# WHITE DWARFS

# ASTROPHYSICS AND SPACE SCIENCE LIBRARY

## VOLUME 214

# WHITE DWARFS

Proceedings of the 10th European Workshop on White Dwarfs,
held in Blanes, Spain,
17–21 June 1996

Edited by

## J. ISERN

*Institut d'Estudis Espacials de Catalunya, CSIC,*
*Barcelona, Spain*

## M. HERNANZ

*Institut d'Estudis Espacials de Catalunya, CSIC,*
*Barcelona, Spain*

and

## E. GARCÍA-BERRO

*Universitat Politècnica de Catalunya,*
*Barcelona, Spain*

SPRINGER SCIENCE+BUSINESS MEDIA, B.V.

A C.I.P. Catalogue record for this book is available from the Library of Congress

ISBN 978-0-7923-4585-5    ISBN 978-94-011-5542-7 (eBook)
DOI 10.1007/978-94-011-5542-7

*Coverpicture:*
*White Dwarf Stars in M4. HST.WFPC2.*
*PRC95-32. ST ScI OPO. August 28, 1995. H. Bond (ST ScI), NASA.*

*Printed on acid-free paper*

# TABLE OF CONTENTS

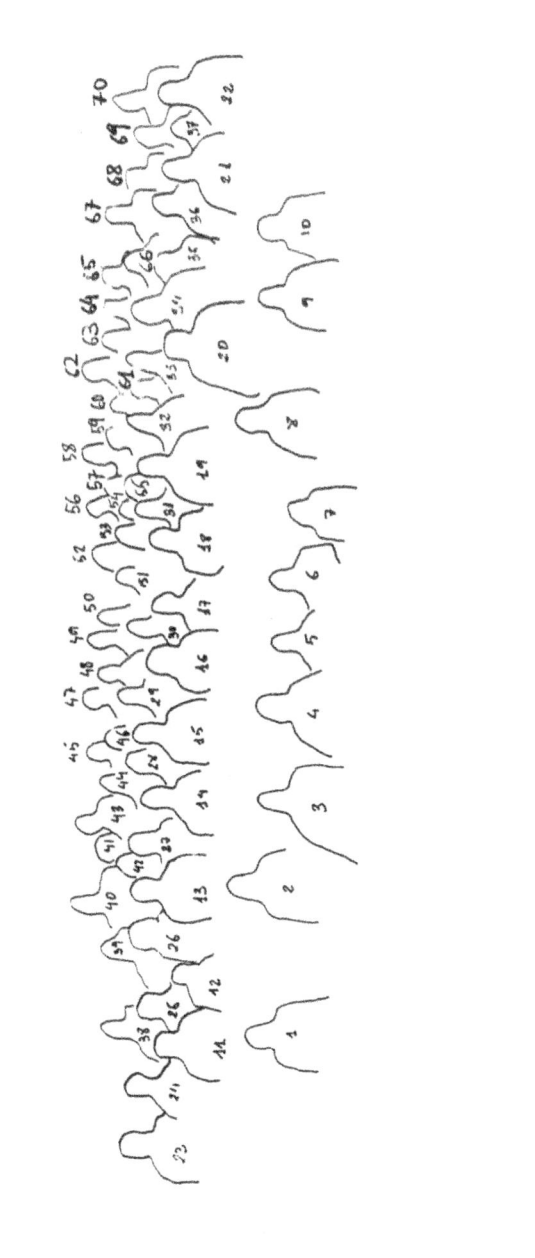

# PREFACE

The study of white dwarfs has been steadily growing during the last years and now is a mature field extending its influence over many others. Since white dwarfs are long lived objects, they can be used to obtain information about the history of the Galaxy. The simplicity of their structures enables them to act as precise particle physics laboratories and the extreme conditions reached at their surfaces allow us to test the equation of state and to study the behavior of matter under conditions impossible to be reached in terrestrial laboratories. Nevertheless, white dwarfs are still challenging astrophysicists. Many questions, ranging from the determination of fundamental parameters to the evolution of their outer layers, are still waiting for a satisfactory answer.

The European Workshop series on White Dwarfs started in 1974 as a consequence of the effort and enthusiasm of Professor Volker Weidemann. The existing proceedings of these meetings, together with those corresponding to the IAU Colloquia held in Rochester (1979) and in Hanover (1988), provide a unique opportunity to follow the development of this field. We hope that the present volume will provide a representative snapshot of the state of the art in 1996. In this sense we are very indebted to all the participants that have sent their contributions according to the instructions (this clearly excludes everybody beyond three sigmas from the standards).

We are very indebted to all of our colleagues of the Scientific Organizing Committee: M. Barstow, F. D'Antona, G. Fontaine, I. Iben, D. Koester, J. Liebert, R. Mochkovitch, G. Shaviv, P. Thejll, H. Van Horn, and G. Vauclair. We want also to express our recognition to Carme Arenillas, Pilar Camps and Evilio del Río for their enthusiastic support. We are specially grateful to our colleagues at the *Centre d'Estudis Avançats de Blanes* for their support anf for showing us that not all in science is Astrophysics. We also thank the *Consell Comarcal de la Selva* for logistic support and the *Blanes City Hall* for their hospitality. Finally we acknowledge the financial support received from the *Dirección General de Investigación Científica y Técnica* (D.G.I.C.Y.T.), the *Comissió Interdepertamental de Recerca i In-*

*novació Tecnología* (C.I.R.I.T.), the *Consejo Superior de Investigaciones Científicas* (C.S.I.C.), the *Universitat Politècnica de Catalunya* (U.P.C.) and the *Fundació Catalana per a la Recerca* (F.C.R.).

Barcelona, January 1997

Jordi Isern and Margarida Hernanz
Institut d'Estudis Espacials de Catalunya
(C.S.I.C. Research Unit)

Enrique García-Berro
Universitat Politècnica de Catalunya

# List of Participants

D. ALVAREZ, (64)     Dept. d'Astronomia i Meteorologia, Universitat de
                     Barcelona, Martí i Franquès, 1, 08028 Barcelona,
                     SPAIN
                     *e-mail*: dalvarez@mizar.am.ub.es


F. ALLARD, (28)      Dept. of Physics, Wichita State University,1845
                     Fairmount, Wichita KS 67260-0032, USA.
                     *e-mail*: allard@eureka.physics.twsu.edu


J.M. APARICIO, (67)  Département de Physique, Université de Montréal,
                     C.P. 6128, Succ. Centre-Ville, Montréal, Quebec,
                     CANADA H3C 3J7
                     *e-mail*: aparicio@astro.umontreal.ca


M.A. BARSTOW, ( 2)   Dept. of Physics and Astronomy, University of
                     Leicester, University Road, Leicester LE1 7RH,
                     UK
                     *e-mail*: mab@star.le.ac.uk


T. BLÖCKER, (52)     Astrophysikalisches Institut Potsdam,
                     Telegrafenberg A27, D-14473 Potsdam,
                     GERMANY
                     *e-mail*: tbloecker@aip.de


P.A. BRADLEY, (46)   Los Alamos National Laboratory, XTA MS
                     B220, Los Alamos NM 87545, USA
                     *e-mail*: pbradley@lanl.gov


A. BRAGAGLIA, (59)   Osservatorio Astronomico di Bologna, Via
                     Zamboni 33, I-40126 Bologna, ITALY
                     *e-mail*: angela@astbo3.bo.astro.it


P. BRASSARD, (61)    Dept. de Physique, Université de Montréal, C.P.
                     6128, Succ. Centre-Ville, Montréal, Québec,
                     CANADA H3C 3J7
                     *e-mail*: brassard@astro.umontreal.ca


I. BUES, (58)        Astronomisches Institut der Universitaet Erlangen,
                     Remeis-Sternwarte Bamberg, Sternwartstrasse 7,
                     D-96049 Bamberg, GERMANY
                     *e-mail*: bues@sternwarte.uni-erlangen.d400.de

M. BURLEIGH, (49)   Physics and Astronomy Department, University of Leicester, Leicester LE1 7RH, UK
*e-mail*: mbu@star.le.ac.uk


R. CANAL, (31)   Dept. d'Astronomia i Meteorologia, Universitat de Barcelona, Av. Diagonal, 647, 08028 Barcelona, SPAIN
*e-mail*: ramon@mizar.am.ub.es


A.M. COOL, (16)   Astronomy Department, University of California, 601 Campbell Hall, Berkeley, CA 94720-3411, USA
*e-mail*: cool@spark.berkeley.edu


L. CUCURULL, ( 7)   IEEC, Edifici Nexus - 104, Gran Capità 2-4, 08034 Barcelona, SPAIN
*e-mail*: cucurull@ieec.fcr.es


G. CHABRIER, (32)   Laboratoire de Physique, Ecole Normale Supérieure de Lyon, 46 Allee d'Italie, 69364 Lyon Cedex 07, FRANCE
*e-mail*: chabrier@cral.ens-lyon.fr


P. CHAYER, (25)   Center for EUV Astronphysics, 2150 Kitteredge Street, Berkeley, California 94720, USA
*e-mail*: chayer@cea.berkeley.edu


P.D. DOBBIE, (57)   X-Ray Astronomy Group, Dept. of Physics and Astronomy, University of Leicester, University Road, Leicester, LE17RH, UK
*e-mail*: pdd@star.le.ac.uk


N. DOLEZ, (10)   Observatoire Midi-Pyrénées, 14 Avenue Edouard Belin, 31400 Toulouse, FRANCE
*e-mail*: dolez@obs-mip.fr


I. DOMÍNGUEZ, ( )   Física teórica y del cosmos, Universidad de Granada, c/ Fuentenueva 18002, Granada, SPAIN
*e-mail*: inma@ugr.es

S. DREIZLER, (24)    Institut für Astronomie und Astrophysik, Universität
                     Kiel, D-24098 Kiel, GERMANY
                     *e-mail*: dreizler@astrophysic.uni-kiel.de


J. DUPUIS, (14)      Center for EUV Astrophysics, 2150 Kittredge
                     Street, Berkeley CA 94720, USA
                     *e-mail*: jdupuis@cea.berkeley.edu


D.S. FINLEY, (38)    Eureka Scientific, Inc, 2452 Delmer St., Oakland
                     CA 94602, USA
                     *e-mail*: david@cea.berkeley.edu


T.A. FLEMING, (36)   University of Arizona, Steward Observatory,
                     Tucson, AZ 85721, USA
                     *e-mail*: taf@as.arizona.edu


G. FONTAINE, ( 8)    Département de Physique, Université de Montréal,
                     C.P. 6128, Succ. Centre-Ville, Montréal, Quebec,
                     CANADA H3C 3J7
                     *e-mail*: fontaine@astro.umontreal.ca


B.T. GÄNSICKE, (15)  Universitäts-Sternwarte Göttingen,
                     Geismarlandstrasse 11, D-37083 Göttingen,
                     GERMANY
                     *e-mail*: boris@uni-sw.gwdg.de


E. GARCÍA-BERRO, ( 6) Dept. de Física Aplicada, Universitat Politècnica
                     de Catalunya, Jordi Girona s/n, Mòdul B-4,
                     08034 Barcelona, SPAIN
                     *e-mail*: garcia@rigel.upc.es


D. GARCÍA-SENZ, (60) Universitat Politècnica de Catalunya, Dept. Física i
                     Enginyeria Nuclear, UPC - Campus Nord, Sor
                     Eulàlia de Anzizu s/n - Mòd. B4, 08034 Barcelona,
                     SPAIN
                     *e-mail*: domingo@polux.upc.es


A. GEMMO, (12)       Dipartimento di Astronomia, Osservatorio
                     Astronomico di Padova, Vicolo dell'Osservatorio,5,
                     I-35122 Padova ITALY
                     *e-mail*: gemmo@astrpd.pd.astro.it

J. GÓMEZ, (  ) — IEEC, Edifici Nexus - 104, Gran Capità 2-4, 08034 Barcelona, SPAIN
e-mail: gomez@ieec.fcr.es

J. GREEN, (66) — C A S A, University of Colorado, 1255 38th Street - Campus Box 593, Boulder, Colorado 80309, USA
e-mail: jgreen@casa.colorado.edu

J. GUERRERO, (54) — IEEC, Edifici Nexus - 104, Gran Capità 2-4, 08034 Barcelona, SPAIN
e-mail: guerrero@ieec.fcr.es

K. GUNDERSON, (68) — Astrophysics Research Laboratory, University of Colorado, 1255 38th Street, Boulder, CO 80303, USA
e-mail: gunderso@casa.colorado.edu

J. GUTIÉRREZ, (63) — Dept. d'Astronomia i Meteorologia, Universitat de Barcelona, Martí i Franquès, 1, 08028 Barcelona, SPAIN
e-mail: jgutier@mizar.am.ub.es

M. HERNANZ, (18) — IEEC, Edifici Nexus - 104, Gran Capità 2-4, 08034 Barcelona, SPAIN
e-mail: hernanz@ieec.fcr.es

F. HERWIG, (43) — Astrophysikalisches Institut Potsdam, Telegraphenberg A 27, D- 14473 Potsdam, GERMANY
e-mail: fherwig@aip.de

J.B. HOLBERG, (39) — Lunar and Planetary Laboratory-west, University of Arizona, Gould-Simpson Building, Tucson AZ 85721, USA
e-mail: holberg@argus.lpl.arizona.edu

I. IBEN, (21) — Astronomy Department, University of Illinois, 1002 West Green St., URBANA, IL 61801, USA
e-mail: icko@astro.uiuc.edu

J. ISERN, ( 1) — IEEC, Edifici Nexus - 104, Gran Capità 2-4, 08034 Barcelona, SPAIN *e-mail*: isern@ieec.fcr.es

S. JORDAN, (44)        Institut für Astronomie und Astrophysik, Universität
Kiel, D- 24098 Kiel, GERMANY
*e-mail*: jordan@astrophysik.uni-kiel.de

S. KAWALER, (34)        Dept. of Physics and Astronomy, Iowa State
University, Ames IA 50011, USA
*e-mail*: sdk@iastate.edu

I.R. KING, ( 9)        Astronomy Department, University of California,
Berkeley, CA 94720-3411, USA
*e-mail*: king@glob.berkeley.edu

S.J. KLEINMAN, (33)        Astronomy Dept., University of Texas, Austin, TX
78712, USA
*e-mail*: sjk@bullwinkle.as.utexas.edu

R. KNOX, (47)        Institute of Astronomy, University of Edinburgh
Royal Observatory, Blackford Hill, Edinburgh EH9
3HJ, UK
*e-mail*: rak@roe.ac.uk

D. KOESTER, (29)        Institut für Astronomie und Astrophysik, Universität
Kiel, D-24098 Kiel, GERMANY
*e-mail*: koester@astrophysik.uni-kiel.d400.de

J. KUBÁT, (40)        Astronomický ústav, Akademie ved Ceské
Republiky, Fricova 1, 251 65 Ondrejov, CZECH
REPUBLIC
*e-mail*: kubat@sunstel.asu.cas.cz

R. LAMONTAGNE, (30)        Département de Physique, Université de Montréal,
C.P. 6128, Succ. Centre-ville, Montréal, Québec,
CANADA  H3C 3J7
*e-mail*: lamont@astro.umontreal.ca

J. LIEBERT, ( 3)        University of Arizona, Steward Observatory,
Tucson, AZ 85721, USA
*e-mail*: liebert@as.arizona.edu

J.A. MARKIEL, (42)    Dept. of Physics and Astronomy, University of
                      Rochester, Rochester, NY 14627, USA
                      *e-mail*: markiel@callisto.pas.rochester.edu


C. MASSACAND, (23)    IMR - Nordlysobservatoriet, University of Tromso,
                      N-9037 Tromso, NORWAY
                      *e-mail*: xtophe@lie.uit.no


R. McLEAN, (70)       CASA, University of Colorado, Boulder, ARL/1255
                      38th St., Boulder, CO, USA
                      *e-mail*: mclean@casa.colorado.edu


R. MOCHKOVITCH, (41)  C N R S, Institut d'Astrophysique de Paris, 98 bis,
                      Blvd. Arago, 75014 Paris, FRANCE
                      *e-mail*: mochko@iap.fr


T.K. NYMARK, (17)     The Auroral Observatory, University of Tromso,
                      9037 Tromso, NORWAY
                      *e-mail*: tanja@lie.uit.no


D. O'DONOGHUE, (19)   Dept. of Astronomy, University of Cape Town,
                      Rondebosch 7700, Cape Town, SOUTH AFRICA
                      *e-mail*: dod@uctvms.uct.ac.za


J. PROVENÇAL, (26)    Dept. of Physics and Astronomy, University of
                      Delaware, Newark, DE 19716, USA
                      *e-mail*: jlp@chopin.udel.edu


A. PUTNEY, (11)       California Institute of Technology, 105-24,
                      Pasadena, CA 91125, USA
                      *e-mail*: axp@astro.caltech.edu


T. RAUCH, (45)        Institut für Astronomie und Astrophysik, Universität
                      Kiel, D-24098 Kiel, GERMANY
                      *e-mail*: rauch@astrophysik.uni-kiel.de


M.T. RUIZ, (27)       Department of Astronomy, Universidad de Chile,
                      Casilla 36-D, Santiago, CHILE
                      *e-mail*: mtruiz@das.uchile.cl

P. RUIZ-LAPUENTE, (35) Dept. Astronomia i Meteorologia, Universitat de
Barcelona, Diagonal 647, 08028 Barcelona, SPAIN
e-mail: pilar@mizar.am.ub.es


R.A. SAFFER, (51)          Space Telescope Science Institute, 3700 San
Martin Drive, Baltimore MD 21218, USA
e-mail: saffer@stsci.edu


M. SALARIS, (65)           IEEC, Edifici Nexus - 104, Gran Capità 2-4, 08034
Barcelona, SPAIN
e-mail: maurizio@mpa-garching.mpg.de


H. SCHMIDT, (50)           Institut für Astronomie und Astrophysik, Universität
Kiel, D- 24098 Kiel, GERMANY
e-mail: schmidt@astrophysik.uni-kiel.d400.de


G. SHAVIV, (22)            Asher Space Research Institute and Dept. of
Physics, Israel Institute of Technology, Haifa,
ISRAEL 32000
e-mail: gioras@physics.technion.ac.il


H. SHIPMAN, ( 4)           Physics and Astronomy Department University of
Delaware Sharp Laboratory Newark - Delaware
19716 USA
e-mail: harrys@udel.edu


E.M. SION, ( )             Dept. of Astronomy and Astrophysics Villanova
University Villanova, PA 19085 USA
e-mail: emsion@ucis.vill.edu


J.E. SOLHEIM, (56)         Institute of Mathematical and Physical Sciences
University of Tromso Auroral Observatory N-9037
Tromso NORWAY
e-mail: janerik@mack.uit.no


S. STARRFIELD, (13)        Physics & Astronomy Arizona State University P.O
Box 871504 Tempe, AZ 85287-1504 USA
e-mail: sumner.starrfield@asu.edu


S. TORRES,( )              Universitat Politècnica de Catalunya EUP Mataró
Av. Puig i Cadafalch, 101-111 08303 MATARÓ
e-mail: santi@benard.fa.upc.es

A. ULLA, (  )  Instituto de Astrofísica de Canarias c/ Vía Láctea s/n 38200 La Laguna SPAIN
*e-mail*: aulla@ll.iac.es

K. UNGLAUB, (62)  Astronomisches Institut der Universitaet Erlangen Remeis-Sternwarte Bamberg Sternwartstrasse 7 D-96049 Bamberg GERMANY
*e-mail*: unglaub@sternwarte.uni-erlangen.d400.de

M. VAN KERKWIJK, (69)  Dept. of Astronomy, m.s. 105-24 California Institute of Technology Pasadena, CA 91125 USA
*e-mail*: mhvk@astro.caltech.edu

H. VÄTH, (48)  Institut für Astronomie und Astrophysik Universität Kiel D-24098 Kiel GERMANY
*e-mail*: supas097@astrophysik.uni-kiel.de

V. WEIDEMANN, (20)  Institut für Astronomie und Astrophysik Universität Kiel D-24098 Kiel GERMANY
*e-mail*: supas058@astrophysik.uni-kiel.de

F. WESEMAEL, (53)  Département de Physique Université de Montréal C.P. 6128, Succ. Centre-Ville Montreal - QUÉBEC CANADA H3C 3J7
*e-mail*: wesemael@astro.umontreal.ca

B. WOLFF, ( 5)  Institut für Astronomie und Astrophysik Universität Kiel D-24098 Kiel GERMANY
*e-mail*: supas089@astrophysik.uni-kiel.d400.de

M. WOOD, (55)  Dept. Physics and Space Sciences Florida Institute of Technology 150. W. University Blvd. Melbourne, FL 32901-6988 USA
*e-mail*: wood@kepler.pss.fit.edu

Y. WU, (37)  Theoretical Astrophysics, 130-33 California Institute of Technology Pasadena, CA 91125 USA
*e-mail*: wyq@tapir.caltech.edu

# Section I:
# Structure and Evolution

# THE EMPIRICAL WHITE DWARF MASS - RADIUS RELATION

HOLGER SCHMIDT

*Institut für Astronomie und Astrophysik der Universität*
*D-24098 Kiel, Germany*

## 1. Introduction

The famous mass - radius relation of white dwarf stars is based on theoretical considerations. However, the few cases in which the masses and the radii of white dwarfs were directly or indirectly determined from observations do not agree well with the theoretical relation.

We present the current situation of the empirical mass - radius relation using new spectroscopic determinations of $T_{\mathrm{eff}}$ and $\log g$ for a large number of DA white dwarfs. The results show a large scattering around the theory, and it seems that the observations do not show the expected correlation between mass and radius. It is shown that the observed masses and radii are, within their errors, in accordance with the theoretical relation. We also discuss the improvements of the empirical mass - radius relation expected from the parallax measurement by the HIPPARCOS satellite.

## 2. Used methods and data

There are two main ways in which we determine the masses und radii for individual white dwarfs :

- The surface brightness method compares the energy fluxes (observed and theoretical) in the visual band. With the trigonometric parallax $\pi$ the radius $R$ can be calculated. The mass $M$ can then be determined either a) via surface gravity $g$ or b) via gravitational redshift $v_{\mathrm{grs}}$
- The gravitational redshift method combines gravitational redshift $v_{\mathrm{grs}}$ of spectral lines with surface gravity $g$, which has the advantage that it does not depend on parallaxes.

All used methods need the atmospheric parameters $T_{\mathrm{eff}}$ and $\log g$. The application of spectroscopic techniques (fitting hydrogen line profiles as predicted by stellar atmosphere models to an observed spectrum) is currently

*I. Isern et al. (eds.), White Dwarfs, 3–9.*
© *1997 Kluwer Academic Publishers.*

the best way to get precise values. We use recently published analyses from Bergeron *et al.* (1995a, 1995b, 1992), Finley *et al.* (1996), Kidder (1991), and Bragaglia *et al.* (1995). Trigonometric parallaxes were taken from the parallax catalog (Van Altena *et al.* 1991) and from the McCook & Sion (1987) white dwarf catalog. Model atmospheres from Koester (1994) were used. We took the values for the gravitational redshift from the literature (published after 1980, only exception: Sirius B).

## 3. The empirical mass - radius relation

Fig. 1 shows the obtained masses and radii using all available data. This figure, and all following mass - radius diagrams, also shows the theoretical zero temperature relation of Hamada & Salpeter (1961) and the evolutionary models of Wood (1994) for a carbon white dwarf with $T_{\text{eff}}$=30000 K with a thick hydrogen layer.

Independent of the methods, the observations do not confirm a mass - radius relation, on the contrary, it appears that the observations contradict a relation! The reasons for this scattering can be traced back to the following causes:

- Mass distribution of the white dwarfs.
  Because the observations reflect the intrinsic distribution around 0.6 $M_{\odot}$, only very few objects are found that have a significantly different mass.
- Influence of observational errors on the position in the mass - radius diagram.
  It can be shown that uncertainties in observational quantities like $T_{\text{eff}}$, $g$, $V$, $\pi$, or $v_{\text{grs}}$ tend to increase the deviation between the theoretically expected masses and radii and those determined from observations.

Since some of the observational quantities and their errors affect the calculation of both, radius and mass (e.g., the gravitational redshift and $\log g$ in the gravitational redshift method) the errors in mass and radius are not statistically independent. The combination of the uncertainties were considered by using a multidimension normal (Gauss) distribution for error propagation. The error ellipse is the graphical representation of this error treatment. The area within the error ellipse corresponds to a 68.3 percent probability (1 $\sigma$ level of a gaussian) of the combined quantities radius and mass to occur (confidence region), just as the standard deviation represents a 68.3 percent probability (confidence interval) of the individual quantities (see Press *et al.* (1992) and Taupin (1988) for more background).

Fig. 2 shows some typical error ellipses and error bars of the gravitational redshift method. One can see the typical dimensions of the error. The orien-

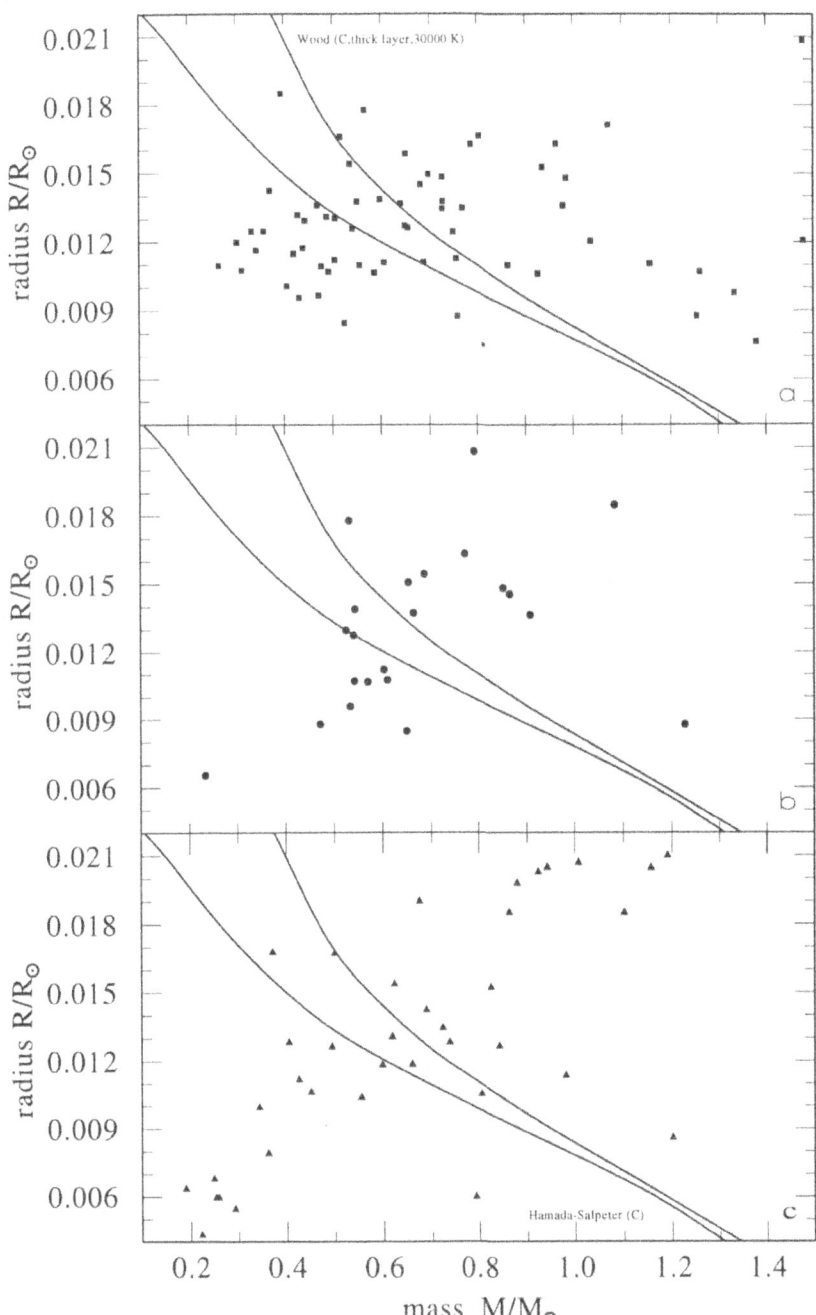

*Figure 1.*   Masses and radii of white dwarfs using all available observational data with values obtained with **a)** the surface brightness method (mass via $\log g$), **b)** the surface brightness method (mass via $v_{grs}$), **c)** with the gravitational redshift method ($\log g$ & $v_{grs}$). If a white dwarf has a parallax measurement and a value for gravitational redshift, then it is displayed in all three figures.

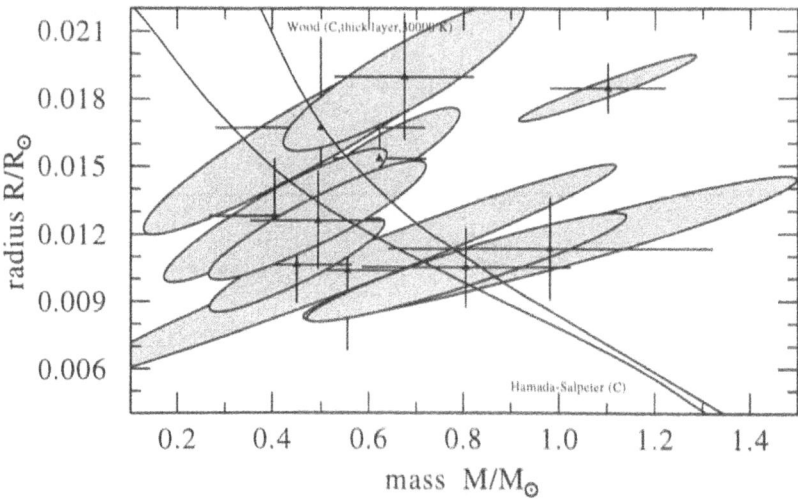

*Figure 2.* Some typical error ellipses and error bars of objects determined with the gravitational redshift method. The area within the error ellipse corresponds to a 68.3 percent probability (1 $\sigma$ level of a gaussian) of the combined quantities radius and mass to occur (confidence region). The orientation of the ellipses reflects the correlation between mass and radius.

tation of the error ellipses reflects the correlation between mass and radius. Therefore almost every uncertainty in observation tends to increase the derivation to a relation. But within the uncertainties the obtained masses and radii are compatible with the theoretical relation.

To select a sample of objects which we expect to have more reliable observational measurements we only use those objects that are near by in order to reduce the error in parallax. Only if the value of the parallax is large ($\pi \geq 0.033''$) the observational error becomes small enough to make the parallax useful for our purpose. Furthermore, we only include those objects that are sufficiently hot such that the uncertainties in the treatment of convection in the atmosphere models do not become important. Therefore we choose objects which have $T_{eff} \geq 15000$ K (for both methods) and $\pi \geq 0.04''$ (for surface brightness method) and depict the resulting mass-radius diagram in Fig. 3.

Again, there is a concentration around 0.6 $M_{\odot}$, but the scattering (compare with Fig. 1 !) is reduced noticeable. In case of Sirius B all three methods show a surprising agreement (with the atmospheric analysis by Kidder (1991), $T_{eff}$=24700 K, $\log g$=8.85), but they lie above the relation.)

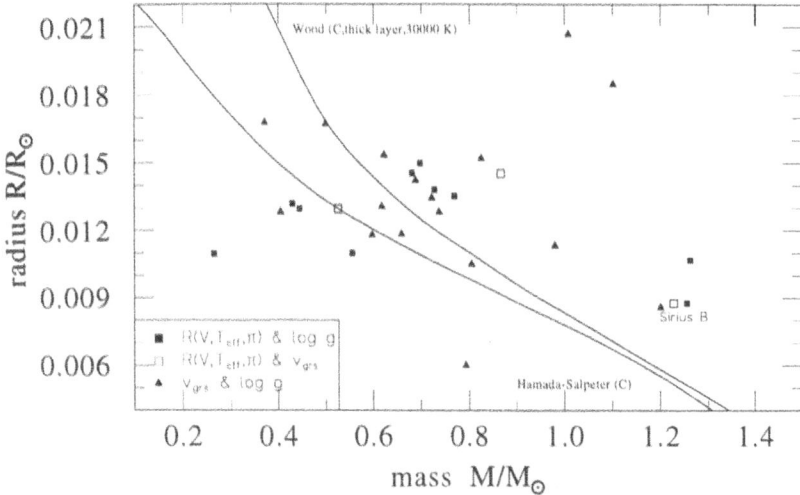

*Figure 3.* Selection of objects with more reliable observational measurements. For surface brightness method (filled and open squares) the distances are lower equal than 25 pc. Futhermore all object have $T_{eff} \geq 15000$ K.

## 4. Possible improvements by HIPPARCOS

The trigonometric parallax is one of the most uncertain quantities. A significant improvement in this respect can be expected from the parallax measurements by the HIPPARCOS satellite which has measured at least for a few white dwarfs parallaxes with an accuracy of the order of milliarcseconds. Because we do not have these measurements yet, we want to discuss the expected improvements. The mass - radius diagram for these stars with the pre-HIPPARCOS parallaxes can be seen in Fig. 4. The influence of the parallax uncertainty on the area of the error ellipse (if we assume an error of 1 mas) is clearly recognizable.

With the more precise values of the HIPPARCOS measurement the resulting masses and radii are no longer compatible with a theoretical relation. Therefore, we expect that the scattering should get smaller and the relation should be confirmed better by the observation.

## 5. Conclusion

The present situation of observations of masses and radii of white dwarfs shows a large scattering around the theoretical mass - radius relation. Within the uncertainties, the observations are compatible with all theories (e.g., Hamada & Salpeter 1961; Wood 1994), but do not favor one over the others. The reason for this unsatisfactory situation lies essentially in the un-

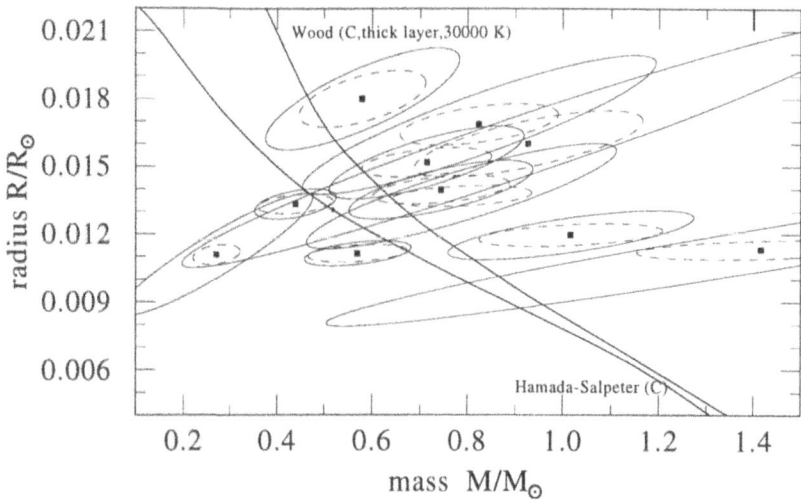

*Figure 4.* Error ellipses of some of the HIPPARCOS white dwarfs with pre-HIPPARCOS parallaxes. For some of the HIPPARCOS objects the error ellipses with the stated error of the observed quantities (solid lines) is shown. The dashed error ellipses show the 68.3 percent probability if the standard deviation of the parallax is artificially reduced to one milliarcsecond.

certainties of the observational parameters.

A large improvement can be expected for the white dwarfs on the HIP-PARCOS input list, when the measurements become available in the near future.

## References

Bergeron P., Saffer R.A., Liebert J., 1992, *ApJ*, **394**, 228

Bergeron P., Liebert J., Fulbright M.S., 1995a, *ApJ*, **444**, 810

Bergeron, P., Wesemael, F., Lamontagne, R., et al., 1995b, *ApJ*, **449**, 258

Bragaglia A., Renzini A., Bergeron P., 1995, *ApJ*, **443**, 735

Finley D.S., Koester D., Basri G., 1996, *in preparation*

Hamada T., Salpeter E.E., 1961, *ApJ*, **134**, 683

Kidder K.M., 1991, *Dissertation, University of Arizona*

Koester D., 1994, *private communication*

McCook, G.P., Sion E.M., 1987, *ApJS*, **65**, 603

Press W.H., Teukolsky S.A., Vetterling W.T., et al., 1992, *Numerical Recipes in FORTRAN*, 2. Ed., Cambridge University Press, Cambridge

Taupin D., 1988, *Probabilities, Data Reduction and Error Analysis in the Physical Sciences*, les éditions de physique, France

Van Altena W.F., Lee J.T., Hoffleit D.E., 1991, *Cat. Trig. Stellar Parallaxes*, Prel. Version on CDROM, Yale Observatory

Wood M.A., 1994, *private communication*

## Discussion

*I. King*: When talking about parallaxes, you should be sure to include those from the U.S. Naval Observatory, which have also achieved milliarcsecond accuracy. Note also that the new Yale Parallax Catalog has finally been published within the last month.

*H. Schmidt*: I used trigonometric parallaxes out of the McCook & Sion White Dwarf Catalog (1987) and out of the Parallax Catalog by Van Altena et al. I used the preliminary version which was published in 1991. Whenever I had more than one value for an object I applied the latest measurement.

*P. Bradley*: When will HIPPARCOS data be publicly available for you to complete your work?

*H. Schmidt*: The HIPPARCOS data will be published in the near future, but I don't know exactly when they will be available.

# WHAT WE HAVE LEARNED ABOUT WHITE DWARFS FROM STUDIES OF NOVAE

SUMNER STARRFIELD

*Department of Physics and Astronomy*
*Arizona State University, Tempe, AZ 85287-1504*

## 1. Introduction

The nova outburst is one consequence of accretion of hydrogen-rich material onto a white dwarf. Although most of our interest has been in trying to understand the cause and evolution of the nova outburst, it is impossible to separate that from an interest in the structure and characteristics of white dwarfs. For example, it is the strong electron degeneracy of the white dwarf that forces the nuclear reactions to temperatures that exceed $10^8$K. Nova studies have also shown that the peak temperature in the nuclear burning region and the peak effective temperature depend on the mass of the white dwarf. It may be the case, in addition, that the lifetime of the nova outburst, the amount of accreted mass, and the ejected mass depend on the mass of the white dwarf. It is also important to realize that there have been numerous determinations of the elemental abundances in nova ejecta and they show, unequivocally, that there are two composition classes for the white dwarfs in nova systems. One class consists of carbon-oxygen (CO) white dwarfs and they must have originated as main sequences stars with masses of around $5M_\odot$ or below. There is a second class of outbursts in which the elemental abundance pattern in the ejected gas can only be explained if it consists of material from the interior of an oxygen-neon-magnesium (ONeMg) white dwarf that has been processed through hot, hydrogen burning. This class must have originated from main sequence stars with masses from $8M_\odot$ to about $12M_\odot$. Finally, the evolutionary studies of the explosion suggest strongly that the white dwarf must be relatively cool which, in turn, implies that nova systems are relatively old objects.

*I. Isern et al. (eds.), White Dwarfs*, 11–17.
© 1997 *Kluwer Academic Publishers.*

## 2. The Thermonuclear Runaway Model of the Outburst

It is now commonly accepted that a classical nova outburst is caused by a thermonuclear runaway (hereafter, TNR) which occurs in the accreted hydrogen-rich envelope of a white dwarf (hereafter, WD) in a close binary system (Truran 1982, 1990; Starrfield 1989, 1993, 1995). The secondary star in the system fills its Roche lobe and loses mass through the inner Lagrangian point into the lobe surrounding the WD. The material enters the Roche lobe of the WD with the angular momentum of the secondary and, therefore, creates an accretion disk before falling onto the WD. Observational determinations of the elemental abundances in nova ejecta have shown that the abundances are very non-solar (see Table 1) and strongly imply that, at some time during the evolution, material from the WD core is mixed into the gas accreted from the secondary (Starrfield et al. 1996a, and references therein). Neither the mechanism responsible for the mixing nor the phase of the outburst in which the mixing occurs is, as yet, known.

Hydrodynamic studies have shown that the consequences of this accretion is a growing layer of hydrogen-rich gas on the WD. When both the WD luminosity and the rate of mass accretion onto the WD are sufficiently low, such that the deepest layers of the accreted material become electron degenerate, a TNR occurs near the base of the accreted layers. For the physical conditions of temperature and density that are expected to obtain in this environment, thermonuclear burning proceeds by means of hydrogen burning, first by means of the proton-proton chain and, subsequently, via the carbon, nitrogen, and oxygen (CNO) cycles, through the peak of the outburst. If there are heavier nuclei present in the nuclear burning shell, they contribute to the nucleosynthesis but, for the range of interior temperatures achieved in classical nova outbursts, they will not give rise to significant energy production. Both compressional heating and the energy released by the nuclear reactions heat the accreted material ($\sim 10^{-6}M_\odot$ to $\sim 10^{-4}M_\odot$, depending on WD mass) until an explosion occurs.

Energy production and nucleosynthesis associated with the CNO hydrogen burning reaction sequences impose interesting constraints on the energetics of the runaway: in particular, the rate of nuclear energy generation at high temperatures ($T > 10^8$K) is limited by the timescales of the *slower and temperature insensitive positron decays*, particularly $^{13}$N ($\tau_{1/2} = 600$s), $^{14}$O ($\tau_{1/2} = 102$s), and $^{15}$O ($\tau_{1/2} = 176$s). The behavior of the positron decay nuclei holds important implications for the nature and consequences of classical nova outbursts. For example, significant enhancements of envelope CNO concentrations are required to insure higher levels of energy release on a hydrodynamic timescale (seconds for WD's) and thus produce a violent outburst (Starrfield, Truran, and Sparks 1978; Truran 1982; Starrfield

1995).

The large abundances of the positron decay nuclei, at the peak of the outburst, have important consequences for the evolution. (1) The rate at which energy is produced, at temperatures exceeding $10^8$K, depends only on the half-lives of the positron decay nuclei and the numbers of CNO nuclei initially present in the envelope. (2) Since convection operates throughout the entire accreted envelope, it brings unburned CNONeMg nuclei into the shell source, when the temperature is rising very rapidly, and keeps the nuclear reactions operating far from equilibrium. (3) The convective turn-over time scale is $\sim 10^2$ sec near the peak of the TNR so that a significant fraction of the positron decay nuclei are able to reach the surface without decaying and the rate of energy generation at the surface can exceed $10^{15}$ erg $gm^{-1}$ $s^{-1}$ (depending upon the enrichment). (4) Finally, the decays of these nuclei provide an intense heat source throughout the envelope that helps eject the material from off the WD. Theoretical studies of this mechanism show that sufficient energy is produced, during the evolution described above, to eject material with expansion velocities that agree with observed values and that the predicted bolometric light curves for the early stages are in reasonable agreement with the observations (Truran 1982; Starrfield 1989, 1995; Starrfield et al. 1992; Politano et al. 1995; Starrfield et al. 1996b).

## 3.  The Oxygen Neon Magnesium Composition Class

An important development in studies of the nova outburst, has been the discovery of the oxygen-neon-magnesium (hereafter, ONeMg) compositional class (Williams et al. 1985; Starrfield, Sparks, and Truran 1986). Our recent studies have shown that it is the outbursts that occur on ONeMg WDs that produce the most interesting nucleosynthesis. The recent members of this class are V838 Her 1991 (Vanlandingham et al. 1996a), QU Vul (Saizar et al. 1992; Austin et al. 1996), and V693 CrA 1981 (Williams et al. 1985; Vanlandingham et al. 1996b).

V1974 Cyg 1992 was the brightest nova seen in outburst since V1500 Cyg 1975, and it was observed from $\gamma$-ray to radio wavelengths. Important data were obtained by ROSAT, which was able to follow it through its entire X-ray outburst (Krautter et al. 1996), and IUE, which was able not only to study it in its fireball stage but also provided a wealth of information about the evolution of its nebular spectrum (Hauschildt et al. 1994a; Shore et al. 1993, 1994, 1996; Austin et al. 1996). In addition, it's outburst was analyzed with two new methods. First, Hauschildt et al. (1992, 1994a, 1994b, 1995, 1996) studied the early fireball stage with a Non-LTE, spherical, expanding, model atmosphere code that was developed to study novae and supernovae.

TABLE 1. Heavy Element Abundances in Novae (mass fraction)

| Object | Year | X | Y | C | N | O | Ne | Z |
|---|---|---|---|---|---|---|---|---|
| T Aur[4] | 1891 | 0.47 | 0.40 | | 0.079 | 0.051 | | 0.13 |
| RR Pic[13] | 1925 | 0.53 | 0.43 | 0.0039 | 0.022 | 0.0058 | 0.011 | 0.043 |
| DQ Her[15] | 1934 | 0.34 | 0.095 | 0.045 | 0.23 | 0.29 | | 0.57 |
| DQ Her[7] | 1934 | 0.27 | 0.16 | 0.058 | 0.29 | 0.22 | | 0.57 |
| HR Del[12] | 1967 | 0.45 | 0.48 | | 0.027 | 0.047 | 0.0030 | 0.077 |
| V1500 Cyg[3] | 1975 | 0.49 | 0.21 | 0.070 | 0.075 | 0.13 | 0.023 | 0.30 |
| V1500 Cyg[6] | 1975 | 0.57 | 0.27 | 0.058 | 0.041 | 0.050 | 0.0099 | 0.16 |
| V1668 Cyg[11] | 1978 | 0.45 | 0.23 | 0.047 | 0.14 | 0.13 | 0.0068 | 0.32 |
| V693 CrA[14] | 1981 | 0.29 | 0.32 | 0.046 | 0.080 | 0.12 | 0.17 | 0.39 |
| V1370 Aql[10] | 1982 | 0.053 | 0.088 | 0.035 | 0.14 | 0.051 | 0.52 | 0.86 |
| GQ Mus[5] | 1983 | 0.37 | 0.39 | 0.008 | 0.125 | 0.095 | 0.0023 | 0.23 |
| PW Vul[8] | 1984 | 0.69 | 0.25 | 0.0033 | 0.049 | 0.014 | 0.00066 | 0.067 |
| QU Vul[9] | 1984 | 0.30 | 0.60 | 0.0013 | 0.018 | 0.039 | 0.040 | 0.10 |
| QU Vul[2] | 1984 | 0.36 | 0.19 | | 0.071 | 0.19 | 0.18 | 0.44 |
| V842 Cen[1] | 1986 | 0.41 | 0.23 | 0.12 | 0.21 | 0.030 | 0.00090 | 0.36 |
| V827 Her[1] | 1987 | 0.36 | 0.29 | 0.087 | 0.24 | 0.016 | 0.00066 | 0.35 |
| QV Vul[1] | 1987 | 0.68 | 0.27 | | 0.010 | 0.041 | 0.00099 | 0.053 |
| V2214 Oph[1] | 1988 | 0.34 | 0 26 | | 0.31 | 0.060 | 0.017 | 0.40 |
| V977 Sco[1] | 1989 | 0.51 | 0.39 | | 0.042 | 0.030 | 0.026 | 0.10 |
| V433 Sct[1] | 1989 | 0.49 | 0.45 | | 0.053 | 0.0070 | 0.00014 | 0.062 |
| V351 Pup[16] | 1991 | 0.37 | 0.25 | 0.0056 | 0.076 | 0.19 | 0.11 | 0.38 |
| V1974 Cyg[17] | 1992 | 0.30 | 0.52 | 0.015 | 0.023 | 0.10 | 0.037 | 0.18 |
| V838 Her[18] | 1991 | 0.80 | 0.093 | 0.018 | 0.019 | 0.0032 | 0.068 | 0.11 |
| Solar | | 0.705 | 0.275 | 0.003 | 0.001 | 0.010 | 0.002 | 0.020 |

References. 1: Andreä et al. 1994; 2: Austin et al. 1996; 3: Ferland & Shields 1978; 4: Gallagher et al. 1980; 5: Morisset & Pèquignot 1996; 6: Lance et al. 1988; 7: Petitjean et al. 1990; 8: Saizar et al. 1991; 9: Saizar et al. 1992; 10: Snijders et al. 1987; 11: Stickland et al. 1981; 12: Tylenda 1978; 13: Williams & Gallagher 1979; 14: Williams et al. 1985; 15: Williams et al. 1978; 16: Saizar et al. 1996; 17: Hayward et al. 1996; 18: Vanlandingham et al. 1996a.

Second, Austin et al. (1996) developed a Monte Carlo-type algorithm to analyze nova nebular spectra and applied this method to V1974 Cyg (Table 1). The analysis of Austin et al. showed, beyond doubt, that the outburst occurred on an ONeMg WD.

Table 1 shows that novae ejecta are enriched in the elements which drive extremes of nuclear energy generation. The mean heavy element mass fraction for these *well-studied* cases is Z = 0.34 (we averaged multiple determinations for the same nova). The need to understand the cause of the large discrepancies between some of the abundance determinations, *for the same nova outburst*, warrants further study (Schwarz et al. 1996, in preparation; Vanlandingham et al. 1996b). We note that for three recent novae: V1370 Aql 1982, V2214 Oph 1988, and V838 Her 1991 (Matheson et al. 1993; Andreä et al. 1994; Vanlandingham et al. 1996a) the ejected material was enriched in sulfur as well as in neon and magnesium. The source of these large abundance enrichments must be matter dredged up from the underlying ONeMg WD and processed through hot, hydrogen burning. The production of sulfur requires, in addition, that the WD be massive (Politano et al. 1995). This implies that there are, likely, mass differences between the WDs in ONeMg novae such as V838 Her 1991 and those in novae such as V1974 Cyg 1992 and V351 Pup 1991 (Starrfield et al. 1992; Politano et al. 1995; Saizar et al. 1996).

## 4. Discussion

In addition to the dependence of the abundance results on WD mass, it has also been shown that the peak $T_{eff}$ and the lifetime of the outburst are dependent on WD mass (Krautter et al. 1996). These results provide a way, therefore, of determining the mass of the white dwarf in classical nova systems independent of dynamical methods. This may prove to be important for statistical studies since there a number of problems with the dynamical determinations. The major problem is that there are additional sources of light in these systems, in addition to the primary and secondary stars, which causes errors in the mass determinations. Finally, all of these studies have shown that the masses of the WD's in classical nova systems exceed $1.0M_\odot$ and, probably, by large amounts. This value is far larger than the value of $\sim 0.55M_\odot$ found for single white dwarfs but this is probably a selection effect since it takes a much larger amount of material to initiate a runaway on a low mass white dwarf than on a high mass white dwarf (Starrfield et al. 1989). In addition, a nova explosion is much brighter on a massive white dwarf than on a low mass white dwarf so we see them at much larger distances.

## References

Andreä, J, Drechsel, H., and Starrfield, S. 1994. A&A, 291, 869.
Austin, S. J., Wagner, R. M., Starrfield, S., Shore, S. N., Sonneborn, G., and Bertram, R. 1996, AJ, 111, 869.
Ferland, G. J., and Shields, G. A. 1978, ApJ, 226, 172.
Gallagher, J. S., et al. 1980, ApJ, 237, 55.
Hauschildt, P. H., Baron, E., Starrfield, S., and Allard, F. 1996, ApJ, 462, 386.
Hauschildt, P. H., Wehrse, R., Starrfield, S., and Shaviv, G. 1992, ApJ, 393, 307.
Hauschildt, P. H., Starrfield, S., Austin, S. J., Wagner, R. M., Shore, S. N., and Sonneborn, G. 1994a, ApJ, 422, 831.
Hauschildt, P. H., Starrfield, S., Shore, S. N., Gonzalez-Riestra, R., Sonneborn, G., and Allard, F. 1994b, AJ, 108, 1008.

Hauschildt, P. H., Starrfield, S., Shore, S. N., Allard, F. and Baron, E. 1995, ApJ, 447, 829.

Hayward, T. L, Saizar, P., Gehrz, R. D., Benjamin, R. A., Mason, C. G., Houck, J. R., Miles, J. W., Gull, G. E., Schoenwald, J. 1996, ApJ, 469, 854.

Krautter, J., Ögelman, H., Starrfield, S., Wichmann, R., and Pfeffermann, 1996, ApJ, 456, 788.

Lance, C. M., McCall, M. L., and Uomoto, A. K. 1988, ApJS, 66, 151.

Matheson, T., Filippenko, A., and Ho, L.C. 1993, ApJL, 418, L29.

Morisset, C. & Pèquignot, D 1996, A&A, in press.

Petitjean, P., Boisson, C., and Pequignot, D. 1990, A&A, 240, 433.

Politano, M., Starrfield, S., Truran, J. W., Sparks, W. M., and Weiss, A. 1995, ApJ, 448, 807.

Saizar, P. Starrfield, S., Ferland, G. J., Wagner, R. M., Truran, J. W., Kenyon, S. J., Sparks, W. M., Williams, R. E., and Stryker, L. L. 1992, ApJ, 367, 310.

Saizar, P., Starrfield, S., Ferland, G. J., Wagner, R. M., Truran, J. W., Kenyon, S. J., Sparks, W. M., Williams, R. E., and Stryker, L. L. 1992, ApJ, 398, 651.

Saizar, P., Pachoulakis, I, Shore, S. N., Starrfield, S., Williams, R. E., and Rotschild, E. 1996, MNRAS, 279, 280.

Shore, S. N., Sonneborn, G., Starrfield, S., Gonzalez-Riestra, R., Ake, T. B. 1993, AJ, 106, 2408.

Shore, S. N., Sonneborn, G., Starrfield, S., Gonzalez-Riestra, R., and Polidan, R. 1994, ApJ, 421, 344.

Shore, S., N., Starrfield, S., Sonneborn, G., and Gonzalez-Riestra, R. 1996, ApJL, 463, L21.

Snijders, M. A. J., et al. 1987, MNRAS, 228, 329.

Starrfield, S., 1988, in *Multiwavelength Observations in Astrophysics*, ed. F. A. Cordova (Cambridge: University Press), 159.

Starrfield, S., 1989, in *The Classical Novae*, ed. M. Bode and A. Evans, (Wiley: NY), 39.

Starrfield, S. 1993, in *The Realm of Interacting Binary Stars*, ed. J. Sahade, G. E. Mc-Cluskey, and Y. Kondo (Dordrecht: Kluwer), 209.

Starrfield, S. 1995, in *Physical Processes in Astrophysics*, ed. I. Roxburgh, and J. L. Masnou, (Springer: Heidelberg), 99.

Starrfield, S., Truran, J. W., Wiescher, M, and Sparks, W. M. 1996a, in *Cosmic Abundances*, ed. S. S. Holt and G. Sonneborn, (Publications ASP: San Francisco), p. 242.

Starrfield, S., Truran, J. W., Wiescher, M, and Sparks, W. M. 1996b, MNRAS, in preparation.

Starrfield, S., Shore, S. N., Sparks, Sonneborn, G., W. M., Politano. M., and Truran, J. W. 1992, ApJ, 391, L71.

Starrfield, S., Sparks, W. M., and Truran, J. W. 1986, ApJ, 303, L5.

Starrfield, S., Truran, J. W., and Sparks, W. M., 1978, ApJ, 226, 186.

Stickland, D. J., et al. 1981, MNRAS, 197, 107.

Truran, J. W. 1982, in *Essays in Nuclear Astrophysics*, eds. C. A. Barnes, D. D. Clayton and D. N. Schramm (Cambridge: Cambridge U. Press), p. 467.

Truran, J. W. 1990, in *The Physics of Classical Novae*, ed. A. Cassatella and R. Viotti, (HD: Springer), p. 373.

Tylenda, R. 1978, Acta Astronomica,28, 333.

Vanlandingham, K., Starrfield, S., Wagner, R. M., Shore, S. N., Sonneborn, G., 1996a, MNRAS, 282, 563.

Vanlandingham, K., Starrfield, S., Wagner, R. M., Shore, S. N., Sonneborn, G., 1996b, MNRAS, in preparation.

Williams, R. E., and Gallagher, J. S. 1979, ApJ, 228, 482.

Williams, R. E., et al. 1978, ApJ, 224, 171.

Williams, R.E., Ney, E.P., Sparks, W.M., Starrfield, S., Truran, J.W. 1985, MNRAS, 212, 753.

## Discussion

*J. Isern*: Is the presence of $^{24}$Mg crucial for the final nucleosynthesis?

*S. Starrfield*: Yes, it's presence or absence is crucial for the nucleosynthesis. We are currently running a new set of evolutionary sequences where we use the abundance distribution of Ritossa, Garcia-Berro, and Iben (1996, ApJ, 460, 489). Their results show that less magnesium is produced in the evolution of massive stars than found in Arnett and Truran (1969, ApJ, 157, 339).

*V. Weidemann*: Can you explain the reasons why we do not see novae with more typical WD masses (around $0.6M_\odot$)? Is this only a selection effect?

*S. Starrfield*: Yes, I think that it is mostly a selection effect. The amount of mass necessary to initiate the runaway is a steeply decreasing function of increasing WD mass. It may take more than $10^{-3}M_\odot$ of gas on a low mass white dwarf while it takes less than $10^{-5}M_\odot$ on a high mass white dwarf. At the observed accretion rates of $10^{-8}M_\odot$ yr$^{-1}$, or less, the time to outburst on a low mass white dwarf could be a 100 times longer than on a high mass white dwarf. In addition, the peak luminosity is an increasing function of WD mass and the outbursts on high mass white dwarfs become much brighter than on low mass white dwarfs so we can see them at much larger distances. It is likely that the masses of the WD's in the DQ Her and T Aur systems are lower than average and may be as low as in the single white dwarfs.

*M. Hernanz*: What prescription have you adopted for convection and how does it influence your results.

*S. Starrfield*: We use a standard mixing length theory as described in the Cox and Giuli book. We have made a minor adjustment so that it is not a local prescription as described in a paper by Peter Wood in the mid- 1970's. We do not assume that the nuclei in the convective region are completely mixed but actually solve a diffusion equation which allows them to move only a specific distance during a given time step. Our results are strongly dependent on convection. For example, if we change the ratio of mixing-length to scale height from 1 to 2 or 2 to 3, we make large changes in our results. This is discussed in detail in Starrfield et al. (1996b)

# THE REDISTRIBUTION OF CARBON AND OXYGEN IN CRYSTALLIZING WHITE DWARFS

R. MOCHKOVITCH

*Institut d'Astrophysique de Paris, CNRS*

J. ISERN AND M. HERNANZ

*Institute for Space Studies of Catalonia (CSIC Research Unit)*

AND

E. GARCIA-BERRO

*Departament de Física Aplicada, Universitat Politècnica de Catalunya*

## 1. Introduction

The evolution of white dwarfs is essentially a cooling process that lasts for $\sim 10$ Gyr. Since the study of white dwarfs allows to obtain information about the age of the Galaxy (Winget et al. 1987, García-Berro et al. 1988, Hernanz et al. 1994), it is important to identify all the sources of energy as well as the mechanisms that control its outflow.

Segretain et al (1994) computed detailed cooling sequences using the most up to date physics both for the equation of state and the phase diagram. The release of gravitational energy induced by the redistribution of carbon and oxygen at crystallization was found to increase noticeably the cooling times. For instance, the time taken by a typical 0.6 $M_\odot$ white dwarf with equal mass fractions of carbon and oxygen to reach a luminosity $\text{Log}(L/L_\odot) = -4.5$ is 11.5 Gyr when the redistribution process is properly taken into account, instead of 9.2 Gyr when it is neglected.

The redistribution process is caused by the removal (by Rayleigh-Taylor instabilities) of the carbon-rich fluid released at the crystallization front. We check below the efficiency of this mixing mechanism and we obtain simple analytical expressions for the extra source of energy which results from redistribution.

19

*I. Isern et al. (eds.), White Dwarfs, 19–25.*

© 1997 *Kluwer Academic Publishers.*

## 2. The physics of the crystallization process

Due to the spindle shape of the phase diagram of C/O mixtures (Segretain and Chabrier, 1993), the solid formed at crystallization is richer in oxygen than the liquid and therefore denser. Using the condition of pressure continuity, the density excess can be estimated to be:

$$\frac{\delta\rho}{\rho} \simeq -\frac{\delta P_i}{\gamma P_e} - \frac{\delta Y_e}{Y_e} \qquad (1)$$

where $P_i$ and $P_e$ are the ionic and electronic pressures respectively, $\gamma$ is the electron adiabatic index and $Y_e$ is the number of electrons per nucleon. For a $0.6\,M_\odot$ white dwarf with equal amounts of carbon and oxygen, $\delta\rho/\rho \simeq 10^{-4}$. Therefore, as the solid core grows from the center of the star the lighter liquid left behind becomes Rayleigh-Taylor unstable and is redistributed by large scale convective motions (Stevenson, 1980; Mochkovitch 1983).

We first estimate the energy released by the redistribution process. The local energy budget of the white dwarf can be written as:

$$-(\frac{dL_r}{dm} + \epsilon_\nu) = C_v\dot{T} + T\left(\frac{\partial P}{\partial T}\right)_{V,X}\dot{V} - l_s\dot{M}_s\delta(m - M_s) + \left(\frac{\partial E}{\partial X}\right)_{T,V}\dot{X} \qquad (2)$$

where $E$, $V$, $X$, $l_s$ and $\dot{M}_s$ are respectively the specific internal energy, the specific volume, the oxygen mass fraction, the latent heat of crystallization and the rate at which the solid core grows, all other symbols having their usual meaning; the delta function indicates that the latent heat is released at the solidification front.

The role of these terms is well known (Lamb and Van Horn, 1975) except the last one and we now examine its significance. Integrating (2) over the whole star, we obtain:

$$L + L_\nu = -\int_0^{M_{WD}} C_v\dot{T}dm - \int_0^{M_{WD}} T\left(\frac{\partial P}{\partial T}\right)_{V,X}\dot{V}dm + l_s\dot{M}_s$$
$$- (X^{sol} - X^{liq})\left[\left(\frac{\partial E}{\partial X}\right)_{M_s} - \left\langle\frac{\partial E}{\partial X}\right\rangle\right]\dot{M}_s \qquad (3)$$

where $(\partial E/\partial X)_{M_s}$ is the partial derivative evaluated at the limit of the solid core and

$$\left\langle\frac{\partial E}{\partial X}\right\rangle = \frac{1}{\Delta M}\int_{\Delta M}\left(\frac{\partial E}{\partial X}\right)_{T,V}dm \qquad (4)$$

$\Delta M$ being the region where the liquid rehomogenizes. In order to obtain this equation we have assumed that the increase in oxygen abundance of the solidifying shell is compensated by its decrease in an homogeneously mixed zone of mass $\Delta M$. From equation (3) we can define the total energy released per gram of crystallized matter due to the change in chemical composition as:

$$\epsilon_g = -(X^{sol} - X^{liq}) \left[ \left( \frac{\partial E}{\partial X} \right)_{M_s} - \left\langle \frac{\partial E}{\partial X} \right\rangle \right] \tag{5}$$

The square bracket is negative since the $(\partial E/\partial X)$ is dominated by the negative Madelung contribution of the ions and essentially depends on the density, which monotonically decreases outwards.

Before computing the energy released by chemical segregation, it is necessary to check the efficiency of convective mixing. Let $v_{crys}$ be the propagation velocity of the solidification front into the C/O mixture. This velocity, which can be obtained from the models, is very small, $\lesssim 0.1$ cm/yr in most of the star. The mass flux of carbon released by the front in the liquid phase is:

$$F_C^{crys} = \rho v_{crys}(X_C^l - X_C^s) = \rho v_{crys}\Delta X_C \tag{6}$$

where $X_C^l$ and $X_C^s$ are the carbon mass fractions in the liquid and the solid respectively. The criterion for convective instability taking into account heat conduction from the convective eddies is (Stevenson and Salpeter, 1977):

$$\chi > k\epsilon \tag{7}$$

with

$$\chi = -\frac{H_P}{\rho c_s^2} \mu \left( \frac{\partial P}{\partial \mu} \right)_{\rho,T} \left( \frac{1}{\mu} \frac{d\mu}{dr} \right) \tag{8}$$

$$\epsilon = \frac{H_P}{\rho c_s^2} T \left( \frac{\partial P}{\partial T} \right)_{\rho,\mu} \left[ \frac{1}{T} \frac{dT}{dr} - (\Gamma_3 - 1) \frac{1}{\rho} \frac{d\rho}{dr} \right] \tag{9}$$

$$k = \frac{\tau_{cond}}{\tau_{cond} + \tau_{conv}} \tag{10}$$

where $H_P$ is the pressure scale height, $c_s$ is the sound velocity, $\mu$ is the mean molecular weight, $(\Gamma_3 - 1) = (\partial \log T/\partial \log \rho)_{ad}$, and $\tau_{cond}$ and $\tau_{conv}$ are the conductive and the convective characteristic times, respectively. If $l$ is the mixing length, the convective velocity can be written as

$$v_{conv} = c_s \left( \chi - k\epsilon \right)^{1/2} \frac{l}{H_P} \tag{11}$$

and the characteristic times are given by:

$$\tau_{conv} = \frac{l}{v_{conv}} \tag{12}$$

$$\tau_{\text{cond}} = \frac{l^2}{K_T} \tag{13}$$

where $K_T$ is the thermal conductivity. If the carbon mass flux released at the crystallization front is to be efficiently carried by convection we have

$$F_C^{\text{cryst}} = F_C^{\text{conv}} = \rho v_{\text{conv}} \left| \frac{dX_C}{dr} \right| l \tag{14}$$

together with a small superadiabaticity

$$\chi - k\epsilon \sim 0 \tag{15}$$

If the superadiabaticity is indeed small (which will have to be checked on the final results), the gradient of carbon mass fraction in the white dwarf is given by

$$\left| \frac{dX_C}{dr} \right| = k \left| \frac{dX_C}{dr} \right|_{\text{ad}} \tag{16}$$

where

$$\left| \frac{dX_C}{dr} \right|_{\text{ad}} = -\frac{48}{\mu^2} \frac{(\Gamma_3 - 1) T \left( \frac{\partial P}{\partial T} \right)_{\rho,\mu}}{\left( \frac{\partial P}{\partial \mu} \right)_{\rho,T}} \left| \frac{1}{\rho} \frac{d\rho}{dr} \right| = Q \left| \frac{1}{\rho} \frac{d\rho}{dr} \right| \tag{17}$$

The value of Q has been computed for a 0.6 $M_\odot$ crystallizing C/O white dwarf and is typically a few $10^{-2}$. Now it is possible to obtain the Peclet number, $P = \tau_{\text{cond}}/\tau_{\text{conv}}$, as

$$\begin{aligned}
\frac{P^2}{P+1} &\simeq \frac{5}{3} \frac{v_{\text{crys}} \Delta X_C H_P}{K_T Q} \\
&\simeq 0.1 \left( \frac{v_{\text{crys}}}{0.1\,\text{cm/yr}} \right) \left( \frac{\Delta X_C}{0.1} \right) \left( \frac{100\,\text{cm}^2\,\text{s}^{-1}}{K_T} \right) \left( \frac{H_P}{10^9\,\text{cm}} \right)
\end{aligned} \tag{18}$$

where $K_T = 100$ cm$^2$s$^{-1}$ is a typical value of the conductivity in the white dwarf. Then $P \lesssim 1$ and the gradient of carbon mass fraction

$$\left| \frac{dX_C}{dr} \right| \simeq 4 \times 10^{-12} \left( \frac{v_{\text{crys}}}{0.1\,\text{cm/yr}} \right) \left( \frac{\Delta X_C}{0.1} \right) \left( \frac{100\,\text{cm}^2\,\text{s}^{-1}}{K_T} \right) \left( \frac{1}{P} \right) \tag{19}$$

is so small that $X_C$ varies by less than 1% in the mixed region.

We now check that the small superadiabaticity hypothesis is correct, i.e.:

$$\chi - k\epsilon \ll \chi \tag{20}$$

We first compute $\chi$ in a crystallizing 0.6 $M_\odot$ C/O white dwarf

$$\chi \sim 10^{-4} \left(\frac{H_P}{10^9 \text{ cm}}\right) \left(\frac{3 \times 10^8 \text{ cm s}^{-1}}{c_s}\right)^{-2} \left(\frac{\left|\frac{dX_C}{dr}\right|}{4 \times 10^{-12}}\right) \tag{21}$$

An upper limit of the superadiabacity can be obtained by assuming that, due to the interaction of convection with rotation, the Rossby number

$$R_0 = \frac{v_{\text{conv}}}{\omega\, l} \tag{22}$$

is equal to unity, where $\omega$ is the angular velocity of the white dwarf (Stevenson and Salpeter 1976). Then,

$$\chi - k\epsilon = \left(\frac{H_P \omega}{c_s}\right)^2 = 3 \times 10^{-7} \left(\frac{H_P}{10^9 \text{cm}}\right)^2 \left(\frac{c_s}{3 \times 10^8 \text{ cm s}^{-1}}\right)^{-2} \left(\frac{\Pi}{10 \text{ h}}\right) \tag{23}$$

where $\Pi$ is the rotation period. We therefore conclude that, even when rotation is considered, convection remains an efficient mechanism to redistribute the carbon rich fluid out from the crystallization front and that the liquid phase can be considered well mixed.

## 3.  The energetics of chemical differentiation

Since $\epsilon_g$ is dominated by the ionic contribution, we can obtain a more transparent expression of $\epsilon_g$ showing that it also corresponds to a release of gravitational energy. We first write

$$\epsilon_g = -\alpha\left(X^{\text{sol}} - X^{\text{liq}}\right)\left(\frac{\partial E}{\partial X}\right)_{M_s} \tag{24}$$

where we have introduced the parameter $\alpha = \left[\left(\frac{\partial E}{\partial X}\right)_{M_s} - \left\langle\frac{\partial E}{\partial X}\right\rangle\right] / \left(\frac{\partial E}{\partial X}\right)_{M_s}$ which is smaller than unity. The ionic pressure associated to the Madelung energy is $P_i = \frac{1}{3}\rho E_i$ so that

$$\epsilon_g \simeq -3\alpha\left(X^{\text{sol}} - X^{\text{liq}}\right)\frac{1}{\rho}\left(\frac{\partial P_i}{\partial X}\right)_{M_s} \simeq -3\alpha\frac{\delta P_i}{\rho} = 3\alpha\gamma\frac{P_e}{\rho}\frac{\delta\rho}{\rho} \tag{25}$$

$\delta P_i$ being the change of ionic pressure at crystallization. Finally, using the virial theorem in the form

$$3\left\langle\frac{P_e}{\rho}\right\rangle \simeq -\frac{\Omega}{M_{\text{WD}}} \simeq \beta\frac{GM_{\text{WD}}}{R_{\text{WD}}} \simeq \beta g_{\text{WD}} R_{\text{WD}} \tag{26}$$

where the average is taken over the white dwarf mass and where $\Omega = -\beta G M_{\mathrm{WD}}^2 / R_{\mathrm{WD}}$, $g_{\mathrm{WD}}$ and $R_{\mathrm{WD}}$ are respectively the white dwarf gravitational energy, surface gravity and radius ($\beta = 6/7$ for $\gamma = 5/3$), we obtain

$$\epsilon_{\mathrm{g}} \simeq \alpha\beta\gamma \frac{P_{\mathrm{e}}/\rho}{\langle P_{\mathrm{e}}/\rho \rangle} \, g_{\mathrm{WD}} R_{\mathrm{WD}} \frac{\Delta\rho}{\rho} \simeq k \times g_{\mathrm{WD}} R_{\mathrm{WD}} \frac{\Delta\rho}{\rho} \qquad (27)$$

where the factor $k$ is of the order of unity except close to the surface of the white dwarf where $\rho$ rapidly decreases.

## 4. Conclusions

We have provided a new formulation of the thermodynamics of phase separation upon crystallization that proves that chemical differentiation results in a net release of heat that is radiated away thus delaying the cooling of the star. This extra heating is not due to compressional work, but to the local changes of chemical abundances.

We have also shown that the hypothesis of perfect rehomogeneization of the liquid is reasonable and can be used to simplify the problem. Notice that if this was not the case, this would result in a decrease of the efficiency of the redistribution process as an energy source but would not invalidate the result that if there is some redistribution there is a release of heat.

Finally, we want to emphasize that the total delay introduced by this extra source of energy depends on the phase diagrams, on the initial chemical compositions and on the transparency of the outer envelope. Any change in these factors can produce noticeable changes in the final outcome.

**Acknowledgements** This work has been supported by the DGICYT grants PB94–0111, PB94–0827–C02–02, by the CIRIT grant GRQ94–8001 and by the AIHF237B.

## References

García-Berro, E., Hernanz, M., Isern, J., Mochkovitch, R. 1988, Nature, 333, 642

Hernanz, M., García-Berro, E., Isern, J., Mochkovitch, R., Segretain, L., Chabrier, G. 1994, ApJ, 434, 632

Lamb, D.Q., Van Horn, H. M. 1975, ApJ, 200, 306

Mochkovitch, R. 1983, A&A, 122, 212

Segretain, L., Chabrier, G. 1993, A&A, 271, L13

Segretain, L., Chabrier, G., Hernanz, M., García-Berro, E., Isern, J., Mochkovitch, R. 1994, ApJ, 434, 641

Stevenson, D.J. 1980, J. Phys. Suppl. No 3, 41, C2–53

Stevenson, D.J., Salpeter, E.E. 1976, in "Jupiter", Ed. T. Gehrels, The University of Arizona Press, p. 85.

Stevenson, D.J., Salpeter, E.E. 1977, ApJS, 35, 239

Winget, D. E., Hansen, H. J., Liebert, J., Van Horn, H. M., Fontaine, G., Nather, R.E., Kepler, S.O., Lamb, D.Q. 1987, ApJ, 307, 659

## Discussion

*I. Iben*: Over a Hubble time, how close does the crystallization front come to the surface? That is, could your effect influence the abundances achieved by novae in cataclysmic variables?

*R. Mochkovitch*: The increase in carbon abundance at the surface due to the redistribution process is important only when the luminosity has decreased to a very low value $< 10^{-4} L_\odot$. I believe that white dwarfs in cataclysmic variables are normally brighter than this.

*I. King*: Now, that we are actually observing cooling sequences in globular clusters, I would like to ask whether your theories of cooling rates apply equally to stars of low metal abundance — [Fe/H] $\sim -2$ (but probably [O/H] $\sim -1.5$).

*R. Mochkovitch*: A low metallicity probably does not affect the cooling rate of a white dwarf of given mass since gravitational settling will lead anyway to a very low heavy element abundance in the envelope. A low metallicity can however affect the relation between the white dwarf mass and the mass of its main sequence progenitor. This could change the luminosity function to an extend which is difficult to evaluate. Finally, our theory essentially deals with the crystallization phase, which corresponds to white dwarfs still too faint to be observed in globular clusters.

*J. Liebert*: Have you calculated a massive ($\sim 1 M_\odot$) case? Would the delay of the cooling time be a larger or a smaller effect?

*R. Mochkovitch*: We find a smaller effect in a $1 M_\odot$ white dwarf compared to the $0.6 M_\odot$ case if we assume that both stars have equal mass fractions of carbon and oxygen, uniformly distributed before crystallization. However, evolutionary calculations indicate this is true for massive white dwarfs only, while $0.6 M_\odot$ white dwarfs are more oxygen-rich in the central regions. If this difference in initial composition is taken into account in the calculations the delays become very similar.

*M.H. Van Kerkwijk*: How does the heat released by the redistribution compare with the latent heat?

*R. Mochkovitch*: The heat released by the redistribution process is typically a factor of two larger than the latent heat.

*M. Wood*: Is the phase diagram a function of density?

*R. Mochkovitch*: If the phase diagram is constructed so as to give the value of the plasma parameter $\Gamma$ at crystallisation as a function of composition its shape is density independent (at least as long as quantum effects on the nuclei can be neglected).

# THE COOLING OF WHITE DWARFS AND THEIR INTERNAL COMPOSITION

MAURIZIO SALARIS, MARGARIDA HERNANZ AND JORDI ISERN

*Institut d'Estudis Espacials de Catalunya (CSIC Research Unit)*
*Edifici Nexus-104, C/ Gran Capità 2-4, 08034 Barcelona, Spain*

INMACULADA DOMINGUEZ

*Departamento de Física Teórica y del Cosmos, Universidad de*
*Granada, Facultad de Ciencias, 18071 Granada, Spain*

ENRIQUE GARCIA-BERRO

*Departament de Física Aplicada*
*Mòdul B4/B5, Campus Nord UPC, 08034 Barcelona, Spain*

AND

ROBERT MOCHKOVITCH

*Institut d'Astrophysique*
*98 bis Boulevard Arago, 75014 Paris, France*

**Abstract.** White dwarfs are the remnants of stars of low and intermediate masses on the main sequence. Their evolution is a gravothermal process, and the energy release only depends on the detailed internal structure, chemical composition and on the properties of the envelope equation of state and opacity. We have computed detailed evolutionary chemical profiles for white dwarfs having progenitors in the mass range 1 to $7\,M_\odot$, and we examine the influence of such profiles in the cooling process. As a result we find that the contribution to the cooling times of the process of separation of carbon and oxygen during crystallization cannot be neglected, since the best fit to the luminosity functions of Liebert et al. (1988) gives an age of the disk of 9.3 Gyr, instead of 8.3 Gyr obtained when the effect of the separation is neglected.

*I. Isern et al. (eds.), White Dwarfs, 27–33.*

## 1. Introduction

The final result of the evolution of low and intermediate mass stars ($M \leq 7$–$8 \, M_\odot$) is a carbon-oxygen white dwarf. White dwarf evolution can be interpreted in terms of a cooling process, and the rate of cooling is determined, among other factors, by the ionic specific heat which depends on the relative proportions of carbon and oxygen. The change of chemical composition between the solid and the liquid at the onset of crystallization and the gravitationally induced redistribution of carbon and oxygen provide an additional source of energy (Mochkovitch 1983, García-Berro et al. 1988), the importance of which depends, among other things, on the shape of the phase diagram.

In light of the recent determinations of the $^{12}C(\alpha, \gamma)^{16}O$ reaction rate, we investigate whether or not the interiors of white dwarfs are stratified before crystallization sets in, and to determine the effect of the actual chemical profile on the cooling times, thereby providing better estimates of the ages of white dwarfs and of the age of the solar neighborhood.

## 2. Input physics and stellar models

The evolutionary stellar models presented in this paper have been computed using the evolutionary code FRANEC (Frascati RAphson Newton Evolutionary Code), as described in Chieffi & Straniero (1989).

The boundaries of convective regions are set by adopting the Schwarzschild criterion and no mechanical overshooting is allowed. Semiconvection during central helium burning is computed according to the method described in Castellani et al. (1985), and the breathing pulses occuring during the last portion of core helium burning have been inhibited. For $T > 10^4$ K, the OPAL radiative opacities of Iglesias et al. (1992) were used, whereas for $T \leq 10^4$ K, the opacities of Kurucz (1991) were adopted. We have assumed a value $Z = 0.02$ for the solar metallicity and the heavy elements distribution as derived by Grevesse (1991). The solar helium abundance $Y_\odot$ and the value of the mixing length parameter $\alpha$ have been derived by matching the luminosity and radius of a stellar model with $Z = 0.02$ and the solar age to their solar values. The values obtained are $Y_\odot = 0.289$ and $\alpha = 2.25$.

Nuclear reaction rates have been taken from Fowler, Caughlan & Zimmerman (1975), and the subsequent modifications have been taken from Harris et al. (1983), Caughlan et al. (1985) and Caughlan & Fowler (1988). The reaction rate for the $^{12}C(\alpha, \gamma)^{16}O$ reaction is crucial for this study, since during the He burning phase, when central helium is depleted down to $Y = 0.10$, the burning mainly occurs through this reaction, and its rate determines the $^{12}C$ and $^{16}O$ profiles in the final white dwarf structure. Several different teams have recently examined the cross section of the $^{12}C(\alpha, \gamma)^{16}O$ reaction; a detailed analysis of the data obtained shows that these experiments are compatible with a total astrophysical S-factor at 300 keV ($S_{300}$) in the range 120-220 keV b (Trautvetter 1996).

There have also been several attempts to constrain the $^{12}C(\alpha, \gamma)^{16}O$ reaction rate from astrophysical data. However, the fractions of $^{12}C$ and $^{16}O$ produced in a typical star depend both on the reaction rate and on the treatment of convection. Thus, since we cannot disentangle both effects, these constraints are only set on an *effective* cross section for the $^{12}C(\alpha, \gamma)^{16}O$ reaction, given the lack of a reliable theory of convection. Woosley, Timmes & Weaver (1993) studied the role of the $^{12}C(\alpha, \gamma)^{16}O$ reaction rate in producing the solar abundance set from stellar nucleosynthesis, and concluded that the effective astrophysical S-factor for the energies involved during core helium burning that best reproduces the observed abundances should be $S_{300} = 170$ keV b, in good agreement with the experimental data. This value for $S_{300}$ corresponds to the value given by Caughlan & Fowler (1988) multiplied by ~1.3. In their models the Ledoux criterion plus an amount of convective overshooting were adopted for determining the extension of the convective regions. Thielemann, Nomoto & Hashimoto (1996) studied the collapse of gravitational supernovae and compared the predicted amount of $^{12}C$ and $^{16}O$ in their ejecta with the abundances observed in SN1987A and SN1993J. They found that the agreement was excellent when the reaction rate of Caughlan et al. (1985), which was computed assuming $S_{300} = 240$ keV b, and the Schwarzschild criterion without overshooting were adopted.

Therefore, given our treatment of convection, we have adopted the rate of Caughlan et al. (1985) for the $^{12}C$ and $^{16}O$ reaction. However, for a sake of comparison, two evolutionary sequences producing the most probable final white dwarf configurations — that is white dwarf masses between 0.55 and 0.65 $M_\odot$ — have also been computed using a lower cross section, namely the rate inferred by Woosley, Timmes & Weaver (1993) ($S_{300} = 170$ keV b, see above), hereinafter "low rate".

With the input physics briefly described above, we have computed evolutionary sequences — neglecting mass loss — from the zero age main sequence to the end of the first thermal pulse, of stellar models with masses in the range $1.0 \leq M/M_\odot \leq 7.0$. The values of the mass on the main sequence ($M_{MS}$), the time spent during the pre-white dwarf phase ($t$), the mass of the final white dwarf ($M_{WD}$), and the oxygen abundance at the center ($X_O$) at the end of the first thermal pulse are displayed in Table 1.

The left panel of Figure 1 displays the oxygen profiles for some of the CO cores obtained, just at the end of the first thermal pulse, the $^{16}O$ profile (dotted line) of the white dwarf resulting from the evolution of a 3.2 $M_\odot$ stellar model, computed adopting the low value of the $^{12}C(\alpha, \gamma)^{16}O$

TABLE 1. Characteristics of the white dwarfs
obtained adopting the rate of Caughlan et al.
(1985) for the $^{12}C(\alpha, \gamma)^{16}O$ reaction

| $M_{MS}(M_\odot)$ | $\log t$ (yr) | $M_{WD}(M_\odot)$ | $X_O$ |
|---|---|---|---|
| 1.0 | 10.0 | 0.54 | 0.79 |
| 1.5 | 9.36 | 0.54 | 0.79 |
| 2.0 | 9.02 | 0.54 | 0.79 |
| 2.5 | 8.88 | 0.55 | 0.83 |
| 3.2 | 8.56 | 0.61 | 0.74 |
| 3.6 | 8.41 | 0.68 | 0.72 |
| 4.0 | 8.27 | 0.77 | 0.71 |
| 5.0 | 8.01 | 0.87 | 0.68 |
| 7.0 | 7.66 | 1.00 | 0.66 |

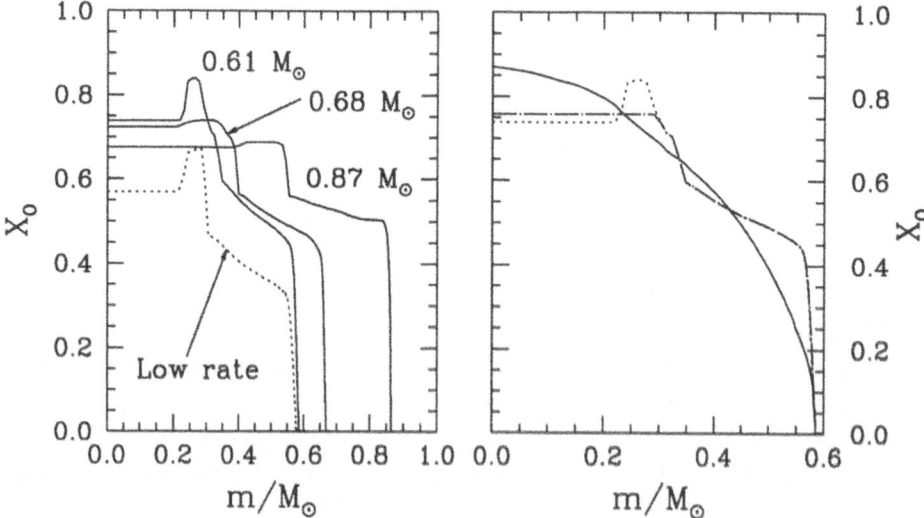

*Figure 1.* Left panel: oxygen profiles for selected white dwarf models. Right panel:
oxygen profile of a 0.61 $M_\odot$ white dwarf at the beginning of the thermally pulsing phase
(dotted line), after the rehomogeneization during the liquid phase (dotted-dashed line)
and after total freezing (solid line).

reaction rate is also shown. The shape of the chemical profile is similar to
that obtained by adopting the rate of Caughlan et al. (1985), but now, due
to the less efficient conversion of $^{12}C$ into $^{16}O$, the two elements have a
more similar abundance in the inner part of the core ($X_C$=0.40, $X_O$=0.57).

The $^{12}C$ and $^{16}O$ profiles at the end of the first thermal pulse are
Rayleigh-Taylor unstable, due to the peak in the oxygen profile, and they

will be rehomogeneized by convection (Isern et al. 1996). The right panel of Figure 1 shows the oxygen profile obtained for the core of the 3.2 $M_\odot$ model at the end of the first thermal pulse (dotted line), and the resulting profile after rehomogeneization (dotted-dashed line). The profiles after Rayleigh-Taylor rehomogeneization are the initial profiles adopted in our cooling sequences.

## 3. Cooling sequences

We have computed cooling sequences for the carbon-oxygen cores previously described according to the method developed by Díaz-Pinto et al. (1994). A helium envelope of mass $10^{-4} M_{\mathrm{WD}}$ was assumed. We have used the equation of state described in Segretain et al. (1994); phase separation during solidification has been included, using the phase diagram of the carbon-oxygen binary mixture of Segretain & Chabrier (1993).

During the crystallization process of the white dwarf interior, the chemical composition of the solid and liquid phases are not equal. A solid, oxygen-rich core grows and the lighter carbon-rich fluid which is left ahead of the crystallization front is Rayleigh-Taylor unstable and is efficiently redistributed by convective motions in the outer liquid mantle. The net effect is a migration of some oxygen towards the central regions which leads to a subsequent energy release (Mochkovitch 1983, Isern et al. 1996). The final profile for a 0.61 $M_\odot$ white dwarf, when the whole interior has crystallized, is shown in Figure 2 as a solid line.

The cooling times as a function of the luminosity for the different models computed are shown in the left panel of Figure 2. The onset of crystallization is clearly marked by the change in the slope of the cooling curves. As an example of the influence of phase separation in the cooling times, the time taken by a 0.61 $M_\odot$ white dwarf to reach $\log(L/L_\odot) = -4.5$ is 9.9 Gyr, to be compared with 8.9 Gyr when phase separation is neglected.

For comparison we have also computed the cooling sequence for the 0.6 $M_\odot$ white dwarf obtained using the low $^{12}C(\alpha, \gamma)^{16}O$ rate (see Figure 1 for the $^{16}O$ profile). The time necessary to reach $\log(L/L_\odot) \simeq -4.5$, is now 10.3 Gyr (to be compared with 9.2 Gyr if separation is neglected). Two aspects of these latter results deserve further discussion. First, either if phase separation is neglected or not, the cooling ages are larger for the low rate model sequence. This is due to its lower oxygen content, which leads to a larger heat capacity and therefore a slower cooling rate. Second, the delay introduced by phase separation during crystallization down to $\log(L/L_\odot) = -4.5$ is practically the same for both cooling sequences ($\sim 1$ Gyr). The reason for this is twofold: on one hand, in the model computed with the low rate of the $^{12}C(\alpha, \gamma)^{16}O$ reaction less oxygen is available for

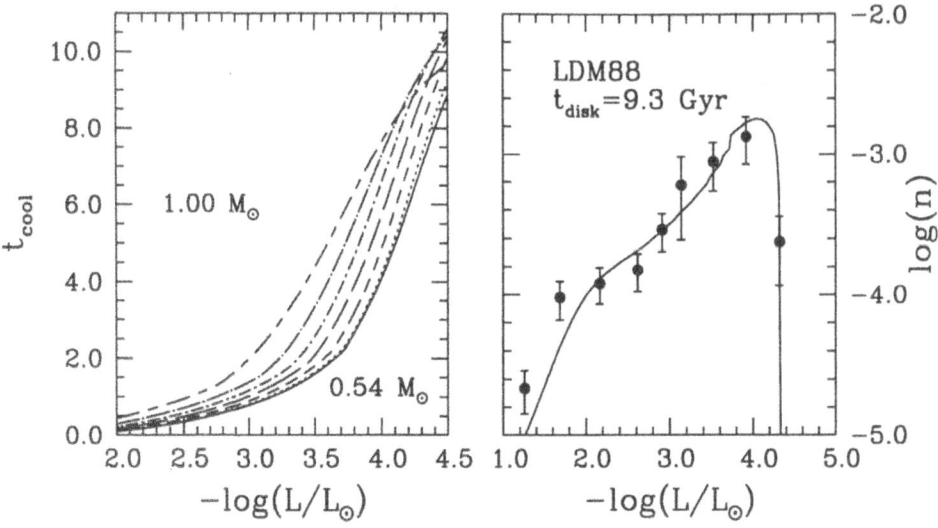

*Figure 2.* Left panel: cooling times for white dwarfs with the masses and chemical compositions quoted in Table 1. Right panel: luminosity function obtained assuming a constant star formation rate per unit volume and an age of the disk of 9.3 Gyr. The observational data come from Liebert, Dahn & Monet (1988).

separation ($M_O = 0.3 M_\odot$ instead of $M_O = 0.4 M_\odot$, see Figure 1) but, on the other hand, the change in its chemical abundance upon crystallization is larger, due to the spindle form of the phase diagram — see Figure 2 in Segretain et al. (1994).

In order to derive an estimate of the age of the solar neighborhood, white dwarf luminosity functions have been computed following the method explained in Hernanz et al. (1994). We have assumed a Salpeter-like initial mass function (Salpeter 1961) and a constant star formation rate per unit volume. The age of the disk that best fits the observational data of Liebert et al. (1988), when adopting blackbody corrections for the cool non-DA white dwarfs (see Figure 4) is 9.3 Gyr. This age of the disk has to be compared with an age of 8.3 Gyr, obtained using the same set of inputs but neglecting phase separation.

## 4. Conclusions

We have examined the influence of the $^{12}C(\alpha, \gamma)^{16}O$ reaction rate on the final structure of carbon-oxygen white dwarfs. For the best choice of the combined effect of convection and the $^{12}C(\alpha, \gamma)^{16}O$ reaction rate, carbon-oxygen profiles showing an enhancement of oxygen in the central regions for all core masses are obtained, whereas for a lower $^{12}C(\alpha, \gamma)^{16}O$ reaction rate, this effect is smaller.

The resulting carbon-oxygen profiles have been used for computing white dwarf cooling sequences, including the effect of phase separation during solidification. This phenomenon leads to a non negligible increase of the cooling ages, which translates into an increase of the age of the disk of the same order. Our best estimate of the age of the disk is 9.3 Gyr, when the data set of Liebert et al. (1988) is used, in contrast with 8.3 Gyr obtained if phase separation is neglected. These values indicate that the effects associated with crystallization should not be neglected when using white dwarfs as a tool to determine the age of the disk.

**Acknowledgements:** This work has been supported by the DGICYT grants PB94-0111 and PB94-0827-C02-02, by the CIRIT grant GRQ94-8001, the AIHF and AIHI. One of us (M.S.) thanks the E.C. for the "Human Capital and Mobility" contract ERBCHGECT920009 of which CESCA (Centre de Supercomputació de Catalunya) is a beneficiary.

# References

Castellani, V., Chieffi A., Pulone, L., Tornambé, A. 1985 *ApJ* **296** 204

Caughlan, G.R., Fowler, W.A., Harris, M.J., Zimmermann, B.A. 1985, *Atomic Data & Nuclear Data Tables*, **32**, 197

Caughlan, G.R., Fowler, W.A. 1988 *Atomic Data & Nuclear Data Tables*, **36**, 411

Chieffi A., Straniero, O. 1989 *ApJS* **71** 47

Díaz-Pinto, A., García-Berro, E., Hernanz, M., Isern, J., Mochkovitch, R. 1994 *Astr. Ap.* **282** 86

Fowler, W.A., Caughlan, G.R., Zimmermann, B.A. 1975 *ARA&A* **13** 69

García-Berro, E., Hernanz, M., Mochkovitch, R., Isern, J. 1990 *Astr. Ap.* **193** 141

Grevesse, N. 1991 *Astr. Ap.* **242** 488

Hernanz, M., García-Berro, E., Isern, J., Mochkovitch, R., Segretain, L., Chabrier, G. 1994 *ApJ* **434** 652

Harris, M.J., Fowler, W.A., Caughlan, G.R., Zimmermann, B.A. 1983 *ARA&A* **21** 165

Iglesias,C. A., Rogers, F. J., Wilson, B. G. 1992 *ApJ* **397** 717

Isern, J., Mochkovitch, R., García-Berro, E., Hernanz, M. 1996 *ApJ* in press

Kurucz, R. L. 1991, in Stellar Atmospheres: Beyond Classical Models, NATO ASI Series C, Vol. 341

Liebert, J., Dahn, C.C., Monet, D.G. 1988 *ApJ* **332** 891

Mochkovitch, R. 1983 *Astr. Ap.* **122** 212

Salpeter, E.E. 1961 *ApJ* **134** 669

Segretain, L., Chabrier, G. 1993 *Astr. Ap.* **271** L13

Segretain, L., Chabrier, G., Hernanz, M., García-Berro, E., Isern, J., Mochkovitch, R. 1994 *ApJ* **434** 641

Thielemann, F.-K., Nomoto, K., Hashimoto, M. 1996 *ApJ* **460** 408

Trautvetter, K 1996 *preprint*

Wood M.A., Winget D.E. 1989, in IAU Colloq. 114, White Dwarfs, ed. G. Wegner (Berlin: Springer), 282

Woosley, S.E., Timmes, F.X., Weaver T.A. 1993, in Nuclei in the Cosmos, Vol. 2, ed. F. Kaeppeler and K. Wisshak (IOP Publishing Ltd), 531

# THEORETICAL MODELS OF WHITE DWARFS WITH HELIUM-RICH INTERIORS

JOSEP M. APARICIO AND GILLES FONTAINE
*Département de Physique, Université de Montréal*
*Montréal, Québec, H3C 3J7 Canada*

**Abstract.** We present the calculation of structure and evolution models of white dwarfs with helium-rich interiors. These are commonly thought to be formed within binary systems with mass exchange, and to be represented by objects in the low-mass wing of the mass distribution function of field stars. We have analyzed models of $0.25 - 0.4 M_\odot$. We present here some sample results for $0.32 M_\odot$.

## 1. Introduction

He-core WD's are a possibility in stellar evolution theory. Stars of low enough mass are supposed to evolve, after its hydrogen burning phase, to a stage where degeneracy appears before helium burning. But such isolated objects are not expected to have yet evolved beyond the main sequence.

However, more massive stars may also become He-core WD's in a time scale smaller than the age of the Galaxy if they belong to a binary system which allows mass loss before He-burning. They can then shorten their main sequence lifetime by means of their higher mass, and then lose most of their envelope. The result will be a very similar object. The loss of the envelope will freeze, or even completely avoid, central helium burning if the final mass is smaller than about $0.45 M_\odot$.

A close look to the mass distribution of field DA white dwarfs suggests that it is not a mediumly wide unimodal distribution, but rather a trimodal one (see Bergeron, Saffer and Liebert 1992). The stars in the low-mass group have masses below the helium burning limit, and thus it is doubtful that they have a C-O core. He-core WD's can thus be recognized by their low mass and the presence of a companion. Interestingly, it has recently been found (Marsh, Dhillon and Duck, 1995) that most of the low-mass DA's in

35

*I. Isern et al. (eds.), White Dwarfs, 35–41.*

the Bergeron, Saffer and Liebert sample are confirmed binaries (they show an oscillating Doppler effect).

The work we present here pretends to modelize the structure of these objects in order to produce a grid of evolutionary sequences. Although many parameters may be necessary to fully describe a binary system, we believe that a 2-parameter representation will suffice to identify all the possible evolutionary sequences of the final helium cores. These are the final mass of the white dwarf after the loss of most of its envelope and the mass of its residual envelope (made up of essentially pure hydrogen due to settling).

## 2.  Input physics used

We have used state of the art constitutive relations for matter in this regime. We mention them in what, we believe, is the order from most to least critical for the final result. The equation of state (EOS) is the most relevant quantity since, a large portion of the envelope being convective, it essentially determines the relationship between central and effective temperatures. Next is the atmosphere model, which provides a realistic upper boundary condition for the interior structure. The importance of the other quantities is, for this kind of object, less significant.

- Equation of state: Helium EOS by Aparicio and Chabrier (1994).
- Atmosphere: Bergeron (see Bergeron, Wesemael and Fontaine 1992).
- Opacity: Radiative opacities from OPAL (Rogers and Iglesias, 1992) with the Grevesse-Noels 93 metal fractions. Conductive opacities from Hubbard and Lampe (1969) for low densities and from Itoh *et al.* (1983, 1984, 1993a,1983,1993b) and Mitake, Ichimaru and Itoh,(1984) for the high densities.
- Reaction rates: Caughlan and Fowler (1988).
- Screening: Partially degenerate Debye, with relativistic corrections for the polarizability of electrons.
- Neutrino cooling: Plasmons from Haft, Raffelt and Weiss (1994). All others from Itoh *et al.* (1990).

## 3.  A sample model

We outline here the properties of a sample model of $0.32M_\odot$ (final mass after mass loss). We have evolved these models from the envelope loss phase until several billion years after that. The amount of hydrogen still present has been left as a parameter, since it is too dependent on details either difficult to modelize or that introduce additional parameters (like the orbital parameters of the binary system, the mass of the companion or the actual physics of the mass loss mechanism).

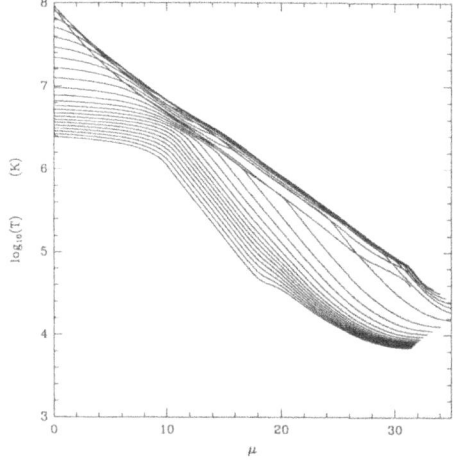

*Figure 1.*    Evolution of the temperature profiles, from $t$=0.1 Gyr (top) to $t = 3.4$ Gyr (bottom) after the loss of the envelope. One of the characteristic features is the region of high *convective* efficiency at about $\mu = 19$ for the cooler models, related to the ionization region of helium (See the gradient picture). The hydrogen envelope begins at about $\mu = 20$ in this example, where the behavior thus changes. The quantity $\mu$ is a lagrangian mass variable defined by $\mu = -\ln(1 - (m/M)^{2/3})$.

As expected, the evolutionary track of a helium WD is qualitatively similar to that of a carbon-oxygen one. After the onset of degeneracy the pressure in both types becomes very similar for equal density and temperature (the number of electrons per baryon is the same). The key points for the difference between them are: the smaller charge per nucleus (which leads to a more uniform, less polarized electron gas, and to smaller Coulomb corrections in the equation of state), and the smaller mass of helium cores (less than about $0.4M_\odot$, as compared to approximately $0.6M_\odot$). The consequences are:

- The limiting central density and surface gravity are smaller (both because of the smaller mass *and* the different EOS). This leaves less gravitational energy available to be released.
- The radius is bigger. It is easier to radiate heat and cooling times are shorter.
- The nuclear charges and the density being smaller, the ion coupling parameter $\Gamma_i$ reaches much smaller values. As can be seen in Figure 3, it is still very far from crystallization after 3.5 Gyr.

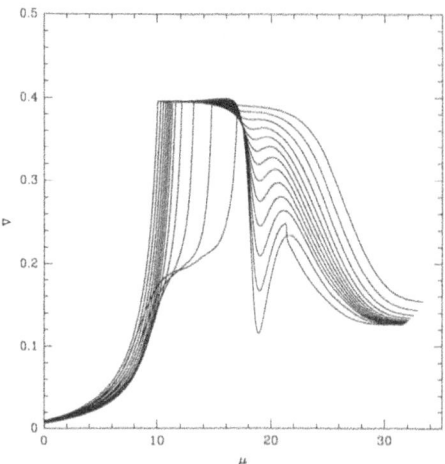

*Figure 2.* Evolution of the gradient profiles, from $t = 0.6$ Gyr to $t = 3.4$ Gyr after the loss of the envelope (The convective envelope deepens with time). The feature that appears in the temperature profile as a quite isothermal region around $\mu = 19$ is shown here as a low adiabatic gradient region where helium becomes recombined. This neutral region is surrounded by ionized material: ionized He below and ionized H above.

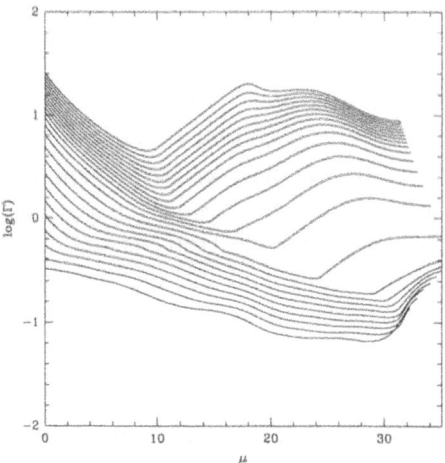

*Figure 3.* Evolution of the Coulomb coupling parameter, from $t = 0.1$ Gyr (bottom) to $t = 3.4$ Gyr (top) after the loss of the envelope. Overall, the plasma becomes more coupled as evolution proceeds. However, even after 3.5 Gyr, it only reaches central values of about 50, far from crystallization (i.e., $\Gamma \simeq 180$). The central, conductive region, is more coupled towards the center, as usual in WD's. In the convective envelope, however, the higher temperature gradient leads to an outwards increase of the coupling. The recombination of He is observable here as a leveling in this progression.

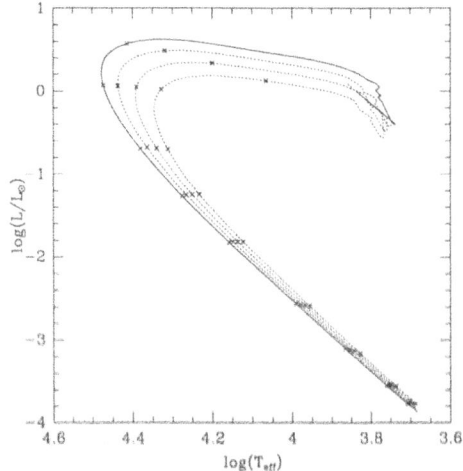

*Figure 4.* Evolution in the HR diagram. The standard track analyzed above of $0.32 M_\odot$ (solid) is compared against other tracks (dotted) of 0.26, 0.28 and $0.30 M_\odot$. The crosses indicate the following ages after the envelope loss (yrs): $1 \cdot 10^7$, $2 \cdot 10^7$, $5 \cdot 10^7$, $1 \cdot 10^8$, $2 \cdot 10^8$, $5 \cdot 10^8$, $1 \cdot 10^9$, $2 \cdot 10^9$ and $3 \cdot 10^9$.

## 4. Discussion

### 4.1. IMPLICATIONS FOR WD COOLING AND THE LUMINOSITY FUNCTION

The presence of structurally different classes of WD may introduce a systematic bias in the analysis of the luminosity function. It must be taken into account that not all observed DA white dwarfs have necessarily carbon-oxygen cores, but some of them may be formed of helium. In order to avoid this bias, one can take measures to select only one class of DA's (e.g. C-O core ones), for example, by making sure that the masses of the stars in the sample is high enough to have C-O cores. Alternatively, the possibility for differences in the structure of a subset of the sample of DA's can be taken into account. Thus, it is also important to consider He-core cooling sequences, not just C-O core ones.

### 4.2. NEXT STEPS

Although these stars differ from a C-O white dwarf in their core structure, their envelope is quite similar to that of a "normal" DA. It is known that the presence of pulsations in DAV's or DBV's is associated to instabilities driven by the *outer* layers, in the partially ionized regions of the envelope. Thus, in a part of the star which is less different from a C-O DA.

We intend to analyze the seismological properties of these objects and search whether or not there is an evolutionary stage where excitation may take place. By contrast to their DA envelope, which favors the onset of instabilities, their surface gravity is relatively low, which is known to unfavor them. In this sense, there is evidence indicating that the red and blue edges of the DAV band converge at smaller surface gravities (see Pierre Brassard's presentation in these proceedings). An extrapolation indicates that under the conditions of He-core DA's (about $\log g = 7.5$) the unstable band is considerably reduced, and possibly closed.

None of the presently known low-mass DA's, however, lies within the DAV temperature range. Thus no luminosity variations should be expected from the currently known objects.

**Acknowledgments** This work has received financial support from Spanish DGICYT project PB94-0111, the NSERC Canada and the Fund FCAR (Québec).

# References

Aparicio, J.M. and Chabrier, G. (1994), Free energy model for fluid atomic helium at high density, *Phys. Rev. E*, **50** #6, pp. 4948-4960.

Bergeron, P.,Saffer, R.A. and Liebert, J. (1992), Spectroscopic determination of the mass distribution of white dwarfs, *Ap. J.*, **394**, pp. 228-247.

Bergeron, P.,Wesemael, F. and Fontaine, G. (1992), On the influence of the convective efficiency on the determination of the atmospheric parameters of DA white dwarfs, *Ap. J.*, **387**, pp. 288-293.

Caughlan, G.R. and Fowler, W.A. (1988), *At. Data Nucl. Data Tables*, **40**, p. 283.

Haft, M., Raffelt, G. and Weiss, A (1994), Standard and nonstandard plasma neutrino emission revisited, *Ap. J.*, **425**, pp. 222-230.

Hubbard, W.B. and Lampe, M. (1969), Thermal conduction by electrons in stellar matter, *Ap. J. Suppl.*, **18**, pp. 297-346.

Itoh, N., Mitake, S., Iyetomi, H. and Ichimaru, S. (1983), Electrical and thermal conductivities of dense matter in the liquid metal phase. I. High temperature results, *Ap. J.*, **273**, pp. 774-782.

Itoh, N., Kohyama, Y., Matsumoto, N. and Seki, M. (1984), *Ap. J.*, **285**, pp. 758-765.

Itoh, N., Adachi, T., Nakagawa, M., Kohyama, Y. and Munakata, H. (1990), Neutrino energy loss in stellar interiors. III. Pair, photo, plasma and bremsstrahlung processes, *Ap. J.*, **285**, pp. 758-765, erratum *Ap. J.*, **360**, p. 741.

Itoh, N. and Kohyama, Y. (1993), *Ap. J.*, **404**, pp. 268-270.

Itoh, N., Hayashi, H. and Kohyama, Y. (1993), *Ap. J.*, **418**, pp. 405-413.

Mitake, S., Ichimaru, S. and Itoh, N. (1983), Electrical and thermal conductivities of dense matter in the liquid metal phase. II. Low temperature quantum corrections, *Ap. J.*, **277**, pp. 375-378.

Marsh, T.R., Dhillon, V.S. and Duck, S.R. (1995), Low-mass white dwarfs need friends: five new double-degenerate close binary stars. *Mon. Not. R. Astron. Soc.*, **275**, pp. 828-840.

Rogers, F.J. and Iglesias, C. (1992), Rosseland mean opacities for variable compositions, *Ap. J.*, **401**, pp. 361-366.

## Discussion

*I. Iben*: Could you describe the initial conditions for your models and explain whether you do or not find the hydrogen shell flashes obtained by Kippenhahn et al. in the 1960's (also by Iben and Tutukov in 1986) for models constructed by abstracting from a red giant model?

*J.M. Aparicio*: The initial conditions are taken from an object in the horizontal branch from which a given portion of the H-rich envelope has been stripped. The mass loss itself is not modeled in any way. Instead, the mass of the remaining envelope is parameterized. In the examples that I have shown this assumed mass was too small to allow H-burning. For thicker envelopes I do indeed find unstable H-burning.

*J. Isern*: How much time takes a $0.45 M_\odot$ He white dwarf to reach a luminosity of the order of $10^{-4.5} L_\odot$?

*J.M. Aparicio*: A mass of $0.45 M_\odot$ is slightly large for this kind of object and close to the He-burning limit. Anyway, objects in this mass range take about 4.5 Gyr to reach this level *after* the mass transfer period.

# THE ELECTROSTATIC SCREENING OF THERMONUCLEAR REACTIONS IN ASTROPHYSICAL PLASMAS

NIR J. SHAVIV AND GIORA SHAVIV
*Department of Physics*
*Technion Israel Institute of Technology*
*Haifa 32000 Israel*

**Abstract.** The classical electrostatic screening is revisited. We find major revisions already in the weak screening limit. Here we discuss the classical expression for the screening and show that the picture used to derive the screening is wrong. When corrected, all the screening factors are multiplied by a factor of 3/2 in the exponent.

## 1. Introduction

The classical expression for the rate of a nuclear reaction between specie $i$ and $j$ with number densities $n_i$ and $n_j$ in a plasma is given by: (e.g., Clayton 1968)

$$R_{ij} = n_i n_j \int f(\vec{v}_i) f(\vec{v}_j) \sigma_{ij} \, |\vec{v}_{ij}| \, dv_{ij}, \qquad (1)$$

where $f(v)$ is the velocity distribution of the relevant particle, $\sigma_{ij}$ is the nuclear cross section and $v_{ij}$ is the relative velocity. This expression assumes that:

1. The particles are randomly distributed and the probability to find the two interacting particles close to each other is independent of the possible existence of a neighboring particle, namely, it assumes that the two particle distribution function can be written as

$$f(\vec{x}_i, \vec{v}_i, \vec{x}_j, \vec{v}_j) = f(\vec{v}_i) f(\vec{v}_j) g_{ij} \quad \text{with} \quad g_{ij}(r, v = |v_i - v_j|) \equiv 1 \qquad (2)$$

and where $g_{ij}(r, v)$ is the two particle correlation function and $v$ the relative velocity between the particles. This assumption is known as molecular chaos.

43

*I. Isern et al. (eds.), White Dwarfs, 43–48.*
© 1997 *Kluwer Academic Publishers.*

2. The interaction between the particles is limited to a small region in space and the interacting particles effectively arrive from infinity where the interaction vanishes and the particles can be considered as moving in free space. The particles feel the self consistent field in the plasma as expressed by the effective interaction potential. This is the *Stosszahlansatz* assumption and it is assumed in the derivation of the Boltzmann equation to provide closure. Generally, this assumption is good whenever the interactions are short range and the interacting particles move a large distance "as free particles on a straight line". In general, it is a good assumption in rarified gases, but not in stellar cores.

The discussion of the above points is deferred to a later communication and we mention them to warn the reader. Here we discuss the validity of the classical picture of electrostatic screening. To this goal we examine first the interacting particles. The energy of the Gamow peak and its width are given in Table 1. The numbers are evaluated for the conditions at the center of the Sun, namely, $T = 1.5 \times 10^7 K$.

The examination of the numbers leads us to conclude that:

1. The pp reaction takes place at almost thermal energies. Namely, the Gamow peak for the pp reaction is almost at thermal energies. Also, the peak for the pp reaction is as wide as 5kT, namely all particles participate in the reaction.

2. The Gamow peak for the other reactions in the pp chain is at about (15-20)kT. The peak is as wide as 10kT.

The electrostatic screening is the extra increase in energy due to the collective effect of other ions and is called the screening energy while its effect on the rate of the reaction is called the screening factor. The relevant parameter is:

$$\Gamma = E_{screening}/kT = \frac{Z^2 e^2}{\langle r \rangle kT}. \tag{3}$$

where $Z$ is the effective charge of the plasma and $\langle r \rangle$ is the mean interparticle distance. The weak screening limit is $\Gamma \ll 1$. In this limit assume *for a moment* that the Debye-Huckel theory is valid. Strong screening is: $\Gamma \gg 1$ and will be discussed elsewhere.

The classical Salpeter picture of the screening is the following: One particle (the target) is at rest and screened and the other, the fast one, is not screened. The screening energy is the energy that the fast particle gains when it falls into the potential well of the target particle.

But a) We note that when the particle distributions are Maxwellian most collisions will be between particles of about equal energy and opposite velocity vectors. Hence, for the pp reaction the relevant energy is 2.5kT.

TABLE 1. The Gamow energy of various relevant reactions

| Reaction | $E_{Gamow}/kT$ | $\Delta/kT$ |
|----------|----------------|-------------|
| p+p | 4.8 | 4.93 |
| $He^3 + He^3$ | 14.0 | 9.40 |
| $He^3 + He^4$ | 14.6 | 9.62 |
| $p + Be^7$ | 11.7 | 8.50 |
| $p + N^{14}$ | 17.3 | 10.47 |

b) Both particles are situated in a potential well. As they interact, the two wells merge to form a new deeper well around the resulting nucleus and hence, the energy gained is equal to the difference between the initial two wells and the final one.

## 2. What about dynamic effects?

Do the screening clouds follow the fast particles? The answer to these questions is given by a Molecular Dynamics simulation that we carried out and we can state already here that for the energies under considerations here say, below 20kT, the screening cloud follows the protons. On the other hand, it is of interest to examine the number of particles in the screening (Debye) cloud.

$$N_D = \frac{4\pi}{3} R_D^3 n = \frac{4\pi}{3} \left( \frac{R_D}{\langle r \rangle} \right)^3 = \frac{1}{4\pi\sqrt{6}\Gamma^{3/2}}, \tag{4}$$

In the core of the Sun $N_D = 1$-3. Hence, expect large fluctuations. These are discussed separately.

## 3. Derivation of the screening energy

In the derivation of the screening energy when the two particles are screened we assume the following assumptions:

1. The two ions move sufficiently slowly to carry with them their screening cloud. In other words, the screening cloud of an ion does not depend on its velocity relative to the plasma.
2. When the ions approach each other, the approach is sufficiently slow so that a new Debye cloud has the time to form around the two ions.
3. The above two assumptions mean that we can assume that the addition to the kinetic energy of the interacting particles, namely the screening

correction, can be derived from the difference in the potential energy between the two following states: the close ions and the two separated ions.

4. The two interacting ions start their relative motion from infinity.

Again, these assumptions are verified with Molecular Dynamics simulations.

## 4.  Derivation of the relevant energies

Two energies are involved in the merger of two protons:

- − The self interaction energy of the charge clouds.
- − The ion-charge cloud interaction.

We assume first that the Debye-Huckel theory is valid. The purpose is to obtain an estimate to the difference between the two pictures discussed above. The DH theory is used to derive the charge distribution. For actual numbers we use the results of the Molecular Dynamics. The complete expressions for the energies are:

$$U_{cloud-cloud} = \frac{1}{2} \int \frac{\rho_{cloud}(\vec{x})\rho_{cloud}(\vec{y})}{|\vec{x}-\vec{y}|} d^3x d^3y = \frac{Z_i^2 e^2}{4R_D} \tag{5}$$

$$U_{ion-cloud} = \int \frac{\rho_{ion}(\vec{x})\rho_{cloud}(\vec{y})}{|\vec{x}-\vec{y}|} d^3x d^3y = -\frac{Z_i^2 e^2}{R_D} \tag{6}$$

where $\rho_{ion} = Z_i e \delta^3(r)$.

The total potential energy of the combined ion is therefore:

$$u_i = -\frac{3(Z_1^2 + Z_2^2)e^2}{4R_D}, \tag{7}$$

and the electrostatic screening energy is:

$$\Delta u = u_{initial} - u_{final} = \frac{3}{2}\frac{Z_1 Z_2 e^2}{R_D}. \tag{8}$$

The new result for the effective interaction energy is therefore:

$$E_{eff} \equiv E_{screening} = E_{kinetic} + \frac{3}{2}\frac{Z_1 Z_2 e^2}{R_D} \tag{9}$$

The old Salpeter result is:

$$E_{eff} \equiv E_{screening} = E_{kinetic} + \frac{Z_1 Z_2 e^2}{R_D}. \tag{10}$$

The differences between the two resuts are due to the following assumptions:

TABLE 2. Parameters of the Models Calculated

| | $n_{26}$ | $T_6$ | $R_D/\langle r \rangle$ | $\Gamma$ | Composition |
|---|---|---|---|---|---|
| case I | 1 | 15 | 0.877 | 0.0517 | Pure H |
| case II | $10^3$ | 15 | 0.277 | 0.517 | Pure H |
| case III | 1 | 15 | 1.132 | 0.103 | H/He mixture |

TABLE 3. Screening Factors for the pp Reaction for $T = 1.5 \times 10^7 K$

| Model | Density(cgs) | Graboske et al. | Present | Screening |
|---|---|---|---|---|
| case I | 166. | 0.041 | 0.0926 | weak |
| case II | $1.66 \times 10^5$ | 1.2964 0.7744 0.8037 | 1.86 | weak intermediate strong |
| case III | 291. | 0.0542 0.02416 | 0.1264 | weak intermediate |

- The neglect of the energy required to form the new electronic cloud around the newly formed nucleous.
- The screening of the projectile.

We note that if we neglect in the new picture the formation of the charge cloud (and include only the screening of the projectile) the factor is 2 not 3/2.

We calculated several models the details of which are given in Table 2. The resulting screening factors for the pp reaction are given in Table 3 where they are compared with the classical approximation of Graboske et al. (1973). The approximation used to derive the Graboske et al. (1973) result is given in the last column. Our results were calculated using the Molecular Dynamic simulations. Recall that the energy enters into the exponent of the correction factor. Thus the pp reaction must be multiplied by the exponent of the screening energy.

### References

Clayton, D.D., 1968, Stellar Structure and Evolution, Academic Press.
Graboske, H.C., De Witt, H.E., Grossman, A.S., Cooper, M.S., 1973, ApJ, 181, 457
Salpeter, E.E., 1954, Australian J. of Physics, 7, 373

## Discussion

*M. van Kerkwijk*: What is the effect on the central temperature of the Sun? And, how does it affect the neutrino problem?

*G. Shaviv*: With the effects discussed here, the central temperature decreases by a few percent. However this is not the complete story because a) not all effects are included and b) we have to look how the effects change with the reaction. We plan to do so shortly.

*J. Kubát*: I am not an expert in this field and maybe I missed something, but one point is not clear to me. Considering two particles in motion should be the same as choosing a reference frame in such a way that one particle is at rest, which is the Salpeter's assumption.

# THE POSSIBLE WHITE DWARF–NEUTRON STAR CONNECTION

R. CANAL AND J. GUTIERREZ

*Department of Astronomy, University of Barcelona*

*Martí i Franqués, 1 - 08028 Barcelona, Spain*

**Abstract.** The current status of the problem of whether neutron stars can form, in close binary systems, by accretion–induced collapse (AIC) of white dwarfs is examined. We find that, in principle, both initially cold C+O white dwarfs in the high–mass tail of their mass distribution in binaries and O+Ne+Mg white dwarfs can produce neutron stars. Which fractions of neutron stars in different types of binaries (or descendants from binaries) might originate from this process remains uncertain.

## 1. Introduction

Gravitational collapse of the Fe–Ni cores of massive stars (initial masses $M \gtrsim 10 - 12 \ M_\odot$) that have reached the end of their thermonuclear evolution is the standard mechanism to form neutron stars (NSs). A supernova explosion ejecting several solar masses of material at high velocities should simultaneously occur in order to get rid of the large mass excess of the object over the maximum possible mass of a NS ($\simeq 2.0 - 2.5 \ M_\odot$). The problem of transferring $\sim 1\%$ of the gravitational energy released in the collapse to the mantle and envelope of the star has not been completely solved yet, but there is little doubt that both isolated pulsars and NSs with massive companions in binaries have formed through this mechanism. NSs, however, are also found in binary systems where the companion is a low–mass ($M \sim 1 \ M_\odot$) star. That raised, already long ago, the question of how such systems might have survived to NS formation by the standard mechanism without being disrupted by the explosive ejection of more than half their total mass.

Canal & Schatzman (1976) (following an earlier suggestion by Schatzman 1974) first proposed that accretion of matter from the companion by a massive white dwarf (WD) in a binary might lead to the formation of a

*I. Isern et al. (eds.), White Dwarfs*, 49–55.

NS with little mass ejection, and presented a preliminary model in which thermonuclear explosion in the stage preceding gravitational collapse of a C+O WD was avoided. The model was further developed by Canal & Isern (1979), Canal, Isern, & Labay (1980), Isern et al. (1983), Hernanz et al. (1988), and Canal et al. (1990). Miyaji et al. (1980), Nomoto (1982, 1987), and Miyaji & Nomoto (1987) considered the case of an O+Ne+Mg WD (see Canal, Isern, & Labay 1990; Canal 1994, for reviews). This possible mechanism of nonexplosive NS formation from WDs is designated in the current literature by the term *accretion–induced collapse* (AIC) (most often without any mention to its original proponents or even with erroneous attribution to other authors: see Verbunt 1993 and reviews by van den Heuvel, for recent examples of it).

AIC of WDs has been proposed in different scenarios to explain the formation of low–mass X–ray binaries, binary pulsars with low–mass companions, binary millisecond pulsars, and single millisecond pulsars. It has even been suggested as a possible mechanism for $\gamma$–ray bursts. In all cases, however, alternative origins for the NSs involved seem also possible (such as capture of the companion by a previously formed, single neutron star). Here we will briefly review the current status of the problem of forming NSs from both C+O and O+Ne+Mg WDs.

## 2. The AIC scenario

The usual AIC *cartoon* (van den Heuvel 1981, 1984) has a WD with a mass below the Chandrasekhar mass accreting material from a low–mass companion which is filling its Roche lobe. When the Chandrasekhar mass is reached, the WD collapses to nuclear matter densities and, the mass of the newly formed object being below the maximum mass for NSs, collapse stops there. Further mass accretion produces X–ray emission and the object becomes a low–mass binary X–ray source.

If the WD were made of completely inert material (no thermonuclear burning, no electron captures), collapse would start due to general–relativistic instability when $M = 1.366\ M_\odot$, $R = 996\ km$, and $\rho_c = 2.495 \times 10^{10}\ g$, for $Z = 6$ material (Canal & Schatzman 1976). WDs, however, are actually made of materials such as He, C+O, and O+Ne+Mg, depending on at which stage of the progenitor evolution its envelope was lost. There the problems start.

— He would definitely explode at much lower densities.
— C typically ignites (explosively) at $\rho \sim (2-3) \times 10^9\ g\ cm^{-3}$.
— O starts to capture electrons at $\rho \sim 2 \times 10^{10}\ g\ cm^{-3}$.
— Ne does the same at $\rho \sim 9.5 \times 10^9\ g\ cm^{-3}$
— Mg begins to capture electrons at even lower densities: $\rho \sim 4 \times 10^9\ g\ cm^{-3}$

Therefore, He WDs are absolutely excluded and in C+O WDs explosive C ignition should always precede collapse. In O+Ne+Mg WDs, $e^-$–captures (ECs) (which produce heating) also precede collapse. However, $M_{Ch} \propto Y_e^2$ ($Y_e$ being the electron mole number), and thus ECs do lower the Chandrasekhar mass. That indicates the main strategy for constructing successful AIC models: to delay explosive ignition (or reduce its effects) until ECs make $M_{Ch} < M_{WD}$.

In the *C+O case*, the preceding means keeping the material cool enough so that the $^{12}C +^{12} C$ reactions take place in the *pycnonuclear* regime (Canal & Schatzman 1976). That is not enough, however, since explosive C ignition still takes place before the start of ECs on O. Nevertheless, explosive C burning transforms the C+O mixture into matter in *nuclear statistical equilibrium* (NSE). Then ECs become fast (especially on free protons) and the Chandrasekhar mass starts decreasing. On the other hand, the explosive burning initiated at the center of the WD propagates outwards and with the energy release the star begins to expand (which would at some point cut–off the ECs). Therefore, only if the burning does not propagate very fast ($v_{burn} \ll c_s$, $c_s$ being the local sound velocity) we can have $M_{Ch} < M_{WD}$ before significant expansion. There, a *solid* WD, in addition to retarding C ignition, would ensure low $v_{burn}$ by suppressing hydrodynamic instabilities and allowing only the slow, purely conductive mode of burning propagation (Canal & Isern 1979).

In the *O+Ne+Mg case*, ECs would start at lower densities. Their effect in heating the material and eventually inducing explosive thermonuclear burning dominates that of decreasing the Chandrasekhar mass and ECs on $^{20}$Ne trigger the explosion. Convective heat transport, however, would delay explosive ignition up to $\rho \gtrsim 2 \times 10^{10}$ $g$ $cm^{-3}$ and the ECs on the NSE material would then rapidly reduce $M_{Ch}$, however fast burning might propagate (Miyaji et al. 1980).

The preceding sets the stage for the current debate on the feasability of AIC from the two types of WDs: C+O and O+Ne+Mg. It involves different aspects of the physics of matter at high densities, the dynamics of thermonuclear burning, mass–accretion processes, and the evolution of intermediate–mass stars in close binary systems.

## 3.  C+O white dwarfs

Nonhomologous heating upon mass accretion (the outer layers of the WD are much more compressible than the central ones) generates a "heat wave" that advances from the surface towards the center and progressively melts the solid core that might have previously formed (Hernanz et al. 1988). Keeping the core solid until C burning starts in the pycnonuclear regime requires either very low or very high $\dot{M}$. The minimum $\dot{M}$, in the "fast" accretion case, increases with decreasing WD masses: larger initial WD masses thus favor collapse, but the maximum mass for C+O WDs is currently estimated to be $M_{C+O} \lesssim 1.1-1.2\ M_\odot$. Taking into account the effects of rotation on the evolution of the cores of AGB stars might, however, increase the limit almost up to the Chandrasekhar mass (Domínguez et al. 1996).

The explosive C ignition density is further sensitive to the approximations adopted for the pycnonuclear reaction rates (*static* vs *relaxed* approximations), to the crystal structure in the solid phase (blocking effects of the O nuclei on the $^{12}C + ^{12}C$ reactions), and even to the way the interpolation between the pycnonuclear and the strong screening reaction rates is made (Isern & Hernanz 1994; Isern & Canal 1994).

Even if the core melts before C ignition, burning still starts propagating conductively and collapse will occur for $\rho_{ign} \gtrsim 8.5 \times 10^9\ g\ cm^{-3}$ (García et al. 1990; Timmes & Woosley 1992). The exact ignition density depends on all of the above factors.

## 4. O+Ne+Mg white dwarfs

Those WDs should come from the evolution of stars in the $8\ M_\odot \lesssim M \lesssim 11 M_\odot$ initial mass range. The debate, here, has been mostly centered on the dependence of the explosive Ne ignition density from the treatment of the convective instability produced by EC heating in the immediately previous stages.

The very high ignition densities ($\rho_{ign} \simeq 2.5 \times 10^{10}\ g\ cm^{-3}$) found by Miyaji et al. (1980) depended on their adoption of the *Schwarzschild criterion* for the start of convection. That neglected the stabilizing effect of the $Y_e$-gradient set up by the electron captures (Mochkovitch 1984). Adoption of the *Ledoux criterion* lowered the ignition density to $\rho_{ign} \simeq (8.0 - 9.5) \times 10^9\ g\ cm^{-3}$ (Miyaji & Nomoto 1987; Canal, Isern, & Labay 1992), leaving some room to the possibility that O+Ne+Mg WDs would explode rather than collapse (Isern, Canal, & Labay 1992).

According to the most recent and accurate model calculations (Gutiérrez et al. 1996a), irrespective of the assumption made as to the development of convection, explosive ONe ignition takes place in the range $9.70 \times 10^9\ g\ cm^{-3} \leq \rho_{ign} \leq 2.12 \times 10^{10}\ g\ cm^{-3}$. That means that gravitational collapse should be the outcome in any case.

Calculations of the evolution of stars in the relevant mass range (Nomoto 1987; Domínguez, Tornambé, & Isern 1994; García–Berro & Iben 1994; Ritossa, García–Berro, & Iben 1996) have consistently shown that a fraction of C is always left unburned at the center after formation of the O+Ne

(+ very little Mg) core. Upon further compression, it might ignite explosively if the C abundance were high enough. Gutiérrez et al. (1996b) have shown, however, that unless the C mass fraction were higher than about 5% (which is distinctly more than the evolutionary model predictions, with the exception of the very low–mass end of the interval) the star should not be disrupted.

## 5.  Conclusions

*C+O WDs* are candidates to form NSs by AIC, provided that $\rho_{ign} \gtrsim 8.5 \times 10^9 \ g \ cm^{-3}$, but if the upper mass limit for those WDs were $M_{C+O} \lesssim 1.1 \ M_\odot$ the initial mass range might become excessively narrow. That, however, still depends on the approximations adopted for the pycnonuclear reaction rates, on the crystal structure in solid WDs, and on the way the interpolation between the pycnonuclear and the strong screening regimes is made in the models. Also, the effects of core rotation during the AGB phase might significantly increase the upper mass limit for C+O WDs.

*O+Ne(+Mg) WDs* can form NSs, unless a fraction of C larger than that predicted by current evolutionary models were left unburned at the end of the C–burning stage. NSs might also result from mergers of C+O WD pairs leading to the formation of a fast spinning O+Ne WDs which later would collapse and produce a single millisecond pulsar (Mochkovitch, Guerrero, & Segretain 1996).

There is an upper limit to the AIC rate: deleptonization of the proto–NS would produce a neutrino–driven wind carrying with it heavily neutronized material, and that should not excessively pollute the Galaxy (Woosley & Baron 1992). Work on the statistics is now still in progress, but the preliminary results indicate that AIC might explain the origin of NSs with low–mass binary companions without violating the nucleosynthesis constraints.

## References

Canal, R. 1994, in *Supernovae*, ed. S.A. Bludman, R. Mochkovitch & J. Zinn–Justin, Amsterdam, North–Holland, 155

Canal, R., García, D., Isern, J., & Labay, J. 1990, *ApJ*, **356**, L51

Canal, R., & Isern, J. 1979, in *White Dwarfs and Variable Degenerate Stars*, ed. H.M. Van Horn & V. Weidemann, Rochester, Univ. Rochester Press, 52

Canal, R., Isern, J., & Labay, J. 1980, *ApJ*, **241**, L33

Canal, R., Isern, J., & Labay, J. 1990, *ARA&A*, **28**, 183

Canal, R., & Schatzman, E. 1976, *A&A*, **46**, 229

Domínguez, I., Straniero, O., Tornambé, A., & Isern, J. 1996, in *Thermonuclear Supernovae*, ed. P. Ruiz–Lapuente, R. Canal, & J. Isern, Dordrecht, Kluwer, 177

Domínguez, I., Tornambé, A., & Isern, J. 1994, *ApJ*, **419**, 268

García, D., Labay, J., Canal, R., & Isern, J. 1990, in *Nuclei in the Cosmos*, ed. H. Oberhummer & W. Hillebrandt, Garching, MPA, 97

García–Berro, E., & Iben, I. Jr. 1994, *ApJ*, **404**, 306

Gutiérrez, J., García–Berro, E., Iben, I. Jr., Isern, J., Labay, J., & Canal, R. 1996a, *ApJ*, **459**, 701

Gutiérrez, J., Canal, R., Labay, J., Isern, J., & García–Berro, E. 1996b, in *Thermonuclear Supernovae*, ed. P. Ruiz–Lapuente, R. Canal, & J. Isern, Dordrecht, Kluwer, 303

Hernanz, M., Isern, J., Canal, R., Labay, J., & Mochkovitch, R. 1988, *ApJ*, **324**, 331

Isern, J., Canal, R., & Labay, J. 1992, *ApJ*, **372**, L83

Isern, J., & Canal, R. 1994, in *The Equation of State in Astrophysics*, ed. G. Chabrier & E. Schatzman, Cambridge, Cambridge Univ. Press, 186

Isern, J., & Hernanz, M. 1994, in *The Equation of State in Astrophysics*, ed. G. Chabrier & E. Schatzman, Cambridge, Cambridge Univ. Press, 106

Miyaji, S., Nomoto, K., Yokoi, K., & Sugimoto, D. 1980, *Publ. Astron. Soc. Japan*, **32**, 303

Miyaji, S., & Nomoto, K. 1987, *ApJ*, **318**, 307

Mochkovitch, R. 1984, in *Problems of Collapse and Numerical Relativity*, ed. D. Banzel & M. Signore, Dordrecht, Reidel, 125

Mochkovitch, R., Guerrero, J., & Segretain, L. 1996, in *Thermonuclear Supernovae*, ed. P. Ruiz–Lapuente, R. Canal, & J. Isern, Dordrecht, Kluwer, 187

Nomoto, K. 1982, *ApJ*, **253**, 798

Nomoto, K. 1987, *ApJ*, **322**, 206

Ritossa, C., García–Berro, E., & Iben, I. Jr. 1996, *ApJ*, in press

Timmes, F.S., & Woosley, S.E. 1992, *ApJ*, **396**, 649

van den Heuvel, E.P.J. 1981, in *Fundamental Problems in the Theory of Stellar Evolution*, ed. D. Sugimoto, D.Q. Lamb, & D.N. Schramm, Dordrecht, Reidel, 155

van den Heuvel, E.P.J. 1983, in *Accretion–Driven Stellar X–Ray Sources*, ed. W. H. G. Lewin & E.P.J. van den Heuvel, Cambridge, Cambridge Univ. Press, 303

Verbunt, F. 1993, *ARA&A*, **31**, 93

Woosley, S.E., & Baron, E. 1992, *ApJ*, **391**, 228

## Discussion

*M.H. van Kerkwijk*: What limits on the mass accretion rate are there for the O+Ne+Mg WDs?

*R. Canal*: They depend on the chemical composition of the accreted material. If it is H or He there should, in principle, be no difference with the C+O WD case: $\dot{M} \gtrsim 5 \times 10^{-8} \ M_\odot \ yr^{-1}$ to avoid both explosive H and He ignitions. There is some evidence that a few WDs in cataclysmic variable binaries are growing in mass from fast accretion of H. Their masses seem to be above 1.2 $M_\odot$, which would be consistent with their being O+Ne+Mg WDs.

# THE IMPACT OF EVOLUTIONARY ENVELOPE MASSES ON THE EVOLUTION OF WHITE DWARFS

T. BLÖCKER

*Institut für Astronomie und Astrophysik, 24098 Kiel, Germany*

AND

F. HERWIG, T. DRIEBE, H. BRAMKAMP, D. SCHÖNBERNER

*Astrophysikalisches Institut Potsdam, 14473 Potsdam, Germany*

## 1. Introduction

Whether white dwarf (WD) models should have "thin" or "thick" envelopes of H and He has been a matter of debate for several years. It is well known that the WD evolution depends sensitively on this question (see e.g. Wood 1995). Evolutionary calculations (e.g. Paczynksi 1971) show that at the tip of the Asymptotic Giant Branch (AGB) the envelope masses are tightly correlated with the mass of the hydrogen exhausted core ($\approx$ total mass). Accordingly the masses of hydrogen, $M_H$, and helium, $M_{He}$, on top of the degenerate C/O interiors decrease by orders of magnitudes with increasing stellar mass. In contrast, many applications of WD calculations consider only single values of $q_{H,He} = \log(M_{H,He}/M_*)$ asuming *either* "thick" *or* "thin" envelopes.

We demonstrate the differential effects of evolutionary envelope masses in comparison with purely thick and thin layers by two typical applications: the mass-radius relation and the mass distribution of DA white dwarfs. Furthermore, we discuss the importance of He white dwarfs for the respective low-mass tails in these relations.

## 2. Evolutionary Envelope Masses

We used the AGB and post-AGB calculations for Pop.I stars between 1 and $7M_\odot$ of Blöcker (1995) which consider mass loss on the AGB by shock- and dust-driven winds (Bowen 1988) and during the post-AGB stage by the radiation driven wind theory of Pauldrach et al. (1988). The resulting

*I. Isern et al. (eds.), White Dwarfs, 57–62.*

T. BLÖCKER ET AL.

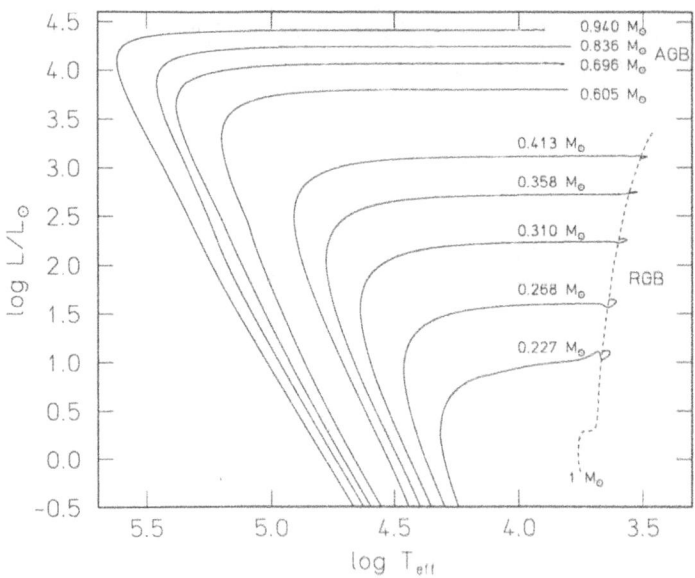

*Figure 1.* Evolutionary tracks for the (pre) C/O und He WDs.

final masses range between 0.53 and 0.94$M_\odot$ and are consistent with the empirical initial-final mass relation of Weidemann (1987).

The lower mass limit for C/O WDs was found to be $\approx 0.45 M_\odot$ since this is the threshold mass for central helium-burning. Thus, WDs below that mass are He white dwarfs. He WDs can be assumed to be products of binary evolution where mass transfer along the Red Giant Branch (RGB) strips off the hydrogen-rich envelope (cf. Kippenhahn et al. 1967, Iben & Tutukov 1986). We simulate this evolutionary channel simply by invoking strong mass-loss rates for a $1 M_\odot$ model at different points on the RGB yielding final masses between 0.227 and 0.413$M_\odot$ (Driebe 1996). Note, that for $M \lesssim 0.3 M_\odot$ the hydrogen shell gets thermally unstable along the cooling track leading to one or multiple loops in the HR diagram.

Fig. 1 illustrates the corresponding evolutionary tracks for selected (pre) C/O und He WDs. In Fig. 2 we show the masses of hydrogen, $M_{\rm H}$, and helium, $M_{\rm He}$ at the beginning of the post-AGB (or post-RGB, resp.) evolution and at $T_{\rm eff} = 30000\,{\rm K}$ on the cooling track as a function of the stellar mass. The evolutionary models show a steep decrease of $M_{\rm H,He}$ with increasing stellar mass, whereas the assumption of single values for $q_{\rm H,He}$ predicts a slight increase. For instance, for $M_* = 0.30 M_\odot \rightarrow 1 M_\odot$, we get $q_{\rm H} = -2.5 \rightarrow -6$ (and for $M_* = 0.45 M_\odot \rightarrow 1 M_\odot$: $q_{\rm He} = -1.0 \rightarrow -3$)! The typically "thick" layer of $q_{\rm H} = -4$ is met at a WD mass of $\approx 0.6 M_\odot$.

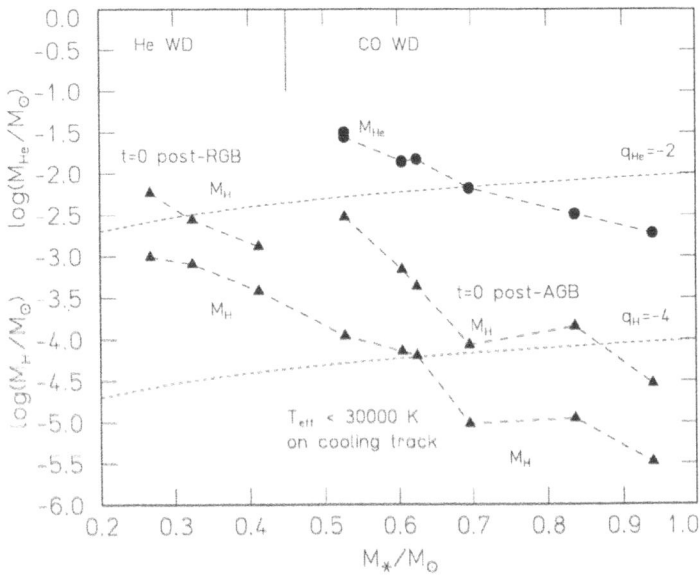

*Figure 2.* Hydrogen and helium layer masses for C/O and He WDs vs. stellar mass. The upper triangles give $M_H$ at the beginning of the post-AGB evolution (i.e. at a pulsational period of $P_0 = 50$ d corresponding to $T_{eff} = 6000....8000$ K), and post-RGB evolution ($T_{eff} = 5000$ K), resp., the lower ones at $T_{eff} = 30000$ K on the cooling track. $M_{He}$ (circles) remains practically constant during the post-AGB evolution. The short-dashed lines refer to the "typically thick" layers of $q_{He} = -2$ and $q_H = -4$ with $q_{H,He} = \log(M_{H,He}/M_*)$.

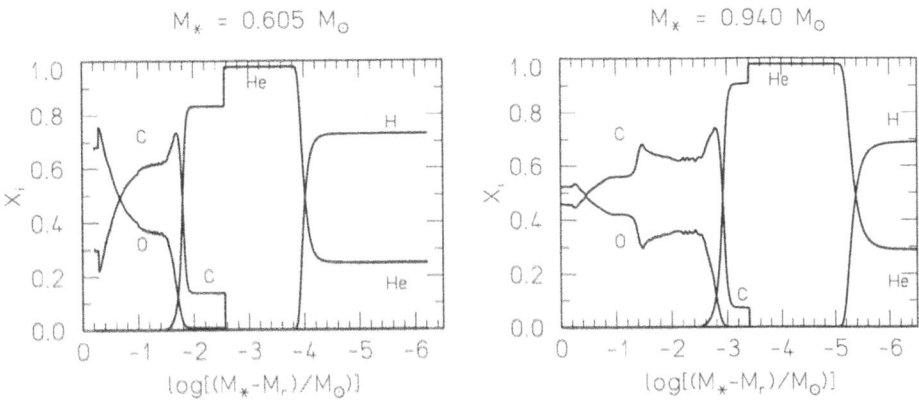

*Figure 3.* Abundance profiles of H,He, C and O for a $0.605M_\odot$ and $0.940M_\odot$ C/O WD.

## 3.  White Dwarf Models

Fig. 3 illustrates the abundance profiles for C/O WDs with 0.605 and $0.940M_\odot$. Whereas the $0.605M_\odot$ model has typically "thick" layers of H and He, i.e. $q_H = -3.9$ and $q_{He} = -1.6$, the massive one can be thought to

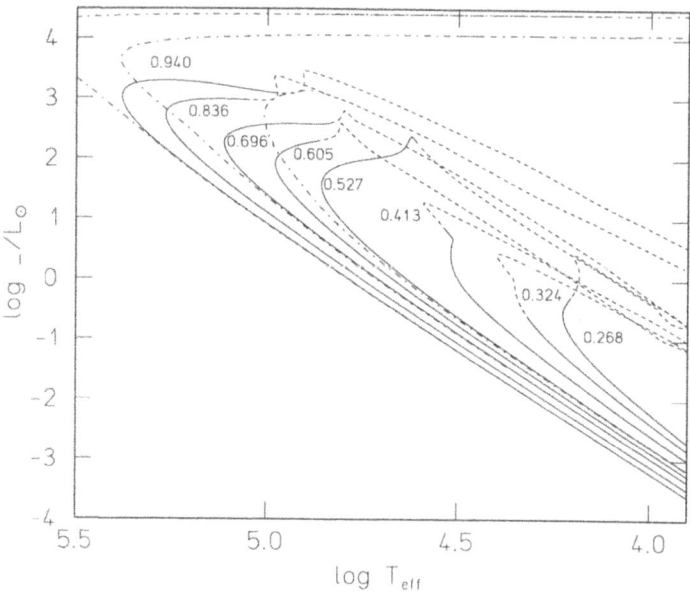

*Figure 4.* Tracks of the contraction WD models. The chemical adaption phase is given by the dashed lines, the cooling phase by the solid ones. The dashed-dotted lines refer to the fully evolutionary calculations for 0.940, 0.696 and $0.527 M_\odot$ (top to bottom).

be typically "thin", i.e. $q_H = -5.5$ and $q_{He} = -2.7$, since there is practically no difference between models with $\approx 10^{-6} M_\odot$ hydrogen and with no hydrogen (Koester & Schönberner 1986).

In order to compare our fully evolutionary models with WD models of the same mass but different H and He layers, we constructed additional WD models. For that purpose we took homogeneous main sequence models and successively adapted a chemical WD profile, forcing the models to a rapid contraction towards the WD domain. Fig. 4 shows the corresponding tracks if one adopts the evolutionary profiles. For the massive C/O WDs the contraction tracks rapidly merge with the evolutionary ones, i.e. at $T_{eff} \geq 10^5$ K, whereas in the case of He WDs convergence is reached only very late, for instance for $M \leq 0.4 M_\odot$ at $T_{eff} \leq 30000$ K.

Finally, we recalculated this grid of "contraction models" with the assumption that *all* models have either "thick" or "thin" layers. For that purpose, we applied the profiles of the $0.605 M_\odot$ and $0.940 M_\odot$ model (cf. Fig. 3).

## 4. Mass-Radius Relation and Mass Distribution

Fig. 5 demonstrates the effects of the different assumptions for the hydrogen and helium layers on the mass-radius relation at $T_{eff} = 30000$ K. The decrease of the layer masses with increasing stellar mass leads to a mass-radius

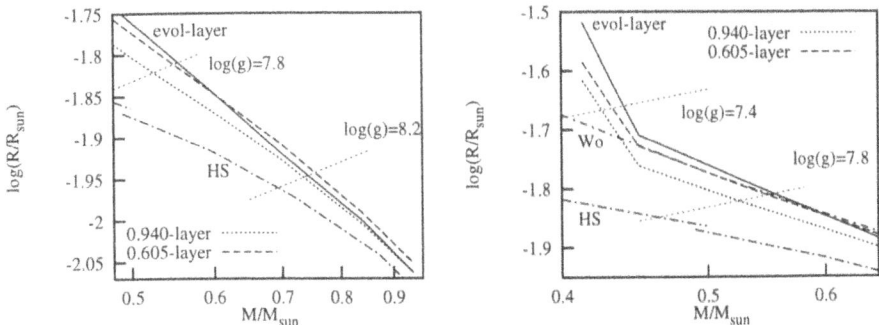

*Figure 5.*      Mass-radius relations (left: whole mass range, right: low-mass tail) at $T_{\mathrm{eff}} = 30000$ K for evolutionary, thick and thin layers of H and He. All models with "thick" layers have the chemical profiles of the evolutionary $0.605 M_\odot$ model, i.e. $q_H = -3.9$ and $q_{He} = -1.6$, those with "thin" layers the ones of the evolutionary $0.940 M_\odot$ model, i.e. $q_H = -5.5$ and $q_{He} = -2.7$. "HS" refers to the zero-temperature results of Hamada & Salpeter (1961), "Wo" to the models of Wood (1995) for C/O WDs with $q_H = -4$ and $q_{He} = -2$.

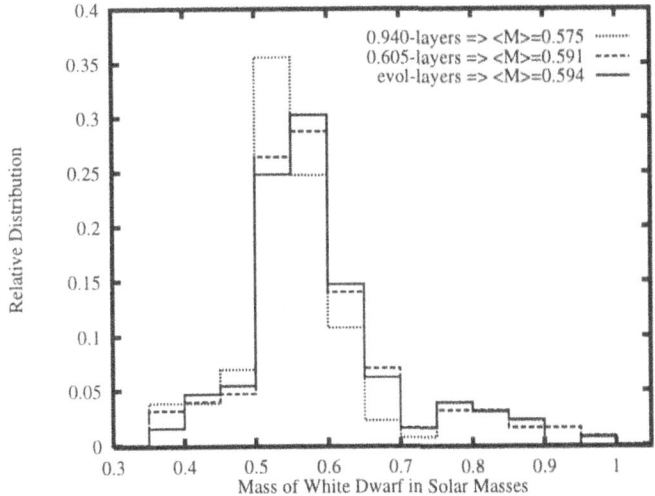

*Figure 6.*      WD mass distribution ($\Delta M = 0.05 M_\odot$) for the sample of Bergeron et al. (1992) with different assumptions for the hydrogen and helium masses, i.e. for evolutionary, thick and thin layers (cf. Fig. 5)

relation which is steeper than the ones based on constant layer masses. Correspondingly, low-mass C/O WDs obey relations given by thick layers whereas the massive remnants follow the ones given by thin layers. The consideration of He WDs leads to a sharp radius increase for $M \leq 0.45 M_\odot$.

The effects of the different layer masses on the WD mass distribution are demonstrated in Fig. 6. For that purpose, we used the sample of Bergeron et al. (1992) consisting of 129 objects. Models with evolutionary profiles give a mean mass of $\langle M \rangle = 0.594 M_\odot$ ($\sigma = 0.12$), whereas those with exclusively "thick" layers lead to $\langle M \rangle = 0.591 M_\odot$ ($\sigma = 0.13$) and those with exclusively "thin" ones to $\langle M \rangle = 0.575 M_\odot$ ($\sigma = 0.13$). Evolutionary-layer and thick-layer models yield almost the same mean mass since both coincide at $M_* \approx 0.6 M_\odot$. The thin-layer models lead to a lower mean mass since they have considerably smaller radii. Since we take He WDs into account, the masses assigned for the lowest gravity objects are not less than $0.35 M_\odot$.

## 5. Summary

Standard evolutionary calculations yield masses of the hydrogen and helium layers which decrease by orders of magnitude with increasing stellar mass. For C/O WDs the hydrogen layer varies from $10^{-4} M_\odot$ ($M_{\mathrm{WD}} = 0.6 M_\odot$) down to $10^{-6} M_\odot$ ($M_{\mathrm{WD}} = 1 M_\odot$).

White dwarfs less massive than $\approx 0.45 M_\odot$ are He WDs, and have been formed by binary evolution. Our simple simulation of this evolutionary channel gives hydrogen layers which are even more massive than those of low-mass C/O WDs (up to one order of magnitude, see Fig. 2). The mass-radius relation of He WDs is considerably steeper than the one given by C/O WDs.

Consequently, the change of the internal composition at $M_* \approx 0.45 M_\odot$ as well as the tight correlation between the H/He layers and the WD mass should be taken into account for the construction of a mass-radius relation.

The resulting WD mass distribution shows no objects below $0.35 M_\odot$ due to the inclusion of He WDs. It gives nearly the same mean mass as one based only on thick layers. However, its shape is even more narrow since the mass-dependence of the layer thickness shifts both the low and high gravity objects towards the center.

## References

Bergeron, P., Saffer, R.A., Liebert, J.: 1992, ApJ 394, 228
Blöcker, T.: 1995, A&A 297, 727; A&A 299, 755
Bowen, G.H.: 1988, ApJ 329, 299
Driebe, T.: 1996, Diploma thesis, in prep.
Hamada, T., Salpeter, E.E.: 1961, ApJ 134, 683
Iben, I. Jr., Tutukov, A.V.: 1986, ApJ 311, 742
Kippenhahn, R., Kohl, K., Weigert, A.: 1967, Zeitschrift f. Astrophys. 66, 58
Koester, D., Schönberner, D.: 1986, A&A 154, 125
Paczyński, B: 1971, Acta Astron. 21, 417
Pauldrach, A.; Puls, J., Kudritzki, R.P., Méndez, R., Heap, S.R: 1988, A&A 207, 123
Weidemann, V.: 1987, A&A 188, 74
Wood, M.: 1995, in *White Dwarfs*, eds. D. Koester & K. Werner, Springer, p. 41.

# THE PROGENITORS OF HIGH-MASS WHITE DWARFS [1]

*The Cluster Age of NGC 2516*

FALK HERWIG

*Astrophysikalisches Institut Potsdam, Germany*
*Telegraphenberg A27, D-14473 Potsdam, fherwig@aip.de*

**Abstract.** Spectroscopic observations of bright main sequence stars and the resulting age redetermination of young open cluster NGC 2516 are presented. The impact on the progenitor masses of white dwarf cluster members is shown.

## 1. Introduction

The semi-empirical cluster method (Weidemann 1987; Herwig 1995) can be employed to derive initial masses of white dwarfs (WDs) in open clusters. In this respect NGC 2516 is of particular interest since its identified WD-members have the highest progenitor masses determined so far by this method.

The long term project to identify white dwarf members in young open clusters by Koester & Reimers (1993) has recently resulted in the determination of stellar parameters of four white dwarf members in NGC 2516 of very good quality (Koester & Reimers, 1996). However, the cluster age determination by Meynet et al. (1993) ($\log(\tau_c) = 8.15 \pm 0.1$) was based on photometric data. We felt that it might be possible to check and improve the age determination for this important cluster.

## 2. Observation and Analysis

We obtained spectra of B- and A-stars in NGC 2516 at the 1.52m telescope of the European Southern Observatory at La Silla (March 28 - April 1, 1996; ESO PID: 56.D-0179). The instrument was the Boller & Chivens spectro-

---

[1] Based on observations collected at the European Southern Observatory, La Silla, Chile.

*I. Isern et al. (eds.), White Dwarfs, 63–66.*
© *1997 Kluwer Academic Publishers.*

graph with the Loral/Lesser 2048x2048 CCD detector. We used grating #32 centered at 4150Å. The resolution was ≈ 1.3Å.

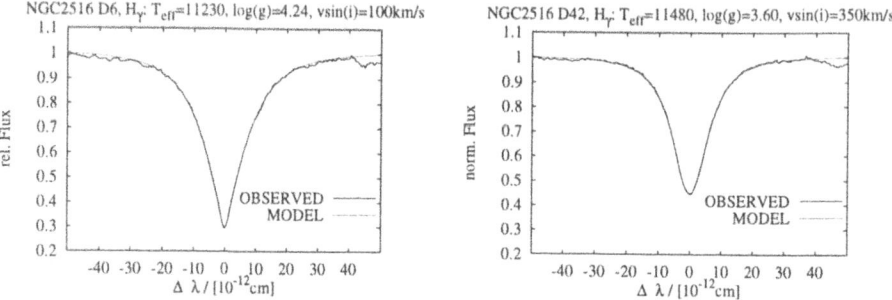

*Figure 1.*   Two spectra of the $H_\gamma$-line from the sample with the respective theoretical line profiles. **Left:** Dachs 6, **right:** Dachs 42.

TABLE 1.   Bright main sequence stars in NGC 2516: $T_{eff}$ determined from Strömgren photometry; $\log(g)$ and $v \sin i$ from analysis of the spectra. The left part displays the more weakly rotating objects with $v \sin i \leq 130 \frac{km}{s}$ while in the right part $v \sin i \geq 180 \frac{km}{s}$.

| Dachs-number | $T_{eff}$ [K] | $\log(g)$ | $v \sin(i)$ $[\frac{km}{s}]$ | Dachs-number | $T_{eff}$ [K] | $\log(g)$ | $v \sin(i)$ $[\frac{km}{s}]$ |
|---|---|---|---|---|---|---|---|
| 4   | 12250 | 4.03 | 115 | 3   | 11130 | 3.95 | 350 |
| 6   | 11230 | 4.24 | 100 | 9   | 11170 | 4.10 | 225 |
| 12  | 12490 | 4.12 | 90  | 13  | 12850 | 3.81 | 180 |
| 22  | 11860 | 4.28 | 115 | 15  | 10490 | 4.10 | 230 |
| 23  | 11200 | 4.31 | 85  | 16  | 11950 | 3.93 | 215 |
| 44A | 13240 | 4.00 | 100 | 29  | 10350 | 4.07 | 250 |
| 47  | 11660 | 4.22 | 130 | 53  | 11300 | 4.15 | 240 |
| 59  | 11930 | 4.21 | 115 | 42  | 11480 | 3.60 | 350 |
|     |       |      |     | 44B | 10450 | 3.48 | 350 |

The standard steps were applied to reduce the data, applying MIDAS procedures. We derived temperatures from Strömgren $uvby\beta$ photometry (Snowden 1975). Napiwotzki et al. (1993) gave a comparison of several calibrations and we followed their recommendations. Thus we assumed an error of ±5% (about ±600K) for $T_{eff}$, which clearly is a rather pessimistic choice and should be improved. The surface gravity and projected rotational velocity (Wenske & Schönberner, 1993) were then determined by fitting theoretical line profiles based on Kurucz's model atmospheres to the observed normalized $H_\gamma$ - line (Fig.1). The error in $T_{eff}$ translates into a

corresponding error in gravity of typically $\Delta \log(g) = 0.13$. The error due
to the fitting is $\Delta \log(g) \approx 0.05$.

## 3. Results and Discussion

As can be seen from Tab.1 half of the objects are rotating strongly and
must therefore be excluded from the age determination. From Fig.2 there
is some evidence that NGC 2516 should be younger than $\log(\tau_c) = 8.15$
as derived from photometry alone. Only one object indicates an age larger
than $\log(\tau_c) = 8.10$. As well the preliminary analysis of the $H_\delta$-line of some
weakly-rotating objects favour a younger age. Thus we conclude the age of
NGC 2516 to be $\log(\tau_c) = 8.08 \pm 0.07$.

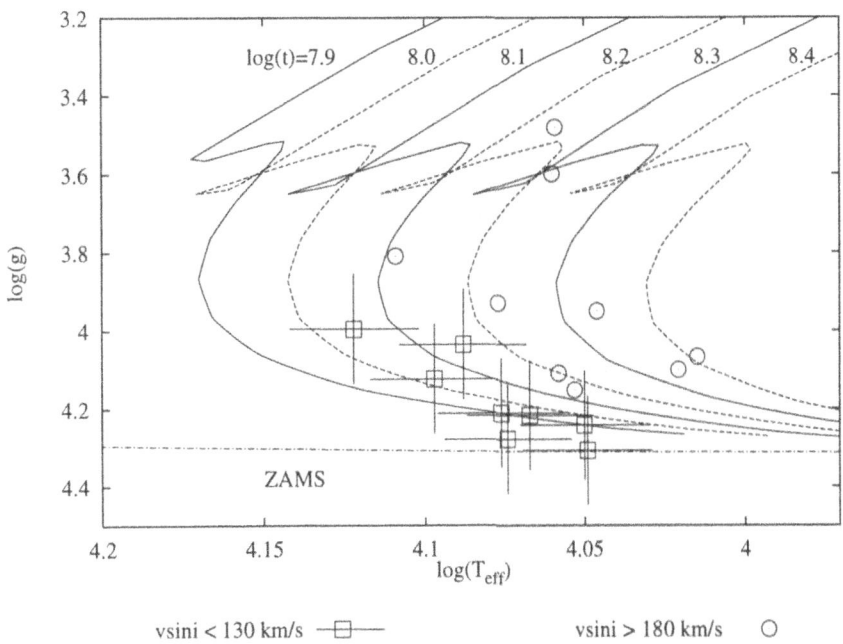

*Figure 2.* Positions of the objects as given in Tab. 1 in the $T_{eff} - \log(g)$ diagram together
with theoretical isochrones from Meynet et al. (1993).

Which consequences follow from this for the progenitors of the four
WD's in the cluster? Generally speaking the initial masses are larger for
a smaller cluster age. Tab.2 gives the numbers, calculated with theoretical
lifetimes from Wood (1994) and Schaller et al. (1992). On average the initial
masses are about $1 M_\odot$ larger due to the younger cluster age.

FALK HERWIG

## 4. Conclusions

Our first results indicate that NGC 2516 is marginally younger than previously assumed. However, we can certainly conclude that according to the cluster age no initial masses smaller than those given by Koester & Reimers (1996) (see Tab.2) are probable.

TABLE 2.   Final and progenitor masses for the four identified WDs in NGC 2516 (Koester & Reimers, 1996). The $2^{nd}$ and $3^{rd}$ column give the masses derived by K&R (with $\log(\tau_c) = 8.15$), the last column gives the progenitor mass determined with $\log(\tau_c) = 8.08$.

| K&R-number | $M_f$ [$M_\odot$] | $M_i$ $\log(\tau_c) = 8.15$ | $M_i$ $\log(\tau_c) = 8.08$ |
|---|---|---|---|
| 1 | 0.914 | 5.60 | 6.51 |
| 2 | 1.050 | 5.45 | 6.31 |
| 3 | 0.975 | 6.97 | 9.56 |
| 5 | 1.019 | 5.95 | 7.11 |

**Acknowledgements** I would like to thank Prof. D. Schönberner for his support. Prof. D. Koester has given me important advice during the observing run and at other times. This research was supported by DFG grant Scho 394/13-1.

## References

Herwig F. 1995, in *Proc. 32$^{nd}$ Liege Int. Astrophys. Coll.*, eds. A. Noels, D. Fraipont-Caro, M. Gabriel, N. Grevesse und P. Demarque, 441

Koester D., Reimers D. 1993, A&A **275**, 479

Koester D., Reimers D. 1996, A&A **313**, 810

Meynet G., Mermilliod J.-C., Maeder A. 1993, A&AS **98**, 477

Napiwotzki R., Schönberner D., Wenske V. 1993, A&A **268**, 653

Schaller G., Schaerer D., Meynet G., Maeder A. 1992, A&AS **96**, 269

Snowden M. S. 1975, PASP **87**, 721

Weidemann V. 1987, A&A **188**, 74

Wenske V., Schönberner D. 1993, in *IAU Coll. 137: Inside the stars*, eds. W. W. Weiss and A. Baglin, 162

Wood M.A. 1994, in *IAU Coll. 147: The Equation of State in Astrophysics*, eds. G. Chabrier and E. Schatzmann, 612

# THE MULTIPERIODIC, SHORT-PERIOD VARIATIONS OF WZ SGE

J. L. PROVENCAL

*Department of Physics and Astronomy, University of Delaware Newark, DE*

AND

R. E. NATHER

*Department of Astronomy and McDonald Observatory University of Texas at Austin Austin, TX*

## 1. Introduction

WZ Sge is an important cataclysmic variable with unique properties that may hold the key to understanding white dwarf interior structure. WZ Sge has one of the shortest orbital periods of any known hydrogen-rich system, undergoing eclipses every 81 minutes 38 seconds. Only the AM CVn stars (Provencal et al. 1995), which suffer from helium, rather than hydrogen, mass transfer, have shorter orbital periods. WZ Sge, originally classified as a recurrent nova, is now classified as a dwarf nova, based on its spectroscopic properties (Warner 1976). This system has one of the largest outburst amplitudes (7 magnitudes), combined with the longest outburst recurrence time (33 yrs) of any known dwarf nova system.

Our interest in WZ Sge lies with the accreting white dwarf, which is exposed during quiescence. Robinson et al. (1978) detected coherent oscillations in WZ Sge with periods of 27.8682 and 28.9596 s, whose simultaneous presence rules out variations arising via the rotation of a spotted white dwarf. We present new high speed photometry of WZ Sge, in which we confirm Robinson's findings and identify a second, complicated band of frequencies near 14 s.

*I. Isern et al. (eds.), White Dwarfs, 67–70.*
© *1997 Kluwer Academic Publishers.*

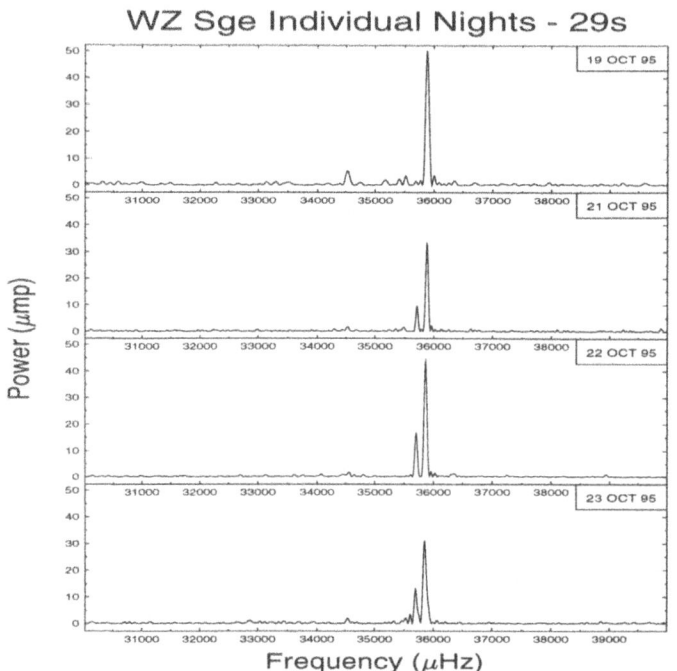

*Figure 1.*   WZ Sge's FT in the region of 29 s for 23-28 Aug.

## 2. Discussion

Our observations of WZ Sge span 23-28 August and 19-23 Oct 1995. The data were taken with the McDonald Observatory 82 inch telescope, in white light, using a three-channel photomultiplier-based photometer, and integration times of 3 s (Figure 1). We employed remote TV guiding, using a dichroic to funnel blue light to the channel 2 (comparision star) PMT and red to the CCD guider. Guiding every 5 seconds or so reduced the star image's motion within the aperture, greatly improving the signal to noise ratio in the 14 s region. We followed the standard reduction techniques outlined by Nather et al. (1990).

Figures 1 and 2 give the sequences of FTs for October in the 29 s and 14 s regions. WZ Sge is a clearly multiperiodic variable, exhibiting complicated behaviour on timescales ranging from night to night and month to month. Such behaviour is observed in single, non-radially pulsating ZZ Ceti white dwarfs (Nather, 1996; Winget et al. 1994). However WZ Sge does not fit the standard ZZ Ceti model. With a quiescent temperature of 15,000 K (Sion

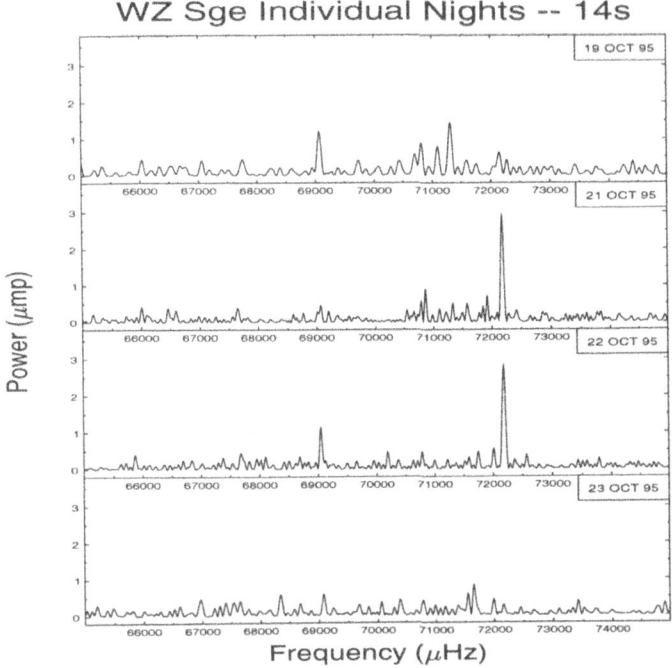

*Figure 2.*  WZ Sge's FT in the region of 29 s for 19-23 Oct.

et al. 1995), WZ Sge is hotter than any theoretical blue edge of the ZZ Ceti instability strip. Furthermore, these periods are a factor of 4 shorter than those observed in the shortest period ZZ Ceti pulsator, and too long for radial modes (Bradley, 1993). The rotation of the accreting white dwarf is expected to be the same order of magnitude as the periods we observe, yet the simultaneous presence of multiple periodicities in WZ Sge's light rules out this possibility as well. We propose, as a tantalizing explanation, torsional oscillations of the degenerate crystalline core, predicted theoretically several years ago (Hansen & Van Horn 1979), but to date unobserved.

If core crystallization does play a major role, WZ Sge holds the possibility of expanding the techniques of asteroseismology (Nather et al. 1990) into a currently unexplored branch, the hidden details of white dwarf core composition. A white dwarf's core contains the byproducts of nucleosynthesis, which theory predicts to be a mixture of carbon and oxygen. Most white dwarf models assume a core of carbon or a carbon/oxygen mixture. Our observational evidence supporting this assumption is sparse. In particular, we know of few white dwarfs with independent measurements of mass and radius, and this knowledge is not sufficiently accurate to determine individual core compositions (Provencal et al. 1996). Understanding the details of white dwarf core compositon, such whether they contain carbon at all, and if so, as the exact ratio of carbon and oxygen, would influence our ideas of such diverse fields as studies of the later stages of stellar evolution, the modelling of supernovae and novae, and energy generation during all phases of stellar evolution.

WZ Sge is a target for an upcoming Whole Earth Telescope run in September 1996, with the goal of mapping the complex behavior of these variations. Identifying, or at least limiting, the timescale of change would provide a major constraint on any model of the process, and a first step towards understanding the underlying driving mechanism.

## References

Bradley, P. A. 1993, Ph.D. thesis, Univ. Texas

Hansen, C. J. & van Horn, H. M. 1979, *ApJ*, **233**, 253

Nather, R. E., Winget, D. E., Clemens, J. C., Hansen, C. J., & Hine, B. P. 1990, *ApJ*, **361**, 309

Nather, R. E. 1996, private communication

Provencal, J. L., Shipman, H. L., Wesemael, F., Bergeron, P., Sion, E., Bond, H., & Liebert, J. 1996, *ApJ*, submitted

Provencal, J. L., et al. 1995, *ApJ*, **445**, 927

Robinson, E. L., Nather, R. E., & Patterson, J. 1978, *ApJ*, **219**, 168

Sion, E. M., Cheng, F. H., Long, K. S., Szkody, P., Gilliland, R. L., Haung, M., & Hubeny, I. 1995, *ApJ*, **439**, 957

Warner, B. 1976, in IAU Symp. 73, *The Structure and Evolution of Close Binary Systems*, ed. P. Eggleton, S. Mitton, & J. Whelan (Dordrecht: Reidel), 85

Winget, D. E., et al. 1994, *ApJ*, **430**, 839

# LIMITS ON GRAVITIES OF EVOLVED STARS, FROM INFRARED PHOTOMETRY AND BINARY COMPANIONS

P. THEJLL

*Nordita, Blegdamsvej 17, DK-2100 Copenhagen Ø, Denmark*

AND

A. ULLA

*Instituto de Astrofísica de Canarias, 38200 La Laguna, Tenerife
and LAEFF-INTA, Apdo. 50727, 28080 Madrid, Spain*

## 1. Hot subdwarf stars as IR emitters?

Thejll et al. (1995; hereafter Paper I) discussed the possible origin for excess infrared fluxes detected from the direction of a hot subdwarf star. As several theoretical explanations compete, they presented $JHK$ photometry for 27 hot objects and corresponding analysis. After discarding the possibility that the IR flux originates in an ejected wind from the hot atmosphere, the authors concluded that the most likely explanation is the presence of a cool binary companion. With the aid of comprehensive literature and databases searches several well known cases were corroborated, suspected cases confirmed and new discoveries uncovered.

72 more hot subdwarfs were observed with the same telescope and instrumentation in June and October, 1994. They are also analysed following the methods described in Paper I but with the aid of the new Kurucz (1993) grid of metal-line blanketed model spectra (see Ulla&Thejll 1996 for details) distributed on CD-ROM discs. The hot subdwarfs in Paper I were reanalyzed. About 40% of all targets display IR fluxes excesses attributable to binary companions.

## 2. Deriving gravities for hot subdwarfs

For extraction of excess IR fluxes we scale an appropriate model spectrum to the reddest possible normalization photometric point that we believe is unaffected by radiation from the companion – a point far into the UV range

*I. Isern et al. (eds.), White Dwarfs, 71–74.*

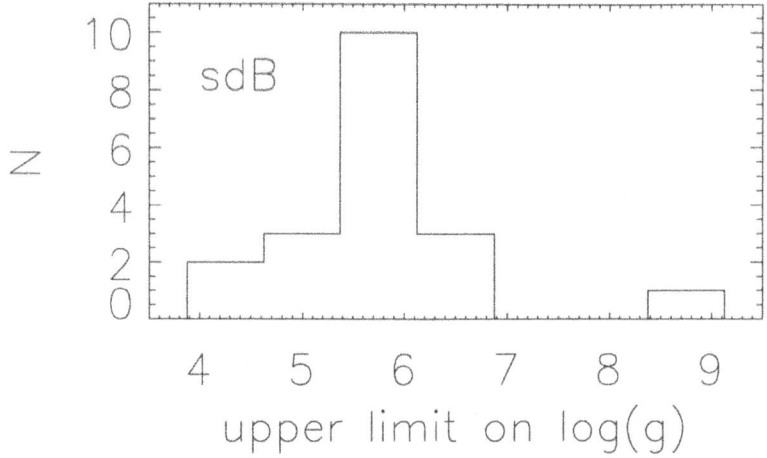

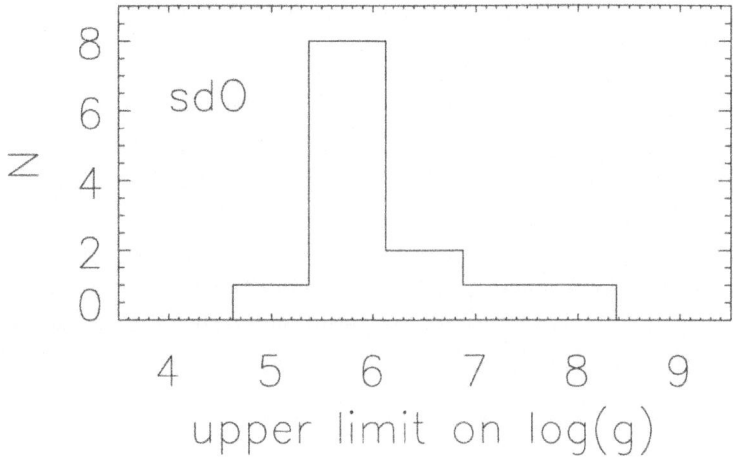

*Figure 1.* Derived upper limits for log(g) of the sample sdB and sdO stars. The few large upper-limit values for log(g) correspond to known cases of evolved companions. The results are consistent with comparisons of sdB atmospheric parameters to theoretical evolution tracks for 0.5 $M_\odot$ He objects. For the sdOs the results do not contradict the suggestion that sdO stars are He Main Sequence objects with masses from 0.25 $M_\odot$ to 1 $M_\odot$.

TABLE 1. Results of fitting Kurucz models to the observed fluxes. The columns are: (1) Object name, (2) Teff of the assumed or fitted hot model, (3) the Teff of the fitted cool model, followed in parenthesis by the dwarf spectral type derived fr om the temperature and standard tables Zombeck (1990), (4) The ratio $R'=R2/R1$, where R2 is the radius of the cool star and R1 the radius of the hot star, and (5) the derived gravity of the hot subdwarf (upper limit) calculated from an assumed mass of 0.55 $M_\odot$ for the hot subdwarf, R' and a mass of the cool companion derived from the mass-radius-Teff relationship for ZAMS stars from Zombeck. Except for BD+34 1543 ([-0.5]), solar metallicity models have been used.

| Object | Teff(1) | Teff(2) | R' | log(g) |
|--------|---------|---------|-----|--------|
| (1) | (2) | (3) | (4) | (5) |
| BD-3 5357 | 40000. | 4500.(K4) | 67. | 8.1 |
| BD-7 5977 | 31000. | 4750.(K2) | 23. | 7.1 |
| BD-11 162 | 35000. | 5250.(K0) | 6. | 5.9 |
| BD +10 2357 | 50000. | 7500.(A8) | 11. | 6.1 |
| BD+28 4211 | 50000. | 3500.(M2) | 2.5 | 5.5 |
| BD+29 3070 | 28000. | 6250.(F7) | 8. | 6.0 |
| BD+33 2642 | 18000. | 6750.(F3) | 1.3 | 4.3 |
| BD +34 1543 | 26000. | 5500.(G8) | 8. | 6.0 |
| Feige 34 | 50000. | 3500.(M2) | 4.6 | 6.1 |
| Feige 80 | 33000. | 5500.(G8) | 6. | 5.7 |
| Feige 108 | 35000. | 3750.(M1) | 6. | 6.0 |
| GD 274 | 24000. | 4750.(K2) | 7. | 6.0 |
| GD 299 | 37500. | 4750.(K2) | 5.0 | 5.8 |
| HD 4539 | 25000. | 10500.(A0) | 2.1 | 4.1 |
| HD 45166 | 40000. | 13000.(B8) | 7. | 4.8 |
| HD 113001 | 47500. | 6750.(F3) | 9. | 6.0 |
| HD 128220 | 40000. | 5250.(K0) | 16. | 6.7 |
| HD 149382 | 34000. | 5750.(G2) | 2.3 | 5.0 |
| HD 185510 | 25000. | 4250.(K8) | 164. | 8.9 |
| HDE283048 | 40000. | 7500.(A8) | 13. | 6.2 |
| MRK509C | 42500. | 3500.(M2) | 8. | 6.6 |
| PG 0110+262 | 21000. | 5000.(K1) | 6. | 5.9 |
| PG 0229+064 | 20000. | 5500.(G8) | 2.1 | 4.9 |
| PG 0232+095 | 21000. | 4750.(K2) | 13. | 6.6 |
| PG 0242+132 | 31000. | 5000.(K1) | 6. | 5.8 |
| PG 2110+127 | 29000. | 6250.(F7) | 8. | 6.0 |
| PG 2118+126 | 25000. | 5250.(K0) | 7. | 6.0 |
| PG 2148+095 | 25000. | 5000.(K1) | 6. | 5.8 |
| PG 2151+100 | 27000. | 3500.(M2) | 10. | 6.7 |
| PG 2219+094 | 21000. | 7000.(F1) | 2.8 | 5.0 |
| PHL 1079 | 28000. | 4500.(K6) | 8. | 6.2 |
| SB 7 | 42500. | 5750.(G2) | 6. | 5.7 |
| TON 139 | 26000. | 5500.(G8) | 7. | 6.0 |

can lead to practical problems if the far-UV flux is heavily depressed by metal opacities. The excess in any band, but in particular the $JH$ and $K$ band is then simply calculated by subtraction of the normalized model flux from the observations.

We next derive gravity estimates for the hot subdwarf by fitting a weighted sum of two Kurucz model spectra to their UV-to-$K$ band energy distributions. The weights used are simply the areas of the stars so after a successful fit we have the ratio of the radii. The fitting procedure is a least-$\chi^2$ method.

The fit also gives the temperature of the best-fitting cool model and if a specific mass-gravity-temperature relationship is then assumed for the companion we can calculate the hot subdwarf gravity with the aid of the following equation

$$log(g_{sd}) = log(g_{comp.}) + log(\frac{M_{sd}}{M_{comp.}}) + 2log(\frac{R_{comp.}}{R_{sd}}), \tag{1}$$

where we need to make an assumption also about the hot subdwarf mass. For purposes of comparing to other work we use the standard hot subdwarf mass value of 0.55 $M_{\odot}$ (see also Paper I). We also use a ZAMS mass-gravity-temperature relationship for the cool companions (Zombeck 1990), which will guarantee upper limits on $log(g_{sd})$ which is desirable.

Our derived upper limits for $log(g)$ values for sdB and sdOs are shown in Table 1 and are also displayed in Figure 1. As seen in Figure 1 both distributions are nearly indistinguishable. Better data is required to test current suggestions that sdOs are evolved sdBs.

**Acknowledgments:** A.Ulla is at present a postdoctoral fellow at the Instituto de Astrofísica de Canarias, within the ISO team there and acknowledges the Spanish "Becas de Reincorporación" grant programme for financial support. P.Thejll acknowledges financial support from Nordita (Copenhagen, Denmark). The IR observations in this work were obtained with the CST operated on the island of Tenerife by the IAC in the Spanish Observatorio del Teide. The staff at the CST are thanked for their skilled assistance during the observations. This work has made use of the SIMBAD databank (CDS, Strasbourg), of the RDAF facility (GSFC, Greenbelt) and of the ULDA archive (LAEFF, Madrid).

### References

Kurucz, R.L., 1993, Kurucz CD-ROM No. 13.
Thejll, P.A., Ulla, A., and MacDonald J. 1995, A&A 303, 773-784
Ulla, A., Thejll, P.A., 1996, in preparation
Zombeck, M.V., 1990, "Handbook of Space Astronomy & Astrophysics", (Cambridge: Cambridge University Press), 2nd ed.

# ROTATION: CO WHITE DWARFS NEAR THE CHANDRASEKHAR LIMITING MASS

INMA DOMINGUEZ
*Departamento de Física Teórica y del Cosmos*
*Universidad de Granada, 18071. Granada (Spain)*

OSCAR STRANIERO
*Osservatorio Astronomico di Collurania*
*I-64100 Teramo (Italy)*

AMEDEO TORNAMBÉ
*Osservatorio di Roma*
*via Osservatorio 2, I-00040 Monteporzio, Roma (Italy)*

AND

JORDI ISERN
*I.E.E.C. (CSIC Research Unit)*
*Edif. Nexus 104. Gran Capita, 2-4, 08034 Barcelona (Spain)*

**Abstract.** The contraction of the CO core at the onset of the AGB phase of intermediate mass stars may induce a significant increase in the angular momentum, even for initially low rotating stars. Following the method described by Kippenhahn et al (1970), we show that a large final CO core mass is the natural consequence of the inclusion of rotation in the computation of stellar models. Reasonable assumptions concerning the redistribution of the angular momentum lead to the production of a final core mass as large as 1.1-1.4 $M_\odot$ for an initial mass of 6.5 $M_\odot$.

## 1. Introduction

It is accepted that intermediate and low mass stars ($M/M_\odot \leq 8$), the precise value depending on various physical parameters) end their life as CO white dwarfs (Reimers and Koester, 1982; Weidemann and Koester, 1983).

Current theory of stellar evolution predicts that the maximum mass of a newly born CO degenerate core cannot exceed 1.08 $M_\odot$ without an off-

75

*I. Isern et al. (eds.), White Dwarfs, 75–78.*
© *1997 Kluwer Academic Publishers.*

center, mildly degenerate, carbon ignition that would convert almost the entire star into oxygen and neon (Domínguez et al. 1993; Iben and Tutukov 1985; Nomoto 1984; Becker and Iben 1979). On the other hand, the growth of the degenerate CO core is inhibited by the second dredge-up, which ends the early AGB phase. Subsequently, during the thermal pulse AGB phase, the mass of the CO core increases. However it is believed that strong mass loss prevents a large number of pulses and the final mass of the CO white dwarf is essentially determined by the value reached at the end of the early AGB phase (Vassiliadis and Wood, 1993; Blöcker, 1995).

On the other hand, the formation of CO white dwarfs with masses close to the Chandrasekhar mass opens interesting possibilities: the occurrence of a thermonuclear explosion (similar to those attributed to Type Ia supernovae), looking like a Type IIL or Type IIP, depending of the mass of the H surrounding envelope. In the case of binaries, new channels could be opened for the accretion induced collapse and Type Ia supernova explosions.

No obvious mechanism can be invoked to induce the formation of massive CO white dwarfs, with the exception of rotation.

## 2. The model

The numerical experiments described here have been performed with the FRANEC (Frascati RAphson Newton Evolutionary Code), slightly modified with respect to the version described in Chieffi and Straniero (1989). We describe in detail the evolution of a 6.5 $M_\odot$ star with Y=0.27 and Z=0.02.

Kippenhahn et al (1970) suggested a method to describe non-spherical rotating stars by spherical models. This method uses a modified equation of hydrostatic equilibrium which includes a term produced by the pressure lifting effect caused by rotation:

$$\frac{dP}{dM_r} = -\frac{GM_r}{4\pi r^4}\left(1 - f\right)$$

the factor f is related to the radial component of the centrifugal force averaged over a sphere, f $= \omega^2/(6\pi r)$, in order to maintain the spherical symmetry.

Three sequences of models have been computed with f $= 0$, 0.1 and 0.25, referred to as cases 0, 1 and 2 respectively. Case 0 obviously corresponds to the non rotating case. The factor f is simply assumed to be constant and is applied just inside the CO core, since our aim is to explore the effect of such a perturbation on the hydrostatic equilibrium equation. In our case, $(\omega/\omega_{cr}) = f^{1/2}$ and gives, 0.316 and 0.5 for cases 1 and 2, respectively.

TABLE 1. Properties of the models at the AGB phase.

| | $\Delta t$ (yr) | $M_{CO}(M_\odot)$ | $\Delta t_{ip}(yr)$ | $M_{CO}$ ($M_\odot$) |
|---|---|---|---|---|
| Case 0 | $3.92\ 10^5$ | 0.986 | 1900 | 0.999 |
| Case 1 | $4.94\ 10^5$ | 1.104 | 2600 | 1.120 |
| Case 2 | $7.41\ 10^5$ | 1.404 | 3000 | 1.423 |

## 3. Stability

To ensure global stability the ratio of the kinetic (T) to gravitational energy (W) must be, for uniform rotating stars, lower than 0.14 (Kippenhahn and Weigert 1991 and reference therein). Our assumed values for the factor f are consistent with this limit: $T/W = 0.05$ for Case 1 and $T/W = 0.125$ for Case 2.

The angular momentum distribution is constrained by local stability criteria. The differential rotation is stable within the CO core against secular (Goldreich-Schubert 1967; Fricke 1968) and dynamical instabilities, except for the inner $10^{-5}$ $M_\odot$.

The existence of meridional currents in the degenerate rotating cores of evolved stars has been studied by Kippenhahn and Molenhöff (1974). In our models the circulation time scale is of the order of a few million years. The currents would tend to reduce the gradients in the angular velocity but not necessarily to lead to solid body rotation. The $\omega$-gradient might be modified by viscosity (Durisen 1973a and 1973b). In our models the time scale needed to influence the angular velocity distribution within the core is order of magnitudes greater than the evolutionary time scale. The situation is different for the convective envelope, where the viscosity time scale is of only a few years. Once rotation affects the envelope, an instantaneous redistribution of angular momentum is to be expected.

## 4. Results and conclusions

The two main effects of rotation, in agreement with previous studies (Kippenhahn et al. 1970; Endal and Sofia 1976 and 1978; Deupree 1995), are the increase in the evolutionary time scales, which is specially relevant in the AGB phase (see Table 1), and the fact that the behavior of rotating models mimics, in terms of the evolution of central temperature and density, the behavior of the non rotating ones with a lower mass, namely about $1 - 2M_\odot$ less massive. Therefore, the limiting mass for C-ignition ($M_{up}$) would be slightly larger when rotation is taken into account.

The physical properties of the models at the AGB phase are shown in table 1. In the first column we show the time interval from the exhaustion of He at the center (when we start to modify the hydrostatic equilibrium equation) to the beginning of the second dredge-up. Notice that this time interval in Case 2 is nearly double that the one obtained in Case 0. As a major result the second column gives the CO core masses at the end of the second dredge-up. The mass of the degenerate CO cores of rotating models may well exceed the canonical value of $1.08M_\odot$ which is the limit for standard models to skip off-center carbon ignition along the E-AGB phase. In the third column we show the elapsed time between pulses at the 16th thermal pulse. This time increases when f increases and since the accretion rate of He is practically the same, the increase of the CO core mass is greater for the rotating models. The fourth column gives the CO core masses at the 16th TP, notice the high value attained in Case 2, $1.42M_\odot$ .

Let us finally note that our computations clearly show that rotation, if present, could affect many observable properties of both AGB stars and their descendants: surface abundances and luminosity, mass loss rate, mass stratification and cooling time of the remnant, and the like.

### References

Blöcker T. 1995, AA, 297, 727

Chieffi A., Straniero O. 1989, ApJSS, 71, 47

Deupree R.G. 1995, ApJ, 439, 357

Domínguez I., Tornambé A., Isern J. 1993, ApJ, 419, 268

Durisen R. H. 1973a, ApJ, 183, 205

Durisen R. H. 1973b, ApJ, 183, 215

Endal A.S., Sofia S. 1976, ApJ, 210, 184

Endal A.S., Sofia S. 1978, ApJ, 220, 279

Fricke K.J. 1968, Z. Astrophys. 68, 317

Goldreich P., Schubert G. 1967, ApJ, 150, 571

Iben I. Jr., Tutukov A.V. 1985, ApJSS, 58, 661

Kippenhahn R., Meyer-Hofmeister, H.C., Thomas, H.C. 1970, AA, 5, 155

Kippenhahn R., Möllenhoff 1974, AA, 31,117

Kippenhahn R., Thomas, H.C. 1970, in Stellar Rotation, ed. A. Stehebak, Dordrecht, Reidel, p.20

Kippenhahn R., Weigert, A. 1991 in Stellar Structure and Evolution, ed. Springer-Verlag

Nomoto K. 1984 ApJ, 277, 791

Reimers D., Koester D. 1982, A&A, 116, 341

Vassiliadis E., Wood P.R. 1993, ApJ, 413, 641

Weidemann V., Koester D. 1983, A&A, 121, 77

# THE ACCRETION-INDUCED COLLAPSE OF WHITE DWARFS REVISITED IN 3D

DOMINGO GARCÍA-SENZ AND EDUARDO BRAVO

*Departament de Física i Enginyeria Nuclear, UPC*

*08034 Barcelona, Spain*

## 1. Introduction

Explosive phenomenae in white dwarfs are commonly associated with Type Ia Supernovae or to Novae. Closely related with the Supernovae events but far less studied is the collapse of a massive white dwarf to a neutron star structure induced by electron captures. Such quiet mechanism of formation of neutron stars has been invoked to explain the origin of, at least, a fraction of the low-mass X-ray sources and millisecond pulsars (van den Heuvel & Habets 1985). There are several possibilities to reach the optimal point to trigger the implosion. The most explored scenario involves those stars which develop a degenerate core made of Oxygen, Neon and Magnesium during the Carbon burning stage and then become dense enough to avoid the explosion once the Oxygen is ignited when the central density is as high as $9\ 10^9 \text{g.cm}^{-3}$ (Gutierrez *et al.* 1996, and references therein). Another possibility is to suppose that a white dwarf composed of Carbon and Oxygen (C-O), with a mass of $1 - 1.1\ M_\odot$ manages to rise its mass by accretion at a low rate, until it reaches the conditions appropiate for explosive carbon ignition when the central density is again close to $10^{10} \text{g.cm}^{-3}$ (Hernanz *et al.* 1988).

In this communication we will explore the consequences of relaxing the constraint of spherical symmetry on the evolution of the deflagration in a C-O white dwarf, corresponding to the last of the above scenarios. Thus, despite the many problems encountered in the pre-ignition stage (allowed accretion rates, uncertainities in the input physics, etc.), we will start from the point in which carbon has ignited explosively in a small volume around the center of the star, when the central density is about $9\ 10^9 \text{g.cm}^{-3}$ . Thereafter the evolution is followed by using a multidimensional hydrocode until

*I. Isern et al. (eds.), White Dwarfs, 79–82.*

densities $\simeq 2\ 10^{10}$g.cm$^{-3}$ , when the way to the complete implosion is irreversible. As several previous works have stated, the fate of the white dwarf after carbon ignition relies on the speed at which combustion propagates through the interior. It is reasonable to think that a slow propagation will favour collapse instead of explosion. The faster mode of propagation comes out if a detonation wave is formed. In that case the star is always disrupted. The minimum value is dictated by a front which expands spherically with a velocity given by its laminar, conductive value, $v_l$. Were this the case, collapse is the outcome if the central density is $\geq 7.5 - 8\ 10^9$g.cm$^{-3}$ (Timmes & Woosley 1992, García-Senz & Bravo 1994). As it is difficult for a detonation to survive in a very high-density medium, the common believe is that not a detonation, but a deflagration with an effective velocity, $v_{def}$, $v_l \leq v_{def} < v_{sound}$ is formed. The actual value of $v_{def}$ is still unknown but its excess over the pure laminar value $v_l$ is related to the geometrical properties of the nuclear flame, in particular its effective surface, through which heat is exhanged.

## 2.  Models and results

Our hydrocode uses the Smooth-Particle-Hydrodynamics (SPH) technique (Hernquist & Katz 1989). The inclusion of a thermal difussive term in the energy equation allows us to handle with deflagrations in degenerate material. The algorithm used to simulate flame advance and other details has been described elsewhere (García-Senz, Bravo & Serichol 1997) and we refer the reader to that work. The value of $h$ is updated every time-step in order to keep the numbers of neighbours constant at around $N_n \simeq 50$, the total number of particles was $N = 30976$. All calculations were made on the whole star, without imposing any geometrical constraint to the calculation. The input physics is quite complete and include an equation of state with contributions of partially degenerate and relativistic electrons with pair corrections, ions considered as an ideal gas plus coulomb and other minor corrections and radiation. The binding nuclear energy and electron capture rates of matter in nuclear statistical equilibrium (NSE) (Fuller, Fowler & Newman 1982) have been calculated interpolating from a table which uses $\rho, T$ and $Y_e$ as an input. Our initial models consist of spherically symmetric isothermal white dwarfs built through a fourth order Runge-Kutta integration. The one-dimensional profile was then mapped to a 3D distribution of $N$ particles and afterwards relaxed using a damping term in the momentum equation. Once a stable enough model was got, the temperature of a small subset of particles was artificially increased up to the self-consistent value corresponding to the NSE temperature for such density, and the ensuing dynamical evolution followed by means of the SPH hydrocode.

An interesting issue is to know to what extent a 1D simulation would be enough to give a qualitatively right picture of the phenomenon. To try to clarify this last point as well as to study the flame dynamics at such high density we carried out several numerical experiments with our SPH code.

The evolution of a white dwarf with initial central density, $\rho_c = 8.5 \; 10^9$ g.cm$^{-3}$ has been followed in three cases, corresponding to the models shown in Table 1. For the first one, Model 1, the flame front is supposed to be born with perfect spherical symmetry. In the second case, Model 2, such symmetry is broken by a combination of random perturbations of different amplitude and wavelength. Finally we have explored the consequences of a more extreme situation: a deflagration starting from a bubble located near the center, Model 3.

TABLE 1. Characteristics of the computed models

| Model | $M_{inc}^o/M_\odot$ | $\rho_c^0$ g.cm$^{-3}$ | t (s) | $M_{inc}/M_\odot$ | $\rho_c$ g.cm$^{-3}$ | Outcome |
|-------|------|------|------|------|------|---------|
| 1 | 0.017 | $8.5 \cdot 10^9$ | 0.65 | 0.24 | $1.9 \cdot 10^{10}$ | Collapse |
| 2 | 0.017 | $8.5 \cdot 10^9$ | 0.60 | 0.26 | $2 \cdot 10^{10}$ | Collapse |
| 3 | 0.017 | $8.5 \cdot 10^9$ | 0.98 | 0.23 | $5.8 \cdot 10^9$ | Explosion |

In agreement to already known results from one-dimensional calculations (see for instance García-Senz & Bravo 1994) Model 1 evolved towards collapse, first slowly but it accelerated once the Chandrasekhar-mass limit was crossed. The comparison between Models 1 and 2 is interesting. As a deformed sphere presents a higher surface for heat exchange and the mean density is slightly lower than in the undeformed case (then, the electron capture rate is also lower) we would expect for the implosion of Model 2 to be slower than it is in Model 1. Moreover, the initial seeding of perturbations will enhance the chance for instabilities to grow, especially the Rayleigh-Taylor instability. All together works to favour explosion against collapse. In spite of all that, Model 2 evolves in a similar way, even faster, than Model 1 does. In contrast to what is seen in deflagrations at lower densities (when $\rho_c \simeq 2 \cdot 10^9$g.cm$^{-3}$ ), we have not observed the growth of any instability during the time covered by the calculation. Even more, due to the high velocity of the laminar flame at these densities the initial perturbations were rapidly erased by the combustion, and the flame front gained spherical symmetry as time went on. Thus, both models evolve similarly, the time spent to reach $2 \cdot 10^{10}$g.cm$^{-3}$ is not very different and the total incinerated mass is similar too. Model 3 exhibits a different behaviour. Owing to the lower density, an off-center ignition leads to a reduction in the amount of electron captures in comparison to a centered ignition. Actually, the pressure deficit caused by the deleptonization is not enough to prevent the explosion of the star. As mentioned in the introduction, previous one-dimensional calculations gave $\rho_c \simeq 8 \cdot 10^9 \mathrm{g.cm}^{-3}$ as a minimum critical density which separates collapse from explosion. Our initial density was slightly higher, $8.5 \cdot 10^9 \mathrm{g.cm}^{-3}$ , thus the evolution of Model 3 towards explosion, although unexpected, was not too surprising. We conclude that an off-center ignition will raise the density threshold necessary to get the accretion-induced collapse. In what extent such threshold is shifted deserves further examination.

## 3. Conclusions

The evolution of a massive white dwarf towards a neutron star has been calculated in 3D by means of a SPH code. A comparison between a calculation which assumes spherical symmetry of the flame front and another which relaxes this last requirement shows that dimensionality effects are not decisive to determine the outcome of a deflagration in the AIC scenario, unless very different initial conditions (like off-center ignitions for example) were considered. For central ignitions it comes out that the value of $8 \ 10^9 \mathrm{g.cm}^{-3}$ obtained (for a $X(^{12}\mathrm{C})=X(^{16}\mathrm{O})=0.5$) using standard one-dimensional hydrocodes could give an adequate idea about the threshold which separates explosion from collapse for these white dwarfs. In these cases the dynamics of the implosion is also very similar when calculated in one or more dimensions.

## Acknowledgements

This work has been supported by the DGICYT grant PB94-0827-C02-02.

## References

Fuller G.M., Fowler, W.A. and Newman, M.J. (1982), *Astrophysical Journal Supplement Series 48*, 279.

García-Senz, D. and Bravo, E. (1994), in *Equation of State in Astrophysics*, Ed. G. Chabrier (Cambridge Univ. Press), 571

García-Senz, D., Bravo, E. & Serichol, N. (1997). Submitted to *Astrophysical Journal Supplement Series*.

Gutierrez, J., García-Berro, Iben, I.Jr., Isern, J., Labay, J. and Canal, R. (1996), *Astrophysical Journal 459*, 701

Hernanz, M., Isern, J., Canal, R., Labay, J. and Mochkovitch, R. (1988), *Astrophysical Journal, 324*, 3

Hernquist, L. and Katz, N. (1989), *Astrophysical Journal Supplement Series 70*, 419

Timmes, F.X. and Woosley, S.E. (1992), *Astrophysical Journal, 396*, 649

van den Heuvel, E.P.J. and Habets, G.M.H. (1985), in *Supernovae, their Progenitors and Remnats*. Ed. G. Srinivasan and V. Radhakrishnan (Bangalore), 129

# Section II:
# Luminosity Functions and Populations

# TOWARDS A LUMINOSITY FUNCTION OF HOT DA WHITE DWARFS

J. LIEBERT , T.A. FLEMING AND J.B. HOLBERG
*Steward Obs., University of Arizona*
*Tucson, AZ 85721 USA*

P. BERGERON
*Université de Montreal*

AND

R. A. SAFFER
*Villanova University*

**Abstract.** This is a progress report on our luminosity function of 329 Palomar Green DA white dwarfs, all reobserved with CCD spectrophotometry and analyzed in the manner of Bergeron, Saffer & Liebert (1992). The combined luminosity and mass functions are presented. Three distinct mass groups of white dwarfs may be identified: (1) the predominant peak group with a median mass just below 0.6 $M_\odot$; (2) the component of likely close binaries where the visible white dwarf is composed of a helium interior and mass < $0.5 M_\odot$; and, (3) massive white dwarfs $\geq 0.9 M_\odot$. This last group includes all three magnetic DA stars for which a measurement or estimate of the mass and radius are available.

## 1. Introduction

The Palomar Green Survey (Green, Schmidt & Liebert 1986) provides a complete, magnitude and $U-B$ color limited survey covering nearly one quarter of the sky at Galactic latitudes $b \geq 30$ degrees. The first estimation of the DA white dwarf formation rate and luminosity function (Fleming, Liebert & Green 1986) utilized photographic and image tube spectra and correspondingly limited analysis techniques - ie. the assumption of a mean mass of $0.6 M_\odot$, and the estimates of $T_{eff}$ using a variety of colors and H$\beta$ equivalent width measurements. Moreover, a significant fraction of stars in

*I. Isern et al. (eds.), White Dwarfs, 85–89.*
© *1997 Kluwer Academic Publishers.*

the PG Survey classified originally as DA white dwarfs turn out to be lower surface gravity stars (and a few migrated the other direction).

In this reanalysis, all DA stars estimated to be hotter than $T_{eff} \sim$ 10,000 K were reobserved with optical CCD spectrophotometry. Hydrogen Balmer line profiles through $H\beta$ were fit with model atmosphere models to determine the effective temperature and gravity of the star and estimate the mass following the procedures of Bergeron, Saffer, & Liebert (1992). Once again the evolutionary models of Wood (1992, and private communication) were used to provide a second relation between radius and mass at the derived $T_{eff}$, so that individual stellar parameters for each star could be obtained.

## 2. Discussion

In Liebert & Bergeron (1995), we emphasized the importance of combining the mass and luminosity functions into one distribution function. Now it is also possible to calculate directly the formation rate of DA white dwarfs in terms of the cooling ages – isochrones in the mass – luminosity distribution function.

In Figure 1, this data set (filled circles) is shown in an $M_V$ vs. $T_{eff}$ "HR" diagram. Diagonal solid curves from the upper left to the lower right are lines of constant mass from the Wood models, from $0.4M_\odot$ (top) to $1.2M_\odot$ in intervals of $0.2M_\odot$. Crossing the mass lines are isochrones of cooling times for $10^6$, $10^7$, $3\times10^7$, $10^8$, $3\times10^8$ and $10^9$ years, respectively from the upper left to the lower right. Also shown as open squares are (only) the Palomar Green DO stars analyzed by Dreizler & Werner (1996), and DB stars by Beauchamp (1994) and Beauchamp et al. (1996).

The ridgeline of stars near the peak of the mass distribution just below $0.6M_\odot$ is an obvious feature. The DB and DO data appear to include very few stars far from this ridgeline. That is, as concluded by Beauchamp et al. 1996, the helium atmosphere white dwarfs appear to lack the components of unusually high and low masses discussed below.

Also apparent in Fig. 1 are the components of low mass near the $0.4M_\odot$ curve, and the high mass "tail" to the distribution. The stars with masses below $0.45M_\odot$ are likely to be composed of helium cores; many have been shown to be binary stars, with either a likely white dwarf or low mass main sequence companion (cf. Marsh, Dhillon, & Duck 1995). The hottest of these include DAO stars.

The isochrones become nearly horizontal for the oldest of these DA stars. Hence, there are two selection effects operating in opposite directions on stars of high and low mass, for a magnitude and color-limited survey: First, the smaller radii of the massive stars means that the available search

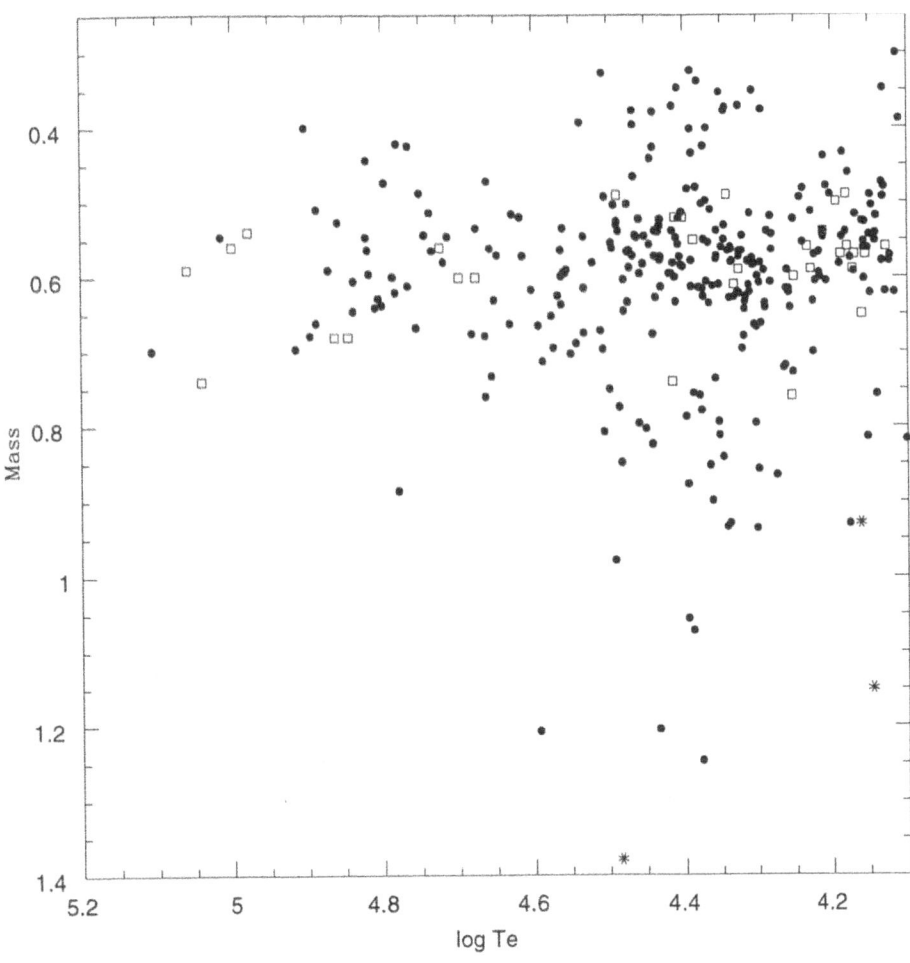

*Figure 1.*  $M_V$ vs. $T_{eff}$, isochrones and mass curves (see text)

volume is much smaller than for ordinary white dwarfs in a magnitude-limited survey. However, their longer cooling time means that they remain hot for a longer period. Of course, each of these effects reverses for the low mass, probably-helium core binary stars. The formation rate of white dwarfs must take into account these important effects.

In Fig. 2 we plot the mass distributions explicitly. The three mass groups of DA stars are perhaps even more apparent, as is the preference of the DB-DO stars for lying near the mass peak.

Three of the stars with masses greater than $0.9M_\odot$ are the magnetic

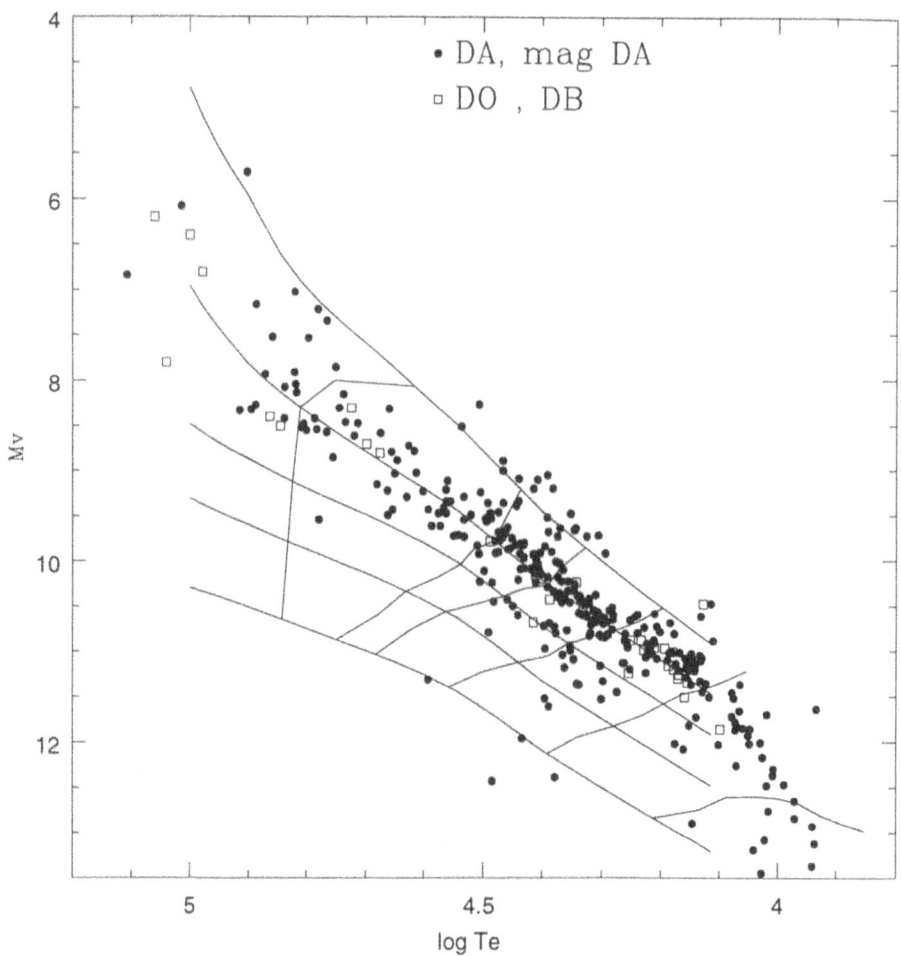

*Figure 2.*   Mass vs. $T_{eff}$, as explained in the text

white dwarfs PG 1658+441 (Schmidt et al. 1992), 0945+246a (Liebert et al. 1993), and 1015+014 (using an unpublished trigonometric parallax courtesy of C. C. Dahn and the U.S. Naval Observatory to estimate $M_V$ and the mass). These three are plotted as asterisks in Fig. 2. The three remaining magnetic DA stars which are part of this sample (for a total magnetic fraction of 2%) could not be included in the plot, since the field strengths are too high to estimate $\log g$ (and hence $M_V$ and the mass) from the Zeeman-split line profiles, and a trigonometric parallax measurement is not available. In any case, this sample contributes to the growing evidence that the magnetic white dwarfs are usually more massive than the average white dwarf, and can have masses not far from the Chandrasekhar limit.

## References

Beauchamp, A. (1996), PhD dissertation, Université de Montreal.

Beauchamp, A., Wesemael, F., Fontaine, G., Lamongtagne, R.,Saffer, R.A., & Liebert, J. (1996), in "Hydrogen-Deficient Stars and Related Objects," Bamberg, Germany, September 1995.

Bergeron, P., Saffer, R.A., & Liebert, J. (1992), *ApJ*, **394**, 228

Dreizler, S., & Werner, K. (1996), *Astr. Ap.*, in press

Fleming, T.A., Liebert, J., & Green, R.F. (1986), *ApJ*, **308**, 176

Green, R.F., Schmidt, M., & Liebert, J. (1986), *ApJS*, **61**, 305

Liebert, J., & Bergeron, P. 1995, in *White Dwarfs*, eds. D. Koester & K. Werner (Heidelberg: Springer–Verlag), p. 12

Liebert, J., Bergeron, P., Schmidt, G.D., & Saffer, R.A. (1993), *Ap.J.*, **418**, 426

Marsh, T.R., Dhillon, V.S., & Duck, S.R. (1995), *M.N.R.A.S.*, **275**, 828

Schmidt, G.D., Bergeron, P., Liebert, J., & Saffer, R.A. (1992), *Ap.J.*, **394**, 603

Wood, M.A. (1992), *ApJ*, **386**, 539

# WHITE DWARFS IN M67

T.A. FLEMING  AND  J. LIEBERT
*Steward Obs., University of Arizona*
*Tucson, AZ 85721 USA*

AND

P. BERGERON  AND  A. BEAUCHAMP
*Université de Montreal*

**Abstract.** We present optical spectra taken of two white dwarfs in the field of the old open cluster M67. One is of type DA, the other of type DB. The DA is hot (68,000 K) and of standard mass (0.57 $M_\odot$), while the DB is much cooler (17,000 K) and less massive (0.49 $M_\odot$), athough the results for the DB are less certain. Should they prove to be members of the cluster, their existence shows that dwarf stars of near solar mass do beget both hydrogen and helium atmosphere white dwarfs.

## 1. Introduction

The open cluster M67 is one of the oldest such clusters known (4.0 ± 0.5 Gyr, turn-off mass of 1.25 $M_\odot$; Dinescu et al. 1995). At a distance of 870 pc, however, it has been one of the more difficult clusters in which to study low luminosity stars like white dwarfs. Recent space-based images taken of M67 in the X-ray (ROSAT) and UV (ASTRO) regimes have helped locate white dwarfs within the field of the cluster (Pasquini & Belloni 1994; Landsman 1995, private comm.) This is a significant find since most other studies of white dwarfs in open clusters have been conducted on much younger clusters with turn-off masses $\geq 2.5 M_\odot$ (e.g. Weidemann 1987, Koester & Reimers 1996). For the first time, we have obtained masses for white dwarfs whose progenitors are known to be close to the Sun in mass.

We have obtained moderate-resolution (3.5 Å), optical spectra for two of these white dwarfs. The first one was brought to our attention by the work of Pasquini, Belloni, & Abbott (1994), who detected it as a very

*I. Isern et al. (eds.), White Dwarfs, 91–96.*
© 1997 *Kluwer Academic Publishers.*

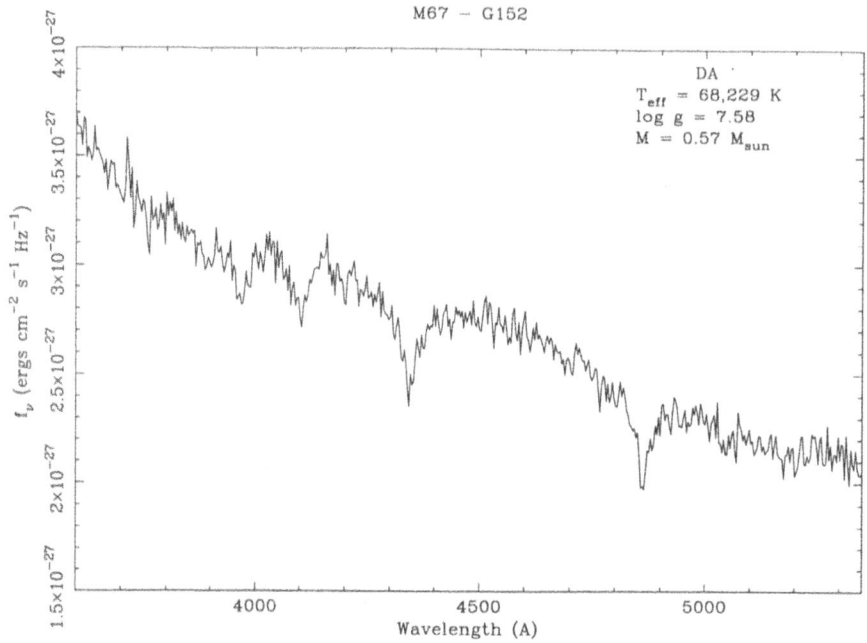

*Figure 1.* Spectrum of star G152 in M67

soft X-ray source using ROSAT and identified the star as a DA with a low-resolution spectrum. This star had originally been discovered by Baade and catalogued by Luyten (1963; LB 6339) and it also appears in the catalogue of Gilliland et al. (1991) as G152.

The second star was brought to our attention by Landsman (1995, private comm.) It had been imaged at 1600 Å by the ASTRO mission aboard the Space Shuttle as a bright UV source which was positionally coincident with star number MMJ 5973 in the catalogue of Montgomery, Marschall, & Janes (1993). To our knowledge, we are the first to take a spectrum of this star and have identified it as a DB white dwarf.

## 2. Data Analysis

The spectra shown in Figures 1 & 2 were taken on 1995 November 21 at the Multiple Mirror Telescope on Mt. Hopkins, AZ. We used the MMT Blue Channel spectrograph with a 500 l/mm grating in first order to achieve a spectral resolution of 3.5 Å. These spectra clearly show that G152 is type DA and MMJ 5973 is type DB.

The spectra were next fit with white dwarf atmosphere models to de-

M67 − MMJ5973

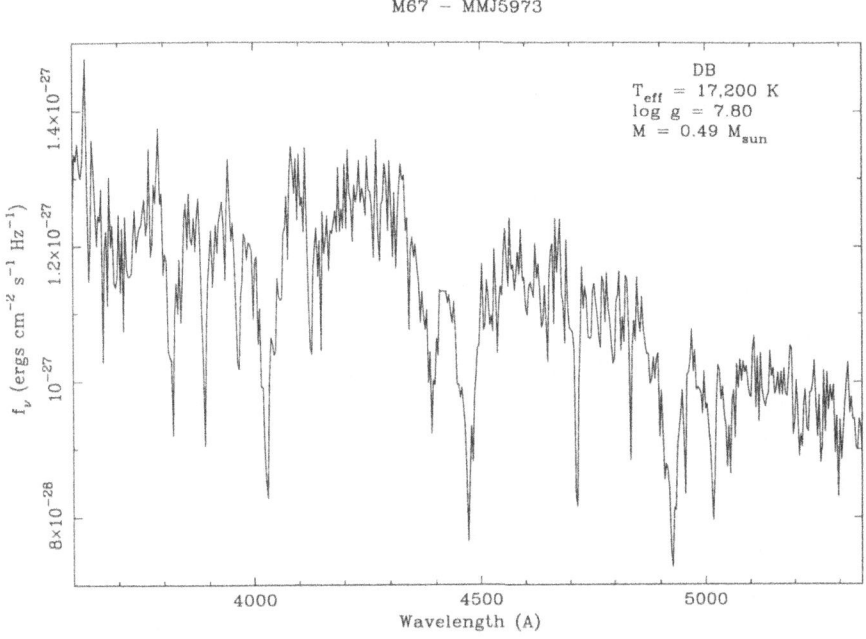

*Figure 2.*  Spectrum of star MMJ 5973 in M67

termine the effective temperature and gravity of the star and estimate the mass in the same way as was done by Bergeron, Saffer, & Liebert (1993). The He atmosphere models which were fit to the DB spectrum come from Beauchamp (1995). The results of these model fits are shown in Table 1.

TABLE 1.  Results of Model Fits

| M67 # | Type | $T_{eff}$ | log g | $M/M_\odot$ | $M_V$ | $\tau_c$ | V |
|---|---|---|---|---|---|---|---|
| G152 | DA | 68,230 K | 7.58 | 0.57 | 8.01 | $7.35 \times 10^5$ yr | 18.71 |
|  |  | $\pm 3,200$ | $\pm 0.16$ | $\pm 0.05$ | $\pm 0.31$ |  |  |
| MMJ5973 | DB | 17,152 K | 7.77 | 0.47 | 10.72 | $1.12 \times 10^8$ yr | 19.66 |
|  |  | $\pm 311$ | $\pm 0.14$ | $\pm 0.07$ | $\pm 0.19$ |  |  |

These model parameters were determined independently from the distance, i.e. the values of $M_V$ given in Table 1 were derived from the model fits, not fixed using the known distance to M67. The distance modulus of M67 is 9.7 (Dinescu et al. 1995) which implies that, should these stars be members, the absolute magnitudes of G152 and MMJ 5973 are 9.0 and

10.0, respectively. We are forced, therefore, to question whether these stars are indeed members of M67. These two stars are too faint to have been included in any of the proper motion surveys done on M67 in order to determine cluster membership. Further work needs to be done in ascertaining the proper motions of these stars.

The ASTRO images of M67 contain 5 hot white dwarf candidates (Landsman 1995, private comm.), including the two stars presented here. At the distance of M67, given the known space density of white dwarfs and the angular extent of the cluster, we would expect 2.5 field white dwarfs to be in the line of sight. Taking into account uncertainties in the models for very hot DAs (e.g. non-LTE effects) and DBs, plus the low SNR ($\sim 20$) of the DB spectrum, we would say that these stars have an even chance of being cluster members.

## 3.  Discussion

Assuming that G152 and MMJ 5973 are members of M67, then their existence is quite significant, which tells us some important things. First, the cooling times listed in Table 1 and the post-main sequence lifetimes of these stars are insignificant compared to the cluster age of 4.0 Gyr. This means, in effect, that the main sequence progenitors for both of these stars had masses of 1.25 $M_\odot$. These white dwarfs represent the endpoint of stellar evolution for stars like the Sun.

The initial-final mass relation has been studied extensively by, e.g., Reimers, Koester, Reid (see the review by Weidemann 1990). But these authors have studied clusters much younger than M67, so that most of the data points in their work have initial masses between 2 and 3 $M_\odot$. Being the oldest cluster for which we have measured white dwarf masses, M67 provides us with white dwarfs for which we have the lowest known initial masses. As expected from theoretical (Iben & Renzini 1983) and semi-empirical (Weidemann & Koester 1983) initial-final mass relations, the white dwarfs in M67 have lower masses (0.57 and 0.47 $M_\odot$) than the white dwarfs in the younger clusters studied (0.6 - 0.8 $M_\odot$).

Furthermore, we see that solar-mass stars can evolve into both DA and DB white dwarfs. This argues against the idea of separate evolutionary channels for H and He atmosphere white dwarfs. Of course, it is possible that the DB is the result of binary star evolution but, given the faintness of the star, this possibility would be difficult to prove.

More spectra need to be taken of the white dwarfs in M67 in order to sort out the questions of membership and evolutionary status. This will be made possible with the next generation of large telescopes. Hopefully, time will be granted since the determination of white dwarf masses for such an old cluster is very important. With M67, we can see what kind of star the Sun will ultimately evolve into.

## References

Bergeron, P., Saffer, R.A., & Liebert, J. (1992), *ApJ*, **394**, 228

Dinescu, D.I., Demarque, P., Guenther, D.B., & Pinsonneault, M.H. (1995), *AJ*, **109**, 2090

Gilliland, R. et al. (1991), *AJ*, **101**, 541

Iben, I. & Renzini, A. (1983), *ARA&A*, **21**, 271

Koester, D. & Reimers, D. (1996), *A&A*. **313**, 810

Luyten, W. (1963), Univ. of Minnesota Obs. Pub. No. 32

Montgomery, K.A., Marschall, L.A., & Janes, K.A. (1993), *AJ*, **106**, 181

Pasquini, L., Belloni, T., & Abbott, T. (1994), *A&A*, **290**, L17

Weidemann, V. (1987), *A&A*, **188**, 74

Weidemann, V. (1990), *ARA&A*, **28**, 103

Weidemann, V. & Koester, D. (1983), *A&A*, **121**, 77

## Discussion

*M. Barstow*: To make the DA a cluster member, does the gravity need to go up or down? Non-LTE seems to give different gravities in these very hot DAs compared to LTE, but the tendency is to get lower values rather than higher.

*T. Fleming*: The gravity needs to go up in order for the DA to be a cluster member, so the effect you suggest would make the situation worse.

*V. Weidemann*: Did you check on the number of expected white dwarfs based on the number of main sequence and giants like we did for Hyades?

*T. Fleming*: No

*J. Liebert*: It is well known that model analyses of DA stars at $T_{\mathrm{eff}} \sim 60000$ K are fraught with serious difficulties and potential systematic errors. If a cluster member, this DA star should be an easier test since, to be such a strong X-ray source, it must have close to a pure H atmosphere. That we seem to miss by $\Delta M_V \sim 1^{\mathrm{mag}}$ is discouraging, isn't it?

*T. Fleming*: Yes it is. Although it is reasonable and even intuitive that the DA should be a cluster member, we have to believe the numbers. There is no strong reason, at the moment, to doubt the model fits.

*I. King*: You should make a real effort to get the proper motion of your DA star. There must surely be good first-epoch plates; Baade's plates are nearly all excellent, since he tended to take them in good seeing. Even if the new Palomar Sky Survey doesn't separate the star from its brighter neighbor (which it should), a quick CCD image should suffice.

*T. Fleming*: We will.

*M.T. Ruiz*: How sure are you about the reddening? Could it be high enough in that spot (where the DA white dwarf is located)? If it is a young object, it could still have some ejecta from the PN phase.

*T. Fleming*: The reddening ($E(B - V) = 0.03$) comes from Dinescu et al. (1995) who fit evolutionary models to color-magnitude diagram. This translates into an extinction at $V$, $A_V \approx 0.1^{\mathrm{mag}}$, which is much less than the $1^{\mathrm{mag}}$ required to get the proper distance modulus for M67.

# MONTE CARLO SIMULATIONS OF THE KINEMATICS AND LUMINOSITY FUNCTION OF WHITE DWARFS

ENRIQUE GARCIA-BERRO AND SANTIAGO TORRES

*Departamento de Física Aplicada*
*Universidad Politécnica de Cataluña*
*Jordi Girona Salgado, s/n, Módulo B4, Campus Norte*
*08034 Barcelona, Spain*

**Abstract.** White dwarf stars are long-lived and well studied objects. Therefore, the white dwarf luminosity function when combined with the kinematical data of white dwarfs in the solar neighborhood can provide useful information about the past history of the local neighborhood. In this work we present the results of Monte Carlo simulations of the kinematics and luminosity function of white dwarfs and compare them with observational results.

## 1. Introduction

The white dwarf luminosity function has become an important tool to determine some properties of the solar neighborhood, such as its age (Winget et al. 1987; García-Berro et al. 1988; Hernanz et al. 1994), the past history of the star formation rate (Isern et al. 1995), or even the properties of the halo (Mochkovitch et al. 1990). This has been possible because now we have observational luminosity functions (Liebert et al. 1988; Oswalt et al. 1996) which seem to be reliable. The most important feature of the luminosity function of disk white dwarfs is the presence of a pronounced cut-off at $\log(L/L_\odot) \sim -4.4$, although the exact position is still today somehow uncertain. All these studies rely on the statistical significance of the cut-off. In this paper we explore the statistical significance of the cut-off using a Monte Carlo method coupled with bayesian inference techniques, together with a consitent model of galactic evolution and improved cooling sequences.

*I. Isern et al. (eds.), White Dwarfs, 97–104.*
© *1997 Kluwer Academic Publishers.*

## 2. Building the sample

We randomly choose two numbers for the galactocentric coordinates $(r, \theta)$ of each star in the sample within approximately 200 pc from the sun. Next we draw two more pseudo-random numbers for the mass $(M)$ on the main sequence of each star — according to the initial mass function of Salpeter (1961) — and for the time at which each star was born $(t_b)$ — according to a given star formation rate. We have chosen an exponentially drecreasing star formation rate per unit time and unit surface: $\psi \propto e^{-t/\tau_s}$. Once we know the time at which each star was born we assign the $z$ coordinate by drawing another random number according to an exponential disk profile. The scale height adopted here decreases exponentially with time: $H_p(t) \propto z_i e^{-t/\tau_h} + z_f$. In order to determine the heliocentric velocities $(U, V, W)$ three more quantities are drawn according to normal laws:

$$
\begin{aligned}
n(U) &\propto e^{-(U-U_0')^2/\sigma_U^2} \\
n(V) &\propto e^{-(V-V_0')^2/\sigma_V^2} \\
n(W) &\propto e^{-(W-W_0)^2/\sigma_W^2}
\end{aligned}
\tag{1}
$$

where $(U_0', V_0', W_0)$ take into account the differential rotation of the disk and derive from the peculiar velocity of the sun $(U_0, V_0, W_0)$. The velocity dispersions and $V_0$ are not independent of the scale height. From main sequence star counts, Binney & Tremaine (1987) obtain

$$
\begin{aligned}
\sigma_V^2/\sigma_U^2 &= 0.32 + 1.67 \ 10^{-5}\sigma_U^2 \\
\sigma_W^2/\sigma_U^2 &= 0.50 \\
H_p &= 6.52 \ 10^{-4}\sigma_W^2
\end{aligned}
$$

$$
\begin{aligned}
U_0 &= 0 \\
V_0 &= \sigma_U^2/120 \\
W_0 &= 0
\end{aligned}
\tag{2}
$$

which is what we adopt here. The values of the free parameters have been taken from Isern et al. (1995), namely: $\tau_s = 24$ Gyr, $\tau_h = 0.7$ Gyr, $z_i/z_f = 485$. Since white dwarfs are long lived objects the effects of the galactic potential in their motion, and therefore in their positions and proper motions, can be potentially large, specially for very old objects which populate the tail of the white dwarf luminosity function. Therefore, the $z$ coordinate is integrated using the galactic potential proposed by Flynn et al. (1996). We do not consider the effects of the galactic potential in the $r$ and $\theta$ coordinates. This is the same to assume that the number of white

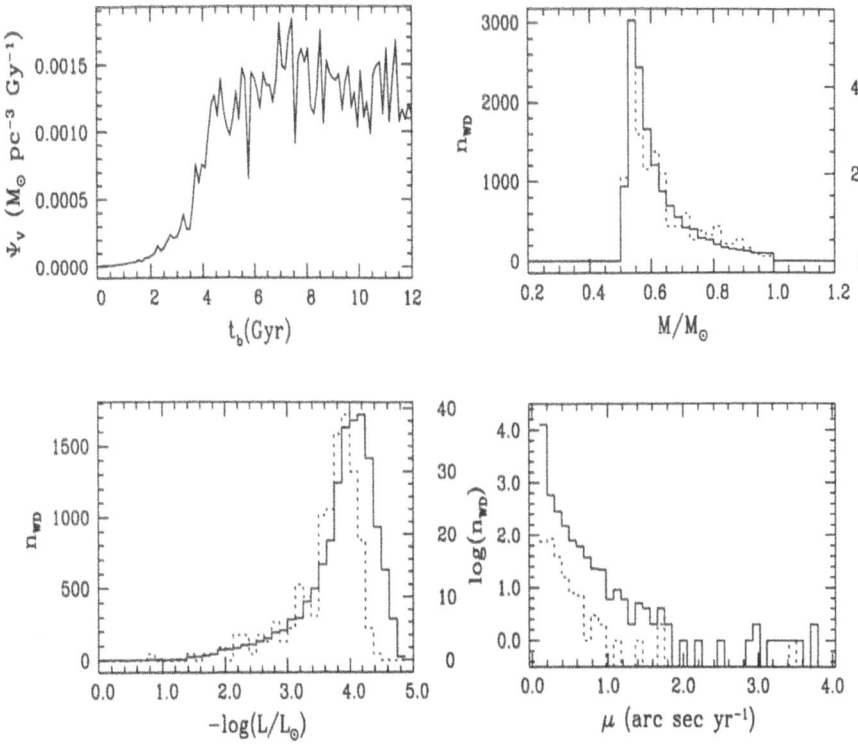

*Figure 1.* Some relevant distributions obtained from a single Monte Carlo simulation. See text for details.

dwarfs that enter into the sector of the disk that we are considering is, on average, equal to the number of white dwarfs that are leaving it. From this set of data we can now compute parallaxes and proper motions for all the stars in the sample ($\sim$ 200000). Given the age of the disk ($t_{\mathrm{disk}}$) we can also compute how many of these stars have had time to evolve to white dwarfs and, given a set of cooling sequences (Ségretain et al. 1994), which is their luminosity. Of course, a relationship between the mass on the main sequence and the mass of the resulting white dwarf is needed. Main sequence lifetimes must be provided as well. For these two relationships have used those of Iben & Laughlin (1989). The size of this sample typically is of $\sim$ 14000 stars (hereinafter "second sample"). Finally, for the white dwarfs belonging to this sample bolometric corrections are calculated by interpolating in the atmospheric tables of Bergeron et al. (1995) and their $V$ magnitude is obtained, assuming that all are DB white dwarfs.

The final step is to compute the white dwarf luminosity function using the $V/V_{\mathrm{max}}$ method (Schmidt, 1968) according to a set of restrictions. The contribution of each white dwarf to the total error budget of each bin is

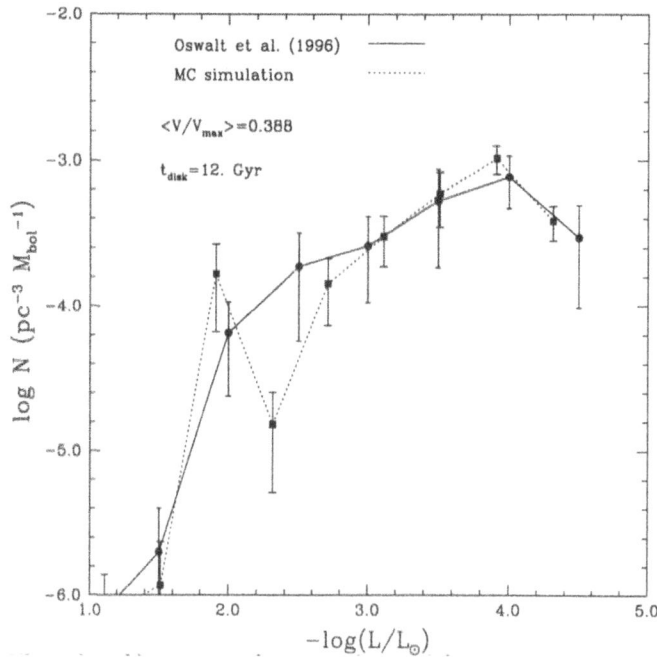

*Figure 2.* White dwarf luminosity function obtained from a single Monte Carlo simulation. See text for details.

conservatively estimated to be equal to the star's contribution itself. We have chosen the following criteria for selecting the sample: $m_V \leq 18.5^{mag}$ and $\mu \geq 0.16'' \, yr^{-1}$ (Oswalt et al. 1996). These restrictions determine the size of the third sample which typically is $\sim$ 200 stars (hereinafter "third sample"). Finally we normalize the total density of white dwarfs to its observed value in the solar neighborhood.

In Figure 1 we show a summary of the most relevant results for a disk age of 12 Gyr. In the upper left panel the *volumetric* star formation rate — before normalizing the resulting density of white dwarfs to its observed value — is shown. In agreement with Isern et al. (1995) the SFR is smoothly increasing during the initial 4 Gyr and remains almost flat during the rest of the life of the solar neighborhood. In the upper right panel, the mass distribution of those stars that have been able to become white dwarfs (solid line, left scale) and of those white dwarfs that are selected for computing the luminosity function (dotted line, right scale) are shown. Both distributions are well behaved and peak at around $0.55 \, M_\odot$ in good agreement with the observations. In the lower left panel we show the raw distribution of luminosities for the stars in the second (solid line, left scale) and third (dotted line, right scale) samples. The differences between both distributions are quite apparent: first, the second sample has a broad peak centered at $\log(L/L_\odot) \sim -4.2$, whereas the third sample is narrowly peaked at a

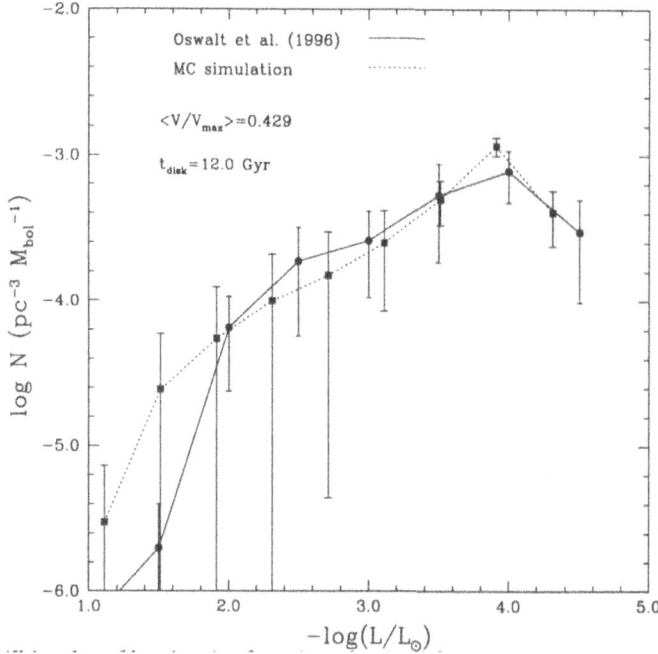

*Figure 3.* White dwarf luminosity function obtained from bayesian inference techniques. See text for details.

slightly larger luminosity. Obviously, since the third sample is selected on a kinematical basis — see the lower right panel of Figure 1 — some very faint and low proper motion white dwarfs are discarded. It is important to realize that only $\sim 2\%$ of the total number of white dwarfs with $\log(L/L_\odot) > -4.0$ are selected for the third sample. This number decreases if we consider white dwarfs with $\log(L/L_\odot) < -4.0$ to $\sim 0.5\%$. That is, the luminosity function is based on relatively poor statistics.

In Figure 2 the white dwarf luminosity function computed with the distributions of Figure 1 is shown as square symbols. For comparison, the luminosity function of Oswalt et al. (1996) is also shown as circles. Both sets of data have been connected to provide a visual reference (dotted and solid lines, respectively). The agreement between both sets of data is excellent at low luminosities and poor at high luminosities. The reason is that at high luminosities the evolution is dominated by neutrino losses and it is fast. Therefore, the probability of finding such white dwarfs is relatively small and the statistical significance of those bins is low. Consequently, any conclusion regarding the SFR coming from data at high luminosities is based on weak grounds. On the other hand, the exact shape of the luminosity function at $\log(L/L_\odot) \geq -3.0$ is strongly dependent on the initial seed of the pseudo-random number generator. Note, however, that $\langle V/V_{\max} \rangle = 0.388$ not far to be considered a complete sample.

## 3.  Bayesian analysis of the luminosity function

Changing the initial seed of the random function provides different realizations of the white dwarf luminosity function. All these realizations are "a priori" equally good. Besides, since the number of objects that is used to compute the white dwarf luminosity function is relatively small, large deviations are expected, specially at relatively high luminosities for which the cooling timescales are short. Consequently, in order to obtain reliable luminosity functions we have used bayesian statistical methods (Press, 1996). First we run our code with different seeds (typically 20) and the same set of restrictions previously described, thus resulting in different samples and luminosity functions. Then we assume constant "a priori" probability, that is, all data are equally good. The problem can be stated as follows: for a given luminosity, $L$, we want to know the most probable value of the white dwarf luminosity function, $N$, given a set of $N_i$ simulations. To compute $N$ one must maximize the probability distribution

$$P(N/N_i) \propto \prod_i \frac{1}{2}\left(P_{G_i} + P_{B_i}\right) \tag{3}$$

where $P_G$ and $P_B$ are the probability of being a good and a bad simulation, respectively. We can calculate them following the expresions:

$$
\begin{aligned}
P_{G_i} &= \exp\left[-\frac{(N_i - N)^2}{2\sigma_i^2}\right] \\
\\
P_{B_i} &= \exp\left[-\frac{(N_i - N)^2}{2S^2}\right]
\end{aligned}
\tag{4}
$$

where $\sigma_i$ is the error bar of each bin of the luminosity function and $S$ is a large but finite number characterizing the maximum expected deviation in $N_i$. We recall that the contribution to the error of each white dwarf is equal to the inverse of its maximum volume and the total error in the bin is equal to the geometric mean of the individual white dwarf errors. We have taken a 90% confidence level in the determination of the resulting error bars (approximately $1.5\sigma$). The result is shown in Figure 3. Now the agreement between both sets of data is substantially improved at high luminosities. The completness of the simulated sample has been also increased to a value of $\langle V/V_{max} \rangle = 0.429$. With this sample we have run a "standard" code to derive an age of the disk, with exactly the same inputs. The age of the disk obtained in this way is 13 Gyr. Thus, the typical uncertainty in the age estimates deriving just from *statistical arguments* is 1 Gyr.

## 4. Conclusions

The most important conclusion of this work is that the white dwarf luminosity function is fully consistent with a standard model of galactic evolution. Moreover, the white dwarf luminosity function when coupled with the kinematics of the sample provides a wealth of information, specially at low luminosities. However, the statiscal significance of the low luminosity bins must be checked since very few white dwarfs are found at these low luminosities. Nevertheless, the observed cut-off in the white dwarf luminosity function seems to be real, although its exact position depends somehow on the details of the survey and on how the data is binned. The statistical error in the position of the cut-off is typically of 1 Gyr. The age of the solar neighborhood obtained from the cut-off in the white dwarf luminosity function depends in the detailed history of the star formation rate. For our choice of both scale heights and star formation rate the age that best fits the observational luminosity function of Oswalt et al. (1996) is $12 \pm 1$ Gyr.

*Acknowledgments.* This work has been supported by the DGICYT grant PB94-0827-C02-02. One of us (EG-B) acknowledges useful discussions with J. Isern and A. Burkert.

## References

Bergeron, P., Wesemael, F., & Beauchamp, A. (1995), *PASP*, 107, 1047

Binney, J. & Tremaine, S. (1987), *Galactic Dynamics* (Princenton: Princenton University Press).

Flynn, C., Sommer-Larsen, J., & Christensen, P.R. (1996), *astro-ph/9603106*

García-Berro, E., Hernanz, M., Isern, J., & Mochkovitch, R., (1988), *Nature*, 333, 642

Hernanz, M., García-Berro, E., Isern, J., Mochkovitch, R., Ségretain, L., Chabrier G. (1994), *ApJ*, 434, 652

Iben, I., & Laughlin, G. (1989), *ApJ*, 341, 312

Isern, J., García-Berro, E., Hernanz, M., Mochkovitch, R., Burkert, A. (1995), in *White Dwarfs*, Ed. Koester, D. & Werner, K., Springer Verlag : Berlin

Liebert, J., Dahn, C.C., & Monet, D.G. (1988), *ApJ*, 332, 891

Mochkovitch, R., García-Berro, E., Hernanz, M., Isern, J., & Panis, J.F. (1990), *A& A*, 233, 456

Oswalt, T.D., Smith, J.A., Wood, M.A., & Hintzen, P. (1996), *Nature*, 382, 692

Press, W.H. (1996) *astro-ph/9604126*

Salpeter, E.E. (1961), *ApJ*, 134, 669

Schmidt, M. (1968), *ApJ*, 151, 393

Segretain, L., Chabrier, G., Hernanz, M., García-Berro, E., Isern, J., Mochkovitch, R. (1994), *ApJ*, 434, 643

Winget, D.E., Hansen, C.J., Liebert, J., Van Horn, H.M., Fontaine, G., Nather, R., Kepler, S.O., & Lamb, D.K. (1987), *ApJ*, 315, L77

## Discussion

*J. Liebert*: Is the kinematics analysis consistent with the conclusions of main sequence stars by Edvardson et al. (1994)?

*E. García-Berro*: I do not know that paper, but I guess so.

*I. Iben*: What physical mechanism is responsible for the increase with time of the $W$ dispersion for white dwarfs in the galactic disk? Does it have anything to do with interactions with giant molecular clouds as postulated long ago by Schwarzschild & Spitzer?

*E. García-Berro*: We do not know for sure. We have used a relationship between the scale height and the $W$ dispersion obtained from number counts of main sequence stars and this is the responsible of the increase in the velocity dispersion for old white dwarfs. The same kind of behaviour is obtained for other types of stars.

*S. Kawaler*: How easy would it be to include a halo white dwarf population?

*E. García-Berro*: It is trivial. You only have to change the kinematics and the star formation rate.

*I. King*: Both you and M. Wood make your point very well: that the observational results are very uncertain. But I would like to ask you about one detail: you assumed gaussian velocity distributions, yet nothing in nature is really gaussian; you always find a bigger tail. If you took this into account, would it make a difference?

*M. Wood*: Possibly, but since I tried to choose my velocity distribution to mimmic the observed distribution, I do not think the effect would be large. I certainly do not think it will change the fundamental result that we have a 1 to 2 Gyr uncertainty resulting from statistical variations.

*E. García-Berro*: In my analysis what is really important is the velocity dispersions, which are directly related to the scale height. I do not expect large changes if we change the shape of the distribution.

# MONTE CARLO SIMULATIONS OF THE WHITE DWARF POPULATION AND LUMINOSITY FUNCTION

MATT A. WOOD

*Department of Physics and Space Sciences*
*Florida Institute of Technology*
*Melbourne, FL 32901-6988*

**Abstract.** The white dwarf luminosity function has been used in recent years as an independent probe of the age and evolution of the local Galactic disk. A long-standing uncertainty of the technique is that the reality of the reported downturn in the luminosity function hinges on just a handful of stars and on statistical arguments that fainter (older) objects would have been observed were they present. Using a Monte Carlo approach, I explore the uncertainties in the derived ages and star formation rates resulting from the small-number statistics of the lowest-luminosity bin. The results suggest that (i) Schmidt's $1/V_{\mathrm{max}}$ technique underestimates by typically 25 to 50% the true space density of white dwarf stars in surveys with proper motion limits $\mu''_{\mathrm{lim}} \lesssim 1.''0 \ \mathrm{yr}^{-1}$, and (ii) there is $\sim 1$ Gyr statistical uncertainty in the age inferred from a sample with $N \lesssim 5$ objects in the lowest-luminosity bin, confirming the conservative error estimates adopted by LDM and OSWH.

## 1. Introduction

The practical application of white dwarf (WD) cosmochronometry begins with Winget et al. (1987) and Liebert, Dahn, & Monet (1988; hereafter LDM). Schmidt (1959) first suggested that observations of the coolest white dwarfs interpreted through Mestel cooling theory could provide an independent age estimate for the local Galactic disk, and D'Antona & Mazzitelli (1978) first laid out the method for calculating theoretical luminosity functions (LFs), but it was not until the late 1980's that the observational determination of the downturn in the LF near $M_V \approx 16.5$ [$\log(L/L_\odot) \approx -4.4$] was sufficiently secure to allow detailed theoretical interpretation. These interpretations followed *en masse* (see Winget et al. 1987, Yuan 1992, Wood

105

*I. Isern et al. (eds.), White Dwarfs*, 105–111.

1992, Hernanz et al. 1994, Wood & Oswalt 1997, and references therein). These efforts gleaned all information available from the LDM data, but one long-standing worry about the observed LF is that there are 3 or less objects in the lowest-luminosity bin, depending upon which binning is used.

Recently, Oswalt et al. (1996, OSWH) published an independent determination of the WDLF using WDs in common proper motion binaries. This sample includes fainter objects than found in LDM, suggesting a local age of ~9.5 Gyr using the thick DA models of Wood (1995) — some 2 Gyr older than the age obtained using these same models to fit the LDM LF. The OSWH sample also suggests a factor of ~2 larger space density than the LDM sample. These discrepancies have reopened the question of the absolute reliability of LFs calculated with small ($N \sim 50$) samples and using the $1/V_{\text{max}}$ method of Schmidt (1968).

To test the behavior of the $1/V_{\text{max}}$ method as applied to the observed WD sample, I wrote a simple Monte Carlo (MC) simulation code. I discuss the code briefly and the results more extensively in the remainder of this paper (see also Wood & Oswalt 1997 for a more detailed discussion).

## 2. The Simulation Program MCGoLF

The code MCGOLF (= Monte Carlo Generator of Luminosity Functions) draws pseudo-random samples from a parent sample which is kinematically similar to the observed sample. The $x_j$ ($j = 1, 2, 3$) positions are drawn randomly in the first octant, and the velocities are drawn from a normal (Gaussian) distribution with center and width of $v_{\text{rms}} = 40$ km s$^{-1}$. Once the phase-space coordinates are assigned, MCGoLF discriminates based on an integrated LF computed with the code LFINT (see Wood 1992), as follows. First, on input the integrated LF is put on a linear scale and normalized to a peak of unity. Then for each trial object, two random numbers are drawn. The first is scaled to provide a value for the trial luminosity $\ell_{\text{test}} \equiv \log(L/L_\odot)_{\text{test}}$ between 0 and $-8$. The spline-interpolated value of the integrated LF at this random trial luminosity, $\Phi_{\text{LFINT}}(\ell_{\text{test}})$, is compared with the value of the second random number $\Phi_{\text{test}}$. If $\Phi_{\text{test}} < \Phi_{\text{LFINT}}(\ell_{\text{test}})$, then the object "exists" in the sample volume $V_{\text{samp}}$ at the location $(x_j, v_j, \ell)$ and contributes to the overall space density $\Phi_{\text{true}} = N_{\text{tot}}/V_{\text{samp}}$, where $N_{\text{tot}}$ is the total number of objects existing in $V_{\text{samp}}$ whether in the "observationally-selected" subsample or not.

Having now a procedure for populating a space with objects that have luminosities drawn from a probability distribution function defined by integrated LFs of various ages, the next step is to determine whether a particular object makes it into the "observationally-selected" sub-sample — i.e., whether the proper motion and $V$ magnitude are within the specified observational limits. Objects are first culled if the calculated proper motion is below the input lower limit $\mu_0''$. For objects still in the running, the code interpolates in a 0.6 $M_\odot$ DA sequence (Wood 1995) to obtain $t_{cool}$, $T_{eff}$, and $\log g$ corresponding to the object's luminosity $\ell_{test}$, and then use these to interpolate in the atmospheric tables of Bergeron, Wesemael, & Beauchamp (1995) to obtain $M_V$ and hence $V$ magnitudes. If the $V$ magnitude is brighter than the input limit, then object becomes the $i$'th member of the observationally-selected sub-sample, and the data are stored ($\vec{r}_i$, $\vec{v}_i$, $v_{rad,i}$, $v_{tan,i}$, $\mu_i''$, $\ell_i$, $T_{eff,i}$, and $V_i$). Roughly $10^2$ objects are in $V_{samp}$ for each object that makes it into the observationally-selected subsample.

Once the observationally-selected sub-sample is populated the luminosity function can be calculated. The $1/V_{max}$ method of Schmidt (1968) is generally regarded as the superior estimator for samples of this kind. In this method, each star's contribution to the luminosity function is weighted by the inverse of the maximum volume in which this star would be observable. Following LDM, I conservatively set the uncertainty of each star's contribution equal to that star's contribution (e.g., $1 \pm 1$), and sum the errors in quadrature within a given luminosity bin.

## 3. Results

### 3.1. KINEMATICS AND SPACE DENSITIES

Because of limited space, I will only report here the results obtained from 2 representative parameter choices, differing only in the adopted proper motion limit $\mu_0''$. Both sets have 50 objects in each subsample, velocities $v_{rms} = 40$ km s$^{-1}$ (see above), ages ranging from 7 to 18 Gyr, and a $V$ magnitude limit $V_{lim} = 19$. Parameter set A has $\mu_0'' = 0.''15$ yr$^{-1}$, characteristic of OSWH, and parameter set B has $\mu_0'' = 0.''80$ yr$^{-1}$, characteristic of LDM.

As expected for samples selected by proper motion and $V$ magnitude, the sample mean distance is small and the sample is biased against objects with $v \sim 0$. This is a strong function of $\mu_0''$: samples from parameter set A have maximum radii of $\sim$60 pc versus $\sim$30 pc for set B.

Observationally, it is no simple matter to insure that WD surveys are complete, yet we can reliably estimate the space density only when the observational limits are chosen so that completeness is either known (assumed) to be 100% or the fractional incompleteness can be quantitatively estimated (see, e.g., Oswalt & Smith 1995). Indeed, LDM chose their lower proper motion survey limit of $\mu_{lim}'' = 0.''8$ yr$^{-1}$ to be secure that their observational database was complete.

Our MC simulations provide an ideal testbed to explore the reliability of the space density estimation in Schmidt's $1/V_{max}$ method, wherein the space density $\Phi_{V_m}$ is given as the sum of the $1/V_{max}$ values for all objects in the sample. For each sample we calculated $\Phi_{V_m}$, and then compared this with the space density of *all* the test objects in $V_{samp}$, $\Phi_{true} = N_{tot}/V_{samp}$. Plotting histograms of the number of samples as a function of $\Phi_{V_m}/\Phi_{true}$ for $\mu_0''$ ranging from 1.0 down to 0.15 shows a clear trend in the statistical incompleteness as a function of proper motion limit (see Wood & Oswalt 1997). A linear-least-squares fit to the median value of $\Phi_{V_m}/\Phi_{true}$ versus $\mu_0''$ yields the following relation for the incompleteness factor $\alpha$:

$$\alpha = 0.334\mu_0'' + 0.469. \tag{1}$$

The uncertainty in $\alpha$ is typically $\pm 0.2$.

## 3.2. TURNDOWN AGES: STATISTICAL UNCERTAINTIES

In Figures 1 and 2, I show LFs computed with MCGOLF for the proper motion limit $\mu_0'' = 0.''15 \text{ yr}^{-1}$ (parameter set A) and $\mu_0'' = 0.''80 \text{ yr}^{-1}$ (parameter set B), respectively. In each Figure, three LFs each are shown for the 4 choices of input age, 7, 10, 13, and 16 Gyr. The numbers of objects in each bin is indicated above the error bars in each case, and the solid lines are the input LFINT curves for the same 4 ages. At the left of each curve are 3 numbers. The first is the age of the input curve in Gyr, the second is $\Phi_{V_m}$, and the third is $\Phi_{true}$. Note that the LF points have been renormalized to the input LFINT curves for purposes of display and also because as "observed" samples we would normalize one to the other to determine an age.

There are several points of interest in these Figures. First, note that in Figure 1 the $1/V_{max}$ space densities consistently underestimate $\Phi_{true}$ by $\sim 50\%$, whereas in Figure 2 the $1/V_{max}$ space densities overestimate $\Phi_{true}$ $\sim 25\%$ of the time, consistent with the trend fitted by Equation 1. Second, note that it is common for pathological objects to cause pronounced "spikes" or "dips" in the LF. These features, if seen in an observed LF, could well be interpreted as reflecting variations in the SFR — the MC results suggest great caution should be used in any detailed interpretation of observed LFs. Third, and related to the previous point, there is significant variation in sample-to-sample ages derived by "re-fitting" the LFINT curves to the MC LFs. The results of these calculations suggest a $\sim 1$ to 2 Gyr uncertainty in the age determination from sampling statistics alone for 50-object samples, and $\sim 0.5$ to 1.0 Gyr uncertainty for 200-point samples (not shown). These uncertainties are consistent with the error estimates adopted by LDM and OSWH. Finally, since the statistical variations are most pronounced for bins with $N \lesssim 5$ objects, these results suggest that future observational determinations of the LF would benefit from a binning choice that puts $N \sim 5$ objects in the lowest-luminosity bin.

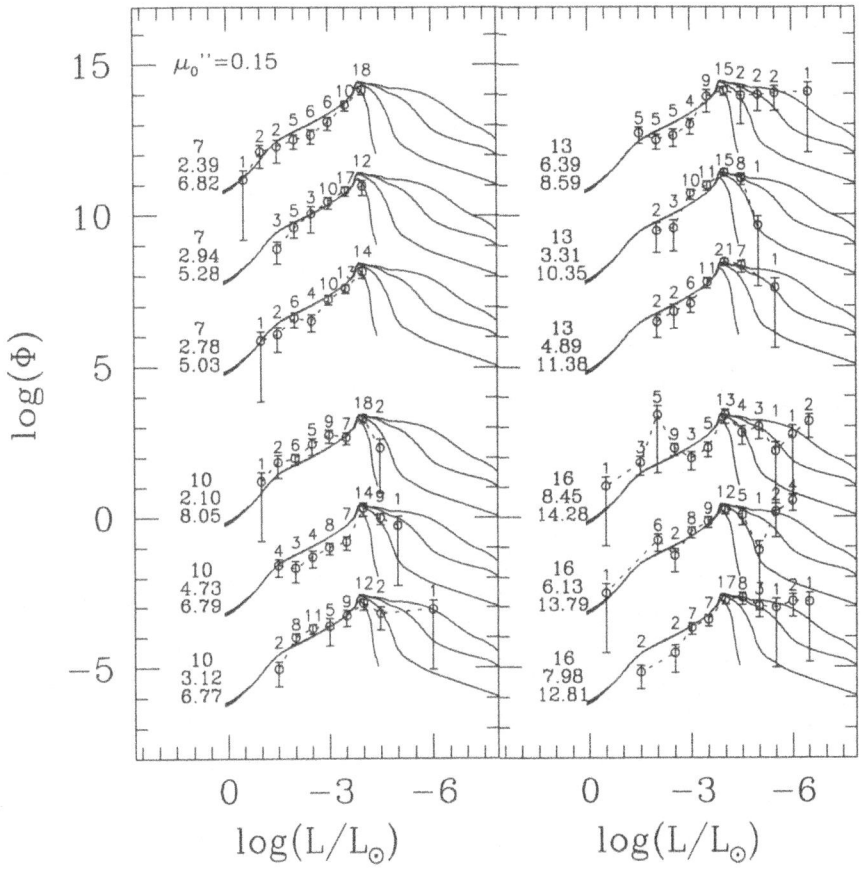

*Figure 1.*   MC LFs with $\mu_0'' = 0.''15 \ \mathrm{yr}^{-1}$. See text.

**Acknowledgments:**   I would like to thank Terry Oswalt for numerous enlightening discussions. This work was supported by NSF grant AST92-17988 and NASA Astrophysics Theory Program grant NAG 5-3103.

# References

Bergeron, P., Wesemael, F., and Beauchamp, A. 1995, PASP, 107, 1047

D'Antona, F., & Mazzitelli, I. 1978, A&A, 66, 453

Hernanz, M. García-Berro, E., Isern, J., Mochkovitch, R., Segretain, L., & Chabrier, G. 1994, ApJ, 434, 652

Liebert, J., Dahn, C. C., & Monet, D. G. 1988, ApJ, 332, 891

Oswalt, T.D., & Smith, J.A. 1995 in *White Dwarfs*, (Berlin: Springer-Verlag), eds. D.Koester & K.Werner, p. 24

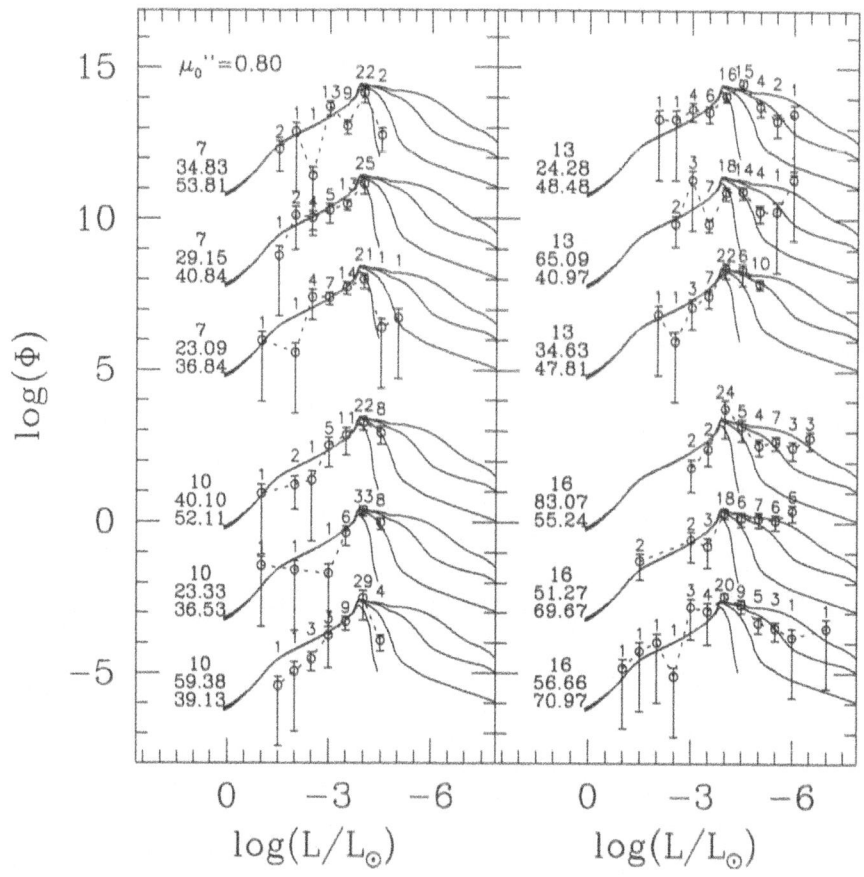

*Figure 2.* MC LFs with $\mu_0'' = 0.''80$ yr$^{-1}$. See text.

Oswalt, T. D., Smith, J. A., Wood, M. A., & Hintzen, P. M., 1996, Nature, 382, 692

Schmidt, M. 1959, ApJ, 129, 243

Schmidt, M. 1968, ApJ, 151, 393

Winget, D. E., Hansen, C. J., Liebert, J., Van Horn, H. M., Fontaine, G., Nather, R. E., Kepler, S. O., & Lamb, D. Q. 1987, ApJ, 315, L77

Wood, M. A. 1992, ApJ, 386, 539

Wood, M. A. & Oswalt, T. D. 1997, ApJ, submitted.

Yuan, J. W. 1992, A&A, 261, 105

## Discussion

*J. Liebert*: I just want to confirm that the formation rate of (hot) white dwarfs currently may be estimated much more accurately using color (magnitude) limited surveys such as Palomar Green.

*M. Wood*: You'll get no argument from me, Jim!

*S. Kawaler*: Do I understand correctly that you and Enrique (García-Berro) find that there's *at least* 1 Gyr uncertainty in the age of the disk that comes solely from the sampling statistics? Am I wrong to be depressed at this news?

*M. Wood*: Yes, the uncertainty is ~1 Gyr from sampling statistics, and yes, you are wrong to be depressed about this since essentially these results simply *confirm* the error estimates given by LDM and OSWH — they do not represent an additional uncertainty in the age estimates.

*S. Kawaler*: How much would it help the situation if the sample was increased in numbers by say, a factor of 10?

*M. Wood*: Considerably: the statistical uncertainty should then be ~0.25 Gyr or less.

*V. Weidemann*: You do not include the inflation effect do to scale height evolution. How did you calibrate your theoretical luminosity functions?

*M. Wood*: For purposes of intercomparison, I normalized the LDM sample and the LFINT curves to the OSWH space density.

*J. Isern*: My models point out that using the Oswalt et al. LF the age of the disk is ~13 Gyr instead of ~10 Gyr for the case of the Liebert et al. LF. What is the present observational situation?

*M. Wood*: It is unfortunate that Terry Oswalt could not be here to answer your question! The OSWH and LDM samples are largely independent, and in fact the disagreement between the two samples was one of the primary reasons that I started this work. The ages for the two samples agree at roughly the $2\sigma$ level, but new results of Allyn Smith with a sample of ~170 WDs from common proper motion binaries — currently under analysis — seem to indicate an even higher age than we found in OSWH!

# HALO WHITE DWARFS: A CONSERVATIVE POINT OF VIEW

J. ISERN & M. HERNANZ
*Institut d'Estudis Espacials de Catalunya (Unitat de Recerca del CSIC), Barcelona, Spain*

E. GARCIA–BERRO
*Departament de Física Aplicada (Universitat Politècnica de Catalunya), Barcelona, Spain*

N. ITOH
*Department of Physics (Sophia University), Tokio, Japan*

AND

R. MOCHKOVITCH
*Institut d'Astrophysique de Paris, (CNRS), Paris, France*

**Abstract.** Recent gravitational microlensing experiments by the MACHO collaboration suggest that the dark matter in the galactic halo could be predominantly in the form of white dwarfs. In this paper we compute the luminosity function of the halo using the most realistic chemical profiles and the most standard hypothesis about the properties of the halo in order to provide a frame to discuss these issues. We also critically analyze the results that are obtained using non–standard initial mass functions.

## 1. Introduction

The interest in halo white dwarfs started when Liebert et al. (1989) provided the first preliminary luminosity function since this opened the possibility of obtaining information about the time elapsed between the formation of the halo and the disk. This interest has recently increased due to the intepretation of the MACHO events (Bennett et al. 1996) as due to objects with an average mass $0.5^{+0.3}_{-0.2}$ $M_\odot$ that could account the 40% of the total halo mass.

White dwarfs could be very attractive microlensing candidates. However, the tight constraints imposed by galactic properties demand the use of

113

*I. Isern et al. (eds.), White Dwarfs, 113–119.*

"ad–hoc" non–standard initial mass functions (Adams and Laughlin 1996; Chabrier, Ségretain and Méra 1996) since it is necessary to avoid the over-production of red dwarfs and to avoid problems with the metallicity (Ryu et al. 1990) and with the luminosity of galactic halos at large redshift (Charlot and Silk 1995). Since the interpretation of the microlensing results has not been yet settled down (Mao and Paczyński 1996; De Paolis et al. 1996), it seems worthwhile to construct a series of standard models of halo white dwarf populations for comparison purposes.

## 2. The Standard Model

The luminosiy function is defined as the number of white dwarfs per unit volume and per unit of bolometric magnitude:

$$n(M_{\rm bol}) = \int_{M_{\rm i}}^{M_{\rm s}} \Phi(M)\, \Psi[T - t_{\rm cool} - t_{\rm MS}]\, \tau_{\rm cool}\, dM$$

where $M$ is the mass of the parent star (for convenience all white dwarfs are labelled with the mass of their main sequence progenitors), $\tau_{\rm cool} = dt_{\rm cool}/dM_{\rm bol}$ is the characteristic cooling time, $M_{\rm s}$ and $M_{\rm i}$ are the maximum and the minimum masses of the main sequence stars able to produce a white dwarf of magnitude $M_{\rm bol}$ at time $T$, $t_{\rm cool}$ is the time necessary to cool down to this luminosity, $t_{\rm MS}$ is the lifetime on the main sequence and $T$ is the age of the population under study.

The cooling times and the characteristic cooling times have been obtained from Salaris et al. (1997) for C–O white dwarfs (white dwarf masses in the range 0.5–1 $M_\odot$ and progenitors in the mass range 0.7–8 $M_\odot$), which take into account the presence of high quantities of oxygen in the central regions produced by the high rates of the $^{12}$C$(\alpha, \gamma)^{16}$O reaction, and from García–Berro et al. (1997) for O–Ne white dwarfs (white dwarf masses in the range of 1–1.4 $M_\odot$ and progenitors in the mass range 8–11 $M_\odot$). In both cases, the adopted envelope was pure helium, $M_{\rm He} = 10^{-4}\ M_\odot$. The main difference from C–O and O–Ne models is that the last ones cool down more quickly. For instance, two white dwarfs of 1 $M_\odot$, one made of C–O and the other of O–Ne take 11.3 Gyr and 8.1 Gyr respectively to reach a luminosity of $10^{-5}\ L_\odot$. This is due to the smaller heat capacity and to the negligible influence of the gravitational settling in the case of O–Ne mixtures.

Beyond luminosities of the order of $\log(L/L_\odot) \lesssim -5$ the structure of the envelope of white dwarfs is very uncertain. At this stage the major part of the star has a specific heat that obeys the Debye law and the bulk of the luminosity is provided by the gravitational contraction of the outer layers (D'Antona and Mazzitelli 1989) which prevents the sudden disappearence of the white dwarf. We have tried different atmospheric models with arbitrary

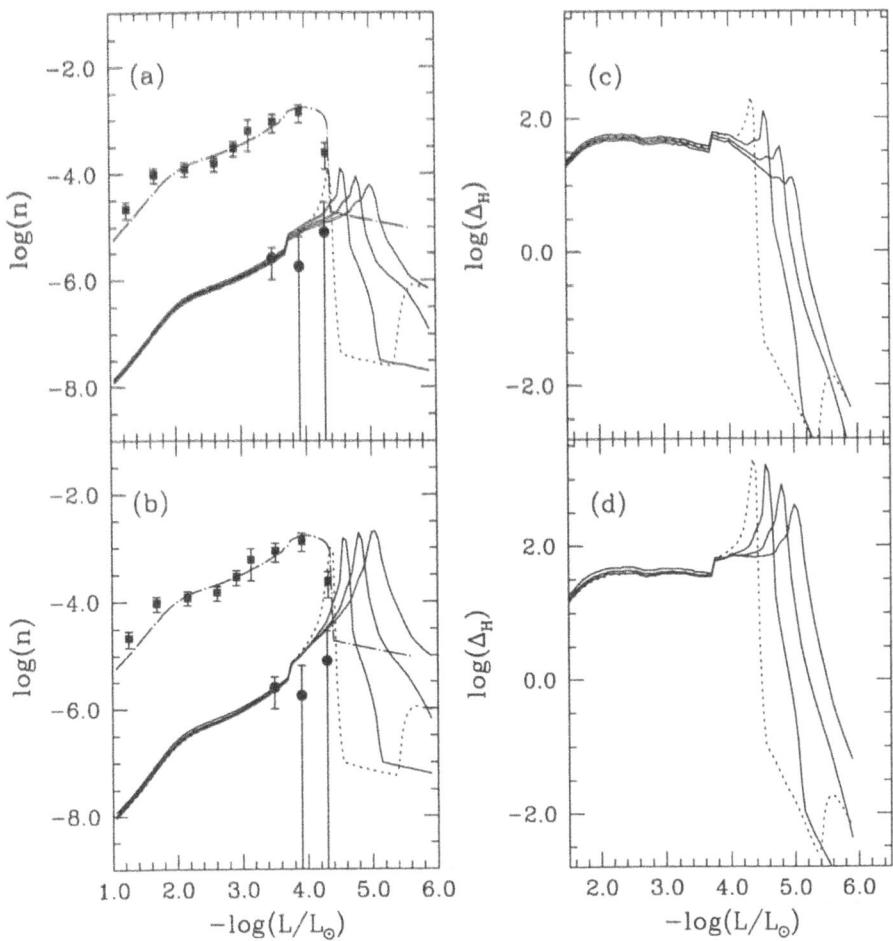

*Figure 1.* Luminosity functions for halo white dwarfs. Panel (a): luminosity function for halo ages 10 (shown as a dotted line for sake of clarity), 12, 14, and 16 Gyr computed using a standard IMF. Panel (b): The same assuming a strongly peaked initial mass function (Adams and Laughlin 1996). In both cases, the luminosity function of disk white dwarfs (dashed–dotted line) for $t_{disk} = 9.3$ Gyr is plotted for comparison purposes. Panels (c) and (d) display the discovery function for both LFs

degrees of transparency. The only consequence was a change in the shape of the luminosity function below $\log(L/L_\odot) \lesssim -5$ which is irrelevant for the problem posed here.

Figure 1a displays the luminosity functions of halo and disk white dwarfs computed with a standard initial mass function (Salpeter 1961). The observational data for both the disk and the halo have been taken from Liebert, Dahn and Monet (1989). The theoretical luminosity functions have been normalized to the points $\log(L/L_\odot) \simeq -3.5$ and $\log(L/L_\odot) \simeq -2.9$ for the halo and the disk respectively due to their smaller error bars. The luminosity function of the disk was obtained assuming an age of 9.3 Gyr and a constant star formation rate per unit volume, and those of the halo assuming a burst that lasted 1 Gyr and started at $t_{halo}$= 10, 12, 14 and 16 Gyr. Due to the higher cooling rate, O–Ne white dwarfs produce a long tail in the disk luminosity function and a bump (only shown in the case $t_{halo}$= 10 Gyr) in the halo luminosity function. The total density of halo white dwarfs obtained in this way ranges, depending on the adopted age of the halo, from $5.3 \times 10^{-5}$ to $7.8 \times 10^{-5}$ white dwarfs per $pc^3$ which can only account for 1% of the total amount of dark matter.

The discovery function gives for the whole sky the number of white dwarfs per interval of magnitude which can be detected in a survey limited to a given apparent magnitude. If we limit ourselves to nearby halo white dwarfs, the available volume can be considered spherical and the discovery function, $\Delta_H(L)$, is readily obtained from the luminosity function

$$\Delta_H(L) = \frac{4\pi}{3}d^3(L)n_H(L)$$

where $d(L)$ is the distance at which a white dwarf of luminosity $L$ has an apparent magnitude $m_\lambda$ (in a photometric band centered at $\lambda$) and can be obtained as

$$d(L) = 1.14\sqrt{\frac{L}{L_\odot}}\, 10^{0.2[m_\lambda+(V-M_\lambda)+BC]}$$

where $(V - M_\lambda)$ is a color index and $d$ is in parsecs.

Since white dwarfs with luminosities $\log(L/L_\odot) \sim -5$ have effective temperatures of $\sim 3000$ K and radiate most of their energy in the red or infrared, we have computed the discovery function for the $I$ band assuming a limiting magnitude $M_I \simeq 19^{mag}$ and that white dwarfs behave as blackbodies. For reasonable ages of the halo ($t_{halo} \sim 12$–14 Gyr), the discovery function has a flat profile ($\sim 70$ stars/magnitude/$pc^3$) with a peak (less pronounced as the age of the burst increases) in the faint end that sharply declines at $\log(L/L_\odot) \sim -5$ (Figure 1c). Finally, the expected number of white dwarfs in the Huble Deep Field (HDF) is $\sim 0.8$.

## 3.  The Non–Standard Models

The use of non–standard IMFs as those proposed by Adams and Laughlin (1996) or Chabrier et al. (1996) not only prevents the formation of red dwarfs, but also allows to invest more mass into white dwarf progenitors. Figure 1b displays the same luminosity functions as those in Figure 1a but using the IMF proposed by Adams and Laughlin (1996). In this case, the local density of white dwarfs increases by an order of magnitude as compared with the standard case ($3.2 \times 10^{-4}$ to $1.3 \times 10^{-3}$ pc$^{-3}$) but it is still below the quantity necessary to fill half of the halo dark matter with white dwarfs ($\sim 7 \times 10^{-3}$ pc$^{-3}$ in the solar neighborhood).

Figure 1d shows the corresponding discovery function. Because of the normalization condition, this function is similar to that found in the previous case in the high luminosity region. However, it rapidly rises to $\sim 1000$ pc$^{-3}$ below $\log(L/L_{\odot}) \sim -4.5$ which could be easily detected by a deep CCD survey in the red or the near infrared. Concerning the HDF survey, the field is so narrow that it does not allow to discriminate among the different hypothesis. In the case of the IMF we are considering this number is just 1.2. The similarity of this value with that obtained with a normal IMF is due to the fact that the main contribution comes from bright white dwarfs.

Chabrier et al. (1996) have also considered similar, but narrower, IMFs as those of Adams and Laughlin (1996). This procedure has the advantage of increasing the number of white dwarfs that are formed per unit mass but it has one inconvenient. If the number of main sequence stars with mases $\leq 1$ $M_{\odot}$ is too small, it is impossible to account for the density of bright halo white dwarfs found by Liebert et al. (1989) as it can be seen from their Figure 2, case IMF2. Nevertheless, since all these IMFs are arbitrary and they play a role of weight function in the calculation of the luminosity function, it is always possible to find the ad hoc shape (for instance, adding low mass tails) to simultaneously fit the Liebert et al. (1989) observations and to increase the number of white dwarfs in the halo.

The problem of using non–standard IMFs is that they are not devoided of secondary effects. For instance, since the shape of the IMF has been modified to only produce intermediate mass stars, it turns out that for each white dwarf, 1 $M_{\odot}$ of gas is returned to the interstellar medium. If the total number of white dwarfs present in the halo is $\sim 10^{12}$, the total mass of returned gas is $\sim 10^{12}$ $M_{\odot}$. Where is this gas? Certainly not in the disk! If we assume an halo with a typical radius of 60 kpc, the total energy necessary to expel this gas is $\sim 10^{60}$ ergs, and it does not seem realistic that AGB stars could provide such energy (remember that gravitational supernovae have been suppressed). Furthermore, intermediate mass stars produce carbon and nitrogen. The mechanism for expelling the gas should be very efficient to erase the traces of such elements in low metallicity stars.

## 4.  Conclusions

We have computed the luminosity functions of white dwarfs in the halo using the standard IMF and assuming bursts of 1 Gyr of duration that started 10, 12, 14, and 16 Gyr ago. The strength of the burst has been adapted to fit the constraints imposed by the observations of Liebert et al. (1989). Under these conditions, the total mass of halo white dwarfs represents the 0.5% of the total mass of the halo and the observed microlensing events have to be interpreted in another way.

The IMF plays the critical role of weighting the contribution of the different categories of stars (low, intermediate and high mass stars) to the galactic evolution. It is obvious that there is no physical reason ensuring the constancy of the IMF, but in the lack of any direct, independent, evidence this function must be kept constant (just because its role of weight function!) in order to ensure the consistency of the galactic models. Therefore, since non-standard "ad hoc" IMFs not only do not solve the problem (except for marginal situations), but they introduce more problems that they solve, they should be discarded until new, independent, evidences would be obtained. In any case, a deep survey in the red or the near infrared ($I \gtrsim 20^{mag}$) could solve the problem.

**Acknowledgements:** This work has been supported by DGICYT grants PB94-0111, PB94-0827-C02-02, by the CIRIT grant GRQ94-8001, by the AIHF 237-B and by the $C^4$ consortium.

## References

Adams, F., Laughlin, G., 1996, *Astrophys. J*, **468**, 586
Bennett, D., Alcock, C., Cook, K., Allsman, R., Alves, D., Axelrod, T., Freeman, K., Peterson, B., Rodgers, A., Griest, K., Guern, J., Lehner, M., Quinn, P., Marshall, S., Pratt, M., Becker, A., Stubbs, C., Sutherland, W., 1996, in *AAS Meeting 187*
Chabrier, G., Ségretain, L., Méra, D., 1996, *Astrophys. J (Letters)* **468**, L21.
Charlot, S., Silk, J., 1995, *Astrophys. J* **445**, 124
D'Antona, F., Mazzitelli, I., 1989 *Astrophys J* **347**, 934
DePaolis, F Ingrosso G., Jetzer Ph. 1996 *Astrophys J* **470** 493.
Fields, B., Mathews, G., Schramm, D., 1996, *Astrophys J (Letters)* submitted
Flynn, C., Gould, A., Bahcall, J. N., 1996, *Astrophys J* **446**, 55
García–Berro, E., Hernanz, M., Isern, J. 1997, *MNRAS*, submitted
Liebert, J., Dahn, C.C., Monet, D.G., 1989, in *"White Dwarfs"*, Eds.: Wegner, G., *IAU Coll. 114*, Springer Verlag, 15
Isern, J., García–Berro, E., Itoh N., Hernanz, M., Mochkovitch, R., 1996 in *White Dwarfs*, ed. J. Isern, M. Hernanz, E. García–Berro (Kluwer), in press.
Mao S., Paczyński B. preprint (astro–ph/9604002) 1 Apr 1996.
Mochkovitch, R., García–Berro, E., Hernanz, M., Isern, J., Panis, J.F., *Astron Astrophys* **233**, 456
Ruiz, M.T., Bergeron, P., Leggett, S.K., Anguita, C., 1995, *Astrophys J (Letters)* **455**, L159
Ryu, D., Olive K.A., Silk, J., 1990, *Astrophys J* **353**, 81
Salaris, M., Domínguez, I., García–Berro, E., Hernanz, M., Isern, J., Mochkovitch, R., 1997, this volume
Salpeter, E.E., 1961, *Astrophys J* **134**, 669

## Discussion

*S. Starrfield*: Comparison of nova observations with our evolutionary sequences suggest that not much Mg exists in ONeMg Novae. This agrees with your study, why is not much Mg produced?

*J. Isern*: In our evolutionary models for the ONe progenitors, the envelope is not treated as a rigid boundary but its allowed to evolve. As a consequence, the temperatures of the core are smaller.

*S. Jordan*: If the assumption that there is a significant number of white dwarfs in the halo were true, one would expect some SNIa to explode in the halo of other galaxies. Has that been observed yet? Or is the number density too low?

*J. Isern*: Certainly. However, the supernova rate would be dependent on the number of massive white dwarfs that are produced. If non–standard IMF are tuned to avoid problems with SNII, there is no reason to do the same to avoid problems with SNIa!

*M.T. Ruiz*: The existence of ESO 439–26, a 1.2 $M_\odot$ white dwarf, found in a small volume $\sim$ 950 pc$^3$ may be an evidence in favor of a different IMF

*J. Isern*: The existence of such star represents a problem but it is hard to decide the exact density of such stars when only one is known. In any case, this star would favor the formation of $\sim$ 9 $M_\odot$ stars, and if they were from the halo they would be much more colder.

*H. Shipman*: What kind of evidence would you find compelling enough to persuade you to appeal to a very unusual IMF?

*J. Isern*: The role of each kind of stars is weighted by the IMF. It is possible to fit almost all the observations by adopting an "ad-hoc" IMF for each problem. Therefore we can only use different IMFs if direct independent evidences are provided or all the standard possibilities have been discarded.

# HALO WHITE DWARFS : A NON CONVENTIONAL SCENARIO

G. CHABRIER, L. SEGRETAIN, D. MERA
*C.R.A.L, Ecole Normale Supérieure*
*69364 Lyon Cedex 07, France*

## 1. Introduction

The MACHO collaboration, which includes one more year LMC data, yields now a total of *seven* microlensing candidates towards the LMC, with *longer* durations, from 30 to 110 days (Alcock et al. 1996). This yields an average mass for the dark objects $< m > \approx 0.5 \, M_\odot$.

We examine the possibility for these objects to be halo stellar remnants, in particular halo white dwarfs (Chabrier, Segretain & Méra, 1996).

## 2. Halo white dwarf Luminosity Function

A stringent constraint on the white dwarf contribution to the halo mass budget comes from the observed white dwarf luminosity function (WDLF) in the solar neighborhood, as considered initially by Tamanaha, Silk, Wood and Winget (1990). However, although pointing the way, these calculations were based on simplified WD interior and a WD cooling theory aimed at describing the *disk* WDLF, thus appropriate for objects younger, and thus warmer, than the expected halo population. In particular these calculations do not include a *complete* treatment of crystallization (see below), which occurs around $\log L/L_\odot \approx -3.5$ in WD interiors. This affects substantially the cooling of halo WDs and will modify significantly the expected halo WDLF. More recent, similar calculations have also been completed by Adams and Laughlin (1996). These calculations, however, are based on simplified (pure carbon) WD interior and a crude treatment of the Debye cooling.

In the present calculations, we use the most updated WD cooling theory for carbon/oxygen WDs, with the appropriate equation of state both in the classical and in the quantum (crystal) regime (Segretain et al., 1994; Chabrier, 1993) and a helium-rich atmosphere (Wood 1992), characteristic

*I. Isern et al. (eds.), White Dwarfs, 121–127.*
© *1997 Kluwer Academic Publishers.*

of most cool ($T_{\rm eff} \lesssim 6000$ K) so-called "DB" WDs. As first suggested by Stevenson (1980), the gravitational energy release due to carbon-oxygen *differentiation* at crystallization affects drastically the subsequent cooling time of the star, thus changing the luminosity for a given age (Segretain & Chabrier 1993). A consistent treatment of the crystallization phase diagram along WD evolution has been derived recently by Segretain et al. (1994). As shown by these authors, the crystallization processes modify appreciably the WD cooling time and then the WDLF for $\log L/L_\odot \lesssim -4$, characteristic of old disk and halo WDs. The LF derived with this theory yields an estimate for the age of the Galactic disk $\tau_{disk} \approx 10.5 - 12$ Gyr, depending on the bolometric correction used for the *observed* LF (Hernanz et al. 1994), about 20% larger than estimates based on cooling theories which do not include the *complete* crystallization process (see Segretain et al. 1994 §4.1 for details).

The calculations proceed as in Hernanz et al. (1994). The WDLF reads:

$$n(L) = \int_{m_{inf}(L)}^{m_{sup}} \tau_{cool}(L, m) \times \psi[t_h - t_{cool}(L, m) - t_{ms}(m)] \times \phi(m)dm \quad (1)$$

Here $\tau_{cool} = dt_{cool}/dM_{bol}$ is the *characteristic* cooling time, where $t_{cool}$ is the WD cooling time. $t_{ms}$ and $t_h$ denote respectively the age spent on the main sequence for the WD progenitor and the age of the halo. The function $\phi(m)$ is the initial mass function and $m_{inf}$ and $m_{sup}$ denote respectively the minimum and the maximum mass of the WD progenitors which contribute at luminosity $L$. Since the age of the halo is much larger than any time associated with star formation, the initial stellar formation rate $\psi(t)$ is well approximated by a burst at $t = 0$, i.e. a $\delta(t = 0)$ function. In that case eqn(1) reduces to :

$$n(L) = \frac{dt_{cool}}{dM_{bol}} \times \nu(t_h - t_{cool}) \times \frac{dm}{dt} \quad (2)$$

where $\nu(t_h - t_{cool})$ represents the number of WDs formed at $t = t_h - t_{cool}$, i.e. the number of stars with a main sequence lifetime $t_{ms} = t_h - t_{cool}$. The progenitor-WD mass relation is $m_{WD} \approx 0.45 + 0.1\,m$ (Iben and Tutukov 1984). The WDLF is normalized to :

$$\int n\,dM_{bol} = -2.5 \int n\,d\log(L/L_\odot) = X_{WD}\,\rho_{dyn}/ <m_{WD}> \ {\rm pc}^{-3} \quad (3)$$

where $X_{WD}$ is the (sought) mass fraction under the form of WDs in the halo of the Galaxy. As shown in eqn(2) the most essential parameter in this calculation is the white dwarf cooling time $t_{cool}$. We use the afore-mentioned WD cooling sequences calculated in Segretain et al. (1994) and Garcia-Berro et al. (1996).

The second important parameter to be determined in Eq.(1) is the IMF $\phi(m)$. A severe constraint arises from the recently determined mass-function (slope and normalization) of halo M-dwarfs (Méra, Chabrier & Schaeffer 1996). The predicted star counts obtained with this MF for a spheroid$(1/r^3)$+halo$(1/r^2)$ density profile are in perfect agreement with the observations of the HST at large magnitude ($I \geq 25$) (Méra et al., 1996). On the other hand, the observed halo metallicity implies that stars above $m \gtrsim 8\,M_\odot$, believed to be type II Supernovae progenitors, represent at most $\sim 1\%$ of the halo initial stellar population (Ryu, Olive & Silk, 1990). These observational constraints show that, for WDs to contribute significantly to the mass of the halo, the IMF must exhibit a strongly *bimodal* behaviour and peak around some characteristic mass in the range $[m_{inf} \sim 8\,M_\odot]$. The min-imum mass corresponds to a main-sequence lifetime of the progenitor equal to the age of the halo, i.e. $m_{inf} \approx 0.9\,M_\odot$ for $t = 10$ to 25 Gyr. We elected a simple cut-off power-law function $\phi(m) = dN/dm = A\,e^{-(\bar{m}/m)^{\beta_1}}m^{-\beta_2}$ (see e.g. Larson 1986). This form mimics adequatly a strongly peaked IMF and is very similar to functional forms based on stellar formation theory (Adams and Laughlin 1996). The IMF is normalized to :

$$\int_{m_{inf}}^{\sim 8\,M_\odot} \phi(m)m_{WD}(m)dm = X_{WD}\rho_{dyn}, \qquad (4)$$

which determines $A$ (for a given $X_{WD}$ and $m_{WD}(m)$ relation). The parameter-space for $(\bar{m}, \beta_1, \beta_2)$ is constrained by the required negligible number of stars outside the mass-range $[\sim 0.9\,M_\odot, \sim 8\,M_\odot]$ but different values yield quantitatively different mass-distributions. A large number of masses $\geq 2\,M_\odot$ would raise severe problems for the fraction of ejected gas and the subsequent helium and metal galactic enrichment (Hegyi and Olive 1986; Ryu et al. 1990). In order to examine the dependence of the IMF on the results, we thus considered two functions, namely ($\bar{m} = 2.0, \beta_1 = 2.2, \beta_2 = 5.15$), peaked around $\sim 1.3\,M_\odot$ (hereafter IMF1), and ($\bar{m} = 2.7, \beta_1 = 2.2, \beta_2 = 5.75$), peaked around $\sim 1.7\,M_\odot$ (hereafter IMF2).

## 3.  Observational constraints

The LF of field WDs has been obtained by Liebert et al. (1988) up to $M_V \approx 19$ (i.e. $L/L_\odot \gtrsim 10^{-5} - 10^{-6}$, depending on the bolometric correction $BC_V$). The LF declines abruptdely for $M_V \approx 16$, which corresponds to $\log L/L_\odot \approx -4.2$ to $-4.6$. As stated by these authors, *no* WD was found at fainter magnitudes, with this or with other proper-motion samples, whereas

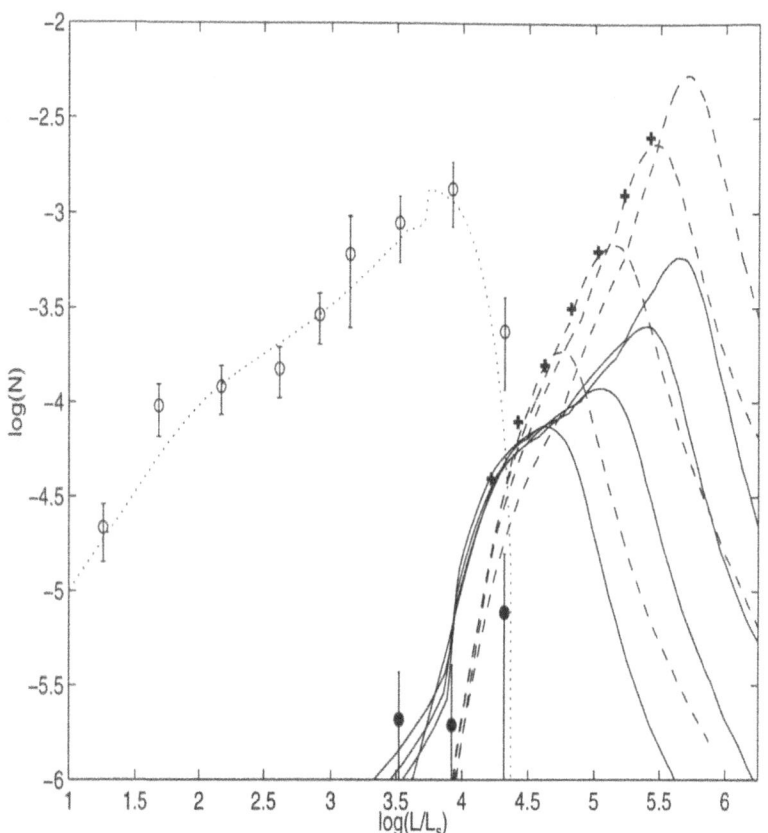

*Figure 1.*  White dwarf luminosity function (pc$^{-3}$ M$_{bol}^{-1}$). Empty circles : Liebert et al. (1998). Filled circles : high-velocity WDs (Liebert et al. 1988). Crosses : limit of detection at the 2-$\sigma$ level (see text). Dotted line : disk WDLF for $t_d = 10.5$ Gyr. Solid lines : halo WDLF for $t_h = 14, 16, 18, 20$ Gyr and $X_{WD} = 1, 2, 4, 8\%$, from left to right, with IMF1. Dashed line : halo WDLF for $t_h = 14, 16, 18, 20$ Gyr and $X_{WD} = 1.7, 8, 25, 50\%$, with IMF2.

stars up to $M_V = 19$, i.e. three magnitudes fainter, have been observed with similar programs. Five WDs in the Liebert et al. sample have tangential velocities $v_{tan} > 250$ kms$^{-1}$ and $M_V \geq 13$ and thus are assignable to the halo sample (shown by filled circles on the Figure).

The observed WDLF is represented on Figure 1, with different halo WDLFs. The dotted line is the *disk* WDLF from Segretain et al. (1994) for an age $\tau_d = 10.5$ Gyr. The crosses correspond to a 90% exclusion confidence level in the limit of detection, i.e. the possibility to see at least two WDs above this line whereas none has been detected is rejected at the 2-$\sigma$ level.

The solid lines show the *halo* WDLFs for halo ages $t_h = 14, 16, 18$ and 20 Gyr, normalized to $X_{WD} = 1, 2, 4$ and 8% respectively, for calculations done with IMF1. For the distribution of progenitors corresponding to this IMF, differentiation at crystallization in the WD interiors leads to a bump in the halo WDLF in the range $-5 \lesssim \log(L/L_\odot) \lesssim -4$, therefore ruling out substantial WD mass fractions. This shows convincingly the importance of a *complete* treatment of crystallization in WD cooling. Calculations with no carbon/oxygen differentiation will underestimate the number of WDs by more than a factor $\sim 5$, for a given age and luminosity. Conversely they will yield halo ages $\sim 2$ Gyr younger for a given LF. In the same vein, an incorrect Debye treatment will change significantly the shape of the WDLF. The dashed lines correspond to the same calculations when using IMF2. The normalizations correspond now to $X_{WD} = 1.7, 8, 25$ and 50%, for the same halo ages. Clearly, for the IMF2, a halo WD mass fraction $\gtrsim 30\%$, in agreement with the MACHO results, can not be excluded, *provided* a halo age $\gtrsim 18$ Gyr.

We have compared the star counts predicted by these WDLF's with the recent HST observations at large magnitudes (Flynn et al. 1996), for a $1/r^3$-spheroid and a $1/r^2$ halo. *All* WDLFs predict *at most* (depending on $BC_V$) $\sim 1.4$ WD in the HST field at the limit magnitude $I = 26$ for a 100% WD halo, and thus are consistent with the HST counts.

## 4. Conclusion

We have examined the possibility for the recent MACHO events towards the LMC (assuming these events are genuine microlensing events) to be due to white dwarfs. The luminosity function of *halo* white dwarfs has been calculated with the most updated white dwarf cooling theory which includes the gravitational energy release due to differentiation at crystallization. This WDLF is confronted to *all* available observational constraints on halo objects. We show that, under the two *necessary conditions* that i) the IMF in the halo differs totally from the one in the disk and exhibits a strongly peaked behaviour around $m \sim 1.5-2\,M_\odot$, and ii) the halo is older than $\sim 18$ Gyr, the white dwarf mass fraction in the halo can represent $\sim 25$ to 50% of the dark matter density, in agreement with the recent MACHO results. This would imply an initial *stellar* mass fraction $> 50\%$ and thus an essentially *baryonic* halo. These results are consistent with the ones obtained from galactic chemical evolution (Ryu et al. 1990), though they are in conflict with the conclusion raised by these authors that the disk must form no later than the halo. However, as stated by these authors, alternative scenarios in the disk formation can be advocated : the left-over gas fraction might have been ejected into the intergalactic medium, as suggested by the recently advocated presence of metal-rich hot gas in the Local Group (Fields et al. 1996). The present results are also consistent with the ones obtained by Charlot and Silk (1995), based on the expected radiation signature in high-redshift galactic halos. These authors considered a Hubble time < 13 Gyr, and solar metallicity (i.e. slowly evolving) stars. The evolution of significantly older, i.e. highly redshifted, low-metallicity stellar populations will certainly be consistent with these observational constraints.

## References

Adams, F.C. and Laughlin, G., 1996, ApJ, september issue

Alcock C. et al., 1996, preprint astroph-9604176

Chabrier, G., 1993, ApJ, 414, 695

Chabrier, G., Segretain, L. & Méra, D., 1996, ApJ, 468, 21

Charlot, S. and Silk, J., 1995, ApJ, 445, 124

Flynn C., Gould A. and Bahcall J.N., 1996, ApJ 466, L55

Garcia-Berro, E., Hernanz, M., Isern, J., Chabrier, G., Segretain, L., and Mochkovitch, R., 1996, A&A Sup., 117, 13

Fields, B., Mathews, G. and Schramm, D., submitted to ApJ

Hegyi, D.J. and Olive K.A., 1986, ApJ 303, 56

Hernanz, M., Garcia-Berro, E., Isern, J., Mochkovitch, R., Segretain, L., and Chabrier, G., 1994, ApJ 434, 652

Iben I. Jr. & Tutukov, A.V., 1984, ApJ 282, 615

Larson, R.B., 1986, MNRAS, 218, 409

Liebert, J., Dahn, C.C., and Monet, D.G. 1988, ApJ, 332, 891

Méra, D., Chabrier, G. and Schaeffer, R., 1996, Europhysics Letters 33, 327

Ryu, D., Olive K.A., and Silk, J., 1990, ApJ 353, 81

Segretain, L. & Chabrier, G., 1993, A&A 271, L13

Segretain, L., G. Chabrier, M. Hernanz, E. Garcia-Berro, J. Isern and R. Mochkovitch, 1994, ApJ 434, 641

Stevenson, D.J., 1980, *Journal de Physique* 41, C2-61

Tamanaha, C.M., Silk, J., Wood, M.A. and Winget, D.E., 1990, ApJ, 358, 164

Wood, M.A., 1992, ApJ, 386, 529

## Discussion

*M. Wood*: First, your comments in your talk on my models are highly misleading and simply wrong. The Lamb & VanHorn (1975) EOS does include quantum effects and Debye cooling, and my sequences evolve down to a typical luminosity of $\log(L/L_\odot) \sim -5.5$, which is well below the cut-off. Indeed, it is not possible to calculate believable models below this point since the physics of the photosphere cannot be calculated at this time. Finally you commented that crystallization proceeds down to $\log(L/L_\odot) = -5.0$, but in fact it is $\sim 99\%$ complete at $\log(L/L_\odot) \sim -4.5$.

*G. Chabrier*: My apologies for my overstatement about the Debye cooling, which indeed is included in the original Lamb's code and thus in yours. It is just extrapolated in the recent Adams & Laughlin calculations. But I disagree with you about the crystallization effect. First of all, a model can evolve down to any luminosity, it does not mean that the *age* which corresponds to this luminosity is correct. It is the very, essential effect of crystallization: it starts occuring in the WD interior at $\log(L/L_\odot) \sim -3.5$ and then delays the cooling, changing the age-luminosity relation for old WDs, which is essential to calculate the halo WDLF. Moreover, the effect of crystallization on the WDLF is a convolution of i) the crystallization itself, which occurs at a given temperature, and ii) the number of WDs at this temperature/luminosity. Example (as shown on the figure): For the *disk*, there is a strong decrease of the number of WDs at $L \sim L_{cryst}$, and thus just a little bump in the WDLF, whereas for the halo (e.g. IMF1), there are *many* WDs at this luminosity, because of the evolution of the progenitors at this age, and the effect (bump) is important. It is what limits the possible number (or conversely the age) of halo WDs in this case, w.r.t. observations. In any case, the effect of crystallization on the *age*, at fixed LF, is always important. At last, the mass fraction crystallized in a WD (for a 50% C/O ratio) is not 99% but 90% (resp. 77%) for a 0.6 $M_\odot$ (resp. 0.5 $M_\odot$) WD at $\log(L/L_\odot) = -4.5$ and is entirely completed only at $\log(L/L_\odot) \sim -5.0$ (98% for the 0.6 $M_\odot$, 95% for the 0.5 $M_\odot$). And the latter the crystallization takes place, i.e. the fainter the star, the larger the induced time-delay ($\Delta\tau \propto \Delta E/L$). For the question of the photosphere, I agree with you that it is certainly the next important question to be adressed to improve the acuracy of the present calculations, and the possibility of detection (e.g. discovery functions in different colors) if halo WDs do represent a significant population.

# HST OBSERVATIONS OF WHITE DWARFS IN THE GLOBULAR CLUSTER NGC 6397[1]

A. M. COOL

*Dept. of Physics and Astronomy, San Francisco State U.*
*1600 Holloway Ave., San Francisco, CA 94132, U.S.A.*

AND

C. SOSIN AND I. R. KING

*Astronomy Department, University of California*
*Berkeley, CA 94720-3411, U.S.A.*

## 1. Introduction

White dwarfs (WDs) in globular clusters are the remnants of known and nearly equal-mass progenitors. The vast majority of those still bright enough to observe are the descendants of main sequence stars with masses not much more than that of the present cluster turnoff mass. This uniformity of origin makes cluster WDs of particular interest, and they have been prime targets for observation with the Hubble Space Telescope (HST). Significant numbers have been detected by several groups in at least six clusters (see, *e.g.*, Cool, Piotto, & King 1996, and references therein).

A prime cluster in which to observe WDs is NGC 6397, just 2.2 kpc distant. Two studies of WDs in this cluster to date were driven in large part by efforts to observe faint main sequence stars, and focused on fields in the cluster envelope, using HST *V* and *I* filters. A WD luminosity function was presented (Paresce, De Marchi, & Romaniello 1995), and an upper limit of 0.05 $\mathcal{M}_\odot$ was placed on the dispersion of cluster WD masses (Cool *et al.* 1996). A limitation of these studies was in the total numbers of WDs observed, which amounted to a few tens of WDs in each case.

Here we present initial results from a new photometric study of NGC 6397 that focuses on the central regions of the cluster where the stellar den-

---

[1]Based on observations with the NASA/ESA *Hubble Space Telescope*, obtained at the Space Telescope Science Institute, which is operated by AURA, Inc., under NASA contract NAS5-26555.

*I. Isern et al. (eds.), White Dwarfs, 129–134.*
© *1997 Kluwer Academic Publishers.*

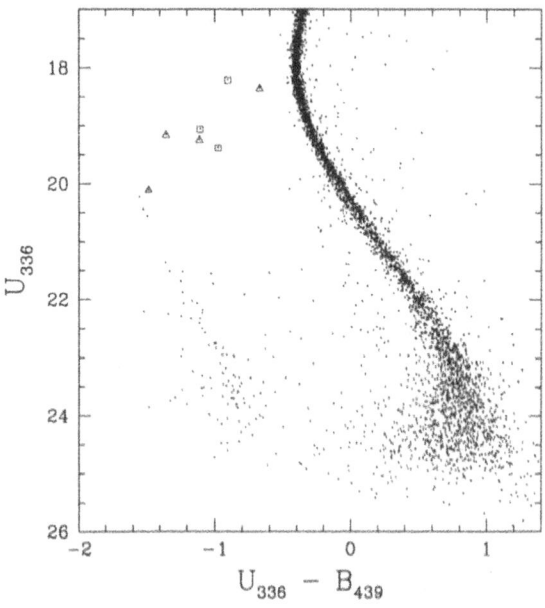

*Figure 1.* CMD of ~4400 stars near the center of NGC 6397. The bend in the upper main sequence is a bandpass effect, not the turnoff (which is closer to $U_{336}$ ~16). The triangles and squares mark cataclysmic variables and other unusual blue stars, respectively.

sities are much higher. The population of WDs sampled is larger than that in the outer fields, and as a result the observed WDs are younger, brighter and bluer on average. We present a color–magnitude diagram (CMD), comparisons with model WD sequences, and a WD luminosity function. The results are preliminary, as the analysis is still under way.

## 2. Data, Analysis, and a Color–Magnitude Diagram

We observed NGC 6397 with the HST/WFPC2 through the F336W and F439W filters ($U_{336}$ and $B_{439}$), with the PC roughly centered on the cluster core. Two overlapping pointings were used so that chip artifacts could be recognized. The total exposure times in $U_{336}$ and $B_{439}$ were 3.9 hrs and 2.2 hrs, respectively. A single combined frame was made for each filter at each pointing and analyzed using DAOPHOT II (Stetson 1987). Stars were identified in each image separately, first using automated routines, and then by eye. Those that appeared in both pairs of combined images are shown in Fig. 1. No rejection of possible artifacts has been made beyond requiring

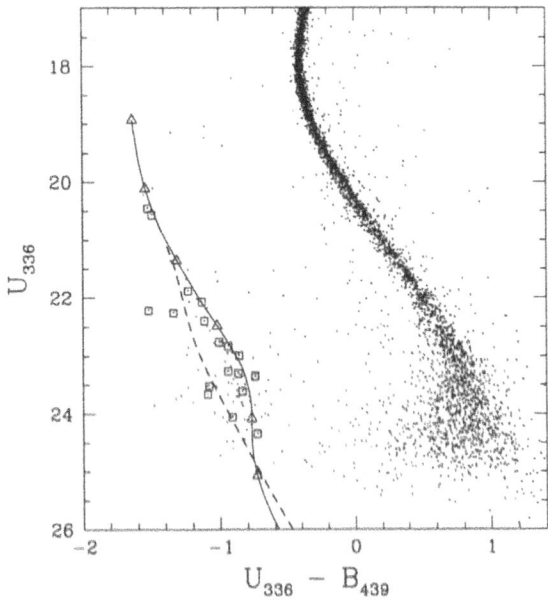

*Figure 2.*    CMD from Fig. 1 with model DA WD sequences overlaid. The apparent distance modulus and reddening were taken to be $A_U = 12.63$ and $E(U - B) = 0.16$, based on distances and reddenings given by Pryor & Meylan (1993). The five curves represent models with $\log g = 7.0$, 7.5, 8.0, 8.5, and 9.0 (top to bottom). The squares mark a subset of relatively uncrowded WDs.

that each object be detected in both filters in each of two pointings.

The WD sequence below and to the left of the main sequence in Fig. 1 contains $\sim$100 stars, with apparent magnitudes in the range $U_{336} \sim 20.5$–25, although a sequence can no longer be distinguished below $U_{336} \sim 24$. A number of stars are also visible between the WD and main sequences, most of which are likely to be field stars, although some portion are potentially WD–MS binaries. In the range $U_{336} \sim 18$–20 to the left of the main sequence are seven specially marked objects. Four of these (the triangles) are cataclysmic variables; the other three (squares) are blue stars but are not cataclysmics (Cool *et al.* 1997).

## 3.  Model Comparisons

Preliminary comparisons of the observed WD sequence to WD model sequences are shown in Figs. 2 and 3. P. Bergeron has kindly performed

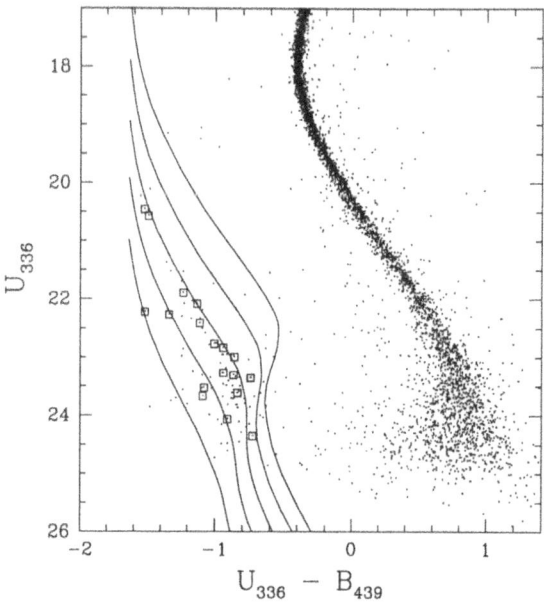

*Figure 3.* CMD from Fig. 1 with $\log g = 8.0$ model WD sequences overlaid. The solid and dashed lines represent DA and DB models, respectively. The six triangles mark DA models with effective temperatures of $T_{\text{eff}} = 100000$, 40000, 25000, 17000, 10000, and 8000 K (top to bottom). The squares are as in Fig. 2.

synthetic photometry, using the HST filter response curves, of a grid of DA and DB WD models (see Bergeron, Wesemael, & Beauchamp 1995). Fig. 2 shows the full grid of DA models, with five values of the gravity. Fig. 3 shows the DA and DB models for $\log g = 8.0$ only. These plots also highlight a subset of the WDs that are sufficiently isolated from neighbor stars that they should have the most reliable magnitudes and colors.

The large majority of the most isolated WDs (and many of the more crowded ones) lie close to the $\log g = 8.0$ model sequence, which corresponds to a WD mass of $\sim 0.6$–$0.7$ $\mathcal{M}_\odot$. While subject to the uncertainties in the assumed values of the distance modulus and reddening, this mass is in keeping with theoretical expectations, and is similar to those determined for a large sample of field WDs (Bergeron, Saffer, & Liebert 1992).

A small but potentially interesting subset of WDs in globulars could have more complex origins in binary stars and/or stellar interactions. WDs formed via these alternate routes would in some cases have unusual masses (low or high, depending on the details of their formation histories). In the

CMD, high and low mass WDs would appear too blue and too red, respectively, for their magnitudes (see Fig. 2). Several such outlying stars appear in the observed sequence. However, considering the crowdedness of the field, it is likely that the measured magnitudes for some of these objects are adversely affected by near neighbors. On the other hand, two of the relatively isolated (and reasonably bright) WDs, which should be well measured, do appear to be outliers to the left side ($U_{336} \sim 22.2$). If these stars' unusual colors are confirmed in more detailed analyses currently under way, it may suggest that the cluster contains a small but not insignificant population of WDs with masses on the order of 0.9–1.2 $\mathcal{M}_\odot$.

The appearance of WDs with unusually blue colors could instead be a result of atmospheric effects, at least in some cases. As can be seen in Fig. 3, in this filter combination, colors for the DA and DB model sequences separate for temperatures in the range $T_{\mathrm{eff}} \sim 10000$–20000 K. At least one of the bright outliers might instead be explained as a DB star. Two of the fainter WDs appear to also be better fit by the DB sequence, although the spread in the WD sequence at these faint magnitudes is large.

## 4. Luminosity Function

Measuring an extended WD sequence in globular clusters also makes it possible to test WD evolutionary models, by comparing the observed numbers of WDs as a function of magnitude to predictions of WD cooling ages from the models. Such measurements require special care, as incompleteness in the star counts can be significant even in only moderately crowded fields. To this end, we performed a series of experiments to assess the completeness along the WD sequence by adding artificial stars with colors appropriate to the WD sequence at $0\overset{\mathrm{m}}{.}5$ intervals from $U_{336} = 20.5$–24. On reanalyzing the images, the recovery rate for the artificial stars indicated that the completeness was $\geq 80\%$ for $U_{336} < 21.5$, $\sim 50\%$ at $U_{336} = 23.5$, and $< 20\%$ at $U_{336} = 24$. Fig. 4 shows the raw and completeness-corrected WD star counts down to $U_{336} = 23.6$. (Note that only the $\sim 70$ WDs that had been found by automated routines are included here, as the others required human intervention to identify, a step not performed in the artificial star experiments. Of these, 55 are brighter than $U_{336} = 23.6$, or $M_{336} = 11.0$.)

Overlaid on the measured star counts in Fig. 4 are predictions based on Wood's evolutionary models for DA WDs with $\log g = 8.0$ (smooth solid line) (Wood 1995). We normalized this curve by counting the number of horizontal branch (HB) stars in the field (20), and then assuming that the number of WDs with ages equal to or shorter than the HB age of $\sim 10^8$ yr should be the same as the number of HB stars. Since Poisson errors are considerable in this normalization procedure, we also include curves

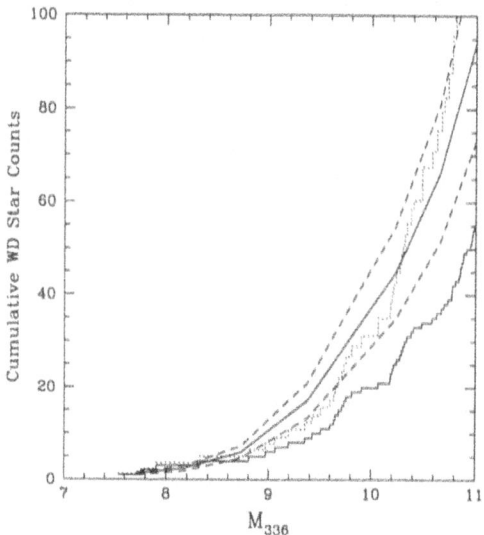

*Figure 4.* A comparison of observed and predicted WD star counts in NGC 6397 (see text). The assumed distance modulus in $U_{336}$ was 12.63. The jagged solid line represents the raw star counts; the jagged dotted line has had completeness corrections applied.

obtained for numbers of HB stars $\pm 1\sigma$ above and below the observed value (smooth dashed curves). Given that the observed star counts are close to being within the $\pm 1\sigma$ curves over the full range of magnitudes, we conclude that the models are in good agreement with the observations.

We thank P. Bergeron for generously sharing his WD models and computing synthetic colors, and J. Liebert for helpful discussions. This work was supported by NASA grant NAG5-1607 and by grant GO-5929 from the Space Telescope Science Institute.

## References

Bergeron, P., Saffer, R. A., & Liebert, J. 1992, *ApJ*, 394, 228
Bergeron, P., Wesemael, F., & Beauchamp, A. 1995, *PASP*, 107, 1047
Cool, A. M., Piotto, G. & King, I. R. 1996, *ApJ*, 468, 655
Cool, A. M., *et al.* 1997, in preparation
Paresce, F., De Marchi, G., & Romaniello, M. 1995, *ApJ*, 440, 216
Pryor, C. & Meylan, G. 1993, in Structure and Dynamics of Globular Clusters, ASPCS 50, eds. S. G. Djorgovski & G. Meylan (San Francisco: ASP), p. 357
Stetson, P. B. 1987, *PASP*, 99, 191
Wood, M. A. 1995, in 9th European Workshop on White Dwarfs, NATO ASI Series, eds. D. Koester & K. Werner (Berlin: Springer), p. 41

# THE DISTANCE AND AGE OF THE GLOBULAR CLUSTER NGC6752 MEASURED BY HST OBSERVATIONS OF CLUSTER AND FIELD WHITE DWARFS

ANGELA BRAGAGLIA, FRANCESCO R. FERRARO
*Osserv. Astronomico di Bologna, I-40126 Bologna, Italy*

ALVIO RENZINI, ROBERTO GILMOZZI
*E.S.O., D-85748 Garching b. München, Germany*

JAY B. HOLBERG
*Univ. Arizona, Lunar&Planetary Lab., Tucson AZ 85721, USA*

JAMES LIEBERT
*Univ. Arizona, Steward Observatory, Tucson AZ 85721, USA*

SERGIO ORTOLANI
*Dipartimento di Astronomia, I-35122 Padova, Italy*

FRANCOIS WESEMAEL
*Département de Physique, Montréal, Québec, Canada H3C 3J7*

AND

RALPH BOHLIN
*STScI, Baltimore, MD 21218, USA*

**Abstract.** We present first results of a new method for finding distances, hence ages, of Galactic Globular Clusters with a precision of about 10 %. We measured the White Dwarfs cooling sequence in NGC6752 with the *Hubble Space Telescope* and, comparing it to a fiducial cooling sequence of field WDs we derive a distance modulus $(m-M)_0 = 13.05 \pm 0.10$, and an age of about 15 Gyr $(\pm 1.5$ Gyr$)$.

## 1. Introduction

The astrophysical importance of a very precise age determination for Globular Clusters (GCs) lies in the fact that they are among the oldest objects in the Universe. They set a hopefully stringent lower limit to the age of the

*I. Isern et al. (eds.), White Dwarfs, 135–141.*
© *1997 Kluwer Academic Publishers.*

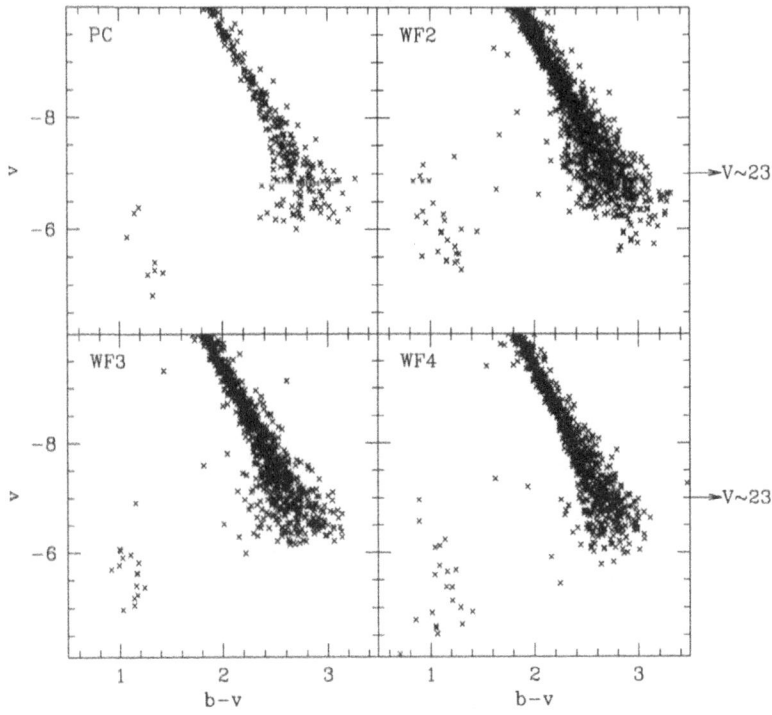

*Figure 1.*    Preliminary reduction of the *median* B and V images for all 4 WFPC2 cameras: notice how the WD cooling sequence is better defined in the WF4 chip

Universe $t_0$ which, togheter with $H_0$, constrains possible values for $\Omega_0$ (or a combination of $\Omega_0$ and $\Lambda$). While $H_0$ is still very controversial (uncertainty $\sim$ factor of 2), there has always been a better consensus on GCs ages (see e.g. VandenBerg, Bolte & Stetson 1996: age $15 \pm 3$ Gyr).

But, *how good* are these ages? and do they really set firm constraints on anything? To date GCs, we make use of the Turn-Off luminosity. Theoretically, this stellar clock is quite sound, but applying it to actual GCs requires knowledge of some observable quantities, each one with its attached uncertainty (Renzini 1991): a) apparent magnitude of the TO ($\sigma$=0.10 mag $\rightarrow$ 9% error in age); b) helium content Y ($\sigma$=0.02 $\rightarrow$ 2% error in age); c) global metallicity [Fe/H] ($\sigma$=0.3 $\rightarrow$ 9% error in age); d) distance modulus ($\sigma$=0.25 mag $\rightarrow$ **22%** error in age). **Precision ages** of GCs then mean **precision distances**. At the moment the techniques usually adopted, based on the Horizontal Branch or on the Main Sequence, cannot guarantee the required accuracy.

## 2. A new way to globular cluster distances

We have then adopted a new idea, i.e. to fit the GC WD cooling sequence to an empirical fiducial sequence composed by field WDs of accurately known parallax and mass (Fusi Pecci & Renzini 1979, Renzini & Fusi Pecci 1988, Renzini et al. 1996). Observation of GC cooling sequences has only recently become feasible, using the *HST* (e.g. Renzini et al. 1996, Cool, Piotto & King 1996, Richer et al. 1995), while accurate masses and parallaxes for field WDs have been accumulating in the last years (see e.g. Bergeron, Saffer & Liebert 1992, Bragaglia, Renzini & Bergeron 1995). We have chosen 10 field WDs having masses similar to GC WDs ($M_{WD}^{GC} = 0.53 \pm 0.02\ M_{\odot}$, Renzini & Fusi Pecci 1988), in the right $T_{eff}$ interval (10000-20000 K), and with a small error on the parallax ($< \sigma(\pi)/\pi > = 0.05$) to build up our fiducial sequence; Table 1 gives names and properties of these WDs.

Our startegy is the following: a) we observe with the WFPC2 on board *HST* the selected GCs (NGC6752 has been our first target, while 47Tuc is presently under analysis) in the UBVI filters; b) we observe each field WD with the same compliment of filters, and in each chip separately. This way we can form fiducial sequences for the four cameras, and do any comparison using only **instrumental magnitudes**. This reduces the uncertainty of our fits, since we avoid transformation of *HST* data to a standard system of magnitudes. More detailed information can be found in Renzini et al. (1996) and Bragaglia et al. (1996).

We have done preliminary reductions in all four filters and for the 4 chips, but have chosen to focus first on the B and V exposures in WF4, since they are the better exposed filters, and WF4 the least crowded camera. Figure 1 shows result of the reduction of the *median* B and V images; those have been used only to detect the candidate WDs. The sequence in WF4 is evidently the better defined, and the one going fainter. We have then reduced each frame separately, using the PSF fitting routines in ROMAFOT (Buonanno et al. 1993) and weight-averaged the measurements, thus greatly reducing the scatter of the cooling sequence. The field stars have been measured with an aperture of 0.5 arcsec, and we have corrected the PSF derived magnitudes for the cluster WDs to the same aperture.

Of the 10 field WDs observed, 8 are DAs and 2 DBs; we have temporarily discarded three of the DAs. Two of them have masses slightly larger than the GC WDs, and we have avoided to correct them using theoretical models (e.g. Wood 1995), so to keep to a strictly empirical procedure, at least for the time being. The other one (1647+591) lies on the cooling sequence of ~0.55 $M_{\odot}$ WDs, while its mass seems to be 0.69 $M_{\odot}$; either the mass or the parallax are in error, or it's a close binary system, and we have decided to defer its use until the Hipparcos parallax is released.

TABLE 1. Literature data on the Field Calibration WDs; references are as follows. a: Bragaglia, Renzini & Bergeron 1995; b: Bragaglia & Bergeron 1996; c: Bergeron, Saffer & Liebert 1992; d: Bergeron et al. 1995; e: Beauchamp 1995; f: Oswalt et al. 1991. All $\pi$'s are in arcsec and come from the 1993 Yale compilation of Van Altena, except for 1647+591, coming from the USNO

| WD | $T_{eff}$ | log g | $M/M_\odot$ | ref. | $\pi$ | err$\pi$ | V | B-V |
|---|---|---|---|---|---|---|---|---|
| DA WDS: | | | | | | | | |
| 0839−327 | 9400 | 7.955 | 0.553 | a | 0.1127 | 0.0092 | 11.862 | 0.222 |
| 1647+591 | 13700 | 8.140 | 0.690 | d | 0.0812 | 0.0046 | 12.218 | 0.170 |
| 1935+274 | 13300 | 7.871 | 0.512 | b | 0.0554 | 0.0029 | 12.980 | 0.163 |
| 2326+049 | 11800 | 8.140 | 0.690 | d | 0.0725 | 0.0048 | 13.027 | 0.160 |
| 2341+322 | 13700 | 7.833 | 0.494 | b | 0.0563 | 0.0019 | 12.929 | 0.143 |
| 1327−083 | 14000 | 7.847 | 0.502 | a | 0.0612 | 0.0028 | 12.312 | 0.085 |
| 2126+734 | 15000 | 7.856 | 0.513 | b | 0.0469 | 0.0024 | 12.828 | 0.030 |
| 0644+375 | 21000 | 8.095 | 0.655 | c | 0.0626 | 0.0018 | 12.076 | -0.080 |
| DB WDS: | | | | | | | | |
| 1917−077 | | | 0.55 | f | 0.0988 | 0.0025 | 12.30 | 0.060 |
| 0002+729 | 13300 | 7.69 | 0.60 | e | 0.0287 | 0.0047 | 14.35 | -0.070 |

## 3. Present results and future plans

Figure 2 shows first results of our work: the dereddened cluster and field sequences for the DA WDs have been fitted, resulting in a distance modulus of 13.05 for NGC6752. We estimate a 0.10 mag error on this value, i.e. a 10 % error in the derived age.

We adopt for NGC6752 the following values: $(m-M)_0$=13.05 (this work, Renzini et al. 1996); $V^{TO}$=17.4, E(B-V)=0.04 (Penny & Dickens 1984); [Fe/H]=−1.54 (Zinn 1995); Y=0.23 (Boesgaard & Steigman 1985). Using the isochrones of VandenBerg & Bell (1985), which assume solar proportions for the metal content Z, an age of 18 Gyr is derived; using insted an enhancements of $\alpha$-elements [$\alpha$/Fe]=0.5 (Salaris, Chieffi & Straniero 1993, Bergbush & VandenBerg 1992), we arrive at an age of 15.3 Gyr. There is also a possible systematic error of about 0.5 Gyr, due to uncertainties in He diffusion computations. In summary, we obtain an age of t = 15 ± 1.5 ± 0.5 Gyr.

In the next future we plan to analyze the other 3 chips and fit their cooling sequences, and do the same for the U and I filters, where anyway the S/N is worse. We will also try to increment the number of field WDs, to enlarge our fiducial sequence and stenghten the fits.

We will then compare our findings with other GCs observed by different groups: as an exemple, in the case of NGC6397 (Cool et al. 1996) we

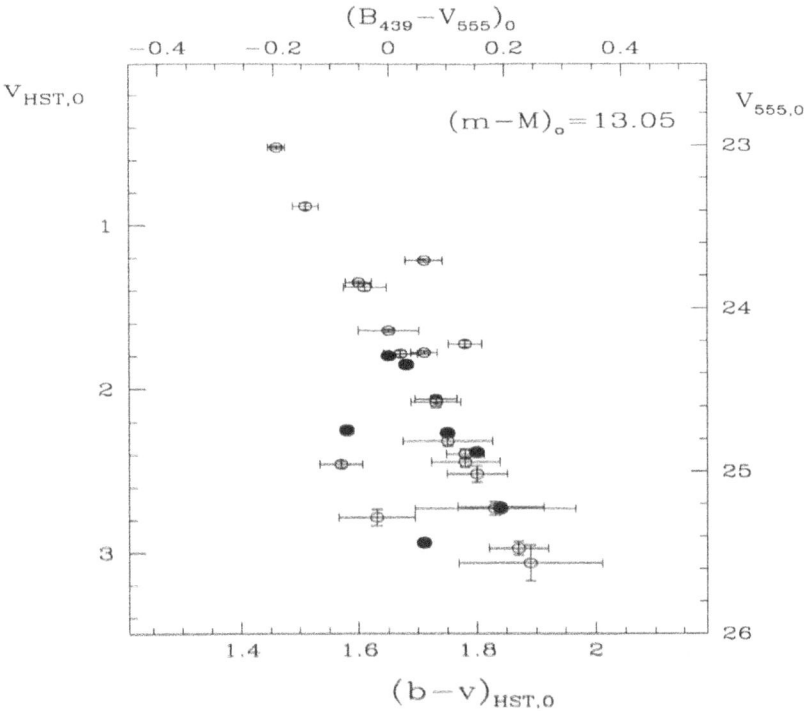

*Figure 2.* Results for WF4: the cluster WD cooling sequence (open circels) and the field fiducial sequence (filled dots) have been fitted using a distance modulus 13.05. Errorbars for magnitudes and colours for the cluster WDs are shown. Errors on the field WDs, exposed to a S/N$\geq$100, come essentially from the parallaxes. We have worked only with instrumental magnitudes, but give also the B and V scale for reference

have done a very preliminary comparison of the V,V-I colour-magnitude diagrams; imposing coincidence of the WD cooling sequences we obtain a differential distance modulus of 1.2 mag.

We will then procede to compare the cooling sequences of NGC6752 and 47Tuc, keeping to instrumental magnitudes. We will derive the differential distance modulus and the (possible) age difference, which has in turn important bearings on the timescales of the formation mechanism of our Galaxy.

## References

Beauchamp, A. 1995, PhD Thesis, Université de Montréal

Bergeron, P., Saffer, R., & Liebert, J. 1992, ApJ, 394, 228

Bergeron, P., Wesemael, F., Lamontagne, R., Fontaine, G., Saffer, R.A., & Allard, N.F. 1995, ApJ, 449, 258

Bergbush, P.A., & VandenBerg, D.A. 1992, ApJS, 81, 163

Boesgaard, A.M., & Steigman, G. 1985. ARA&A, 23, 319

Bragaglia, A., Renzini, A., & Bergeron, P. 1995, ApJ, 443, 735

Bragaglia, A. et al. 1996 (in Preparation)

Bragaglia, A., & Bergeron, P. 1996 (in Preparation)

Buonanno, R., Buscema, G., Corsi, C.E., Ferraro, I., & Iannicola, G. 1983, A&A, 126, 278

Cool, A.M., Piotto, G., & King, I.R. 1996, ApJ, 468, 655

Fusi Pecci, F., & Renzini, A. 1979, *In* Astronomical Uses of the Space telescope, eds. F.Macchetto, F.Pacini and M.Tarenghi, (ESO), p. 181

Oswalt, T.D., et al. 1991, AJ, 101, 583

Penny, A.J., Dickens, R.J. 1986, MNRAS, 220, 845

Renzini, A. 1991. *In* Observational Tests of Cosmological Inflation, ed. T. Shanks et al. (Dordrecht: Kluwer), p. 131

Renzini, A., & Fusi Pecci, F. 1988, ARA&A, 26, 199

Renzini, A., Bragaglia, A., Ferraro, F.R., Gilmozzi, R., Ortolani, S., Holberg, J.B., Liebert, J., Wesemael, F., Bohlin, R.C. 1996, ApJ, 465, L23

Richer, H.B. et al. 1995, ApJ, 451, L17

Salaris, M., Chieffi, A., & Straniero, O. 1993. ApJ, 414, 580

Van Altena, W.F., Lee, J.T., & Hoffleit, D. 1991, The General Catalogue of Trigonometric Stellar Parallaxes, Preliminary Version (Yale University Observatory)

VandenBerg, D.A., & Bell, R.A. 1985, ApJS, 58, 561

VandenBerg, D.A., Bolte, M., Stetson, P.B. 1996, ARAA, 34, in press

Wood, M.A. 1995, in "White Dwarfs", eds. D. Koester, K. Werner (Springer), p. 41

Zinn, R. 1985, ApJ, 293, 424

## Discussion

*I. King*: From comparison of your WDs with ours in NGC6397, you get $(m–M)_0=11.85$. When you correct for extinction how does this compare with the $(V–M_V)_{apparent}=12.3$ that we used for NGC6397?

*A. Bragaglia*: $(V–M_V)_{apparent}=13.42$ if an extinction $A_V=0.57$ is used, so I think it compares quite well, given the fact that our reductions are only preliminary.

# DA WHITE DWARFS IN THE MONTRÉAL–CAMBRIDGE–TOLOLO SURVEY

R. LAMONTAGNE, F. WESEMAEL, G. FONTAINE, S. DEMERS
*Département de Physique, Université de Montréal*

P. BERGERON
*Lockheed Martin Electronic Systems Canada*

M.J. IRWIN
*Royal Greenwich Observatory*

AND

W.E. KUNKEL
*Las Campanas Observatory, Carnegie Institution of Washington*

## 1. Introduction

Large–scale colorimetric surveys, such as the Palomar–Green survey (Green *et al.* 1986), are an important source of new hot subluminous objects which can form the basis of many different types of investigations. Thus the PG survey has, over the years, yielded a complete sample of DA white dwarfs from which the luminosity function of hot white dwarfs can be derived (Fleming *et al.* 1986), increased considerably the number of objects in some sparsely populated classes (e.g., the DBA stars), as well as revealed the existence of entirely new kinds of objects (e.g., the very hot PG 1159 stars). Comparable strides are now being made in studies of other kinds of subluminous objects, such as the hot B and O subdwarfs, which appear in large numbers in the PG and the Kitt Peak–Downes (Downes 1986) surveys.

We have recently witnessed a flurry of activity in the field of colorimetric surveys, culminating with at least four analogs of the PG survey proceeding concurrently: the Kiso survey (Noguchi *et al.* 1980; Kondo *et al.* 1984); the Montréal–Cambridge–Tololo (MCT) survey (Demers *et al.* 1986); the Edinburgh–Cape survey (Stobie *et al.* 1987, 1992); and the Homogeneous Bright Quasars Survey, a Key Project currently being carried out at ESO (Gemmo *et al.* 1993; Gemmo *et al.* 1995).

*I. Isern et al. (eds.), White Dwarfs, 143–147.*

Of course, a mere colorimetric identification is not sufficient, and a substantial effort must also be invested in follow–up spectroscopic observations of the color–selected candidates. Thus Wegner and his collaborators have undertaken a thorough spectroscopic study of the Kiso survey at the Michigan–Dartmouth–MIT (MDM) Observatory (e.g. Wegner & Dupuis 1993, and references to seven earlier papers within), while follow–up spectroscopy is proceeding as well in the Edinburgh–Cape survey (e.g., O'Donoghue 1995).

As far as the MCT survey is concerned, we have as of this year identified a sample of 158 DA white dwarfs. Most of these stars are new spectroscopic discoveries; only 33 figure in the Third Edition of the McCook & Sion (1987) white dwarf catalog. More recently, after the spectroscopic identification of most of the new DA white dwarfs presented here had been made, several of these objects were identified independently as X–ray sources in the ROSAT experiment, which provided optical identifications (Pounds *et al.* 1993).

## 2. Observations

A detailed description of the photographic part of the MCT survey has been presented by Demers *et al.* (1986). Follow–up spectroscopy of the blue candidates in the MCT survey has been ongoing since 1985. Most of the observations were obtained at the CTIO 1.5m and 4m telescopes, with a variety of instrumental setups. As a general rule, we concentrate on the bluest objects, those with $(U - B) \leq -0.6$, although some redder objects, with $-0.6 \leq (U - B) \leq -0.4$ have been occasionally observed.

The 158 objects selected in this first list of degenerates are based on our own spectroscopic identifications and/or cross–correlations with several catalogs: the *HK* objective–prism/interference–filter survey of Beers *et al.* (1992); the *ROSAT Bright Source Catalog* (Pounds *et al.* 1993); the *Catalogue and Luminosity Function of White Dwarfs in the ROSAT All-sky Survey* (Fleming *et al.* 1996); the recent *Second Extreme Ultraviolet Explorer Source Catalog* (Bowyer *et al.* 1996); the Third Edition of the *Catalog of Spectroscopically Confirmed White Dwarfs* (McCook & Sion 1987);

## 3. Analysis of the sample

We have carried out a model atmosphere analysis of the subsample of 113 DA stars for which the optical spectra showed no signs of peculiarities (obviously composite spectrum, hybrid spectral type, etc ...). For our analysis, we have followed the procedure discussed by Bergeron *et al.* (1992; hereafter BSL), to which the reader is referred for further details. The grid of synthetic spectra used is the one used for the extensive tabulations of Bergeron *et al.* (1995a). Below 17,000 K, convective energy transport is included, with an efficiency characterized by the ML2/$\alpha$ = 0.6 parametrization (Bergeron et al. 1995b).

The determination of surface gravity for a large sample of DA stars permits, in principle, a determination of the mass distribution of stars in the sample. This technique was applied to a large sample of northern DA stars by Kidder (1991), by BSL, and by Bragaglia et al. (1995) to various samples selected from the McCook & Sion catalog. There are two substantial differences between this analysis and that of BSL, which we use as a point of comparison. Firstly, the BSL data base is much more homogeneous than ours; second of all, the BSL investigation focused explicitly on stars above 20,000 K, where uncertainties associated with the efficiency of convective mixing could be safely avoided[1].

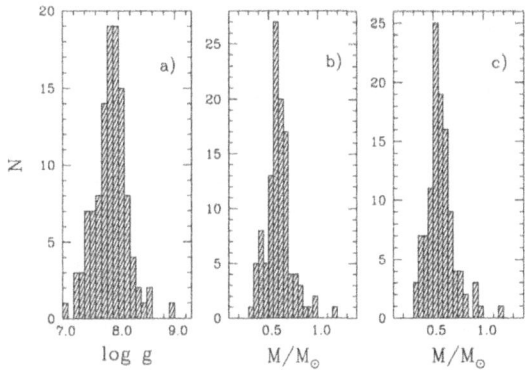

Figure 1. Histogram of the distribution of our sample of 113 objects. a) Surface gravity. b) Mass, obtained from stellar models with thick hydrogen layers. c) Mass, obtained from stellar models with thin hydrogen layers.

Here, our colorimetric selection criterion has allowed many stars below 20,000 K to be included in our sample, and the uncertainties in the effective temperature determinations of cool stars do have some impact on the assigned gravities. We have thus determined a mass distribution based on our complete sample of 113 stars, as well as one for a restricted sample of 80 objects with temperatures above 20,000 K. In each case, we have considered stellar models covering a spread in hydrogen–layer thickness. Thus we use Wood's (1990) carbon–core models with fractional helium layers $q_{He} = 10^{-4}$ and no hydrogen (essentially equivalent to models with very thin fractional hydrogen layers, $q_H \sim 10^{-10}$). These are the models considered by BSL. But we use, as well, the more recent Wood (1995) carbon–core sequences with thicker fractional helium and hydrogen masses, $q_{He} = 10^{-2}$ and $q_H = 10^{-4}$, respectively. The surface gravity and mass

---

[1]Note that much progress on this topic has been accomplished since, mostly within the detailed spectroscopic investigation of ZZ Ceti stars of Bergeron et al. (1995b).

distributions are presented in Figure 1 for the complete sample of 113 objects. The surface gravity distribution has a mean value of $\log g = 7.860$ with a standard deviation of $\sigma(\log g) = 0.296$, while the mass distribution has a mean of $M = 0.604\ M_\odot$ with $\sigma(M) = 0.131$ using thick models, and $M = 0.574\ M_\odot$ with $\sigma(M) = 0.144$ for thin models (BSL obtained $M = 0.562\ M_\odot$, $\sigma(M) = 0.137$ using the same thin models).

For the subsample of 80 stars with temperatures above 20,000 K, we obtain a mean value of $\log g = 7.826$ with $\sigma(\log g) = 0.301$ for the surface gravity distribution. The mass distributions yield means of $M = 0.608\ M_\odot$ $(\sigma(M) = 0.126)$ and $M = 0.576\ M_\odot$ $(\sigma(M) = 0.144)$ using thick and thin models respectively.

We are grateful to the CTIO Time Allocation Committee for its unswerving support of this project, and to the CTIO staff for its help and technical support over the years. This work was supported in part by the NSERC Canada, by the Fund FCAR (Québec), and by NATO.

## References

Beers, T.C., Preston, G.W., Shectman, S.A., Doinidis, S.P., & Griffin, K.E. 1992, *A.J.*, **103**, 267

Bergeron, P., Saffer, R.A., & Liebert, J. 1992, *A.J.*, **394**, 228 (BSL)

Bergeron, P., Wesemael, F., & Beauchamp, A. 1995a, *PASP*, **107**, 1047

Bergeron, P., Wesemael, F., Lamontagne, R., Fontaine, G., Saffer, R.A., & Allard, N. 1995b, *Ap.J.*, **449**, 258

Bowyer, S., Lampton, M., Wu, X., Jelinsky, P., & Malina, R.F. 1996, *A.J.Suppl.*, **102**, 129

Bragaglia, A., Renzini, A., & Bergeron, P. 1995, *Ap.J.*, **443**, 735

Demers, S., Kibblewhite, E.J., Irwin, M.J., Nithakorn, D.S., Béland, S., Fontaine, G., & Wesemael, F. 1986, *A.J.*, **92**, 878

Downes, R. 1986, *Ap.J.Suppl.*, **61**, 569

Fleming, T.A., Liebert, J., & Green, R.F. 1986, *Ap.J.*, **308**, 176

Fleming, T.A., Snowden, S.L., Pfeffermann, E., Briel, U., & Greiner, J. 1996, *Astr.Ap*, in press

Gemmo, A.G., La Franca, F., Cristiani, S. & Barbieri, C. 1993, in *White Dwarfs: Advances in Observation and Theory*, ed. M.A. Barstow, p. 23

Gemmo, A.G., Cristiani, S., La Franca, F. & Andreani, P. 1995, in *White Dwarfs*, eds. D. Koester & K. Werner, p. 31

Green, R.F., Schmidt, M., & Liebert, J. 1986, *Ap.J.Suppl.*, **61**, 305

Kidder, K. 1991, Ph.D. thesis, Univ. of Arizona

Kondo, M., Maehara, H., & Noguchi, T. 1984, *Ann. Tokyo Astron. Obs.*, **20**, 130

McCook, G.P., & Sion, E.M. 1987, *A.J. Suppl.*, **65**, 603

Noguchi, T., Maehara, H.,, & Kondo, M. 1980, *Ann. Tokyo Astron. Obs.*, **18**, 55

O'Donoghue, D. 1995, in *White Dwarfs*, eds. D. Koester & K. Werner, p. 297

Pounds *et al.* 1993, *MNRAS*, **260**, 77

Stobie, R.S., Morgan, D.H., Bhatia, R.K., Kilkenny, D., & O'Donoghue, D. 1987, in *The Second Conf. on Faint Blue Stars*, eds. A.G. Davis Philip, D.S. Hayes, & J. Liebert, p.493

Stobie, R.S., Chen, A., O'Donoghue, D., & Kilkenny, D. 1992, in *Variable Stars and Galaxies*, ed. B. Warner, p.87

Wegner, G. & Dupuis, J. 1993, *A.J.*, **106**, 390

Wood, M. A. 1990, Ph.D. Thesis, Univ. of Texas at Austin
Wood, M. A. 1995, in *White Dwarfs*, eds. D. Koester & K. Werner, p. 41

# COOL WHITE DWARF SURVEY USING STACKED DIGITISED SCHMIDT PLATES

RICHARD KNOX
*Institute for Astonomy*
*University of Edinburgh*

AND

M R S HAWKINS
*Royal Observatory*
*Edinburgh*

## 1. Introduction

A survey for cool degenerates in the 5° field centred around RA = 21h 28m and Dec = −45° is being undertaken using digitally scanned Schmidt plates. The database consists of around 200 plates in several colours over a 15 year baseline. At present the survey's photometric limits are $B_J \sim 23$ and $R_F \sim 21.2$. A reduced proper motion diagram has been produced for objects with proper motion ≥60 mas/yr. A distinct white dwarf locus is apparent, providing a sample of cool degenerate candidates for spectroscopic analysis.

The observed cut-off in the white dwarf luminosity function yields a valuable independent estimate of the age of the Galactic Disk(Oswalt *et al* 1996). The ultimate aim of this project is to identify the degenerates within the survey limits and construct a luminosity function from them.

## 2. Project Database

The Schmidt plate database consists of:

- 64 IIIa-J ($B_J$ photometric band) plates
- 64 corresponding IIIa-F ($R_F$ band) plates
- others (including 30 IV-N) providing additional U,V and I photometry

All plates were digitally scanned with Edinburgh's COSMOS scanning machine(MacGillivray 1984).

*I. Isern et al. (eds.), White Dwarfs, 149–152.*

Once scanned, the plate images can be 'stacked' into another pixel array to gain greater depth. For this project the $B_J$ and $R_F$ plates were stacked into 16 arrays, each consisting of 4 stacked plates. The criterion defining 'an object' was the discovery of 3 adjacent pixels on an array above a certain limiting transmission threshold. Every object then had its x and y plate co-ordinates, ellipticity, area(in pixels) and photographic transmission recorded.

Transformations converting object co-ordinates in the 2nd to 16th array into the 1st(master) array co-ordinate frame allow objects to be traced through the 16 stacks. The final dataset includes every object's master frame co-ordinates in each array in which it was detected, thus providing up to 16 epochs of positional information. Photometric magnitudes are also derivable from the COSMOS transmission measurements.

A proper motion was calculated for every object simply by fitting a straight line to a plot of object displacement vs time. Plots such as figure 1 were fitted using a standard least squares routine to calculate x and y

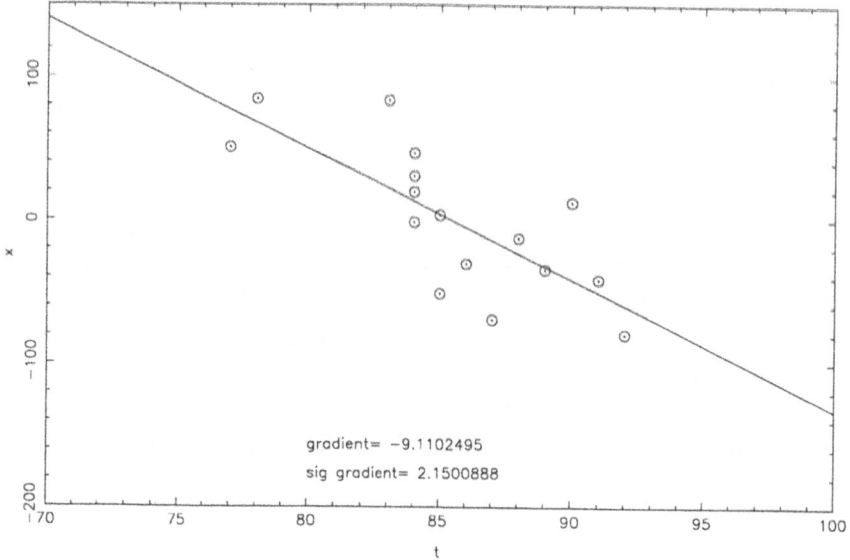

*Figure 1.* Example of a least squares fit to a time against position plot(the number on the time(t) axis denotes year this century)

proper motions, which were combined to yield total proper motions and position angles. The proper motions derived from the $B_J$ and $R_F$ stacks are totally independent.

It is essential for the population discrimination technique used here that the vast majority of proper motions in a sample are genuine. Since proper motions are calculated for all objects, position measurement error will inevitably lead to the time vs displacement plots of some zero proper motion objects mimicking a high proper motion. This source of contamination was combatted firstly by cross-correlation of the independent proper motion measures from the $B_J$ and $R_F$ stacks and subsequent rejection of incompatible motions. It was then simply a matter of choosing a proper motion limit high enough to ensure the remaining motions were real. The survey limits used for the sample displayed here were:

- Proper motion $\geq$60 mas/yr
- $B_J \leq 23.0$
- $R_F \leq 21.2$

## 3. Reduced Proper Motion Diagrams

The 'reduced proper motion'(RPM) is defined by

$$H = m + 5 \log_{10} \mu + 5 \tag{1}$$

where m is apparent magnitude and $\mu$ proper motion. A reduced proper motion diagram(RPMD) is a plot of colour against RPM. It is a powerful way of combining proper motions and photometry to distinguish stellar population groups. Equation 1 can be re-written using the relationships $m = M - 5 + 5 \log d$ and $\mu = V_t/4.74d$ to give

$$H = M + 5 \log_{10} V_t - 3.379 \tag{2}$$

where M is the absolute magnitude, $V_t$ the transverse velocity (in $kms^{-1}$) and d the distance. Since M and $V_t$ are both intrinsic properties of the star, so too is H.

To see the significance of H, suppose that every star had an identical $V_t$; H would then clearly be simply M plus a constant and the distribution of a particular population group in H at a particular colour would depend solely on the spread of the populations' colour-magnitude relation at that colour. Of course there is a distribution in $5 \log_{10} V_t$ for each population, but since the tangential velocities are distributed around a most probable value the RPM serves as an estimate of M, $i.e.$ $M = a + bH$(a and b constants). The resulting locus for each population in the RPMD is then the convolution of its colour magnitude distribution with its $5 \log_{10} V_t$ distribution over the diagrams colour range. To allow population discrimination in some colour region we therefore only require that the various population loci do not overlap significantly in that region of the RPMD. In effect the RPMD is analogous to the Hertsprung-Russell diagram, and in both plots the white dwarf population is quite distinct in most colours.

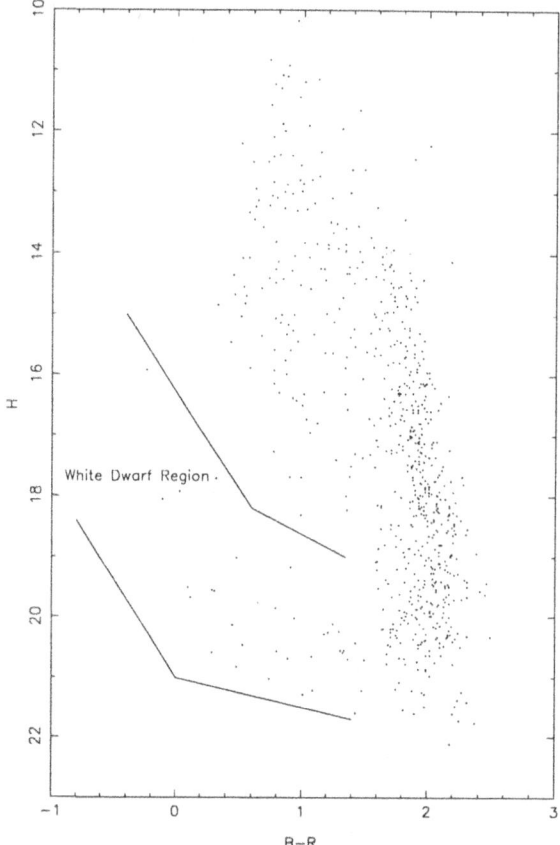

*Figure 2.*   RPMD for objects with proper motion $> 60 mas/yr$

## 4.  A Reduced Proper Motion Diagram from our Data

The sample consists of $\sim$ 700 objects drawn from a database of several hundred thousand. The RPMD obtained is shown in Fig 2. The bulk of stars, around $B_J - R_F \sim 2$, are M dwarfs; the main sequence locus is visible and the white dwarf locus is quite distinct and has been labelled. While no firm conclusions can be drawn without spectroscopic confirmation of the candidates status as bona fide degenerates, crude estimates derived from the photographic colour indices indicate the sample should be extremely cool. Spectra of these objects are to be taken in the near future.

## References

MacGillivray,H.T.,Stobie,R.S (1984) *Vistas in Astronomy*,**27**,433
Oswalt, T. D., Smith , J. A, Wood, M. A., Hintzen, P (1996) *Nature*,**382**,692

# Section III:
# Atmospheres

# CHEMICAL EVOLUTION OF COOL WHITE DWARF'S ATMOSPHERES

MARÍA TERESA RUIZ

*Departamento de Astronomía, Universidad de Chile, Casilla 36-D, Santiago, Chile;*
*mtruiz@das.uchile.cl*

P. BERGERON[1]

*Département de Physique, Université de Montréal, C.P. 6128, Succ. Centre-Ville,*
*Montréal, Québec, Canada, H3C 3J7; bergeron@astro.umontreal.ca*

AND

S. K. LEGGETT

*NASA Infrared Telescope Facility, P.O. Box 4729, Hilo, HI 96720;*
*skl@galileo.ifa.hawaii.edu*

**Abstract.** Spectroscopy and photometry (B,V,R,I,J,H,K) of 110 cool white dwarfs have been analyzed with state-of-the-art model atmospheres appropriate for cool white dwarfs with pure hydrogen, pure helium, as well as mixed H/He compositions. The observed energy distributions are obtained from a combination of both optical $BVRI$ and infrared $JHK$ photometric data, and used to derive the effective temperature and the atmospheric composition of each star. Masses for 60 white dwarfs with known trigonometric parallaxes, are also derived.

Most cool white dwarfs turned out to have energy distributions which are well reproduced by either pure hydrogen or pure helium models, with very few objects showing mixed atmospheric compositions. Our results reveal an inhomogeneous temperature distribution of hydrogen-rich and helium-rich white dwarfs, with the presence of a non-DA gap in the range $5000 \lesssim T_{\mathrm{eff}} \lesssim 6000$ K. Hydrogen-rich white dwarfs reappear below 5000 K, and are present down to the lowest temperatures covered by this sample.

[1]Current address: Loral Canada, 6111 Royalmount ave., Montréal, Québec, Canada H4P 1K6.

*I. Isern et al. (eds.), White Dwarfs, 155–163.*
© *1997 Kluwer Academic Publishers.*

## 1. Introduction

The empirical white dwarf luminosity function determined by Liebert, Dahn, & Monet (1988, hereafter LDM) using the Luyten Half-Second Catalog (Luyten 1979), indicate that the luminosity function peaks around log $L/L_\odot = -4.3$, and has a sudden drop towards lower luminosities which has been interpreted as a consequence of the finite age of the Galactic disk. Winget et al. (1987) using theoretical cooling sequences and the empirical white dwarf luminosity function by LDM determined an age of the Galactic disk of $9.3 \pm 2.0$ Gyr. Wood (1992) has improved this analysis studying the effect on the derived cooling ages of various parameters in the white dwarf evolutionary models like stellar masses core and atmospheric compositions.

Model atmospheres appropriate for the study of cool white dwarfs with both pure hydrogen and pure helium compositions have been recently developed by Bergeron et al. (1995b). In parallel with the development of these new theoretical models, we started a few years ago, an extensive spectroscopic and photometric survey aimed at determining more precise atmospheric parameters for a large sample of cool ($T_{\rm eff} < 8000$ K) white dwarfs. We have derived accurate values of chemical composition, effective temperature, surface gravity and hence radius, mass, luminosity and cooling age for these stars. The complex relationships that exists between composition and temperature stresses the need for an improved understanding of the chemical evolution of cool white dwarfs.

## 2. Observations

Most of our program objects were selected from the catalog of McCook & Sion (1987) but we also included 14 ESO white dwarfs from the common proper motion survey in the southern hemisphere (Ruiz et al. 1993, Ruiz & Takamiya 1995, Ruiz 1996), and two additional stars from Monet et al. (1992; LHS 1093 and LHS 1405). In order to test our models and obtain a better understanding of the chemical evolution of white dwarfs atmospheres, we included in our sample cool degenerates of all spectral types (DA, DZ, DQ, and DC white dwarfs). The total sample consists of 110 white dwarfs for which we have optical CCD photometry (B, V, R, I), IR photometry (J, H, K), and spectroscopy mostly near H$\alpha$.

The observations have been obtained during the period 1991 to 1996, using the CTIO 4m, 1.5m and .9m telescopes, the ESO 3.6m telescope, the KPNO 0.9m telescope, the UKIRT and IRTF (extensive details of the observations and analysis will be published elsewhere).

Our high signal-to-noise spectroscopy revealed the presence of H$\alpha$ in 20 white dwarfs previously classified as DC stars, 4 of which turned out to be magnetic, displaying the characteristic Zeeman splitting of the H$\alpha$ line.

## 3. Analysis

In order to fit the model atmospheres to the observed photometry, making use of all available measurements at different filters simultaneously, we convert the magnitudes into observed fluxes, and compare the resulting energy distributions with those predicted from our model atmosphere calculations. The first step is accomplished by transforming every magnitude $m$ into an average flux $f_\lambda^m$ using the equation

$$m = -2.5 \log f_\lambda^m + c_m ,\tag{1}$$

where

$$f_\lambda^m = \frac{\int_0^\infty f_\lambda S_m(\lambda) d\lambda}{\int_0^\infty S_m(\lambda) d\lambda} ,\tag{2}$$

and where $S_m(\lambda)$ is the transmission function of the corresponding bandpass, $f_\lambda$ is the flux from the star received at Earth, and $c_m$ is a constant which remains to be determined. The transmission functions are taken from Bessell (1990) for the $BVRI$ filters on the Johnson-Cousins photometric system, and from Bessell & Brett (1988) for the $JHK$ filters on the Johnson-Glass system. Since the infrared magnitudes obtained are on the CIT system, they are first transformed onto the Johnson-Glass system using the transformation equations given by Leggett (1992). The constants $c_m$ for each passband are determined using the Vega fluxes as discussed in detail by Bergeron et al. (1995c). The calculations yield $c_B = -20.5072$, $c_V = -21.1158$, $c_R = -21.6726$, $c_I = -22.3764$, $c_J = -23.7854$, $c_H = -24.8671$, and $c_K = -26.0017$.

For each star, we obtain a set of seven (or less) average fluxes $f_\lambda^m$ which can now be compared with the model fluxes. Since these observed fluxes correspond to averages over given bandpasses, the monochromatic fluxes from the model atmospheres need to be converted into *average fluxes* as well, $H_\lambda^m$, by substituting $f_\lambda$ in Eq. [2] for $H_\lambda$, the monochromatic Eddington flux. As discussed by Ruiz et al. (1995), this procedure is to be preferred to that of using the monochromatic fluxes when strong lines are present or when the energy distribution varies considerably over the filter bandpass.

The average observed fluxes $f_\lambda^m$ and model fluxes $H_\lambda^m$ — which depend on $T_{\text{eff}}$, $\log g$, and $N(\text{He})/N(\text{H})$ — are related by the equation

$$f_\lambda^m = 4\pi \ (R/D)^2 \ H_\lambda^m ,\tag{3}$$

where $R/D$ is the ratio of the radius of the star to its distance from Earth. Our fitting technique relies on the nonlinear least-squares method of Levenberg-Marquardt (Press et al. 1986), which is based on a steepest descent method. The value of $\chi^2$ is taken as the sum over all bandpasses of the difference between both sides of Eq. [3], properly weighted by the corresponding observational uncertainties. We consider only $T_{\text{eff}}$ and the solid angle free parameters, and the uncertainties of both parameters are obtained directly from the covariance matrix of the fit. Even though it is also possible, in principle, to constrain the value of log $g$ from fits to the energy distributions, we found that in practice this procedure yields meaningless results both because of the limited accuracy of the observations, and because the models do not necessarily perfectly describe the data. Consequently, for white dwarfs with no parallax measurement, we simply assume log $g = 8.0$.

For stars with known trigonometric parallax measurements, we start with log $g = 8.0$ and determine $T_{\text{eff}}$ and $(R/D)^2$, which combined with the distance $D$ obtained from the trigonometric parallax measurement yields directly the radius of the star $R$. The radius is then converted into mass using the evolutionary models of Wood (1990) for the pure helium and mixed hydrogen/helium compositions, and those of Wood (1995) for the pure hydrogen models. The former models have carbon-core compositions, helium layers of $q(\text{He}) \equiv M_{\text{He}}/M_\star = 10^{-4}$, and no hydrogen layers, while the latter have carbon-core compositions also, but helium layers of $q(\text{He}) = 10^{-2}$, and thick hydrogen layers of $q(\text{H}) = 10^{-4}$. In general, the log $g$ value obtained from the inferred mass and radius ($g = GM/R^2$) will be different from our initial guess of log $g = 8.0$, and the fitting procedure is thus repeated until an internal consistency in log $g$ is reached. The parameter uncertainties are obtained by propagating the error of the trigonometric parallax measurements into the fitting procedure.

The spectroscopic observations (line profiles), are not used directly in the fitting procedure, however, they served as an internal check of our photometric solutions, or they are sometimes used to derive limits on the chemical abundances, or to infer the value of log $g$ in DA stars when no trigonometric parallax measurements are available. The theoretical line profiles are thus simply interpolated at the values of $T_{\text{eff}}$ and log $g$ derived from the energy distribution fits and compared with the observed line profiles (usually $\text{H}\alpha$).

## 4. Results and discussion

The surface gravity, and hence the mass of 60 stars in our sample has been constrained from trigonometric parallaxes. Masses thus obtained are displayed in Figure 1 together with the mass distribution of DA white dwarfs determined by Bergeron et al.(1992) and revised by Bergeron et al.(1995a)

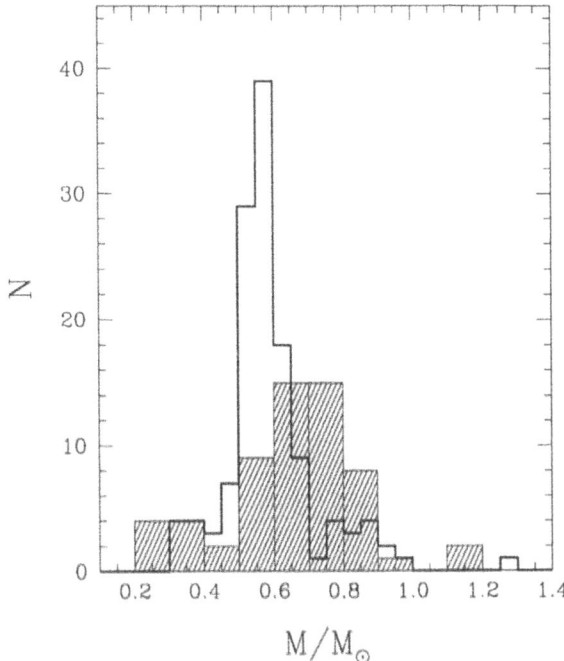

*Figure 1.* Mass distribution of our trigonometric parallax sample (hatched histogram) compared with that determined spectroscopically by Bergeron et al.(1992) for DA stars only (thick solid line) and revised by Bergeron et al.(1995a). Since our parallax sample is considerably smaller than the Bergeron et al. sample, we use mass bins twice as large.

by using the thick hydrogen models of Wood (1995). The latter mass distribution has a mean of 0.59 $M_\odot$ and a standard deviation of 0.13 $M_\odot$. In contrast, the mass distribution of degenerates in our sample (cooler white dwarfs) has both a higher mean mass of 0.66 $M_\odot$, and a broader dispersion of 0.19 $M_\odot$, with both distributions containing clear cases of low and high-mass white dwarfs ( both containing unresolved double degenerates). This difference is readily understood if we consider that the coolest and therefore oldest white dwarfs in the Galactic disk are more massive than their hotter counterparts since they were formed from more massive progenitors. However, as both methods used for determining masses are fundamentally different, and the systematic effects of both techniques (hydrogen line profiles and trigonometric parallaxes) are not fully understood yet, it is hard to assess the significance of the observed difference in the mass distributions.

The stellar masses of all white dwarfs in our trigonometric parallax ensemble are displayed in Figure 2 as a function of effective temperature; stars with no trigonometric parallax measurements are shown at the bottom of the plot with $M < 0$. In the top and bottom panels, the different symbols reflect differences in spectral types (DA and non-DA) and atmospheric composition (hydrogen-rich, helium-rich, and also mixed compositions for

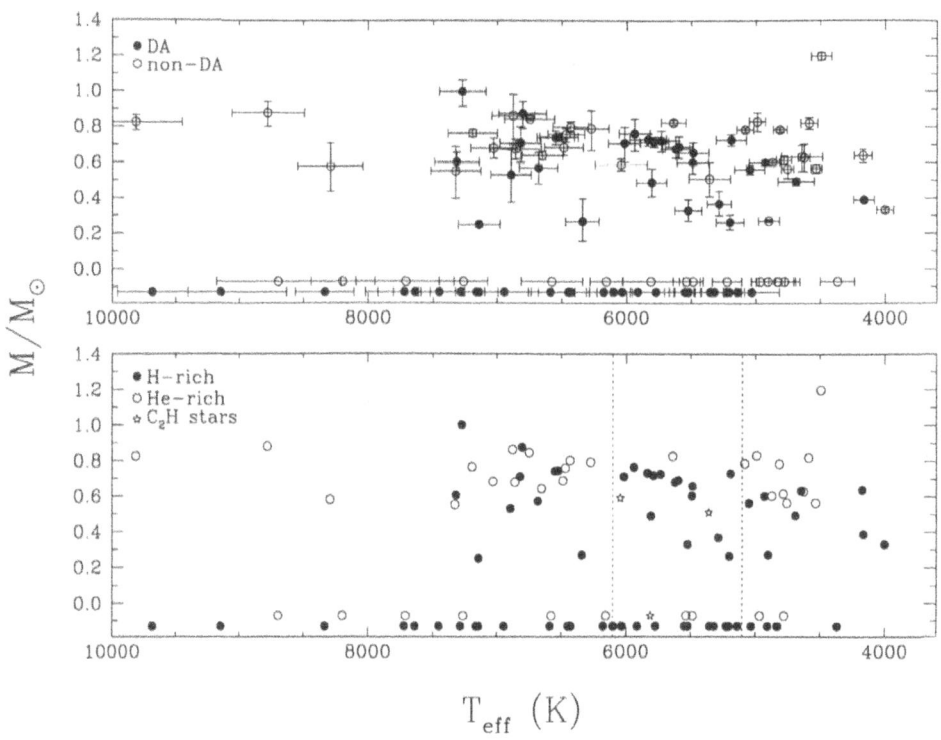

*Figure 2.* Masses of white dwarfs in our trigonometric parallax sample as a function of $T_{eff}$ ; stars with no measurements are shown at the bottom of each panel with $M < 0$ . In the top panel, white dwarfs are divided in DA and non-DA stars, and individual uncertainties are shown for both M and $T_{eff}$ . In the bottom panel, different symbols are used to distinguish the main atmospheric constituent of each object, while the dotted lines delineate the non-DA gap discussed in the text.

the $C_2H$ stars), respectively. Note that some of the coolest non-DA stars in the top panel actually have hydrogen-rich compositions in the bottom panel. Our sample concentrates on stars with $T_{eff} < 7500$ K. There are a few objects at higher temperatures, but not enough to allow us to draw any conclusions about the chemical evolution of hotter white dwarfs.

In the top panel, we also show the uncertainties of M and $T_{eff}$. The mass uncertainties depend mostly on the precision of the trigonometric parallax measurements. The temperature uncertainties decrease at lower $T_{eff}$ simply because the shape of the energy distributions of cooler white dwarfs varies more rapidly with $T_{\text{eff}}$. However, the relative uncertainties remain some-what constant at 1–3% through the entire temperature range considered here. These uncertainties are about twice as large for stars with no infrared photometric measurements, a result which underlines the importance of infrared photometry for studying cool white dwarfs.

The results presented in the bottom panel of Figure 2 hides crucial information about the chemical evolution of cool white dwarfs. Certainly the most striking feature of the temperature distribution of hydrogen-rich and helium-rich atmospheres is the existence of a temperature range, $6100 \gtrsim T_{\text{eff}} \gtrsim 5100$ K, in which the number of helium-rich stars is greatly reduced, a non-DA gap, where we find only three helium-rich objects. Our sample is bias (by design) against DA stars, therefore, further observations of an unbiased sample can only increase the statistical significance of this non-DA gap. Above and below the gap, helium-rich stars are found in large numbers, which leads us to conclude that some physical process is at work to alter the chemical composition of these stars as they cool off turning most, if not all, helium-rich white dwarfs above the non-DA gap into hydrogen-rich stars as they reach $T_{\text{eff}} \sim 6000$ K, some of which will become helium-rich again at $T_{\text{eff}} < 5000$ K.

Further analysis of the distribution of white dwarfs' spectral types with $T_{\text{eff}}$, for stars with masses accurately determined and selected from an unbias sample, may reveal important clues regarding the physical processes involved in the change in composition of their atmospheres as they cool off.

This work was supported in part by FONDECYT grant 1950588, by the NSERC Canada, by the Fund FCAR (Québec), by the NSF grant 93-15372, and by a Chrétien International Research Grant.

## References

Bergeron, P., Saffer, R. A., & Liebert, J. 1992, ApJ, 394, 228

Bergeron, P., Wesemael, F., Lamontagne, R., Fontaine, G., Saffer, R. A., & Allard, N. F. 1995a, ApJ, 449, 258

Bergeron, P., Saumon, D., & Wesemael, F. 1995b, ApJ, 443, 764

Bergeron, P., Wesemael, F., & Beauchamp, A. 1995c, PASP, 107, 1047

Bessell, M. S. 1990, PASP, 102, 1181

Bessell, M. S., & Brett, J. M. 1988, PASP, 100, 1134

Leggett, S. K. 1992, ApJS, 82, 351

Liebert, J., Dahn, C. C., & Monet, D. G. 1988, ApJ, 332, 891 (LDM)

Luyten, W. J. 1979a, The LHS Catalogue, Second Edition (Minneapolis: University of Minnesota)

McCook, G. P., & Sion, E. M. 1987, ApJS, 65, 603

Monet, D. G., Dahn, C. C., Vrba, F. J., Harris, H. C., Pier, J. R., Luginbuhl, C. B., & Ables, H. D. 1992, AJ, 103, 638

Press, W. H., Flannery, B. P., Teukolsky, S. A., & Vetterling, W. T. 1986, Numerical Recipes (Cambridge: Cambridge University Press)

Ruiz, M. T. 1996 AJ, 111, 1267

Ruiz, M. T., Bergeron, P., Leggett, S. K., & Anguita, C. 1995, ApJ, 455, L159

Ruiz, M. T., & Takamiya, M. Y. 1995 AJ, 109, 2817

Ruiz, M. T., Takamiya, M. Y., Mendez, R. A., Maza, J., & Wischnjewsky, M. 1993, AJ, 106, 2575

Winget, D. E., Hansen, C. J., Liebert, J., Van Horn, H. M., Fontaine, G., Nather, R. E., Kepler, S. O., & Lamb, D. Q. 1987, ApJ, 315, L77

Wood, M. A. 1990, Ph. D. thesis, Univ. of Texas at Austin

——. 1992, ApJ, 386, 539

——. 1995, in 9th European Workshop on White Dwarfs, NATO ASI Series, ed. D. Koester & K. Werner (Berlin: Springer), 41

## Discussion

*H. Shipman*: I don't understand how $H^-$ opacity is going to change the observed surface composition of very cool white dwarf stars. Could you explain where this comes from?

*M.T. Ruiz*: There is no change in surface composition, we just say that at about 6000 K the opacity due to $H^-$ in a He dominated atmosphere suddenly becomes very important and produces a change in the pressure of the atmosphere, from a high pressure (low opacity) atmosphere to a low pressure (high opacity) atmosphere but with the same amounts of He and H. When the pressure in the atmosphere decreases, then the levels of H are no longer perturbed and the Balmer lines are observed.

*J. Liebert*: Just a comment. Your last figure shows dramatically how two-dimensional the problem of analyzing the luminosity function has become. You need good parallaxes to get the mass-luminosity function and calculate the cooling age of each star (and estimate the age of the disk, etc.).

# THE CHEMICAL EVOLUTION OF WHITE DWARF STARS: WHAT THE OBSERVATIONS TELL US

HARRY L. SHIPMAN
*University of Delaware*

**Abstract.** This brief review presents an overview of our current picture of the evolution of the surface chemical composition of white dwarf stars. Different investigators have converged remarkably well on a consensus picture regarding white dwarf spectral evolution, especially considering some of the differences and controversies which have provided some spice and heat in the literature and conference discussion in the past several years. Here, I provide some brief summaries of the current picture and list some outstanding problems which are ripe for further research.

## 1. Introduction

The chemical evolution of white dwarf stars has been one of the central problems of our field since Schatzman's seminal book (Schatzman 1958). Everywhere else in the HR diagram, the chemical composition of stars is pretty much the same. Compositional differences are quite rightly described as "peculiarities," in the sense that a "chemically peculiar" star has the same basic composition of mostly hydrogen. Even in Schatzman's time, it was clear that the white dwarf stars were fundamentally different in several ways. First, there is not a single white dwarf that even comes remotely close to the "cosmic composition" of $X = 0.7$, $Y = 0.28$, $Z = 0.02$. Clearly processes operating prior to or during the white dwarf stage have altered the composition of white dwarf photospheres. Second, the white dwarf stars are divided into what we now call DA's and non-DA's. The H/He ratio varies by nearly 15 orders of magnitude between extreme members of the two classes. And third, within these two broadly defined compositional classes, trace elements come and go as the star cools and accretes material from the interstellar medium.

*I. Isern et al. (eds.), White Dwarfs*, 165–171.

Many scientists in the white dwarf community pondered over these differences, and some of us even tried our hand at explaining the surface chemical diversity of white dwarf stars. We began to explore some of the physical processes which altered the surface compositions of white dwarf stars, such as convective mixing (e.g., D'Antona and Mazzitelli 1978, radiative levitation (Fontaine and Michaud 1979, Vauclair, Vauclair, and Greenstein 1979), and accretion from the interstellar medium (Alcock and Illarionov 1980). But it was not until the mid 1980s that the first really comprehensive hypotheses began to emerge. Iben and MacDonald (1985) came up with a concrete proposal which explained the main divergence between the DA's and the non-DA's as a product of the timing of the final thermal pulse in red giant evolution. A somewhat contrasting view was presented at various meetings by Liebert, Fontaine, and Wesemael (1987; see also Fontaine and Wesemael 1989), wherein white dwarf evolution was confined to a single channel and DA's turned into non-DA's and vice versa. Another important development of this time was the recognition of convective mixing as the source of the carbon in DQ white dwarf stars (Fontaine et al. 1984, Pelletier et al. 1986).

These divergent views generated some vigorous debate at white dwarf conferences during the late 1980s and early 1990s. On the scientific side, theorists who analyze stellar pulsations argued vigorously over whether the pulsational properties of white dwarf stars were most easily interpreted in terms of thin or thick hydrogen layers. But perhaps more colorful were the unofficial debates at conferences which extended through the night. A group of astronomers enthusiastically sang the "thin layer blues" in Hanover, New Hampshire, in the wee hours of the morning in the summer of 1988. The temperature range of the DBV stars was deemed sufficiently important by two members of our community so as to put a case of wine at stake. The wager was settled at the European Workshop in Toulouse, and the winner and loser of this bet were gracious enough to permit the conference participants to join together in consuming the stakes.

A number of review articles appeared during these years. Sion (1986) valiantly tried to make sense of it all in one picture, a picture which was seen by at least one other person (Shipman 1988) as being too complex to fit on a T-shirt and thus too complex to be correct, using some principles which were enunciated in a different form many years ago (Ockham 1492). In retrospect, my attempt in Hanover to make sense of it all (Shipman 1989) contained many ideas which have since then turned out to be right, though the observational support for them is considerably richer now than it was at the time. A few other recent reviews, selected somewhat arbitrarily, are those by D'Antona and Mazzitelli (1990). Fontaine and Wesemael (1991), and Weidemann (1990).

A rather contrasting picture emerged in the summer of 1996 on Spain's Costa Brava, where Blanes is located. There was considerable consensus regarding our current ideas on white dwarf evolution. I do not have space in this brief review to do much more than depict the consensus, identify a few key observations which support it, and list a few open questions. Longer treatments can be found in many of the key articles described here, and hopefully in a forthcoming review article (Shipman 1997). The general picture is that the distinction between DA's and non-DA's comes from events which occurred well before a star entered the white dwarf region of the HR diagram. Thus there are at least two independent channels of white dwarf evolution. I will describe each in turn.

## 2. The DA Stars

Figure 1 describes our current understanding of the evolution of the DA stars. For the hotter DA stars, our overall picture seems reasonably complete. It begins at the top with planetary nebula central stars, which are not yet degenerate. Earlier, many investigators expressed a concern had been that there was a gap between the hottest DA stars and the PNN. Several stars have recently been discovered which fill in this gap (EGB 1 - Napiwotzki and Schonberner 1993; WDHS1 - Liebert, Bergeron, and Tweedy 1994; and RE 1738+665 - Barstow et al. 1994) and so it is now clear that this gap no longer exists (Shipman 1994). Many of the DA's which are hotter than 40,000 K show signs of heavy elements, particularly in the extreme ultraviolet spectral region which has only recently been opened up by EUVE. Radiative acceleration pushing heavy ions up through the photosphere has long been suspected as the source of the heavy elements. A comprehensive comparison of state-of-the-art calculations with observations (Chayer, Fontaine, and Wesemael 1995) reveals that there is general agreement between theory and observation. The abundances are about right. The correlation between the presence of heavy elements and stellar parameters like $T_{eff}$ and $\log g$ corresponds pretty well to what is observed. I would characterize this theory as a semi-quantitative success, in that there is general and quantitative agreement between theory and observation but some discrepancies remain (e.g., the Si abundance is predicted to be too low and some stars like HZ 43 and W 1346 are hard to fit; e.g., Holberg 1996)

In the recent past, a very serious difficulty with the single-channel scenario for the DA stars came from observations of the pulsating ZZ Ceti stars. Pulsation is an extraordinarily powerful tool for probing the structure of stellar interiors. However, there have been two barriers to the successful interpretation of data on pulsating stars. Until the advent of the Whole Earth Telescope (WET), it was extremely difficult to identify par-

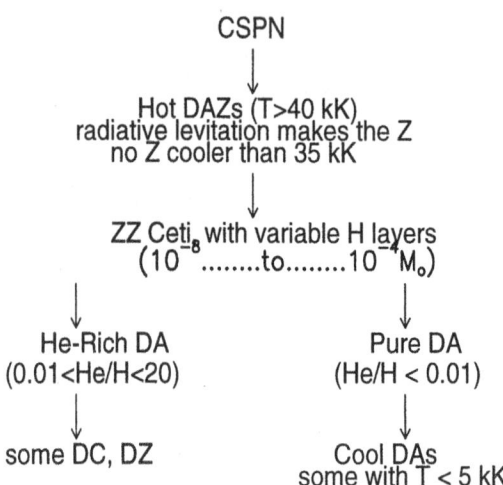

*Figure 1.* A proposed scheme for the chemical evolution of the DA stars. It is simple enough to fit on a T-shirt, as was experimentally demonstrated at the conference.

ticular pulsation modes in an individual pulsator. Thus interpretation of the pulsation data used quantities which are less closely coupled to the envelope structure of a given model, such as the location of the edges of the ZZ Ceti instability strip. With the advent of WET and the development of several state-of-the-art codes (Brassard et al. 1992, 1995; Bradley 1993) these ambiguities have been cleared up.

The most straightforward interpretation of the statistics of the cool DA's (see, e.g., Shipman 1989) has been that there is a variety of layer masses among the DA's. As a DA star cools, its convection zone deepens, and it may reach a point where He will be dredged up from beneath and a DA will turn into a non-DA. The non-DA/DA ratio increases as stars become cooler, but the DA's never completely disappear. The easy way to explain this changing ratio is to propose that some DA's have a relatively thin ($10^{-9}M_\odot$) H layer and mix at high temperatures, while other DA's have a thick ($10^{-4}M_\odot$) layer and never mix at all. While some points remain to be cleared up, the pulsation data appear to be consistent with this interpretation (Bradley 1993).

## 3. The Non-DA Stars

The picture for the non-DA stars remains more complex than that for the DA stars, though it is a bit simpler than we thought a few years ago. An outline of this picture is provided in Figure 2 below, and it too fits on a T-shirt, so it is simple enough to be correct:

Observationally, the single most significant complication in the evolution

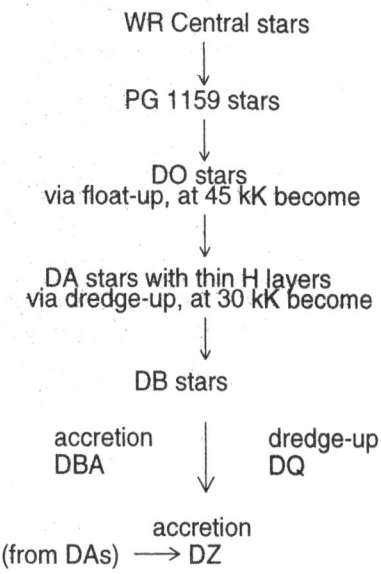

*Figure 2.* A proposed scheme for the spectral evolution of the non-DA stars. Like Fig. 1, it is simple enough to fit on a T-shirt, as was demonstrated at the conference.

of the non-DA stars is the existence of the "DB gap," an absence of DB stars with 30,000 K $< T_{\mathrm{eff}} <$ 45,000 K. This gap was first identified by Liebert et al. (1986). Several surveys of faint blue stars have identified a lot more white dwarf stars than were known ten years ago. The only stars that even remotely resemble DB stars to populate the gap are two hybrid DBA stars (HS 0209+0832, described in Jordan et al.1994; GD 323, described most recently in Koester, Liebert, and Saffer 1994). These objects are predominantly hydrogen, and it is likely that whatever their evolutionary status is, it is not related to the spectral evolution of non-DA stars. The statistical significance of the gap is undoubtedly greater than the ... sigma claimed by Liebert et al. (1986) and is ripe for re-determination. It seems reasonable to assume, as a working model, that the DB gap is real.

It is still true that the best explanation of the DB gap is the float-up and dredge-up scenario first proposed as part of a comprehensive theory of white dwarf spectral evolution (Fontaine, Liebert, and Wesemael 1987) and comprehensively investigated by MacDonald and Vennes (1991). What happens is that trace amounts of H which are present in the DO stars (also called PG 1159 stars) float up to the surface when the star cools to 50,000 K, making the star look like a DA, but with an ultrathin layer of hydrogen. When the helium convection zone thickens at 30,000 K, helium is then brought to the surface and the star once again turns into a DB. Spectroscopy of the PG 1159 stars by Werner and co-workers (Dreizler and Werner 1996; Werner et al. 1996) is consistent with this picture. A question is whether the thin-layered DA stars which are really part of the non-DA sequence of spectral evolution can be identified somehow as a sub-population of the DA stars.

Cooler DA stars also undergo some spectral changes as convective dredge-up brings C to the surface, producing the DQ stars (Pelletier et al. 1986).

## References

Alcock, C. & Illarionov, A. 1980, ApJ, 235, 534
Barstow. M., et al. 1994, MNRAS 271, 175
Brassard, P., Fontaine, G., Wesemael, F., & Tassoul, M. 1992, ApJS, 81, 747
Brassard, P., Fontaine, G., & Wesemael, F. 1995, ApJS, 96, 545
Bradley, P. 1993, *The Encyclopedia Seismologica* (Ph.D. Thesis, Univ. of Texas at Austin)
Chayer, P., Fontaine,G., & Wesemael,F. 1995, ApJS, 99, 189
D'Antona, F., & Mazzitelli, I. 1978. A&A, 66, 433
D'Antona, F., & Mazzitelli, I. 1990, ARAA, 28, 139
Dreizler, S. & Werner, K. 1996, A&A, (in press)
Fontaine, G. & Michaud, G. 1979, ApJ, 231, 826
Fontaine, G., Villeneuve, B., Wesemael, F., & Wegner, G. 1984, Ap.J. Letters 227, L61.
Fontaine, G., & Wesemael, F. 1987, in *The Second Conference on Faint Blue Stars*, eds. A.G.D. Philip, D.S. Hayes, and J. (Schenectady: L. Davis Press), 319
Fontaine, G., & Wesemael, F. 1991, in *Evolution of Stars: the Photospheric Abundance Connection*, eds. G. Michaud & A. Tutukov, (Dordrecht: Reidel), 421-434
Holberg, J., Barstow, M., Bruhweiler, F., & Collins, J. 1996, AJ, in press.
Iben, I., & MacDonald, J. 1985
Jordan, S., et al. 1994, A&A 273, L27
Kawaler, S., & Bradley, P. 1994, ApJ, 427, 415
Koester,D., Liebert,J., & Saffer,R. 1994, ApJ 422, 783
Liebert J., Bergeron P., & Tweedy R., 1994, ApJ, 424,817
Liebert, J., Fontaine, G., & Wesemael, F. 1987, MSAI, 58, 17
Liebert, J., et al. 1986, ApJ, 309, 241
MacDonald, J., & Vennes, S. 1991, ApJ, 371, 719
Napiwotzki & Schonberner 1993, in *White Dwarfs: Advances Observation and Theory* (Proceedings of the NATO Workshop on White Dwarf Stars, ed. M. Barstow (Dordrecht: Kluwer, 1993), 99
Pelletier, C., Fontaine, G., Wesemael, F., Michaud, G., & Wegner, G. 1986, ApJ, 307, 242.
Provencal, J., Shipman, H., Thejll, P., Vennes, S., & Bradley, P. 1996, ApJ, 466, 1011
Schatzman, E. 1958, White Dwarfs, (Amsterdam: North-Holland)
Shipman, H. 1988, in *The Second Conference on Faint Blue Stars* eds. A.G.D. Philip, D.S. Hayes, and J. Liebert, eds., (Schenectady: L. Davis Press),
Shipman, H. 1989, in *White Dwarfs: Proc. IAU Colloquium 114* in Lecture Notes in Physics, 328 (New York: Springer-Verlag), 220
Shipman, H., 1997, PASP, in preparation
Shipman, H. 1994, Nature, 372, 318
Sion, E. 1986 PASP, 98, 821
Vauclair,G., Vauclair, S., & Greenstein, J. 1979, A&A, 80, 79
Weidemann. V. 1990, ARA,. 28, 103
Werner, K., Dreizler, S., Heber, U., & Rauch, T. 1996, in *Hydrogen-Deficient Stars* eds. U. Heber & C.S. Jeffrey, (San Francisco, CA: ASP Conference Series).

## Discussion

*I. Iben*: You have invoked a number of mechanisms such as radiative levitation mixing, accretion, etc. but I didn't hear "winds". It doesn't take much of a wind to beat accretion from the ISM.

*H. Shipman*: Winds may well play a role in modifying the compositions of the hottest white dwarf stars, and they probably should play some role in a comprehensive theory, again for the hottest stars.

*S. Dreizler*: I don't agree that the DO's and DB's are found in different ways. The majority of the DO's are also found in blue star surveys.

*H. Shipman*: That's true for the majority of the DO's. But there are some, and some PG 1159's, which have been found in other ways. To determine the statistical significance of the DB gap, you need to deal with a homogeneous sample.

*J. Holberg*: I disagree with your characterization of Wolf 1346 as an example of a cool DA with a circumstellar shell. We (Holberg, Barstow, Bruhweiler and Collins, June 1996 AJ) determined 1) That the Si features in Wolf 1346 have the same Doppler velocity as the Balmer lines, and 2) that Si II and Si III features imply a consistent Si photospheric abundance. We conclude the Si is photospheric not circumstellar.

*H. Shipman*: You're quite right about the nature of Wolf 1346. However, the question is whether it is a major clue to the spectral evolution of white dwarf stars or whether it is an unusual star which can be treated as such.

# A CRITICAL LOOK AT THE QUESTION OF THICK VS THIN HYDROGEN AND HELIUM ENVELOPES IN WHITE DWARFS

G. FONTAINE AND F. WESEMAEL

*Département de Physique, Université de Montréal*

## 1. Introduction

The question of the possible existence of relatively thin hydrogen and helium envelopes in white dwarfs has certainly enlivened the scientific debate in our community over the last decade. It has led to vigorous exchanges –one of them ending as a long, low-down blues song (!)– as well as some misconceptions and even mild disputes at times, but it has also certainly forced us, collectively, to think seriously about the spectral evolution of white dwarfs. Among the large number of papers devoted to the topic, we single out those with a relatively broad perspective: Sion (1986), Fontaine & Wesemael (1987), Shipman (1989), Koester (1989), Fontaine & Wesemael (1991), Shipman (1993), and Bergeron, Ruiz, & Leggett (1996). The more specialized papers devoted to one aspect or another of the subject are too numerous to mention here, but, for a good sample, we refer the reader to the Proceedings of the various white dwarf meetings held in Frascati (1986), Tucson (1987), Hanover (1988), Toulouse (1990), Leicester (1992), and Kiel (1994). This intense research activity is justified by the fact that the question of the thickness of the outer hydrogen and helium layers is critically important in understanding the weird atmospheric compositions of white dwarfs, in calibrating cooling calculations aimed at estimating the age of the Galactic disk and halo, and in constraining the conditions encountered in pre-white dwarf evolution, which, by all accounts, still constitutes a gray area in the theory of stellar evolution.

Since the Leicester meeting, it appears that the pendulum of wisdom has swung back from an "all-thin" to an "all-thick" scenario, the latter becoming fashionable again in recent years. Although the original proponents of the idea that a *majority* of white dwarfs should have relatively thin hydrogen and helium outer layers, we, along with our students, contributed

173

*I. Isern et al. (eds.), White Dwarfs, 173–192.*
© 1997 *Kluwer Academic Publishers.*

significantly to this shift of thinking by examining carefully, over the last few years, many of the various pieces of evidence backing up the original model. Taking into account the independent contributions of many other researchers, it is now safe to affirm that several of these pieces of evidence did not hold true upon closer scrutiny. We believe, however, that the current "all-thick" tendency is too extreme, and we demonstrate this in the present paper. To this end, we review critically the question of thin *vs* thick layers in the light of several recent developments in the field.

## 2. The case for thin layers (circa $\sim$ 1987)

Standard evolution theory predicts that a typical $\sim 0.6$ $M_\odot$ isolated white dwarf should be made of a C/O core surrounded by a He mantle of mass $q(He) \equiv M(He)/M_* \simeq 10^{-2}$, itself surrounded by a H envelope of mass $q(H) \equiv M(H)/M_* \simeq 10^{-4}$ (e.g., Iben & Tutukov 1984; Koester & Schönberner 1986; D'Antona & Mazzitelli 1987). It is possible to adjust in a clever way the conditions in an AGB model, so that, in one case out of four, it will end up as a white dwarf totally devoid of hydrogen (Iben 1984). In this model, sometimes referred to as primordial scenario (Shipman 1989), a H-rich (DA) white dwarf is born and remains as such throughout its cooling history. Likewise, a He-atmosphere white dwarf is born that way and cools as a non-DA star. In the absence of mechanisms such as mass loss and accretion, this model predicts that the non-DA/DA ratio is fixed at birth (1:3), and does not change throughout the evolution.

This scenario is clearly at odds with the *observed* distribution of the two spectral types as summarized, for example, in Figure 3 of Fontaine & Wesemael (1987), which is still partly relevant to the present discussion. This figure suggests, in contrast to the primordial scenario, that spectral evolution *does* occur along the cooling sequence, i.e., a given star may show two different faces (DA or non-DA), depending on its effective temperature. Of particular interest are the paucity of DA stars at very high effective temperatures, the total absence of He-rich white dwarfs in the range 45,000-30,000 K (DB gap), and the turnover in favor of non-DA stars below $T_{\rm eff} \sim$ 10,000 K.

To understand these observed phenomena, it is necessary to postulate nonstandard structures allowing convective dilution and mixing, and this requires thin H layers, at least in a fair fraction of the DA's. By "thin", we mean here $q(H) \ll 10^{-4}$, i.e., much smaller than the amount predicted by standard evolution theory. A working model was first proposed by Liebert, Fontaine, & Wesemael (1987) who postulated a two-channel feeder to white dwarf cooling sequences: the first one with thick hydrogen layer stars entering and remaining DA objects throughout (akin to the DA models of

Iben 1984) and which we will refer to as channel 1, and the second one containing stars entering as non-DA objects but containing small residual amounts of hydrogen, thus allowing spectral evolution to occur (channel 2).

Fontaine & Wesemael (1987) went one step further, and suggested that a *majority* of white dwarfs go through channel 2. Unfortunately, this has often been confused with a "one-channel scenario". In their original model, Fontaine & Wesemael (1987) did allow for a small, but nonzero contribution from channel 1. In this scenario, most white dwarfs descend from PG1159/DO stars which, it is assumed, contain some minute amounts of hydrogen still mixed in their envelopes. With cooling, hydrogen diffuses upward until an atmosphere's worth of DA material ($q(H) \sim 10^{-16}$) is accumulated at the surface. At that point, the evolving star appears as a DA for the first time. Separation still goes on, however, and the hydrogen outer layer continues to thicken. Depending on the total amount of hydrogen leftover from previous evolutionary phases, He-rich stars turn gradually over to DA objects from $T_{\mathrm{eff}} \sim 80,000$ K until about $T_{\mathrm{eff}} \sim 45,000$ K, where nearly all of them must have done so. This is the blue edge of the DB gap, a clear indication that no He-rich star has been able to survive the overpowering process of diffusion. From the blue edge of the gap, a star cools down to 30,000 K as a DA object. At that temperature, the helium convection zone underlying the thin hydrogen radiative atmosphere may be allowed to break into the latter and dilute it. For those DA stars where the accumulated hydrogen layer mass is less than $q(H) \sim 10^{-15}$ (at $T_{\mathrm{eff}} \sim 30,000$ K), mixing thoroughly dilutes the hydrogen within the relatively massive underlying helium convection zone and DB stars are born. (We note that the basic validity of this model for the DB gap has since been verified quantitatively by MacDonald & Vennes 1991.) The others retain their DA character until the outer hydrogen layer itself becomes fully convective and mixes with the underlying helium convection zone below $T_{\mathrm{eff}} \sim 10,000$ K, a process described several years before by Koester (1976), Vauclair & Reisse (1977), and D'Antona & Mazzitelli (1979). This process has been invoked to explain the observed turnover in the white dwarf population in favor of non-DA objects at low effective temperatures.

In the face of the evidence available to them at the time, Fontaine & Wesemael (1987) reasoned that channel 1 (the one with standard thick hydrogen and helium layer stars) could not be a *major* contributing channel because, otherwise, there would be no significant paucity of DA objects at very high effective temperatures, and no substantial turnover in favor of non-DA stars at low effective temperatures. Moreover, they relied on three independent pieces of circumstantial evidence which, taken together, strenghtened considerably the case for thin outer layers in white dwarfs.

On the first account, it was proposed that the bulk of the available

EUV/soft X-ray broadband observations of hot DA stars could be explained in terms of *stratified* H/He atmospheres made of a thin hydrogen layer ($q(H) \sim 10^{-13} - 10^{-15}$) overlaying He-rich regions still visible in this spectral window. This model was originally articulated by Vennes et al. (1988) and was subsequently invoked by several other authors. We note that one of the misconceptions that has crept in the literature about the suggestion of Fontaine & Wesemael (1987) concerns the apparent conflict between the values of $q(H)$ suggested by Vennes et al. (1988) in hot DA's and those (much larger) required to produce mixing below $T_{\text{eff}} \sim 10,000$ K. Because separation still goes on (depending on the total amount of hydrogen initially hidden in a white dwarf), there is, in fact, no conflict: the buildup of the outer hydrogen layer in a DA star could amount to $q(H) \sim 10^{-14}$ at $T_{\text{eff}} \sim 50,000$ K, but could be much larger at lower effective temperatures where separation is more advanced.

The second piece of evidence came from Pelletier et al. (1986) who carried out detailed time-dependent calculations of the separation of helium and carbon in He-rich white dwarfs in order to explain the DQ phenomenon (i.e., the presence of traces of carbon observed in the atmospheres of such stars with $T_{\text{eff}} \lesssim 15,000$ K). The model they developed, now referred to as the carbon pollution model, hinges on the interaction of the helium superficial convection zone (which develops as a consequence of cooling) with traces of carbon diffusing near the core edge. Under appropriate circumstances, the helium convection zone dredges up some of the carbon, and these traces pollute the atmosphere and become spectroscopically visible. Pelletier et al. (1986) found that the models best representing the observations have $M_* \simeq 0.6 \ M_\odot$ and helium layer masses in the range $10^{-4.0} \lesssim q(He) \lesssim 10^{-3.5}$ (for ML1 convection). These are significantly smaller than the value ($\sim 10^{-2}$) expected from standard evolution theory.

The third piece of evidence came from the *nonadiabatic* pulsation analyses of the ZZ Ceti stars carried out by Winget et al. (1982) and Winget & Fontaine (1982; see also Bradley, Winget, & Wood 1989). These authors uncovered and exploited a sensitivity of the location of the blue edge of the ZZ Ceti instability strip on the hydrogen layer thickness in their models. By identifying the theoretical blue edge with the observed one, they were able to constrain the hydrogen layer mass in these $\sim 11,500$ K DA stars to $10^{-12} \lesssim q(H) \lesssim 10^{-7}$, again much less than expected from standard models. This result was disputed by Cox et al. (1987) who claimed that, on the contrary, the theoretical blue edge temperature is not sensitive to the hydrogen layer mass in ZZ Ceti stars. However, Brassard et al. (1989, 1991) demonstrated convincingly that the results of Cox et al. (1987) were unreliable and, consequently, could not be used to decide on the issue of the sensitivity of the blue edge temperature on $q(H)$.

## 3. Some early objections and answers

A significant objection was immediately raised by Richard Green (see discussion at the end of the Fontaine & Wesemael 1987 paper) who pointed out that, by considering the space density of DO stars and cooling time intervals from models, only $\sim 1/8$ of the cooler white dwarfs could be accounted for. This estimate was taken from the analysis of Wesemael, Green, & Liebert (1985) which used an unpublished evolutionary track of a pure C, 0.6 $M_\odot$ model calculated by Winget, Lamb, & Van Horn. Along with the above result, that analysis led to a predicted number of 39 DB white dwarfs as compared to the 48 DB objects actually discovered in the Palomar-Green (PG) survey.

To answer Green's objection, we repeated the calculations following the same procedures as Wesemael et al. (1985), except that we used more realistic compositionally-stratified models taken from Koester & Schönberner (1986; a 0.598 $M_\odot$ model with $q(He) = 10^{-1.36}$, and no hydrogen) and Iben & Tutukov (1984; their Case B model with 0.6 $M_\odot$, $q(He) = 10^{-1.36}$, and no hydrogen). These were the most appropriate models available at the time. Because the outer helium layer in these latter models is more opaque than pure carbon, the corresponding cooling timescales are significantly longer. We found, accordingly, that the predicted number of DB's descending directly from the DO's was larger than estimated by Wesemael, Green, & Liebert (1985): 92 for the Koester-Schönberner timescales and 159 for the Iben-Tutukov timescales. Taken at face value, this result implied that DO stars evolve into something else, in addition to the 48 DB's found in the PG survey. We then reasoned that this "something else" had to be DA stars.

We point out that we were actually able to go one step further in this game of statistics. Indeed, we estimated how many cooler white dwarfs would descend from the 4 DO's above $T_{eff} \simeq 70,000$ K and the 7 PG1159's appearing in Table 7 of Wesemael et al. (1985). Collectively, these 11 objects have a space density of $\sim 1.2 \times 10^{-7}$ pc$^{-3}$ (Wesemael et al. 1985). Using this in conjunction with the Koester-Schönberner timescales, we found that 148 objects descending from the above very hot He-rich degenerate stars should be present in the PG survey in the interval 70,000 K $\gtrsim T_{eff} \gtrsim$ 12,000 K. Interestingly, this number grew to 240 objects when using the Iben-Tutukov timescales, quite close to the total number of DA white dwarfs in the sample discussed by Fleming, Liebert, & Green (1986), and implying that channel 2 dominates completely! While supportive of our 1987 model, we nevertheless concluded that these numbers should not be taken too seriously. Indeed, following these statistical estimates, it appeared clear to us that their uncertainties were sufficiently large that, on their basis, the real importance of channel 2 as a feeder to the cooler white dwarf population could not be inferred.

A second significant objection was raised by Volker Weidemann who pointed out that a (then) recent estimate of his indicated that some 80 % of the pre-white dwarf population consists of H-rich central stars of planetary nebulae (CPN's), the rest being PG1159 stars (see again the discussion at the end of the Fontaine & Wesemael 1987 paper). For proponents of primordial scenarios, this naturally suggests that the atmospheric composition of a white dwarf is determined at birth: the H-rich CPN's evolving into DA's and the PG1159 stars evolving into non-DA stars. However, we reasoned (and the argument is still valid) that CPN's are still obviously losing mass and it is not at all clear that they necessarily evolve into DA stars with thick hydrogen envelopes or even into stars with H-rich atmospheres for that matter. This presumably would be the case if the CPN's had all thick H-burning envelopes. However, there is a result by Kawaler (1988), which, we feel, has not been given enough attention. According to him, if CPN's with thick H-burning envelopes really exist, they should *all* pulsate through a very potent $\epsilon$ mechanism. This is far from the actual observed situation as demonstrated in a survey of 50 objects in which none has been found to be unstable (Hine & Nather 1987). The implication is that CPN's do not have the structure expected from standard evolutionary calculations. Hence, if Kawaler (1988) is right, these progenitors must evolve into something else than thick hydrogen envelope DA's.

## 4.  A new look at the circumstantial evidence

Progress on several fronts has, of course, been made since the original suggestion of Fontaine & Wesemael (1987). In the light of these advances, we first review critically in this section the three pieces of circumstantial evidence discussed above.

Firstly, the new insights brought upon us by the *ROSAT* and *EUVE* experiments (in particular) have forced us to give up the stratified H/He atmosphere model in favor of an alternative model originally proposed by Vennes et al. (1989) to account for the *EXOSAT* EUV spectrum of the hot DA white dwarf Feige 24. These authors were able to rule out the stratified atmosphere hypothesis for that star, and demonstrated instead that its EUV spectrum could be explained in terms of a host of heavy elements with relatively low individual abundances and possibly supported by radiative levitation in the photospheric layers. The latter idea was further developed by Vennes (1992), and now appears to be widely accepted as the basic explanation for the presence of the well-documented EUV opacity source in hotter DA stars (e.g., Vennes & Fontaine 1992; Barstow et al. 1993;

Tweedy 1993). The detailed calculations carried out by Chayer et al. (1994, 1995a,b) lend very strong support to the idea that the atmospheres of hot white dwarfs are contaminated by complexes of heavy elements supported in part by radiative levitation and possibly subjected to weak stellar winds. These findings do not formally rule out the possibility that *some* hot DA stars could have stratified atmospheres, but, to our knowledge, the signature of such an object is yet to be found in the current *EUVE* spectroscopic archives.

The idea *itself* of stratified H/He atmospheres in white dwarfs took a serious hit with the realization by Bergeron et al. (1994) that even DAO stars do not show compositional stratification in their atmospheres (see also Napiwotzki & Schönberner 1993). Contrary to hot DA stars for which helium abundances had to be *inferred* (under the assumption that this element provided most of the EUV opacity), helium is spectroscopically visible in DAO stars. In order to explain its presence, it was suggested that a thin hydrogen layer would rest in diffusive equilibrium on top of helium. Hence, it was believed by many, including ourselves, that DAO's were the epitomes of stratified atmosphere stars. This is not the case as demonstrated by Bergeron et al. (1994).

The second piece of evidence to be discussed anew is the constraints on $q(He)$ inferred by Pelletier et al. (1986) to account for the DQ spectral type. The carbon pollution model proposed by these authors appears to be the only successful one to date and is still very much viable (see, e.g., Weidemann & Koester 1995). However, the results are model dependent. In particular, there has been considerable progress made in recent years on the constitutive physics of dense plasmas, and these new developments were discussed at length during IAU Colloquium No. 147 (Chabrier & Schatzman 1994). Of direct interest here is the actual extension of the helium superficial convection zone in relatively cool He-rich white dwarfs, which depends primarily on the opacity and equation of state used (given a convective efficiency). In order to test if the results of Pelletier et al. (1986) are significantly altered by the new physics, we computed envelope models in which the equation of state of Fontaine, Graboske, & Van Horn (1977; FGV) for H, He, and C was replaced by those of Saumon, Chabrier, & Van Horn (1995; SCV) for H and He and Fontaine (1993; F) for C. Likewise, we replaced the Los Alamos radiative opacity data (Cox & Stewart 1970; LAO) by the more recent 1995 OPAL data (see Rogers & Iglesias 1992 and Iglesias & Rogers 1996) for the same Iben V (X(He) = 0.999, Z = 0.001) and Weigert V (X(C) = 0.999, Z = 0.001) mixtures considered previously by Pelletier et al. (1986). Furthermore, we included upgrades in the computations of the conductive opacities (Itoh et al. 1983, 1984, 1993a,1993b; Mitake et al. 1984).

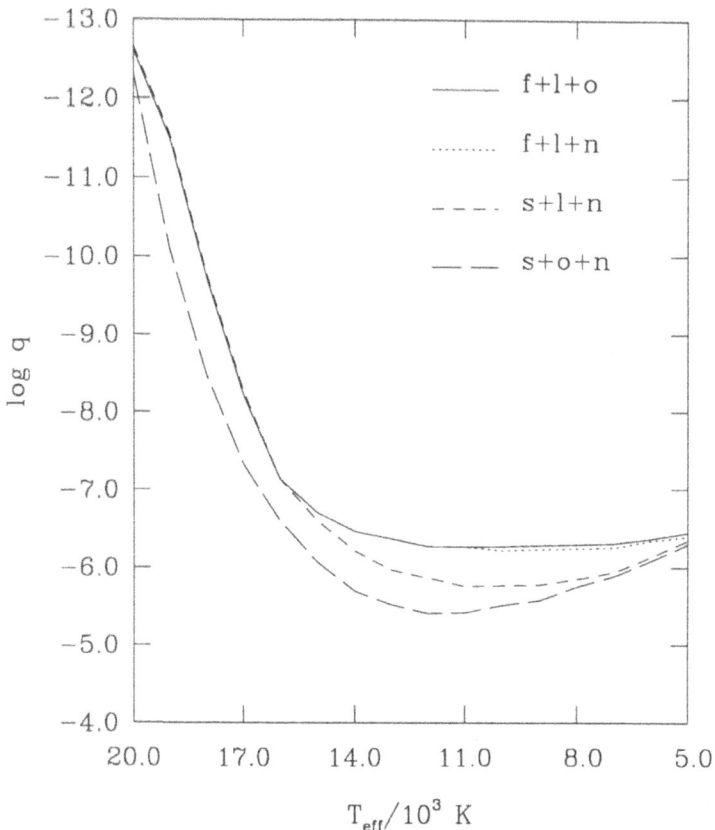

*Figure 1.* Effects of the constitutive physics on the location of the base of the helium convection zone in a typical 0.6 $M_\odot$ He-rich white dwarf model. The calculations assume ML1 convection and there is a trace of heavy elements, Z = 0.001. The symbols f(s) stand for the FGV (SCV+F) equation of state, l (o) for the LAO (OPAL) radiative opacities, and o(n) for the old and new conductive opacities.

The solid curve in Figure 1 corresponds to the location of the base of the superficial helium convection zone in a typical 0.6 $M_\odot$ model computed with ML1 convection and with input physics identical to that used by Pelletier et al. (1986). The other curves illustrate how the successive improvements lower further the base of the convection zone. The effects of improved conductive opacities are shown by the dotted curve; those are combined with the upgrades in the equation of state and lead to the short–dashed curve; finally, the addition of the new radiative opacity data leads to the long–dashed curve. At maximum depth, the convection zone is now $\sim 0.85$ dex deeper than before in log $q$. Using a number a scaling arguments based on the work of Pelletier et al. (1986), it can be shown that this implies a difference of $\Delta \log q(He) \simeq 0.7$ to produce the same amount of carbon pollution. Hence, on the basis of these estimates, we find that helium layers in DQ white dwarfs should have $-3.3 \lesssim \log q(He) \lesssim -2.8$, larger than in Pelletier et al. (1986), but still significantly smaller than the standard value ($\log q(He) \simeq -2$).

We find a puzzling result, however, when using the OPAL radiative opacity tables for *pure* helium (and carbon) instead of those with a trace of heavy elements (Z = 0.001) as above. Figure 2 illustrates an unexpected behavior in that the base of the convection zone keeps sinking with decreasing effective temperature when Z = 0.000, all other things being the same. This is contrary to what we have observed in all white dwarf models so far. Furthermore, the base of the convection zone is deeper than in the models with Z = 0.001, which suggests that thicker helium layers than before are required to produce the same amount of carbon pollution. We estimate, for instance, that the maximum amount of carbon seen in DQ stars could be maintained in a 8000 K (11,000 K) model if $\log q(He) \simeq -1.95(-2.37)$. These estimates are essentially compatible with the canonical value (and see the paper by MacDonald, Hernanz, and Jose in these Proceedings). We note, however, that the OPAL tables for pure helium corresponds to a *minimal* opacity since, for example, helium molecules are not included (F. Rogers 1995, private communication). The same is true for the pure carbon data, so that, taking also into account the absence of metals, the latter values of $q(He)$ may be too extreme. Until this unexpected sensitivity of the carbon pollution model on the metal content in the envelopes of DQ stars is thoroughly investigated, the constraints derived on $q(He)$ will remain uncertain.

Thirdly, the controversy around the possibility of using nonadiabatic results to constrain $q(H)$ in ZZ Ceti stars seems to have subsided. It now appears that numerical inadequacies plagued *all* early nonadiabatic calculations as demonstrated by Brassard et al. (1991), Fontaine et al. (1992), Bradley & Winget (1994a), and Fontaine et al. (1994). The breakthrough on this front came from the utilization by Fontaine et al. (1994) of a new, robust, linear nonadiabatic pulsation code based on a Galerkin finite-element algorithm to analyze the *exact same models* as Winget et al. (1982), Winget & Fontaine (1982), and Bradley et al. (1989). The use of this new pulsation code, in conjunction with older ones, clearly revealed that past nonadiabatic studies suffered from spurious results. Fontaine et al. (1994) established firmly that the Tassoul et al. (1990) models lead to a blue edge which is *not* sensitive to the hydrogen layer mass. More recent calculations, incorporating the latest improvements on the constitutive physics front, confirm this result (Brassard, Fontaine, & Bergeron 1996). Consequently, nonadiabatic results cannot provide constraints on the thickness of the hydrogen

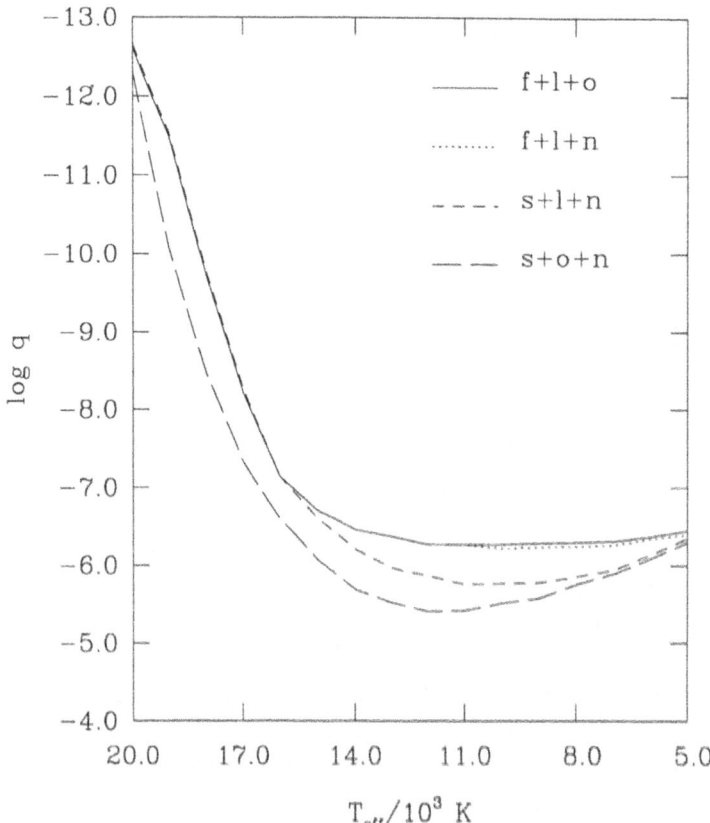

*Figure 2.* Location of the base of the helium convection zone in a typical 0.6 $M_\odot$ pure He white dwarf model. The calculations assume ML1 convection and there is *no* heavy elements, Z = 0.000.

layer in pulsating DA white dwarfs.

It thus appears that the circumstantial evidence that we invoked, back in 1987, to strenghten our case in favor of a dominant channel 2 has either evaporated or must be further investigated. However, asteroseismology based on period-fitting arguments has started to pay dividends, and values of $q(H)$ and $q(He)$ in pulsating white dwarfs *have* been inferred. Lest the reader remains skeptical in view of the past saga concerning nonadiabatic results, we point out that pulsation periods depend essentially on the mechanical structure of a star and can be calculated quite accurately in the adiabatic approximation. In contrast, nonadiabatic calculations are sensitive to the thermal structure and are characteristically much more uncertain.

The best studied ZZ Ceti star is G117-B15A for which a complete aster-oseismological analysis has been carried out by Fontaine & Brassard (1994) and Fontaine et al. (1995; see also Robinson et al. 1995 and Fontaine et al. 1996). The values of $q(H) \simeq 10^{-5.9}$ and $q(He) \simeq 10^{-2.7}$ have been inferred for that star. Other results for ZZ Ceti stars are $q(H) \simeq 10^{-4.0}$ in G226-29 (Fontaine et al. 1992), $q(H) \gtrsim 10^{-6.4}$ in GD 165 (Bergeron et al. 1993), $q(H) \simeq 10^{-9.7}$ in GD 154 (Pfeiffer et al. 1996), and $q(H) \simeq 10^{-4.0}$ in G29-38 (Kleinman et al. 1996). A detailed analysis of the pulsating DB star GD 358 has also been carried by Bradley & Winget (1994b; see also Winget et al. 1994) who inferred a value $q(He) \simeq 10^{-5.7}$. Contrary to the appre-hensions expressed by these authors, this last result is not at variance with the constraints on $q(He)$ derived from the carbon pollution model because He/C separation is not necessarily completed at the relatively high effective temperature of GD 358. It does conflict, however, with the idea of a fixed value of $q(He) \simeq 10^{-2}$ at birth. We note that the claim of Clemens (1995) that *all* ZZ Ceti stars have a thick hydrogen layer $q(H) \simeq 10^{-4}$ is based on an *assumption* of his, and is not backed by the few detailed asteroseismo-logical analyses that currently exist. To summarize, asteroseismology does suggest a *spectrum* of hydrogen and helium layer masses in white dwarfs, from thick to thin, at odds with primordial scenarios.

## 5.  And what about the main evidence?

In this section, we reexamine the main evidence in favor of interactions between spectral types; in order, the turnover in favor of non-DA stars at low effective temperatures, the existence of the DB gap, and the paucity of hot DA's at very high effective temperatures.

The apparent turnover in the white dwarf population in favor of non-DA stars at the cool end of the cooling sequence has been interpreted by Sion (1984; see also Greenstein 1986) as evidence in favor of convective mixing transforming DA's into He-rich stars. This is based on the calculations of Koester (1976), Vauclair & Reisse (1976), and D'Antona & Mazzitelli (1979) which all indicate that, for this process to work, thin hydrogen layers are required. For instance, since the base of the hydrogen convection zone in a $\sim 0.6 \, M_\odot$ DA white dwarf reaches a maximum depth $q \sim 10^{-6}$ during the evolution (see below), it follows that, for mixing to occur in such a star, the hydrogen layer mass *must* be smaller than this limit.

We provide, in Figure 3, an updated picture of the relative distribution of the two types of white dwarfs as a function of effective temperature be-low the DB gap based on substantially improved estimates of the effective temperatures (Bergeron et al. 1996). In our presentation, the dashed hori-zontal line corresponds to the standard (1:3) ratio predicted by primordial

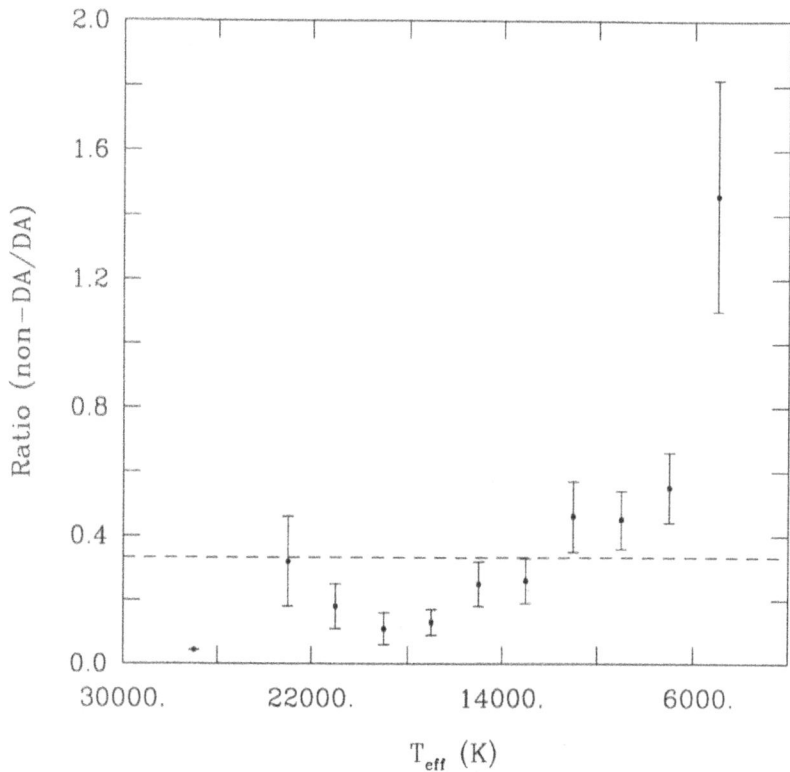

*Figure 3.* Observed distribution of the number ratio of non-DA to DA stars at effective temperatures lower than the red edge of the DB gap.

scenarios. It is not clear whether or not all the variations exhibited by the non-DA/DA number ratio about this value are significant, except for the obvious deficiency at high effective temperatures which is due to the influence of the DB gap. If we adopt a $2\sigma$ criterion, then Figure 3 suggests a deficiency in the two bins at 19,000 K and 17,000 K, and an enhancement in the coolest bin. The former case is quite puzzling as no mechanism is known that would transform DA stars into DB stars as they cool from 19,000 K to 16,000 K (see Bergeron et al. 1996). On the other hand, the enhancement in the coolest bin *could* be due to mixing. If one insists on this interpretation, then the bulk of DA stars would mix around $T_{eff} \sim 6,000$ K.

In order to compare with theoretical results, we also provide, in Figure 4, new values of the fractional mass depth, log $q$, at the base of the hydrogen convection zone in a DA white dwarf as a function of the effective temperature and surface gravity. These are upgraded results taking into account the improvements in the constitutive physics already alluded to in the construction of our Figure 1, as well as the important effects of using detailed atmosphere models as boundary conditions. ¿From Figure 4, one can read, for example, the value of $q(H)$ corresponding to a given surface gravity and mixing temperature by identifying it to the fractional mass depth at the base of the convection zone. Thus, if the bulk of DA stars indeed mix around $T_{\text{eff}} \sim 6,000$ K, then a typical value of the hydrogen layer mass is $q(H) \sim 10^{-7.7}$ (assuming stars with $\sim 0.6\ M_\odot$, $\log g = 8$). We recall, however, that there are some doubts as to the real efficiency of convective mixing in turning cool DA stars into He-rich white dwarfs as advocated by Sion (1984) and others since (see Bergeron et al. 1990 and Fontaine & Wesemael 1991).

We note, in the present context, the important work of Bergeron et al. (1996), dedicated to the chemical evolution of very cool ($T_{\text{eff}} \lesssim 7,500$ K) white dwarfs. These authors have unveiled a picture of cool white dwarfs that is complex and is at odds with the simple mixing scenario discussed above. They show that, when viewed at higher resolution, the distribution of the non-DA/DA number ratio illustrated in Figure 3 on the cool side must be reinterpreted. In particular, they find that the temperature distribution of H- and He-rich white dwarfs is inhomogeneous and is characterized by the puzzling presence of a *non-DA gap* in the range 6,000 K $\gtrsim T_{\text{eff}} \gtrsim 5,000$ K. Their investigation leaves no doubt as to the reality of interactions and changes between the two main atmospheric compositions of white dwarfs on the cool side of the cooling sequence. In order to account for these changes, Bergeron et al. (1996) propose an interplay of convective mixing with a new mechanism by which hydrogen is accreted onto the surface of He-rich white dwarfs while remaining spectroscopically invisible. This would be feasible, according to the authors, through a phase transition between the neutral hydrogen atom and the metallic state. Of prime importance in the present discussion, Bergeron et al. (1996) further infer that their observations can be explained only if $q(H)$ in DA stars covers a *spectrum* of values, from thick to thin.

One piece of evidence in favor of spectral evolution that has remained unshaken over the last decade is the existence of the DB gap. Of course, its existence, proper, does not constrain channel 1 as such, but is a proof that spectral evolution does occur in channel 2. However, by examining the distribution of DA white dwarfs, we uncovered a suggestive excess of DA stars right in the range of effective temperatures corresponding to the DB gap. This can be easily observed in Figure 2 of Liebert (1986), but has apparently remained unnoticed. Taken at face value, this excess must correspond to those stars born through channel 2, and which have turned

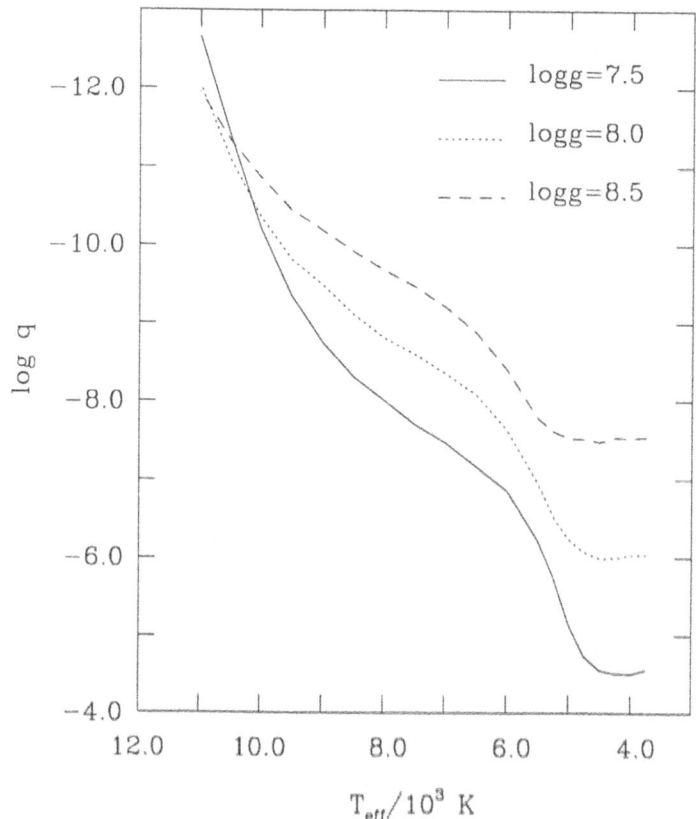

*Figure 4.* The location of the base of the hydrogen convection zone in typical pure hydrogen (DA) white dwarf models of various surface gravities. The calculations assume ML2 convection and incorporate detailed model atmospheres as upper boundary conditions.

into DA stars during the course of their evolution at high effective temperatures. If this excess is proven significant, it implies that channel 2 is an important feeder channel to the white dwarf population as a whole. Conversely, channel 1 cannot be the dominant channel, at least not in the (3:1) proportions suggested by primordial scenarios.

We readdress also the question of the paucity of very hot DA white dwarfs. On the basis of the observations and analyses available in 1987, no DA white dwarf was known with $T_{\text{eff}} \gtrsim 70{,}000$ K (Holberg 1987). Thanks to the work of Napiwotzki & Schönberner (1993) and Napiwotzki (1995), new insights at this problem were provided. It now appears that the paucity of very hot DA's may not be real at all! These authors make a strong case for the evolutionary connection between the H-rich CPN's and the DAO stars (see also Bergeron et al. 1994). They find a continuous sequence between these types of stars. Furthermore, they also infer a (1:4) ratio between He-rich and H-rich CPN's, a value consistent with primordial scenarios. Taken at face value, these findings constitute the strongest argument against that part of the spectral evolution model proposed by Fontaine & Wesemael (1987) in which *most* cooler white dwarfs would descend from PG1159/DO stars. However, as indicated above, there remains the question of whether or not all H-rich CPN's end up *necessarily* as thick hydrogen envelope DA white dwarfs.

In this connection, the recent results of Vennes et al. (1996) are particularly interesting. These authors have analyzed a sample of 110 EUV-selected DA stars in order to determine their space density, population age, and mass distribution. Because their sample is significantly hotter (70,000 K $\gtrsim T_{eff} \gtrsim$ 25,000 K) than the well-established BSL sample of 129 DA stars discussed by Bergeron, Saffer, & Liebert (1992; 40,000 K $\gtrsim T_{eff} \gtrsim$ 14,000 K), it is more sensitive to the puffing up effect exhibited by the outer hydrogen layer in DA white dwarfs at high effective temperatures. Vennes et al. (1996) have exploited this sensitivity and, by requiring that their inferred mass distribution matches that of the cooler BSL sample, they determined that hot DA white dwarfs, in bulk, must have thin hydrogen layers. At the level of accuracy reached, this means any value from $q(H) \simeq 10^{-6}$ and smaller. In addition, Vennes et al. (1996) have uncovered a population of ultramassive ($M \gtrsim 1.1 M_{\odot}$) DA stars in their sample. These 10 objects would exceed the Chandrasekhar mass in the presence of a standard, thick hydrogen layer ($q(H) \simeq 10^{-4}$), and, consequently, must have thin layers. The findings of Vennes et al. (1996) not only constitute a true shot in the arm to the general idea of thin layers in white dwarfs, but also provide possible answers to the question posed above as to the fate of H-rich CPN's and the problem raised by Kawaler (1988) as to the lack of observed pulsational instabilities in these objects.

## 6. Conclusion

Our survey indicates that, in contrast to the suggestion of Fontaine & Wesemael (1987), white dwarf evolution is *not* dominated by channel 2. In retrospect, our model may now appear somewhat naive since, after all, it has been known for a long time that there are many different ways to form white dwarfs. Nevertheless, back in 1987, it had the merit of providing an unified picture and, on the lighter side, of fitting on a normal-size (!) T-shirt (see Shipman 1987). On the other hand, the case for thin layers in at least a substantial fraction of the white dwarfs can still be defended with vigor. Indeed, asteroseismological results based on adiabatic calculations suggest a spectrum of hydrogen layer masses $10^{-10} \lesssim q(H) \lesssim 10^{-4}$ in the ZZ Ceti stars, and nonstandard (smaller than predicted by primordial scenarios) values of $q(He)$ in at least a ZZ Ceti star (G117-B15A) and a DB pulsator (GD 358). In addition, if one insists in interpreting the apparent turnover in favor of non-DA stars at low effective temperatures in terms of a mixing event due to the coalescence of two convection zones, one finds that the bulk of DA stars should mix around $T_{eff} \sim 6,000$ K and they should typically have $q(H) \sim 10^{-7.7}$. As indicated above, there are problems with this simple picture and, in particular, the work of Bergeron et al. (1996) implies a much more complicated situation. Nevertheless, the latter authors still advocate a spectrum of values of $q(H)$ in cool DA white dwarfs. Moreover, the existence of the DB gap remains the proof "par excellence" that spectral changes do occur in channel 2. If confirmed, the excess of DA stars observed in the range of effective temperatures of the DB gap suggests strongly that channel 2 is not a negligible avenue for forming DA stars. Furthermore, these DA's must have relatively thin hydrogen layers. Irrespective of the actual channels of formation, the study of Vennes et al. (1996) of a EUV-selected sample of hot DA stars strenghtens considerably the case for thin hydrogen layers in these stars.

We find that the most important result that emerges from the recent developments summarized in this review is the realization that white dwarfs show a *spectrum* of hydrogen and helium layer masses. This is incompatible with the current tendency of using only thick envelope models in evolutionary calculations. Unfortunately, and despite the significant progress made in the study of CPN's (Napiwotzki 1995), PG1159 stars (Rauch et al., these Proceedings), and DO stars (Dreizler et al., these Proceedings), the relative importance of the various feeder channels to white dwarfs is still not known with any certainty, and this may cloud our detailed understanding of white dwarf evolution for many years to come. The task ahead for all of us will be to determine this relative importance.

We wish to acknowledge the essential contributions of numerous friends and collaborators to our understanding of the spectral evolution of white dwarfs, a most fascinating and challenging subfield of stellar physics. Special thanks are due to our former and current students and associates. This work was supported in part by the NSERC Canada and by the Fund FCAR (Québec).

# References

Barstow, M. A., Fleming, T.A., Diamond, C.J., Finley, D.S., Sanson, A.E., Rosen, S.R., Koester, D., Marsh, M.C., Holberg, J.B., & Kidder, K. 1993, *M.N.R.A.S.*, **264**, 16

Bergeron, P., Saffer, R.A., & Liebert, J. 1992, *Ap.J.*, **394**, 228

Bergeron, P., Ruiz, M.T., & Leggett, S.K. 1996, *Ap.J.Suppl.*, in press

Bergeron, P., Wesemael, F., Fontaine, G., & Liebert, J. 1990, *Ap.J.(Letters)*, **351**, L21

Bergeron, P., Wesemael, F., Beauchamp, A., Wood, M. A., Lamontagne, R., Fontaine, G., & Liebert, J. 1994, *Ap.J.*, **432**, 305

Bergeron, P. et al. 1993, *A.J.*, **106**, 1987

Bradley, P.A., & Winget, D.E. 1994a, *Ap.J.*, **421**, 236

Bradley, P.A., & Winget, D.E. 1994b, *Ap.J.*, **430**, 850

Bradley, P.A., Winget, D.E., & Wood, M.A. 1989, in IAU Colloq. 114, *White Dwarfs*, ed. G. Wegner (New York: Springer), p. 286

Brassard, P., Fontaine. G., & Bergeron, P. 1996, in preparation

Brassard, P., Fontaine, G., & Wesemael, F. 1989, in IAU Colloq. 114, *White Dwarfs*, ed. G. Wegner (New York: Springer), p. 263

Brassard, P., Fontaine, G., Wesemael, F., Kawaler, S.D., & Tassoul, M. 1991, *Ap.J.*, **367**, 601

Chabrier, G., & Schatzman, E. 1994, IAU Colloqium 147, *The Equation of State in Astrophysics* (Cambridge: Cambridge Univ. Press)

Chayer, P., LeBlanc, F., Fontaine, G., Wesemael, F., Michaud, G., & Vennes, S. 1994, *Ap.J.(Letters)*, **436**, L161

Chayer, P., Fontaine, G., & Wesemael, F. 1995a, *Ap.J.Suppl.*, **99**, 189

Chayer, P., Vennes, S., Pradhan, A. K., Thejll, P., Beauchamp, A., Fontaine, G., & Wesemael, F. 1995b, *Ap.J.*, **454**, 429

Clemens, J.C. 1995, in *White Dwarfs*, eds. D. Koester & K. Werner (Berlin: Springer), p. 294

Cox, A.N., & Stewart, J.N. 1970, *Ap.J.Suppl.*, **19**, 261

Cox, A.N., Starrfield, S.G., Kidman, R.B., & Pesnell, W.D. 1987, *Ap.J.*, **317**, 303

D'Antona, F., & Mazzitelli, I. 1979, *Astr.Ap.*, **74**, 161

D'Antona, F., & Mazzitelli, I. 1987, in IAU Colloq. 95, *The Second Conference on Faint Blue Stars*, eds. A.G.D. Philip, D.S. Hayes, & J. Liebert (Schenecdaty: Davis), p. 635

Fleming, T.A., Liebert, J., & Green, R.F. 1986, *Ap.J.*, **308**, 176

Fontaine, G., 1993, unpublished

Fontaine, G., & Brassard, P. 1994, in ASP Conf. Ser. 57, *Stellar and Circumstellar Astrophysics*, eds. G. Wallerstein & A. Noriega-Crespo (San Franscisco: ASP), p. 195

Fontaine, G., & Wesemael, F. 1987, in IAU Colloq. 95, *The Second Conference on Faint Blue Stars*, eds. A.G.D. Philip, D.S. Hayes, & J. Liebert (Schenecdaty: Davis), p. 319

Fontaine, G., & Wesemael, F. 1991, in IAU Symp. 145, *Evolution of Stars: The Photospheric Abundance Connection*, eds. G. Michaud & A. Tutukov (Dordrecht: Kluwer), p. 421

Fontaine, G., Graboske, H.C., Jr., and Van Horn, H.M. 1977, *Ap.J.Suppl.*, **35**, 293

Fontaine, G., Brassard, P., Bergeron, P., & Wesemael, F. 1992, *Ap.J.(Letters)*, **399**, L91

Fontaine, G., Brassard, P., Bergeron, P., & Wesemael, F. 1996, *Ap.J.*, **469**, 328

Fontaine, G., Brassard, P., Bergeron, P., Wesemael, F., Vauclair, G., Pfeiffer, B., & Dolez, N. 1995, in Proc. 4th CFHT Users' Meeting, ed. M. Azzopardi (Kamuela: CFHT Corp.), p. 89

Fontaine, G., Brassard, P., Wesemael, F., & Tassoul, M. 1994, *Ap.J.(Letters)*, **428**, L61

Greenstein, J.L. 1986, *Ap.J.*, **304**, 334

Hine, B.P., & Nather, R.E. 1987, in IAU Colloq. 95, *The Second Conference on Faint Blue Stars*, eds. A.G.D. Philip, D.S. Hayes, & J. Liebert (Schenecdaty: Davis), p. 619

Holberg, J.B. 1987, in IAU Colloq. 95, *The Second Conference on Faint Blue Stars*, eds. A.G.D. Philip, D.S. Hayes, & J. Liebert (Schenecdaty: Davis), p. 285

Iben, I. Jr. 1984, *Ap.J.*, **277**, 333

Iben, I. Jr., & Tutukov, A.V. 1984, *Ap.J.*, **282**, 615

Iglesias, C.A., and Rogers, F.G. 1996, *Ap.J.*, **464**, 943

Itoh, N., Hayashi, H., and Kohyama, Y. 1993a, *Ap.J.*, **418**, 405

Itoh, N., and Kohyama, Y. 1993b, *Ap.J.*, **404**, 268

Itoh, N., Kohyama, Y., Matsumoto, N., and Seki, M. 1984, *Ap.J.*, **285**, 758

Itoh, N., Mitake, S., Iyetomi, H., and Ichimaru, S. 1983, *Ap.J.*, **273**, 774

Kawaler, S.D. 1988, *Ap.J.*, **334**, 220

Kleinman, S. et al. 1996, preprint

Koester, D. 1976, *Astr.Ap.*, **52**, 415

Koester, D. 1989, in IAU Colloq. 114, *White Dwarfs*, ed. G. Wegner (New York: Springer), p. 206

Koester, D., & Schönberner, D. 1986, *Astr.Ap.*, **154**, 125

Liebert, J. 1986, in IAU Colloq. 87, *Hydrogen-Deficient Stars and Related Objects*, eds. K. Hunger, D. Schönberner, & N.K. Rao, (Dordretch: Reidel), p. 367

Liebert, J., Fontaine, G., & Wesemael, F. 1987, *Mem. Soc. Astron. Ital.*, **58**, 17

MacDonald, J., & Vennes, S. 1991, *Ap.J.*, **371**, 719

Mitake, S., Ichimaru, S., and Itoh, N. *Ap.J.*, **277**, 375

Napiwotzki, R. 1995, in *White Dwarfs*, eds. D. Koester & K. Werner (Berlin: Springer), p. 176

Napiwotzki, R., & Schönberner, D. 1993, in *White Dwarfs: Advances in Observations and Theory*, ed. M. Barstow, NATO ASI Ser., Vol. 403 (Dordrecht: Kluwer), p. 99

Pelletier, C., Fontaine, G., Wesemael, F., Michaud, G., & Wegner, G. 1986, *Ap.J.*, **307**, 242

Pfeiffer, B. et al. 1996, *Astr.Ap.*, **314**, 182

Robinson, E.L. et al. 1995, *Ap.J.*, **438**, 908

Rogers, F.G., & Iglesias, C. 1992, *Ap.J.*, **401**, 361

Saumon, D., Chabrier, G. and Van Horn, H.M., 1995, *Ap.J.Suppl.*, **99**, 713

Shipman, H.L. 1987, in IAU Colloq. 95, *The Second Conference on Faint Blue Stars*, eds. A.G.D. Philip, D.S. Hayes, & J. Liebert (Schenecdaty: Davis), p. 273

Shipman, H.L. 1989, in IAU Colloq. 114, *White Dwarfs*, ed. G. Wegner (New York: Springer), p. 220

Shipman, H.L. 1993, in *White Dwarfs: Advances in Observations and Theory*, ed. M. Barstow, NATO ASI Ser., Vol. 403 (Dordrecht: Kluwer), p. 555

Sion, E.M. 1984, *Ap.J.*, **282**, 612

Sion, E.M. 1986, *P.A.S.P.*, **98**, 821

Tassoul, M., Fontaine, G. and Winget, D.E. 1990, *Ap.J.Suppl.*, **72**, 335

Tweedy, R.W. 1993, in *White Dwarfs: Advances in Observations and Theory*, ed. M. Barstow, NATO ASI Ser., Vol. 403 (Dordrecht: Kluwer), p. 317

Vauclair, G., & Reisse, C. 1977, *Astr.Ap.*, **61**, 415

Vennes, S. 1992, *Ap.J.*, **390**, 590

Vennes, S., & Fontaine, G. 1992, *Ap.J.*, **401**, 288

Vennes, S., Pelletier, C., Fontaine, G., & Wesemael, F. 1988, *Ap.J.*, **331**, 876

Vennes, S., Chayer, P., Fontaine, G., & Wesemael, F. 1989, *Ap.J.(Letters)*, **336**, L25

Vennes, S., Thejll, P.A., Galvan, R.G., & Dupuis, J. 1996, *Ap.J.*, in press

Weidemann, V., & Koester, D. 1995, *Astr.Ap.*, **297**, 216

Wesemael, F., Green, R.F., & Liebert, J. 1985, *Ap.J.Suppl.*, **58**, 379

Winget, D.E., & Fontaine, G. 1982, in *Pulsations in Classical and Cataclysmic Variable Stars*, eds. J.P. Cox & C.J. Hansen (Boulder: Univ. Colorado Press), p. 46

Winget, D.E., Van Horn, H.M., Tassoul, M., Hansen, C.J., Fontaine, G., & Carroll, B.W. 1982, *Ap.J.(Letters)*, **252**, L65

Winget, D.E. et al. 1994, *Ap.J.*, **430**, 839

## Discussion

*I. Iben*: You have emphasized the importance of a jump in the non-DA/DA ratio at low surface temperature and interpreted this as in terms of mixing processes. On the other hand, Ruiz in her talk pointed out that there are no non-DA stars in her sample in the surface temperature range 5000-6000 K. How do you explain this?

*G. Fontaine*: What I ment to say is that if one insists (as has been done customarily) in interpreting the observed turnover in favor of non-DA stars at low effective temperatures in terms of mixing events, on is forced to the conclusion that the bulk of DA stars should mix around $T_{eff} \sim 6000$ K and they should typically have $q(H) \sim 10^{-7.7}$. However, the work of Bergeron, Ruiz, and Leggett reveals that there are problems with this simple picture. As summarized by Maria-Teresa, the "fine structure" of spectral evolution at low $T_{eff}$ suggests a much more complicated situation. Thus, when grouped in narrow enough effective temperature bins, cool white dwarfs appear to be dominated by H-rich atmospheres in the 5000-6000 K range. Nevertheless, in order to explain this peculiar distribution, Bergeron et al. do invoke a spectrum (from thick to thin) of values of $q(H)$ in cool DA stars.

*V. Weidemann*: Two years ago Clemmens convinced us about thick-layer models. You did not mention this?

*G. Fontaine*: Although I have not seen the results of Clemens in published form, I understand that he assumed that all ZZ Ceti stars have the same amount of hydrogen (the "canonical" thick value $q(H) = 10^{-4}$), and that used the observed periods to infer "asteroseismological" values of the mass and effective temperature. Currently, there is only a handful of asteroseismological studies of ZZ Ceti stars which can be considered reliable, but the few that exist contradict the assumption of Clemens. At the moment, the available evidence suggests a spectrum of hydrogen layer masses in ZZ Ceti stars, $10^{-4} \geq q(H) \geq 10^{-10}$.

*H. Shipman*: To determine the hydrogen layer masses, you need to identify the individual modes in ZZ Ceti stars. How well identified are these modes? Specifically, can you exclude the possibility that all DA white dwarfs have the same envelope mass?

*G. Fontaine*: Mode identification is indeed the key issue. The best studied case is G117-B15A for which we have identified (on the basis of CFHT observations) three pulsation modes. They all have $l = 1$ and they are consecutive radial order modes ($k=$ 1, 2, and 3). The single mode observed in G336-39 is also well identified ($l = 1, k = 1$). While similar results are being obtained for other simple pulsators, it is true to say that "mode identification" in other ZZ Ceti stars has been based, so far, on circumstancial evidence only. Nevertheless the cases of G117-B15A and G336-39 already demonstrate that ZZ Ceti stars do not have the same hydrogen envelope mass. For the former object, an analysis leads to $q(H) \sim 10^{-5.9}$, while for the latter, we find $q(H) \sim 10^{-4.0}$.

# POLYATOMIC CARBON MOLECULES IN VERY COOL HELIUM-RICH WHITE DWARFS

IRMELA BUES AND TURGUT ASLAN

*Astronomisches Institut der Universität Erlangen-Nürnberg*
*Dr. Remeis Sternwarte,*
*Sternwartstr. 7, 96049 Bamberg, Germany*

**Abstract.** The structure of helium-rich very cool white dwarf atmospheres is investigated if the opacity in the visible and infrared region of the spectrum is mainly due to diatomic and polyatomic species of carbon and hydrogen molecules. For various abundance ratios of H/He, C/He, O/He and N/He the gradients of $C_3$, $C_2H$ and $C_2H_2$ in the outer layers of model atmospheres with $T_{eff} \leq 5000K$, $\log g = 8.0$ have been computed as well as fluxes for comparison with observed spectra. It is shown that even a reduced oxygen abundance is far from negligible for the overall run of partial pressures.

## 1. Introduction

Improved observations of instrinsically faint objects have increased the number of very cool white dwarfs. Photometry as well as spectroscopy by Ruiz et al. (1992,1996) revealed a variety of spectral shapes ranging from pure continuum to broadened and shifted features of hydrogen, neutral helium or of carbon bands when they could be identified. Several objects show strange depressions of infrared colours which could not be accurately reproduced by hotter model atmospheres and quasicontinuous spectra. Ruiz et al. (1992) tried the assumption of hydrogen-rich composition and included collision-induced absorption of $H_2$, $H_2^+$ and $H_3^+$ in the overall opacity and the equation of state. Positive results have been obtained for objects showing the $H\alpha$-profile. For objects with stronger features, Schmidt et al. (1995) proposed a composition of hydrogen and carbon in order of $C_2H$ to be formed. Our experience with helium-rich model atmospheres with various ratios of carbon, oxygen and hydrogen enabled us to calculate conditions

*I. Isern et al. (eds.), White Dwarfs, 193–198.*

for the formation of $C_2H$, $C_3$ and $C_2H_2$ in helium-rich atmospheres. New spectra in the red region were taken for comparison.

## 2. Model atmospheres

Earlier helium-rich model atmosphere calculations for cool white dwarfs showed already (Bues, 1973) for $T_{eff}$=6000K, and a ratio of He/C=300, C/O=300 the occurence of $C_3$ in the outer layers, a factor of 2 more abundant than $C_2$ and neutral carbon. Comparison with observed spectra of G99-37, the magnetic white dwarf with mixed H/He/C/O composition gave a satisfactory fit (Bues, Karl-Dietze, 1995) for a ratio of He/H=1000, C/H=35.5, C/O=300 and confirmed the shifted blue feature of $C_3$ around 4000Å. That is why we started computation for even lower $T_{eff}$'s, where the equation of state as well as pressure effects on the broadening and shift of spectral bands would increase. Aslan and Bues (1997) showed the structure of a $T_{eff}$=5000K, log $g$=8.0 model atmosphere with increased pressure and comparison with LHS1126. $C_2$ can still be responsible for spectral features observed around 5000Å, yet the accurate position is determined by a pressure shift.

New flux constant model atmospheres have been calculated for extremely helium-rich compositions in the range 5000K> $T_{eff}$ >4000K, log $g$=8.0, with varied relative abundances of H/He, C/He, O/He and N/He, where oxygen is most important for gas pressure.

The possible H, C, O reaction processes were critically reviewed. According to the presence of high gas pressure and a ratio of C/H$\geq$1 in agreement with $C_2$ and $C_3$ features some formation processes have been preferred compared to those in hydrogen-rich objects:

$C_2+C\rightarrow C_3$
$C_2+H\rightarrow C_2H$
$C_2+H_2 \rightarrow C_2H_2$

The other processes for formation of $C_2H$ and $C_2H_2$ like

$C_2+H_2 \rightarrow C_2H+H$
$C_2H+H\rightarrow C_2H_2$
$C_2H+H_2 \rightarrow C_2H_2+H$   have been omitted.

The constants of dissociation for $C_3$ and $C_2H_2$ were taken from Tsuji (1964), for $C_2H$ from Perić et al. (1990). The computation of dissociative equilibria was solved by an iterative scheme. For absorption, the molecules were included in a smeared line approximation, relevant for white dwarfs with high pressure. For $C_2H$, transitions calculated by Reimers et al. (1985) and by Perić et al. (1992) with ab initio methods have been included in our calculations. They are at 9500 cm$^{-1}$ (= $1.05\mu m$) and 5500 cm$^{-1}$ (= $1.82\mu m$) and of Perić et al. (1992) at $1.78\mu m$, $1.81\mu m$ and $2.61\mu m$.

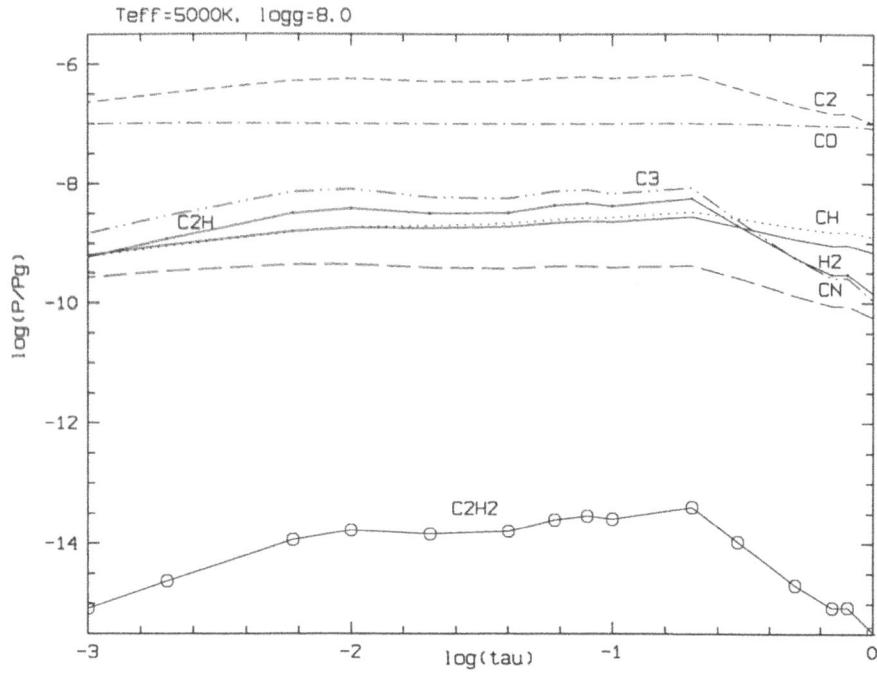

*Figure 1.* Pressure fractions of various diatomic and polyatomic molecules versus the optical depth for $T_{eff}$=5000K, log $g$=8.0, log(H/He)=-5, log(C/He)=-4.5, log(O/He)=-7, log(N/He)=-9

Fig. 1 shows relative partial pressures of the molecules $H_2$, $C_2$, CN, CO, CH, $C_3$, $C_2H$ and $C_2H_2$ compared to the total gas pressure versus the optical depth for a model atmosphere with the indicated parameters. The depth scale in the line and band forming region is $\tau$-Rosseland.

For these reduced abundances of C, N and O, CO and $C_2$ are the main constituents of carbon and most important for the structure of the atmosphere, the gas pressure being due to neutral helium. $C_3$ and $C_2H$ are a factor of 100 less abundant, but have the same order of magnitude as $H_2$ and CH. $C_2H_2$ is negligible. For increasing oxygen abundance the polyatomic species are favoured, $C_2H$ is the most abundant molecule.

Fig. 2 shows the structure for the indicated and slightly changed and reduced abundance ratios, yet the lower $T_{eff}$=4500K. Although the ratio of C/O is only a factor of 3, CO is not the most abundant molecule, the polyatomic ones $C_2H$ and $C_3$ become more efficient, even for a steep gradient of temperature and gas pressure in the line- and band-forming region. The ratio C/H is sensitive for the preference of CH over $C_2H_2$. Both figures can

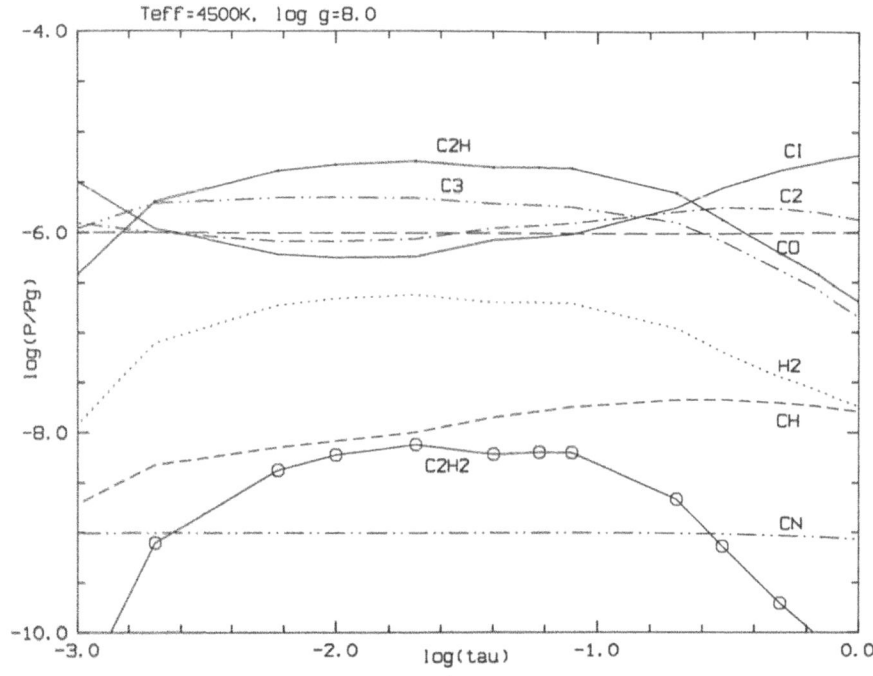

*Figure 2.* Same as Fig. 1 for $T_{\text{eff}}$=4500K, log $g$=8.0, log(H/He)=-5.5, log(C/He)=-5, log(O/He)=-6, log(N/He)=-9

be compared to Fig. 4 of Schmidt et al. (1995) where for a fixed value of $P_g$ hydrocarbons in their relative abundances dependent on temperature only are shown just for a mixture of H, He and C. Our stratification from a flux constant model atmosphere includes the effects on opacity as well, where the O abundance is responsible for the formation of polyatomic molecules due to the formation of CO and the corresponding high pressure.

## 3. Application

Fluxes from a sequence of our model atmospheres have been applied to 8 white dwarfs at the lower end of the cooling sequence. New low resolution spectra in the wavelengths range 4000-9500Å were taken at the ESO 1.52m telescope. For LHS1126, the object with a 12 per cent molecular absorption band at 4990Å±100Å and broad absorption bands at 5450Å±200Å, 6050Å±200Å and 6680Å±100Å in addition to $C_3$ $\lambda$ 4050Å, the analysis, compared to Bergeron et al.(1994) and Scmidt et al. (1995), will be published elsewhere (Aslan and Bues, 1997).

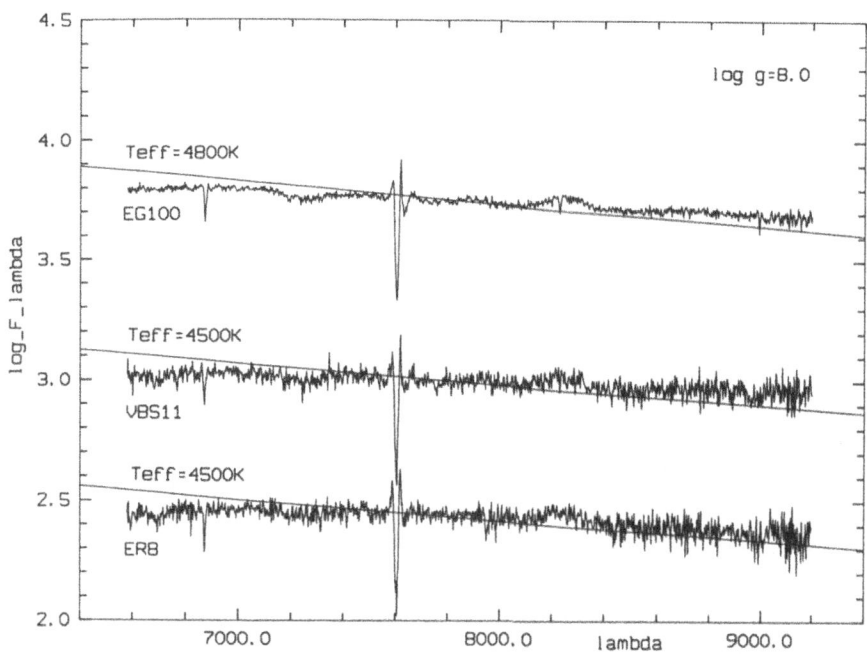

*Figure 3.* Comparison of EG100, VBS11 and ER8 with calculated model atmospheres $T_{eff}$ = 4500K and 4800K, log $g$ = 8.0

With $T_{eff}$ = 5000K, log $g$ = 8.0, log (H/He) = -7, log (C/He) = -6.5, log (O/He) = -7, log (N/He) = -9 the atmosphere consists of nearly pure helium (Aslan et al. 1996) with an immense increase in gas pressure. That is why the features in the visible can be attributed to the high pressure bands of the Swan band and do not need a large magnetic field in connection with a larger carbon abundance. The flux deficiency in the infrared region can be due to $C_2H$ at $1.05\mu m$, $1.78\mu m$ and $1.82\mu m$.

In Fig.3 fluxes of three cooler objects are shown compared to the indicated model atmosphere fluxes taken for abundance ratios log (H/He) = -6.5, log (C/He ) = -7. For EG100, Stancil (1994) determined a helium-rich atmosphere from the nonvisibility of Hα and $T_{eff}$ = 5150K. Our gradient for $T_{eff}$ = 4800K is steeper than his and, in addition, a weak feature around 9100Å  can be identified with a neutral carbon line, thus indicating the presence of metals at all.

ER8, identified as a white dwarf and discussed by Ruiz et al. (1986), and VBS11 have been compared up to now to black body distributions, which indicate effective temperatures less than 4000K. Our flux gradients, however, enable hotter models to reproduce the red region of the spectrum as a consequence of the very weak feature of $C_3$ and the change of slope in the infrared due to the $C_2H$ features.

Thus our computations strengthen the idea that extremely helium-rich white dwarfs remain present at the lower end of the cooling sequence and that traces of carbon, hydrogen and oxygen determine the structure of the atmosphere as well as the corresponding flux gradient by the formation of their diatomic and polyatomic molecules.

## References

Aslan, T., Bues, I. and Karl-Dietze, L. 1996, in *Hydrogen-Deficient Stars*, ed. C.S.Jeffery and U.Heber, CASP, 96, p. 325

Aslan, T., Bues, I. 1997, IAU Symp. 177, in press

Bergeron, P., Ruiz, M.-T., Leggett, S. K., Saumon, D. and Wesemael, F. 1994, Ap. J., 423, 456

Bues, I. 1973, A&A 28, 181

Bues, I. and Karl-Dietze, L. 1995, in *White Dwarfs*, ed. D. Koester and K. Werner, Lec.Not.Phy., 443, p. 201

Perić, M., Peyerimhoff, S. D., Buenker, R. J. 1990, Mol. Phys., 71, 693

Perić, M., Peyerimhoff, S. D., Buenker, R. J. 1992, in *Atoms, Molecules and Clusters*, Z.Phys.D., 24, 177

Reimers, J. R., Wilson, K. R., Heller, E. J. and Langhoff, S. R. 1985, J.Chem Phys., 82, 5064

Ruiz, M.-T., Bergeron, P., Leggett, S.K. 1992, NATO ASI, 403, 245

Ruiz, M.-T., this volume

Ruiz, M.-T., Maza, J., Wischnjewsky, M., Gonzalez, L.E., 1986, Ap. J., 304, L25

Schmidt, G. D., Bergeron, P. and Fegley, B. Jr. 1995, Ap. J., 443, 274

Stancil, P. C. 1994, Ap. J., 430, 360

Tsuji, T. 1964, Ann. Tokyo Astr. Obs. Sec. Ser., Vol. IX, 1

# ANALYSIS OF OPTICAL, UV, AND EUV OBSERVATIONS OF G 191-B2B

BURKHARD WOLFF AND DETLEV KOESTER
*Institut für Astronomie und Astrophysik der Universität*
*D-24098 Kiel, Germany*

AND

ALFRED VIDAL-MADJAR
*CNRS Institut d'Astrophysique*
*98 bis Boulevard Arago, F-75014 Paris, France*

## 1. Introduction

The major tools for the determination of the photospheric composition of DA white dwarfs are observations in the UV, EUV, and X-ray parts of the electromagnetic spectrum. In the case of G 191-B2B carbon, nitrogen, oxygen, silicon, phosphorus, sulphur, iron, and nickel could be detected in UV and FUV spectra (Bruhweiler & Kondo 1981, Sion et al. 1992, Vennes et al. 1992, Holberg et al. 1994, Vidal-Madjar et al. 1994, Werner & Dreizler 1994, Vennes et al. 1996). The presence of these elements results in a strong flux decrease in the EUV and X-ray regions (e.g. Barstow et al. 1993, Dupuis et al. 1995, Wolff et al. 1996).

This decrease causes a problem with the determination of effective temperatures from Balmer lines. Although metal lines are not visible in the optical part the opacity at short wavelengths may cause a redistribution of flux so that an analysis of the Balmer lines results in a higher effective temperature (blanketing effect). Bergeron et al. (1994) could show that – for example – a DA white dwarf of $T_{\rm eff} = 52000$ K would look like a 60000 K object if the helium abundance is He/H $= 10^{-4}$.

In this paper we analyze observations with the Extreme Ultraviolet Explorer, with the Goddard High Resolution Spectrograph on the Hubble Space Telescope, and optical spectra obtained at the Calar Alto Observatory. From the HST observations we determine metal abundances and use these values for a fit of the EUVE spectrum. This approach is neces-

*I. Isern et al. (eds.), White Dwarfs, 199–205.*

sary since it is very difficult to determine metal abundances on the basis of EUVE observations alone (e.g. Jordan et al. 1996a,b). We also demonstrate the effect of metal line-blanketing on the temperature determination with Balmer lines.

## 2. Observations and analysis

The primary aim of the observations with the HST GHRS was a study of the interstellar medium. In the spectral ranges observed photospheric lines of NV, SiIV, FeV, and NiV could also identified. For a detailed description of these observations we refer to Vidal-Madjar et al. (1994).

The EUVE spectra were obtained from the public archive. We use three dithered observations with a total exposure time of about 100000 seconds. The spectra were extracted, flux calibrated, and higher order contributions subtracted with the standard procedures of the IRAF/EUV software package (Version 1.6).

For the analysis we used LTE model atmospheres. Several million metal lines are included which were taken from the Kurucz (1994) CD-ROM. The bound-free opacities were taken from the Opacity Project calculations (Seaton et al. 1992) as provided by TOPBASE (Cunto & Mendoza 1992).

As a first step we analyzed the optical and ultraviolet spectra without considering the blanketing effect caused by metals. With pure hydrogen atmospheres we determined $T_{\text{eff}} = 60800\,\text{K}$ and $\log g = 7.59$ from a fit to the Balmer lines. For the analysis of the UV spectra we included metal lines in the calculation of the synthetic spectra. The effective temperature and gravity were held fixed at the values from the optical determination. The results are $N/H = 4 \cdot 10^{-6}$, $Si/H = 6 \cdot 10^{-7}$, $Fe/H = 4 \cdot 10^{-6}$, and $Ni/H = 1 \cdot 10^{-6}$.

For the analysis of the EUVE spectra we added $C/H = 2 \cdot 10^{-6}$ – derived from IUE observations – and $O/H = 1 \cdot 10^{-6}$ (Vennes et al. 1996) to these abundances. The synthetic spectra were normalized to the visual magnitude and the interstellar absorption was calculated with the IRAF/EUV package according to the model of Rumph et al. (1994). The values for the hydrogen column density and the HeI/HI ratio were taken from Dupuis et al. (1995). With $T_{\text{eff}}$ being the only free parameter a good fit to the EUVE spectra could be obtained with 56000 K.

After this first step we reanalyzed the HST spectra. This time the atmospheric structure was calculated with the previously determined abundances and metal line-blanketing was taken into account. With $T_{\text{eff}} = 56000\,\text{K}$ we derived $N/H = 4 \cdot 10^{-6}$, $Si/H = 1 \cdot 10^{-6}$, $Fe/H = 5 \cdot 10^{-6}$, and $Ni/H = 1 \cdot 10^{-6}$. Most abundances are virtually identical with the first results.

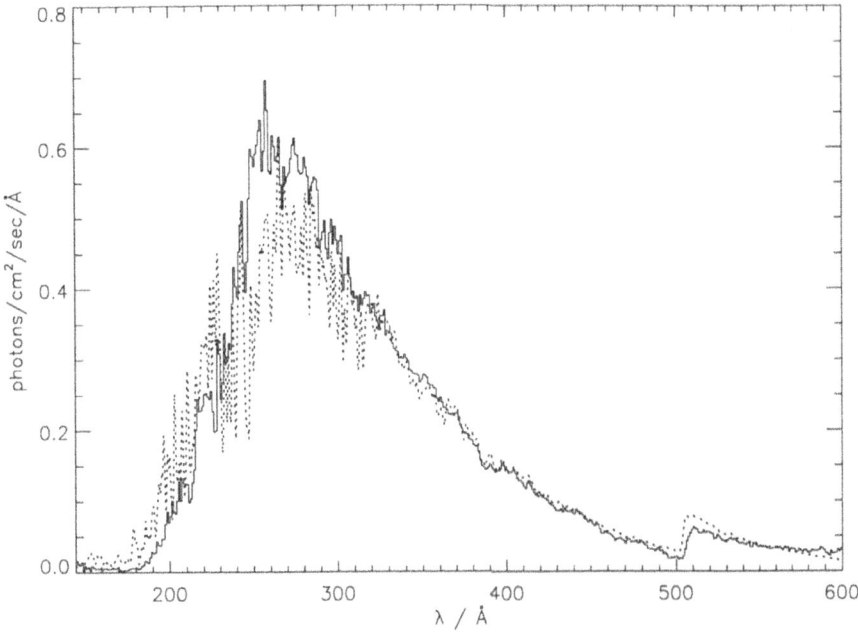

*Figure 1.* Combined medium and long wavelength EUVE spectrum of G 191-B2B (continuous line) compared to a model with $T_{\text{eff}} = 56000\,\text{K}$, $\log g = 7.6$, $C/H = 2 \cdot 10^{-6}$, $N/H = 4 \cdot 10^{-6}$, $O/H = 1 \cdot 10^{-6}$, $Si/H = 1 \cdot 10^{-6}$, $Fe/H = 5 \cdot 10^{-6}$, and $Ni/H = 1 \cdot 10^{-6}$ (dotted line)

Again, these abundances were used for a fit of the EUVE spectra. In Fig. 1 the synthetic spectrum for $T_{\text{eff}} = 56000\,\text{K}$ and $\log g = 7.6$ is compared with the observed medium and long wavelength spectrum. With these parameters the overall shape of the spectrum can be well reproduced. However, there are too strong absorption features at $\lambda \approx 250$–$320\,\text{Å}$, and at $\lambda < 230\,\text{Å}$ the model flux is somewhat too high.

The range of effective temperatures can be limited to $56000 \pm 2000\,\text{K}$ from the EUVE observations. For this conclusion the spectral region at $\lambda > 300\,\text{Å}$ is used because the absorption by heavy elements is rather small in this part. If the metal abundances are high enough to account for the blanketing effect at $\lambda < 250\,\text{Å}$ then $T_{\text{eff}}$ can be very well determined from the EUVE spectra.

In the following we will discuss implications on the metal abundances which can be drawn from the EUVE spectra. For an estimate of the importance of individual elements we have calculated model atmospheres with several element combinations. The main result is that the overall shape of the model spectrum is determined by Iron and Nickel alone. Both are

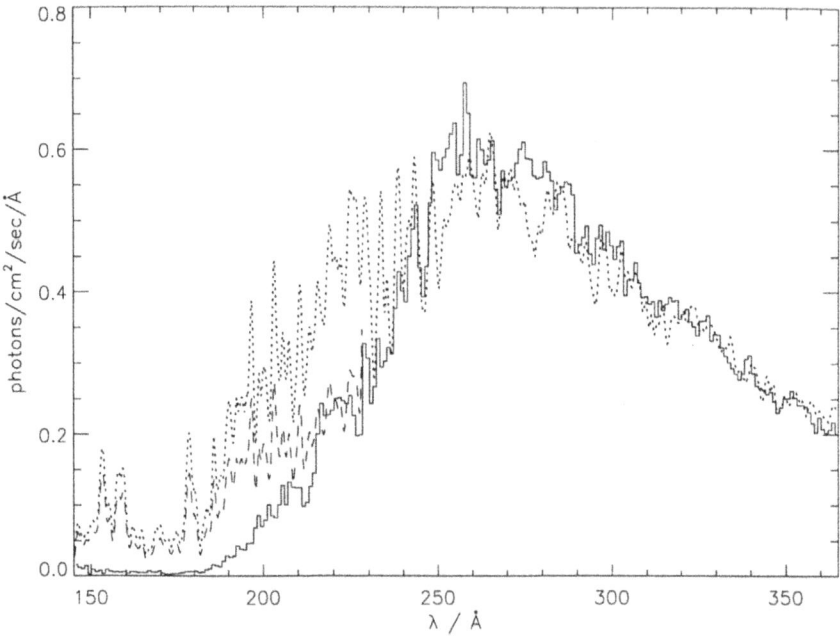

*Figure 2.*    Influence of interstellar HeII. The model parameters are $T_{\text{eff}} = 56000\,\text{K}$, Fe/H $= 2 \cdot 10^{-6}$, Ni/H $= 1 \cdot 10^{-6}$, HeII/HI $= 0.0$ (dotted line), and HeII/HI $= 0.2$ (dashed line), respectively. The observed spectrum is plotted with a continuous line. For clarity, only the medium wavelength region is shown

responsable for the strong decrease in flux shortward of 250 Å. Other elements exhibit several strong lines but their contribution to the flux decrease is rather small.

As can be seen from Fig. 1 the absorption features are somewhat too strong at $\lambda \approx 250$–$320$ Å. Therefore, we have reduced the iron abundance to the lower boundary of the HST analysis (Fe/H $= 2 \cdot 10^{-6}$) and left out the light metals (C, N, O, Si). The result is shown in Fig. 2. Now, the model better fits the observation at 250–320 Å. However, at smaller wavelengths there is some opacity missing. Since this region is just below the HeII absorption edge at 228 Å we tested the possible influence of interstellar HeII. The result for HeII/HI $= 0.2$ is also shown in Fig. 2. It can be seen that HeII can contribute some of the opacity below 228 Å, but the flux at very small wavelenghts is still too high. Nevertheless, we can set an upper limit for interstellar HeII since the absorbtion at the 228 Å edge would be too strong with HeII/HI $> 0.2$ . We can also set an upper limit for photospheric helium of about $10^{-6}$ because at higher abundances HeII lines would be visible.

We have also tested the possible influence of sulphur and phosphorus since these elements were discovered in ORFEUS spectra by Vennes et al. (1996). However, the abundances are far too low for a significant contribution to the opacity in the EUV.

The final step in our analysis was a fit to the Balmer lines using metal-blanketed atmospheres with abundances taken from the HST analysis. This results in an effective temperature of 50400 K, about 10000 K lower than the value obtained with pure hydrogen atmospheres and 5600 K lower than the EUV result. The temperature discrepancy between the optical and EUV values cannot be removed if the abundances are changed to the lower boundaries of the HST analysis. In this case the temperature is only 600 K higher.

## 3. Discussion

We have two main problems with the interpretation of the observations. First, it is not possible to obtain a consistent fit for all optical, ultraviolet, and extreme-ultraviolet observations. A re-analysis of the HST spectra with $T_{\text{eff}} = 50400 \, \text{K}$ results in higher metal abundances making the blanketing-effect even stronger. A higher temperature from the analysis of the Balmer lines can only be achieved if the blanketing effect can be reduced. But this is not possible since a significant amount of absorbing metals is necessary to reproduce the EUVE spectra.

The second problem is that there seems to be some opacity missing at short wavelengths as can be seen from Fig 2. In the following we will discuss possible solutions to these problems: non-LTE effects and non-uniform distribution of trace elements.

One main non-LTE effect is a change of the ionization equilibria. In the case of iron the most abundant ion in the line forming region of an LTE model atmosphere for a DA white dwarf of 56000 K is FeV, followed by FeIV, whereas FeVI is unimportant. This picture changes if a non-LTE atmosphere code is used. Then the abundance of FeIV is strongly decreased whereas the FeVI abundance is increased (e.g. Dreizler & Werner 1993). This explains the fact that FeIV lines are not visible in our HST spectra whereas the LTE calculations predict the presence of this ion.

For a test of the non-LTE influence we calculated model atmospheres using the Accelerated Lambda Iteration code by Werner & Husfeld (1985), Werner (1986), and Dreizler & Werner (1993). Our first results show that it is possible to achieve a somewhat better fit to the EUVE spectrum with these models. Since the FeV lines are less abundant it is possible to increase the iron abundance to $1 \cdot 10^{-5}$ without an increase in the photospheric absorption at 250-320 Å. Then at $\lambda < 250$ Å a better fit to the observed spectrum is possible than in the LTE case. We were also able to fit the optical spectra with $T_{eff} \approx 54000\,K$ which resolves the problem of the discrepant temperature determinations.

A similar result with non-LTE model atmospheres was obtained by Lanz et al. (1996) and Barstow et al. (1996). They could fit the Balmer lines with $T_{eff} = 55200\,K$ which was also consistent with their analysis of the EUVE spectra.

Another assumption made in our analysis is the uniform distribution of metals in the atmosphere. Since the support of trace elements by radiative forces results in a non-uniform distribution the absorption features in the UV and EUV may differ in strength from uniform calculations. Inspite of this remaining uncertainty the good agreement of the non-LTE model with EUV, UV, and optical observations makes us confident that we have uniquely identified the heavy elements present in hot DA white dwarfs and their abundances.

*Acknowledgements.*    This work was supported by the DARA under grant 50 OR 96173.

## References

Barstow M.A., Fleming T.A., Diamond C.J. et al., 1993, MNRAS 264, 16
Barstow M.A., et al., 1996, these proceedings
Bergeron P., Wesemael F., Beauchamp A., et al., 1994, ApJ 432, 305
Bruhweiler F.C., Kondo Y., 1981, ApJ 248, L123
Cunto W., Mendoza C., 1992, Rev. Mexicana Astron. Astrofis., 23, 107
Dreizler S., Werner K., 1993, A&A 278, 199
Dupuis J., Vennes S., Bowyer S., Pradhan A.K., Thejll P., 1995, ApJ 455, 574
Holberg J.B., Hubeny I., Barstow M.A., et al., 1994, ApJ 425, L105 253
Jordan S., Koester D., Finley D., 1996. In: Bowyer S., R.F. Malina (eds.), Astrophysics
    in the Extreme Ultraviolet, Kluwer, Dordrecht, p.235
Jordan S., Koester D., Finley D., 1996, these proceedings
Lanz T., Barstow M.A., Hubeny I., Holberg J.B., 1996, ApJ, in press
Rumph T., Bowyer S., Vennes S., 1994, AJ 107, 2108
Seaton M.J., Zeipper C.J., Tully J.A., Pradhan A.K., Mendoza C., Hibbert A., Berrington
    K.A., 1992, Rev. Mexicana Astron. Astrofis., 23, 19
Sion E.M., Bohlin R.C., Tweedy R.W., Vauclair G.P., 1992, ApJ 391, L29
Vennes S., Chayer P., Thorstensen J.R., Bowyer S., Shipman H.L., 1992, ApJ 392, L27
Vennes S., Chayer P., Hurwitz M., Bowyer S., 1996, ApJ 468, 898
Vidal-Madjar A., Allard N.F., Koester D., et al., 1994, A&A 287, 175
Werner K., 1986, A&A 161, 177
Werner K., Dreizler S., 1994, A&A 286, L31
Werner K., Husfeld D., 1985, A&A 148, 417
Wolff B., Jordan S., Koester D., 1996, A&A 307, 149

## Discussion

*H. Shipman*: Did you use the intensity of the continuum radiation in the UV spectral range to constrain the effective temperature?

*B. Wolff*: No, we did not. It may be useful to do specially if one uses the new ORFEUS observations (see talk by Martin Barstow).

*S. Starrfield*: The missing EUV opacity is probably caused by the fact that the available line lists are very incomplete.

*B. Wolff*: I agree. The line lists may be incomplete and the atmospheric data may also be uncertain in the EUV region. This may influence the predicted opacity.

*J. Kubát*: Do you calculate the partition function with occupation probabilities?

*B. Wolff*: Yes, we do.

# NEW RESULTS ON PG 1159 STARS
# AND ULTRAHIGH-EXCITATION DO WHITE DWARFS

K. WERNER
*Universität Tübingen, Institut für Astronomie und Astrophysik*
*D-72076 Tübingen, Germany*

S. DREIZLER
*Universität Kiel, Institut für Astronomie und Astrophysik*
*D-24098 Kiel, Germany*

U. HEBER
*Dr.-Remeis-Sternwarte Bamberg, D-96049 Bamberg, Germany*

AND

T. RAUCH
*Universität Potsdam, D-14469 Potsdam, Germany*

**Abstract.** We address four topics describing our latest progress since the last white dwarf workshop. 1. We report on new HST UV spectra of selected PG 1159 stars which were taken for fine analyses to constrain the GW Vir instability strip. 2. We describe the results of an analysis of a new ROSAT detected PG 1159 star which turns out to be the hottest known object of this spectral type. 3. Latest HST observations of the so-called ultra-high ionization white dwarfs give surprising insights into the "O VIII" phenomenon. 4. Finally we report on results of a recent effort to search for faint nebulae around hot white dwarfs.

## 1. HST-GHRS Cycle 5 observations of eight PG 1159 stars

During August/September 1995 we have taken a set of UV medium resolution (0.7Å) spectra with HST using the G140L grating in the GHRS instrument. We observed eight PG 1159 stars covering the wavelength range 1170–1460Å. Exposure times were between 20 and 30 minutes and the resulting spectra are of good quality. Our prime motivation was to confine the edges of the GW Vir instability strip by observing pulsators and non-

*I. Isern et al. (eds.), White Dwarfs, 207–212.*
© *1997 Kluwer Academic Publishers.*

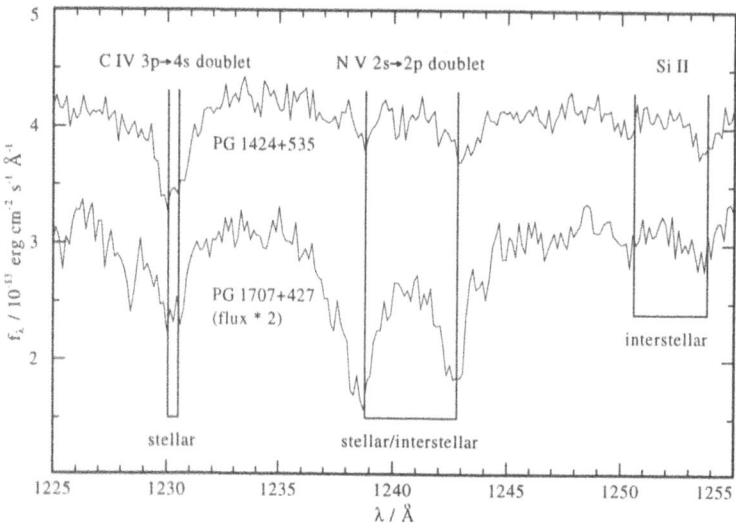

*Figure 1.* Detail of HST spectra from the pulsator PG 1707+427 and the non-pulsator PG 1424+535. The pulsator has a very strong N V resonance line which is probably photospheric. The spectra were shifted by +1.2 and -0.2Å, respectively, in order to roughly bring the C IV line to the rest wavelength. Rest wavelengths are indicated by bars.

pulsators which have similar optical spectra, and to perform precise model atmosphere analyses on hand of sensitive UV metal lines. The data analysis is in progress so that we cannot yet give final results here. A preliminary fit to the non-pulsator PG 1520+535 confirms the recent analysis of an EUVE spectrum (Werner et al. 1995a) that its $T_{\text{eff}}$ (=150 kK) is slightly higher than that of the prototype pulsator PG 1159−035 (=140 kK), hence the blue strip boundary is confined by these two stars. The other seven observed stars are relatively cool ($T_{\text{eff}} \leq 100$ kK) and are located near the red edge of the strip. The case of the red edge is probably more complicated, but one striking feature is immediately evident from coarse inspection of the spectra. There is a remarkably strong N V resonance doublet in three out of the seven objects, and these three are pulsators, while the others are not. If this is a pure accident or if the presence of nitrogen is essential for pulsation driving at the red edge needs to be studied by model calculations. Still, some doubt remains on this finding because we cannot rigorously exclude the possibility that the N V lines are of interstellar origin, though we believe they are stellar because of their strength. Additional observations in other wavelength regions are necessary for a strict conclusion. As an example we display in Fig. 1 details from the spectra of PG 1424+535 and PG 1707+427 for which we deduced equal parameters in a previous analysis of optical spectra ($T_{\text{eff}}$=100 kK, log $g$ =7.0, Werner et al. 1991).

## 2. ROSAT detects the hottest known PG 1159 star

The ROSAT supersoft X-ray source RX J0122.9−7521 is located towards the SMC and was known for some years, but only optical spectroscopy performed by Cowley et al. (1995) revealed its nature as a hot PG 1159 star. Based on these optical spectra and the X-ray data we have recently presented a model atmosphere analysis of this star (Werner et al. 1996). RX J0122.9−7521 turned out to be the hottest PG 1159 star found so far ($T_{eff}$=180 kK, $\log g$ =7.5). In contrast to other members, this object has an unusually low carbon and oxygen abundance: the He/C/O mass fractions (in %) are 68/21/11 in comparison to what we found for the prototype, 33/50/17. The high $T_{eff}$ and the low metal abundance are the reason for the surprisingly high X-ray luminosity of RX J0122.9−7521.

## 3. HST-GHRS Cycle 5 observations of two ultra-high ionization white dwarfs: HE 0504−2408 and HS 0713+3958

Among the most surprising observations was the recent detection of two hot DO white dwarfs (HS 0713+3958 and HE 0504−2408) which show broad and asymmetric absorption lines from ultrahigh ionized metals: C V/VI, N VI/VII, O VII/VIII, and even Ne IX/X (Werner et al. 1995b). Excitation temperatures of the order of one million K are necessary to produce such high ionization stages. Our first suspicion, that we might have detected extremely hot post-AGB stars could be excluded by model atmosphere calculations (Werner et al. 1995c). Instead we proposed that the metal lines are signatures from a very hot stellar wind, which also explains their asymmetric shape. The idea is based on the fact that e.g. O VIII lines were already detected in other hot post-AGB objects, albeit in emission, namely in the hot DO KPD 0005+5106 (Werner & Heber 1992) and in early-type Wolf-Rayet central stars like Sanduleak 3 (Barlow et al. 1980). In both cases, the high temperatures required are possibly generated by shock fronts in a stellar wind. In the case of KPD 0005+5106 this is corroborated by the detection of coronal X-rays (Fleming et al. 1993).

The ultrahigh-ionization absorption lines were first discovered in optical spectra, and we could identify every line transition between levels with principal quantum numbers Δn=1 and 2 in the CNO and Ne ions named above. These objects are strange enough to merit closer inspection in the UV by HST. In Sept. 1995 spectra of HS 0713+3958 and HE 0504−2408 were taken with the GHRS instrument using grating G140L (resolution 0.7Å) in the range 1140–1435Å. Exposure times were 71 and 24 minutes, respectively. Two other spectra from HE 0504−2408 were taken, a 44 minute exposure covering the range 1480–1775Å and a 40 minute exposure covering 2954–3000Å, now with the G270M grating (0.1Å resolution). The vast

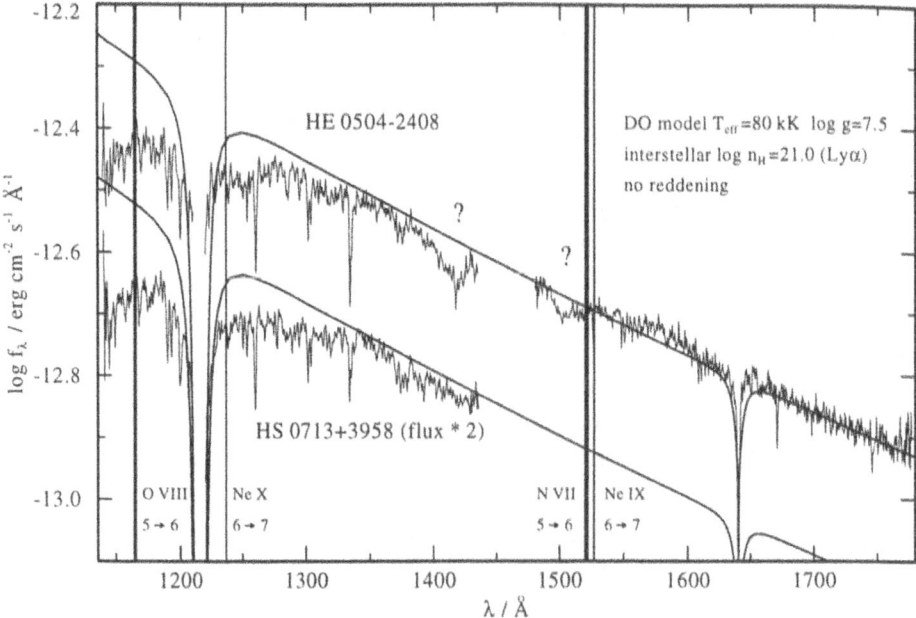

*Figure 2.* HST-GHRS spectra of the two protoype hot-wind DOs. Vertical bars indicate laboratory wavelengths of metal lines we expected to be present. They cannot be identified, however, two other, unknown broad absorptions (question marks) are detected. He II 1640Å is photospheric whereas most other features are of interstellar origin. The data are smoothed with a 0.5Å Gaussian.

majority of the narrow line features (Fig. 2) is of interstellar origin, but some remain unidentified (e.g. 1155Å and 1160Å in both stars). The only certain photospheric feature is He II 1640Å, another candidate is a weak O V 1371Å line. Quite surprisingly, the spectra do *not* show any ultrahigh-ionization absorption line which would be in analogy to the optical spectra! The strongest lines we expected to see are O VIII 1165Å (n=5 → 6) and N VII 1521Å (n=5 → 6) with wings as broad as 20–40Å. Instead, two such broad features, at 1420Å and 1500Å are detectable in HE 0504−2408, but we do not know any plausible identification as CNO or Ne lines. The 1500Å feature could be the blue-shifted N VII 1521Å line, but such a high velocity shift is not expected from the optical spectra, where systematic blue-shifts are measurable, too. In particular the 1420Å absorption has an asymmetric shape similar to that of the optically identified metal absorptions, hence we might see the a line from another metal ion. On the other hand, this feature is much stronger in HE 0504−2408 while in the optical the metal features are stronger in HS 0713+3958. The 1500Å absorption has a similar counterpart in H 1504+65 (one of the hottest PG 1159 stars), which was tentatively identified as a complex of Al V lines (Shipman et al. 1995). Al V

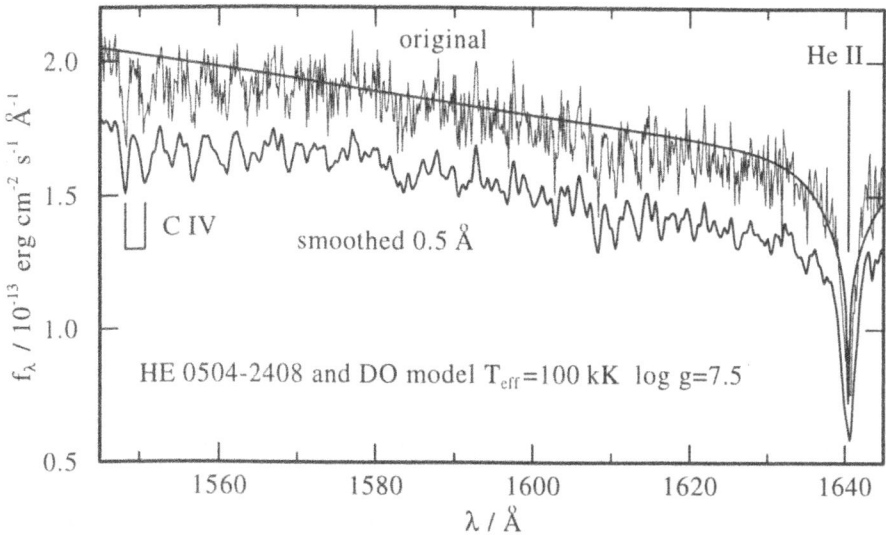

*Figure 3.* Detail of the HE 0504−2408 spectrum. Note the weak lines redward of the C IV resonance line and the occurrence of a similar line pattern near 1610Å.

lines can be expected at photospheric temperatures in hot WDs, but for a ultrahot wind we would expect much higher ionization stages.

Other line features are remarkable. Two bunches of narrow and weak absorption lines spanning about 20Å and reminding of molecular band heads are located near 1550Å and 1610Å (Fig. 3). In the case of 1550Å this might be an indication of circumstellar material (C IV) with different relative velocities, but this seems implausible because the redshifted components would indicate material falling onto the star. The continuum shape of the stars is unusual. It is very close to the hot DO model down to 1350Å, but at shorter wavelengths the observed continuum fluxes do not rise as steeply as the model (which runs essentially Rayleigh-Jeans like), but remain almost at a certain level. This might be the consequence of interstellar reddening, but we did not succeed to obtain a reasonable model fit with the standard reddening law. Another explanation is that the continuum flux is depressed by very strongly broadened wind lines, two candidates are O VIII 1165Å (n=5 → 6) and Ne X 1237Å (n=6 → 7), which we failed to identify as individual lines. Confusion with the interstellar Lyα profile complicates the situation. Future EUV observations as planned with ORFEUS are needed to study the continuum flux to shorter wavelengths.

Further progress in our understanding of what is going on in these strange objects certainly requires detailed modeling of the expanding hot shells. It remains to be mentioned that several other DOs were detected showing similar spectral lines, though less pronounced, and even one case of a DA was found (Dreizler et al. 1995). Hence a large fraction of the hot DO white dwarfs does show this unexplained phenomenon.

## 4. Search for planetary nebulae around hot white dwarfs

Post-AGB evolutionary time scales are still unclear, e.g. the PG 1159 type central star RX J 2117.1+3412 is luminous, i.e. young but its planetary nebula (PN) is among the largest, i.e. eldest known. We have looked for old, faint PN around a number of hot white dwarfs, most of them were recently discovered by the Hamburg-Schmidt-Survey (Hagen et al. 1995). We performed direct Hα narrow band imaging with a new wide angle CCD camera (20'x 20') which was built by the Bonn Observatory. The camera was attached to the 1.23m telescope on Calar Alto. In total we studied 16 objects of different spectral type with typical exposure times of one hour. The search was entirely negative, with one possible exception. An asymmetric nebulosity extending at least 5' was detected around the hot DO white dwarf PG 0109+111. If this is an evolved PN or ionized ambient interstellar gas needs to be clarified. PG 0109+111 was recently analyzed (Dreizler & Werner 1996) and found to be the most massive and among the hottest DO white dwarfs ($T_{eff}$=110 kK, $\log g$ =8, M=0.74 M$_\odot$). No nebula was found e.g. around the hot-wind DO HS 0713+3958 which was discussed above. More details are given elsewhere (Werner et al. 1997).

**Acknowledgement.** ROSAT and HST data analysis in the authors' institutes is supported by DARA under grants 50 OR 94091, 50 OR 96029.

## References

Barlow, M.J., Blades, J.C., Hummer, D.G. (1980), ApJ, L27

Cowley, A.P., Schmidtke, P.C., Hutchings, J.B., Crampton D. (1995), PASP 107, 927

Dreizler, S., Werner, K. (1996), A&A 314, 217

Dreizler, S., Heber, U., Napiwotzki, R., Hagen, H.-J. (1995), A&A 303, L53

Fleming, T.A., Werner, K., Barstow, M.A. (1993), ApJL 416, 79

Hagen, H.J., Groote, D., Engels, D., Reimers, D. (1995), A&AS 111, 195

Shipman, H.L., Provencal, J., Roby, S.W., et al. (1995), AJ 109, 1220

Werner, K., Heber, U. (1992), *The Atmospheres of Early-Type Stars*, eds. U. Heber & C.S. Jeffery, LNP 401, p. 291

Werner, K., Heber, U., Hunger, K. (1991), A&A 244, 437

Werner, K., Rauch, T., Dreizler, S., Heber, U. (1995a), *Astrophysical Applications of Stellar Pulsation*, eds. R.S. Stobie & P.A. Whitelock, ASP Conference Series 83, 96

Werner, K., Dreizler, S., Heber, U., Rauch, T., Wisotzki, L., Hagen, H.-J. (1995b), A&A 293, L75

Werner, K., Rauch, T., Dreizler, S., Heber, U., (1995c), *White Dwarfs*, eds. D. Koester & K. Werner, LNP 443, Springer, Berlin, p. 171

Werner, K., Wolff, B., Pakull, M., Cowley, A.P., Schmidtke, P.C., Hutchings, J.B., Crampton, D. (1996a), *Supersoft X-ray Sources*, ed. J. Greiner, LNP 472, Springer, p. 131

Werner, K., Bagschik, K., Rauch, T., Napiwotzki, R. (1997), *Planetary Nebulae*, eds. H. Habing & H. Lamers, IAU Symp. 180, Kluwer, in press

# NON–LTE ANALYSES OF DO WHITE DWARFS

S. DREIZLER
*Universität Kiel, Institut für Astronomie und Astrophysik*
*D-24098 Kiel, Germany*

AND

K. WERNER
*Universität Tübingen, Institut für Astronomie und Astrophysik*
*D-72076 Tübingen, Germany*

**Abstract.** We review our current knowledge about hot helium rich (DO) white dwarfs. For most known DOs detailed NLTE model atmosphere analysis have now become available. These stars represent the non-DA white dwarf cooling sequence from the hot end ($T_{\rm eff} \approx 120\,000$ K) down to the DB gap ($T_{\rm eff} \approx 45\,000$ K). Determination of effective temperature, surface gravity and metallicities allow a comprehensive comparison with predictions of diffusion calculations.

## 1. Introduction

Among the white dwarfs two distinct spectroscopic sequences exist: The hydrogen rich white dwarfs are called DA and can be found all along the white dwarf cooling sequence. The hottest DAs are closely connected to the hydrogen rich central stars of planetary nebulae (CSPN, Napiwotzki & Schönberner 1995). The helium rich sequence comprises DO ($T_{\rm eff} > 45\,000$ K), DB ($11\,000$ K $< T_{\rm eff} < 30\,000$ K) and DC ($T_{\rm eff} < 11\,000$ K) white dwarfs. Their spectral appearance is determined by the ionization balance of the helium plasma depending on the effective temperature of the star. The DOs display a pure He II line spectrum at the hot end and a mixed He I/II spectrum at the cool end while DBs are characterized by a pure He I line spectrum. At $T_{\rm eff}$ below $\approx 11\,000$ K the temperature is too low to excite He I lines making the DC featureless white dwarfs. Those DOs with detectable traces of metals are denoted DOZ. At the highest effective temperatures the DOs are connected to the helium, carbon and oxygen rich PG 1159 stars (also denoted as DOZ by Wesemael et al. 1985 [WGL]) which are the proposed

*I. Isern et al. (eds.), White Dwarfs*, 213–219.

precursors of the DO white dwarfs. There is an suggestive link from the [WC] CSPNe via the PG 1159 stars to the DOs. At lower effective temperatures the helium rich sequence is obviously interrupted by the so-called "DB gap" between 30 000 K and 45 000 K (Liebert et al. 1986), a fundamental problem in the understanding of the white dwarf evolution.

Recently we started a complete investigation of all known white dwarfs (Dreizler & Werner 1996 [DW96]) which is a resumption of the pioneering work of WGL. However, in the recent ten years the observation facilities and the modeling techniques improved drastically so that it seems well justified also to repeat their analyses. The work of WGL coincided with the publication of the Palomar–Green survey (PG, Green et al. 1986) since a reasonable number of DOs were known then. Our investigation was triggered by the Hamburg–Schmidt survey (Hagen et al. 1995). This survey aims at a complete sample of bright quasars but it is also a rich source of faint blue stars, especially for hot subdwarfs and white dwarfs (Heber et al. 1991). In a collaborative project between the institutes in Hamburg, Kiel and Bamberg follow-up spectroscopy of 320 hot stars has been carried out up to now (see Dreizler et al. 1994 for details). One of our goals was to identify more white dwarfs from rare subtypes, e.g. DO white dwarfs. This could be done very successfully since we doubled the number of DO white dwarfs (Heber, Dreizler, Hagen 1996). We could also detect a new class of hot white dwarfs probably suffering an extremely hot, fast and dense wind (Werner et al. 1995a, Dreizler et al. 1995, see also Werner et al. these proceedings). Within that project we obtained medium resolution spectra (1.5–3.5 Å) of all known DO white dwarfs covering the entire optical wavelength range. Several analyses of individual DOs were motivated by ROSAT, EUVE, ORFEUS, HST and optical high resolution observations (Barstow et al. 1994, 1996, Napiwotzki et al. 1995, 1996, Werner et al. 1995b, 1996 [WDW], Werner 1996). All these analyses made profit of the enormous improvement of the Non–LTE model atmosphere technique in the recent years (Dreizler & Werner 1993, Werner & Dreizler 1996, Hubeny & Lanz 1995) making highly sophisticated models available.

## 2. Results of Non–LTE analyses

The results of our analyses are summarized in Table 1. Most results were already published in previous papers and are therefore only shortly annotated here. All PG DOs – except PG 0038+199 which was identified as DO only recently by Wesemael et al. (1993) – have been analysed previously by WGL using LTE model atmospheres. The newly determined NLTE values are up to 20 000 K higher. This is due to systematic differences between the LTE and NLTE model atmospheres. The latter ones reveal deeper He I

TABLE 1. Atmospheric parameters for DO white dwarfs determined from our NLTE analyses. Preliminary results were derived differentially by comparison with analysed DOs.

| star | type | $T_{\text{eff}}$ | $\log g$ | $M/M_\odot$ | comment |
|---|---|---|---|---|---|
| KPD 0005+5106 | DOZ | 120 000 | 7.0 | 0.46 | WHF |
| PG 0038+199 | DO | 115 000 | 7.5 | 0.59 | DW96 |
| PG 0109+111 | DOZ | 110 000 | 8.0 | 0.74 | DW96 |
| PG 1034+001 | DOZ | 100 000 | 7.5 | 0.56 | WDW |
| HS 1830+7209 | DO | 100 000 | 7.2 | 0.47 | DW96 |
| PG 0108+101 | DOZ | 95 000 | 7.5 | 0.54 | DW96 |
| MCT 2148−294 | DOZ | 85 000 | 7.5 | 0.52 | prelim. |
| PG 0046+078 | DOZ | 73 000 | 8.0 | 0.68 | DW96 |
| PG 0237+116 | DO | 70 000 | 8.0 | 0.68 | DW96 |
| RE 0503-289 | DOZ | 70 000 | 7.5 | 0.49 | DW96 |
| HS 0111+0012 | DOZ | 65 000 | 7.8 | 0.58 | DW96 |
| Lanning 14 | DO | 58 000 | 7.9 | 0.61 | DW96 |
| PG 0929+270 | DO | 55 000 | 7.9 | 0.61 | prelim. |
| HZ 21=PG 1211+332 | DO | 53 000 | 7.8 | 0.56 | DW96 |
| PG 1057−059 | DO | 50 000 | 7.9 | 0.60 | prelim. |
| HS 0103+2947 | DO | 49 500 | 8.0 | 0.65 | DW96 |
| HD 149499 B | DOA | 49 500 | 8.0 | 0.65 | N95a,b |
| PG 1133+489 | DO | 46 000 | 8.0 | 0.64 | prelim. |
| | | | | mean: 0.59 | |

lines resulting in a higher effective temperature. Stellar masses can be determined from the comparison of the effective temperature and gravity with those of evolution calculations (DW96). From the derived masses we conclude that DOs are normal white dwarfs in the sense that their mean mass amounts to $0.6\,M_\odot$. We have, however, to admit that this determination could suffer from non negligible systematic errors. We used the tracks of Wood (1995) to derive the stallar masses. There is a caveat, though, because theses tracks are not evolved consistently through the AGB phase and the hotter parts of the tracks are therefore uncertain. Due to the lack of consistent post–AGB evolution model with helium rich surfaces there is no better alternative at the moment.

Another important difference between the WGL and DW96 results is the non–detection of hydrogen in all DOs. While HZ 21 and PG 0929+270 were supposed to have 30% hydrogen in their atmospheres DW96 found no evidence for trace amounts of hydrogen and derived an upper limit of 10% for HZ 21. Due to the similarity of PG 0929+270 to HZ 21 we suppose that this star has also a no detectable amount of hydrogen left in its atmosphere.

These new results can also be understood as an NLTE effect, since the He II lines are also deeper in the NLTE models. Additional equivalent width from the blending Balmer lines is not required in the NLTE fit. An exploratory search for hydrogen in hot DOs and PG 1159 stars was also unsuccessful (Werner 1996). Due to the importance of our understanding of the evolution of helium rich post–AGB stars this aspect is investigated in more detail in upcoming high resolution observations. Undetected traces of hydrogen could well be responsible for the DB gap. Very small amounts arc sufficient to cover the helium rich envelope with a thin hydrogen atmosphere when element separation is not disturbed by competing processes, e.g. convection or mass loss. Above the DB gap the He II/He III convection zone and below it the He I/He II convection zone could prevent element separation.

Spectra of the very recently detecteded DOs are compared to already analysed ones to estimate their parameter. We find two interesting results. MCT 2148–294 is very similar to PG 0108+101 not only from effective temperature and gravity but also from its carbon abundance. It also clearly shows the C IV complex at 4560 Å in a similar strength. It therefore belongs to the most carbon rich DOs. As already noticed by WGL, PG 1133+489 is the coolest DO. In the optical spectrum only the strongest He II line (4686 Å) remains visible. As in all other DOs – except HD 149499 B (Napiwotzki et al. 1995) – there is no evidence for trace amounts of hydrogen. Since the He II lines overlapping with the Balmer lines are no longer visible, this star will provide the chance to derive a very low upper limit for its hydrogen abundance. PG 1133+489 currently defines the hot end of the DB gap. Despite intensive search in the course of the Hamburg–Schmidt survey no helium rich white dwarf in the DB gap could be found up to now and its existence is therefore confirmed.

In some cases metal abundances of DOs could be determined from optical spectra. Since the lower detection limit is quite high ($\sim 1\%$) this is only applicable for the more metal rich ones. Existing HST observations were utilized by Werner, Heber & Fleming (1994) [WHF] and WDW. Upcoming HST observations will enable us to determine metal abundances of more DOs representing the entire DO cooling sequence.

## 3. Equilibrium abundances

The available metal abundance or upper limits are displayed in Fig. 1. In this figure we overplotted the available equilibrium abundances calculated by Chayer et al. (1995) and Chayer (priv. comm.). In general, the agreement is not very promising. The abundances are derived under the assumption that a steady state in the diffusion process is reached and the gravitational downward forces are balanced by radiative acceleration. In these

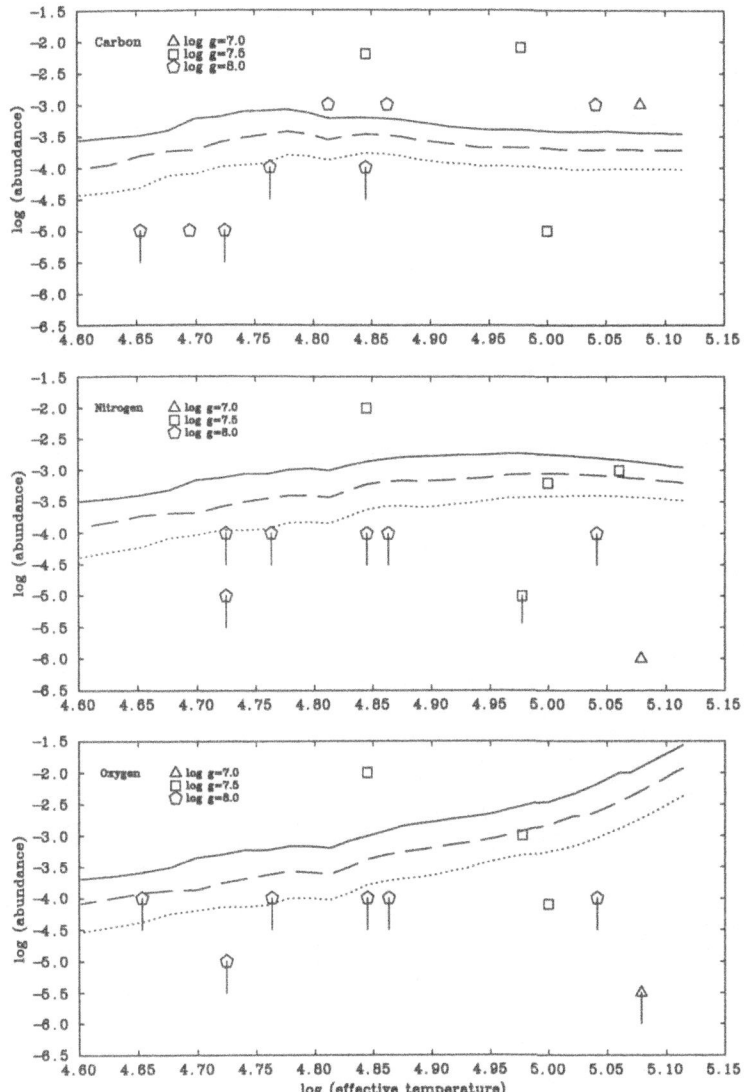

*Figure 1.* Observed metal abundances by number ratio relative to He (symbols) of DO white dwarfs (DW96) compared to equilibrium abundances of Chayer et al. (1995) and Chayer (priv. comm.): Solid log g=7.0, dashed log g=7.5, dotted log g=8.0. Upper limits are marked with a solid vertical line.

calculations an LTE radiation field is used to calculate the radiative forces. As demonstrated above NLTE effects are important in DO atmospheres and we therefore decided to check the influence of a NLTE radiation field on the derived equilibrium abundances. Further more, no feedback on the abundance stratification on the radiation field is taken into account. We therefore adapted the approach of Chayer et al. (1995) and implemented it in our NLTE model atmosphere code. This enables us to check the NLTE effect and by iteration within our atmosphere code also the feedback of the abundance stratification. The results are, however, also not very encouraging. We compared LTE versus NLTE equilibrium abundances in a typical DO atmosphere and found that the differences in the line forming region are very small. Compared to the results of Chayer et al. no significant difference can be found. We therefore confirm that equilibrium abundances are unlikely to explain the observed abundance patterns of DOs. However, the carbon and oxygen rich PG 1159 stars which mark the transition from the CSPN to the white dwarf phase must at some point turn into white dwarfs, very probably into DOs. This transition is marked by the onset of element separation removing the heavy elements out of the atmosphere. Only small traces can be supported by radiative acceleration as long as the effective temperature is high enough to produce a sufficient intense radiation field. From the available results we do, however, not know when and how this transition takes place.

## References

Barstow M.A., Holberg J.B., Werner K., et al. 1994, MNRAS 267, 653

Barstow M.A., Hubeny I., Lanz T., Holberg J. B., Sion E.M. 1996, in Astrophysics in the Extreme Ultraviolet, IAU Coll. 152, Kluwer, Dordrecht, p. 203

Chayer P., Fontaine G., Wesemael F. 1995, ApJS 99, 189

Dreizler S., Werner K. 1993, A&A 278, 199

Dreizler S., Werner K. 1996, A&A 314, 217 (DW96)

Dreizler S., Heber U., Jordan S., Engels D. 1994, in Hot Stars in the Galactic Halo, eds. S.J. Adelman, A.R.Upgren, C.J. Adelman, Cambridge University Press, p. 228

Dreizler S., Heber U., Napiwotzki R., Hagen H.J. 1995, A&A 303, L53

Green R.F., Schmidt M., Liebert J. 1986, ApJS 61, 305

Hagen H.J., Groote D., Engels D., Reimers D. 1995, A&AS 111, 195

Heber U., Jordan S., Weidemann V. 1991, in White Dwarfs, NATO ASI Series C Vol. 336, eds. G. Vauclair and E. Sion, Kluwer, p. 109

Heber U., Dreizler S., H.-J. Hagen 1996, A&A 311, L17

Hubeny I., Lanz T. 1995, ApJ 439, 875

Liebert J., Wesemael F., Hansen C.J., et al. 1986, ApJ 309, 241

Napiwotzki R., Schönberner D. 1995, A&A 301, 545

Napiwotzki R., Hurwitz M., Jordan S., et al. 1995, A&A 300, L5

Napiwotzki R., Jordan S., Bowyer S., et al. 1996, in Astrophysics in the Extreme Ultraviolet, IAU Coll. 152, Kluwer, Dordrecht, p. 241

Werner K. 1996, A&A, 309, 861

Werner K., Dreizler S. 1996, in Computational Astrophysics Vol. II (Stellar Physics), eds. R.P. Kudritzki, D. Mihalas, K. Nomoto, F.-K. Thielemann, submitted

Werner K., Heber U., Fleming T.A. 1994, A&A 284, 907 (WHF)

Werner K., Dreizler S., Heber U., et al. 1995a, A&A 293, L75

Werner K., Dreizler S., Wolff B. 1995b, A&A 298, 567 (WDW)

Werner K., Dreizler S., Heber U., et al. 1996, A&A 307, 860

Wesemael F., Green R.F., Liebert J. 1985, ApJS 58, 379 (WGL)

Wesemael F., Greenstein J.L., Liebert J., et al. 1993, PASP 105, 761

Wood M.A. 1995, in White Dwarfs, eds. D. Koester and K. Werner, Lecture Notes in Physics 443, Springer, Berlin, p. 41

## Discussion

*S. Kawaler:* I recommend caution when using evolutionary models to assign masses to hot white dwarfs. At the hot end ($T_{\mathrm{eff}} > 70\,000\mathrm{K}$) WD models like Wood (1995) are still sensitive to their (usually unrealistic) initial conditions.

*S. Dreizler:* We are aware of that point. There are, however, no better tracks available which are evolved consistently through the AGB phase and are He rich.

*M. Barstow:* I could add a comment on your analysis of RE 0503-289. At the temperature you (and we) derived from the optical data it is not possible to get a consistent measurement of the photospheric abundances across all wavelength ranges, including optical, UV, and EUV.

# NEW SPECTRAL ANALYSES OF PRE-WHITE DWARFS

T. RAUCH

*Universität Tübingen, Institut für Astronomie und Astrophysik*
*D-72076 Tübingen, Germany*

S. DREIZLER

*Universität Kiel, Institut für Astronomie und Astrophysik*
*D-24098 Kiel, Germany*

AND

K. WERNER

*Universität Tübingen, Institut für Astronomie und Astrophysik*
*D-72076 Tübingen, Germany*

**Abstract.** The very hot, H-deficient PG 1159 stars are found in the post-AGB region of the Hertzsprung-Russell diagram. Most likely the "born-again" post-AGB scenario is valid for their evolution. It predicts a final He-flash which brings back the star to the AGB, followed by a second (He burning) descent from the AGB where a superwind may take off the entire H-rich and most of the He-rich envelope and lay bare intershell layers, where Ne is the fourth abundant element after He, C, and O. In order to support this hypothesis, we present a search for the Ne VII $\lambda$ 3644Å absorption line in several of these stars.

The two presently known O(He) central stars of planetary nebulae, K 1-27 and LoTr 4, exhibit almost pure He II absorption line spectra. They are found in the post-AGB region of the Hertzsprung-Russell diagram just amongst the PG 1159 stars which in contrast show additional strong C IV lines. We present an analysis of three recently discovered O(He) stars, HS 0742+6520, HS 1522+6615 and HS 2209+8229, for which no associated planetary nebulae have been detected. It is likely that there exists an evolutionary channel for H-deficient post-AGB stars, "O(He) stars → DO white dwarfs", which is separate from to the well established "Wolf-Rayet [WC] central stars → PG 1159 stars → DO white dwarfs" evolutionary sequence.

*I. Isern et al. (eds.), White Dwarfs, 221–227.*

## 1.  A search for Ne VII $\lambda 3644$Å in PG 1159 stars

The PG 1159 stars are found in the post-Asymptotic Giant Branch (post-AGB) region of the Hertzsprung-Russell diagram (HRD). Spectral analyses by means of NLTE model atmosphere techniques have shown that they cover a wide parameter range of effective temperatures ($T_{eff} = 180 - 65$kK) and surface gravities ($\log g = 5.5 - 8.0$; Dreizler et al. 1995). Due to these wide parameter ranges, a variety of spectral appearance and chemical composition is found: Abundance analyses revealed a "typical" abundance pattern He:C:O of 33:50:17 by mass. H 1504+65 (Werner 1991) is the most extreme example, exhibiting a naked C-O stellar core. The discovery of "hybrid" PG 1159 stars (Napiwotzki & Schönberner 1991) which have a significant amount of H in their photospheres shows that there is no sharp delimitation from the H-rich pre-white dwarfs.

All these exotic objects are direct progenitors of the white dwarfs, presently close to their hottest stage of evolution. It is worthwhile to note here that planetary nebulae (PN) have been detected around about every other PG 1159 star and eight out of 28 presently known PG 1159 stars show multi-periodic, low-amplitude, non-radial g-mode pulsations.

The so-called "born-again post-AGB" star (or "final flash") scenario of Iben et al. (1983) predicts that a star can experience a late He flash at already decreasing luminosity after its first departure from the AGB. It returns to the AGB and in a second, He-burning post-AGB phase, the entire H-rich and most of the He-rich envelope may be expelled from the star and former intershell matter becomes visible. This scenario is able to explain the observed photospheric abundances of PG 1159 stars (Werner et al. 1991). Evolutionary calculations of Iben & Tutukov (1985) show that Ne, with ongoing mass-loss, can become the fourth (He, C, O, Ne) or even third (if the He envelope is completely blown off) abundant element at the star's surface.

In an analysis of the PN NGC 246 and its exciting (PG 1159 – type) central star (CSPN), Heap (1975) found a stellar absorption line at 3644 Å which remained unidentified until Werner & Rauch (1994) realized that a Ne VII line (2s3s-2s2p singlet) at 3643.6 Å is commonly used for the calibration of laboratory plasmas (König et al. 1993). They could identify Ne VII $\lambda 3644$Å in optical spectra of three PG 1159 stars, RX J2117+3412 ($T_{eff} = 170$kK, $\log g = 6.1$), NGC 246 (150, 5.7), and K 1-16 (140, 6.1). These objects have in common very high $T_{eff}$ and relatively low $\log g$ (Werner et al. 1995, Rauch & Werner 1995). Test calculations have shown that three conditions have to be fulfilled to reproduce the observed strong Ne line: $T_{eff} \gtrsim 120$ kK, $\log g \lesssim 6$, and the Ne abundance has to be $\gtrsim 2\%$ by mass! This high value is in good agreement with the calculations of Iben & Tutukov

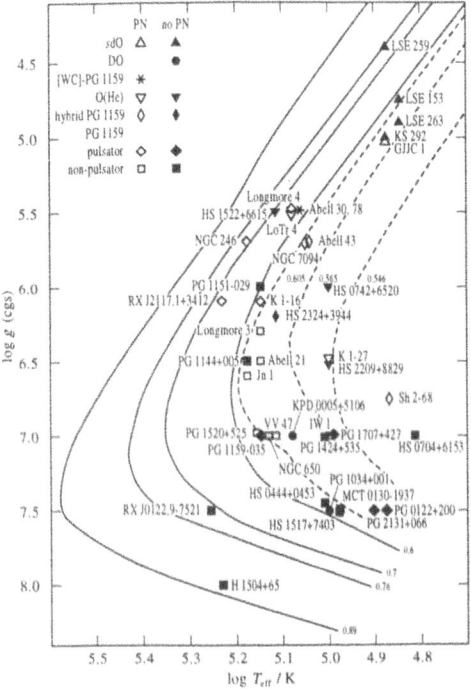

*Figure 1.*    Positions of the presently known PG 1159 stars and related objects in the $\log T_{\text{eff}} - \log g$ diagram compared to theoretical evolutionary tracks of H- (dashed, Schönberner 1983, Blöcker & Schönberner 1990) and He-burning post-AGB stars (Wood & Faulkner 1986). The tracks are labeled with the respective stellar masses (in $M_\odot$). The shaded area indicates the approximate position of the GW Vir instability strip

(1985) and supports the validity of the "born-again" scenario for these stars.

Consequently, we have selected four PG 1159 stars (Fig. 1) which fulfil our Ne detection criteria: Longmore 4 (120, 5.5), PG 1151-029 (140, 6.0), Longmore 3 (140, 6.3), PG 1144+005 (150, 6.5), and for comparison the prototype PG 1159-035 (140, 7.0). They were observed with the ESO 3.5m NTT and EMMI (Apr. 8-10 1995; resolution $\approx 2.5$ Å, S/N $\approx 30\text{-}50$).

Ne VII $\lambda 3644$Å is found only in the spectrum of Longmore 4. This has already been detected by Werner & Rauch (1994) with some uncertainty due to the poor quality of the spectrum available at that time. In the case of PG 1144+005 and PG 1159-035, the higher $\log g$ may be the reason for the non-detection in our spectra. In the case of PG 1151-029 and Longmore 3 which have similar parameters like K 1-16, however, a lower Ne abundance is likely to be the reason.

All four PG 1159 stars with positive Ne detection (Longmore 4, NGC 246, RX J2117+3412, K 1-16) are pulsators (and CSPNe) — this may be an accident because the "Ne detection zone" is just overlapping the "low gravity"

part of the GW Vir instability strip (Fig. 1). However, a relation between a
more efficient mass-loss phase (and thus, a higher Ne abundance) and the
pulsations cannot be excluded.

## 2. The O(He) evolutionary channel to the DO white dwarfs

The spectral sub-type O(He) was introduced by Méndez (1991) for CSPN
showing predominantly absorption line spectra dominated by He. The only
two presently known O(He) CSPN are K 1-27 (100, 6.5) and LoTr 4 (120,
5.5) which have been analyzed by Rauch et al. (1994, 1996). From their
photospheric composition both fit into the "born-again" scenario. In the
case of LoTr 4 hydrogen could be clearly identified in the optical spectrum
(H/He = 0.5 by number), while in the case of K 1-27 (with a higher $\log g$)
only an upper limit (H/He < 0.2) could be determined.

Jeffery et al. (1996) have classified GJJC 1 (Gillet et al. 1989) as an
O(He) CSPN: Harrington & Paltoglou (1993) determined equivalent widths
of H, He, C, and N lines in its optical spectrum (taken by the HST) by
Gaussian fits and compared them to results of an NLTE model atmosphere
analysis of the sdO star KS 292 (Rauch et al. 1991). From the good agree-
ment they concluded that both stars have similar parameters ($T_{\mathrm{eff}} = 75\mathrm{kK}$,
$\log g = 5.0$, H/He =2, C and N strongly enriched). However, GJJC 1 exhibits He II, C IV, N IV, and N V lines of similar strengths and hence, might
better be classified as a "normal" He-rich sdO star. The presence of a PN
around GJJC 1 itself is remarkable and makes GJJC 1 and KS 292 ("PN-
free") an interesting pair in analogy to similar pairs within the PG 1159
group.

Three objects which fit into the O(He) classification were recently discovered within the framework of NLTE analyses of the stellar component in
the Hamburg-Schmidt quasar survey (Heber et al. 1996): HS 0742+6520,
HS 1522+6615 (ROSAT source RX J1522+6604), and HS 2209+8229. For
none of these stars a PN has been detected so far.

Optical spectra (res. $\approx 3$ Å) were taken with the MPG 3.5m telescope
at Calar Alto. A coarse inspection shows that in the case of HS 0742+6520
the He II $\lambda\lambda$ 4686, 5411, 6560 Å absorptions lines are much stronger (Fig. 2)
than those of HS 1522+6615 and HS 2209+8229. It is impossible to reproduce these extraordinarily strong He II lines with theoretical profiles
calculated with our plane-parallel, hydrostatic NLTE code (Werner 1986).
A similar phenomenon was found by Werner et al. (1995) in the case of
hot DO white dwarfs showing ultrahigh-excitation lines. But in the spectrum of HS 0742+6520 none of these lines is identified. In the spectrum
of HS 1522+6615 there is a significant contribution of H $\alpha$ to the blend
with He II $\lambda$ 6560 Å while there is no indication of H in HS 2209+8829. N V

emission lines are found only in the O(He) CSPN K 1-27 and LoTr 4.

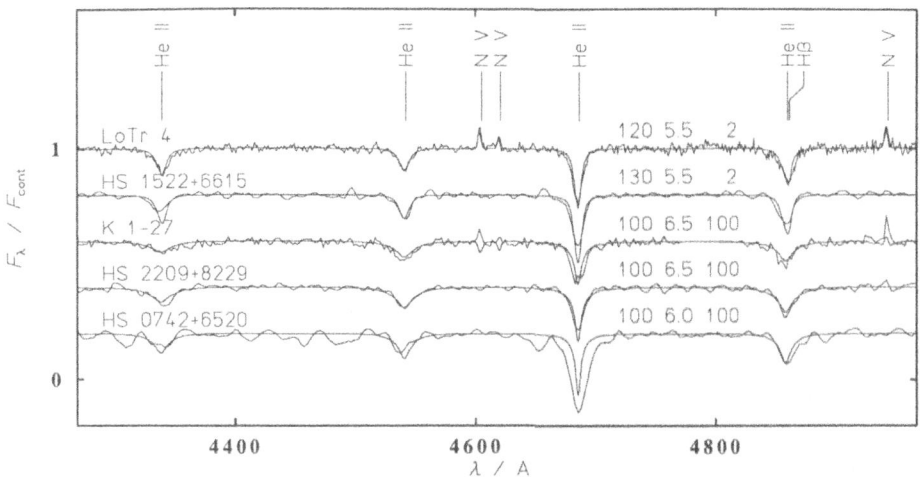

*Figure 2.* Optical spectra of LoTr 4, HS 1522+6615, K 1-27, HS 2209+8229, and HS 0742+6520 compared to theoretical spectra of H+He models, labeled with respective $T_{eff}$ (kK), $\log g$ (cgs), and He/H (number ratio). The models for LoTr 4 and K 1-27 contain N in addition, 0.07% and 0.5%, respectively

We carried out a NLTE analysis (H+He models, e.g. Rauch et al. 1994) for HS 1522+6615 and HS 2209+8829. In the case of HS 2209+8829 we determined $T_{eff} = 100\pm20$kK, $\log g = 6.5\pm0.5$ dex, H/He$\lesssim0.1$. These parameters are similar to those of the CSPN K 1-27. In the case of HS 1522+6615 the observation is reproduced with a theoretical spectrum calculated from a model with $T_{eff} = 130$kK, $\log g = 5.5$, and H/He = 2. These are almost the same parameters like those of the CSPN LoTr 4. In addition, we determined C and O abundances of about 1% from the optical spectrum using the C IV $\lambda\lambda$ 5806Å doublet and the O VI $\lambda$ 5290Å complex, respectively.

In the ROSAT archive two PSPC spectra of RX J1522+6604 are available. We determined a total PSPC count rate of 0.35±0.01 cts/sec. Our final H+He+C model ($T_{eff} = 130$kK, $\log g = 5.5$, H/He = 2, C/He = 1) yields a flux twice as high at energies $\gtrsim0.22$ keV. This might indicate the presence of additional absorbers (besides H, He, and C) in the photosphere. An addition of 1% O reduces the flux strongly at these energies (e.g. Werner et al. 1996) resulting in a flux which is already too low.

Although some fine tuning is still necessary in order to improve the fit of the synthetic spectrum to the observation, it is clear that HS 1522+6615 resembles the CSPN LoTr 4. The additionally identified C IV and O VI lines may indicate that HS 1522+6615 is in a later stage of evolution than the CSPN LoTr 4. The determined C and O abundances are similar to those found typically in DO white dwarfs (Dreizler & Werner 1996).

However, if we assume that the "born-again post-AGB" scenario is valid for both, for the PG 1159 stars as well as for the O(He) stars, there are two possibilities for the further evolution of the O(He) stars: Firstly, significant mass-loss due to a wind is still going on and the He-rich envelope will be blown off. Then the O(He) stars can be progenitors of the PG 1159 stars. However, nothing is known about the present mass of the He-rich surface layers of the O(He) stars. Secondly, the mass-loss phase was simply less efficient than in the case of PG 1159 stars. Then the O(He) stars display the transition of post-AGB stars to the DO white dwarfs without passing the PG 1159 stage.

From the presently known numbers of PG 1159 (28) and O(He) (5) objects one can estimate that a significant fraction of the H-deficient post-AGB stars does not pass the PG 1159 stage. These objects can be assigned to a separate DO white dwarf feeder channel "O(He) $\rightarrow$ DO", which is overlapping the PG 1159 region in the $\log T_{\text{eff}} - \log g$ diagram.

**Acknowledgements** We like to thank Burkhard Wolff (Kiel) who extracted the ROSAT data. This work has been supported by the DARA under grant 50 OR 9409 1.

# References

Blöcker, T., Schönberner, D. (1990), A&A 240, L11

Dreizler, S., Werner, K. (1996), A&A 314, 217

Dreizler, S., Werner, K., Heber, U. (1995), in: White Dwarfs, eds. D. Koester, K. Werner, Lecture Notes in Physics 443, Springer, Berlin, p. 160

Gillet, F.C., Jacoby, G.H., Joyce, R.R., et al. (1989), ApJ 338, 862

Harrington, J.P., Paltoglou, G. (1993), ApJ 411, L103

Heap, S.R. (1975) ApJ 196, 195

Heber, U., Dreizler, S., Hagen, H.J. (1996) A&A 311, L17

Iben, I. Jr., Tutukov, A.V. (1985), ApJS 58, 661

Iben, I. Jr., Kaler, J.B., Truran, J.W., Renzini, A. (1983), ApJ 264, 605

Jeffery, C .S., Heber, U., Hill, P.W., et al. (1996), in: White Dwarfs, eds. D. Koester, K. Werner, Lecture Notes in Physics 443, Springer, Berlin, p. 471

König, R., Kolk, K.-H., Kunze, H.-J. (1993), Physica Scripta 48, 9

Méndez, R.H. (1991), IAU Symp. 145, Kluwer, Dordrecht, p. 375

Napiwotzki, R., Schönberner, D. (1991), A&A 249, L16

Rauch, T., Werner, K. (1995), in: White Dwarfs, eds. D. Koester, K. Werner, Lecture Notes in Physics 443, Springer, Berlin, p. 186

Rauch, T., Köppen, J., Werner, K. (1994), A&A 286, 543

Rauch, T., Köppen, J., Werner, K. (1996), A&A 310, 613

Rauch, T., Heber, U., Hunger, K., Werner, K., Neckel, T. (1991), A&A 241, 457

Schönberner, D. (1983), ApJ 272, 708

Werner, K. (1986), A&A 161, 177

Werner, K. (1991), A&A 251, 147

Werner, K., Rauch, T. (1994), A&A 284, L5

Werner, K., Heber, U., Hunger, K. (1991), A&A 244, 437

Werner, K., Dreizler, S., Heber, U., et al. (1995), A&A 293, L75

Werner, K., Wolff, B., Pakull, M.W., et al. (1996), in: Supersoft X-ray Sources, ed. J. Greiner, Lecture Notes in Physics 472, Springer, Berlin, p. 131

Wood, P.R., Faulkner, D.J. (1986), ApJ 307, 659

# THE SPHERICITY EFFECTS IN THE NLTE MODEL ATMOSPHERES OF WHITE DWARFS

JIŘÍ KUBÁT

*Astronomický ústav, Akademie věd České Republiky,*
*251 65 Ondřejov, Czech Republic*

**Abstract.** We have calculated a sample grid of plane-parallel and spherically symmetric model atmospheres of white dwarfs for different $T_{eff}$ and $\log g$ and with different helium abundances. The model atmospheres consist only of hydrogen and helium. The NLTE model atmospheres are calculated with the help of our own computer code that enables calculation of both plane-parallel and spherically symmetric model atmospheres, and consequently, an easy and reliable estimate of the sphericity effects. The sphericity effects are not drastic, but neglecting them introduces a systematic error into model atmosphere analysis.

## 1. Introduction

A common assumption in the analysis of atmospheres of white dwarfs is that the atmosphere is "thin" relative to radius; it means that the atmosphere may be assumed to be plane-parallel. NLTE plane-parallel model atmospheres have been successfully used for the analysis of white dwarfs by a number of authors. For a brief critical review of previous model atmosphere calculations (both LTE and NLTE) see Hubeny & Lanz (1995) and Lanz & Hubeny (1995). Recently, the white dwarf model atmosphere calculations were brought to a very high degree of sophistication by including NLTE line blanketing effects by Lanz & Hubeny (1995) and Dreizler & Werner (1993, 1994). Here I would like to spend some words on the validity of the plane-parallel approximation and make an estimate of a limit of its applicability as well as a comparison with the spherically symmetric approximation.

*I. Isern et al. (eds.), White Dwarfs, 229–235.*
© *1997 Kluwer Academic Publishers.*

## 2. Basic equations for static atmospheres

For the calculation of static plane-parallel NLTE model atmospheres we need the equations of hydrostatic, radiative, and statistical equilibrium, and the equation of radiative transfer. For the calculation of spherically symmetric model atmospheres we need to add an equation for radius.

### 2.1. DIFFERENCES BETWEEN PLANE-PARALLEL AND SPHERICALLY SYMMETRIC ATMOSPHERES.

Let us have a look on the above mentioned equations to see how they change when we switch from the plane-parallel geometry to the spherically symmetric one.

It is obvious that the radiative transfer equation changes significantly. The method of solution in both geometries is described in detail e.g. in Mihalas (1978, 1985) and I will not repeat it here. It should be noted that the solution for spherical geometry is much more time consuming. Nevertheless, the radiative transfer equation is the only equation that changes significantly.

The equations of statistical equilibrium, and the equation of radiative equilibrium in its integral form do not change at all. The equation of radiative equilibrium in the differential form and the equation of hydrostatic equilibrium change only a little if we introduce in the spherical geometry the generalized column mass depth $dm = -\rho(R^2/r^2)dr$, where $R$ is the reference stellar radius. Detailed form of all these equations were presented in Kubát (1996a).

## 3. The model atmosphere code

Our model atmosphere code is able to calculate both plane-parallel and spherically symmetric static NLTE model atmospheres. It is based on the accelerated lambda iteration method and has been described in Kubát (1994, 1996a). In addition, the population numbers of atomic levels are calculated using the occupation probability formalism of Hummer & Mihalas (1988) in its NLTE form (Hubeny et al. 1994).

## 4. Results

### 4.1. HYDROGEN WHITE DWARFS

In order to test my spherically symmetric code also for the case of hot high gravity stars I calculated (initially only as a joke) a pure hydrogen NLTE model atmosphere for $T_{\text{eff}} = 100000$K, $\log g = 8.0$, and $M = 0.6M_\odot$, and compared the results with the plane-parallel model (see Kubát 1995a). The temperature structures did not differ too much, the continuum fluxes were practically the same. On the other hand, I found a difference about 0.5% of the continuum flux in the core of $H_\alpha$ line. It is not a drastic difference, but since I had expected *no* difference due to a large gravity, it forced me to study this effect more systematically. I calculated a coarse grid of plane-parallel and spherically symmetric pure hydrogen NLTE model atmospheres for effective temperatures $T_{eff}$ = 60000K, 80000K, 100000K, 120000K, 140000K and surface gravities $\log g$ = 6.0, 7.5 (Kubát 1995b). I found differences in temperature structure which were rising with the effective temperature and which were larger for lower gravity. Temperature for spherically symmetric models was always lower, especially in the outer regions. Similar words about differences can be said about the continuum flux. For the coolest models with higher surface gravity the differences disappeared. Differences in Balmer line profiles were much more pronounced, up to 2% for the hottest model with lower gravity for the $H_\alpha$ line. For higher members of the Balmer series they tended to disappear. Detailed results (including figures) were presented in Kubát (1995b) and they will not be repeated here. An important idea that supported the sense of our calculations was to test the possibility of solving the so-called Balmer line problem (see Napiwotzki 1992, 1993). Although the differences found were lowering the Balmer line problem, they were too small to solve it. The Balmer line problem was then successfully solved by Werner (1996a,b) with the help of a detailed treatment of the opacity in the wings of lines of C, N, and O.

## 4.2. HYDROGEN DEFICIENT WHITE DWARFS

Recently, I extended my calculations of pure hydrogen model atmospheres to atmospheres consisting of hydrogen and helium in order to investigate the influence of increasing helium abundance on the sphericity effects (Kubát 1996b,c). I calculated a grid of plane-parallel and spherically symmetric models for $T_{eff}$ = 100000K and $\log g$ = 7.5 for different abundances of helium from a pure hydrogen to a pure helium atmosphere. Generally speaking, the differences are similar to the differences found in pure hydrogen atmospheres, i.e. lower temperature and deeper profiles for spherically symmetric models. Detailed discussion is presented in Kubát (1996c).

Here I would like to present calculations of pure helium models for $T_{eff}$ = 100000K and $\log g$ = 6.0 of an atmosphere of a white dwarf having a typical mass of $0.6 M_\odot$. Our helium model atom has 15 levels of He II and 1 level of He III. He I was neglected due to the large value of temperature, and, consequently, its negligible influence on the atmospheric structure. Atomic data (oscillator strengths, photoionization cross sections, collisional rates, line profiles) are the same as in Kubát (1996c) and I refer the reader there.

MODEL ATMOSPHERE                    CONTINUUM FLUX

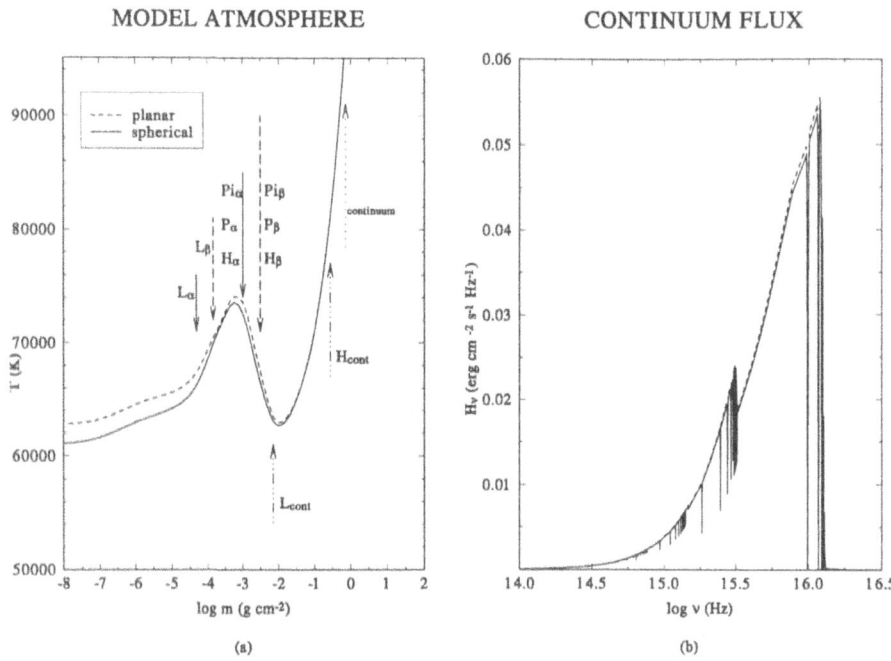

*Figure 1.* The NLTE model atmosphere (a) and emergent continuum flux (b) for a pure helium atmosphere with $T_{\text{eff}} = 100000$K, $\log g = 6.0$, and $M = 0.6M_\odot$. Full lines are spherically symmetric models, dashed lines denote the plane-parallel models. $T$ stands for temperature, $H_\nu$ for the flux, and $m$ for the column mass depth. Arrows indicate the depth of formation of selected He II lines.

The temperature structure of this model is presented on Fig.1a. As in all other models, the spherically symmetric model yields lower temperature in the outer regions than the plane-parallel one. In order to have a better insight into the line formation process, I indicated the depths of formation of particular lines and continua with arrows. The emergent flux (Fig.1b) is only a little lower for the spherically symmetric model. Larger differences can be found in the cores of lines, especially for the $\alpha$ lines. Particular examples of $P_\alpha$ and $Pi_\alpha$ lines are plotted on Fig.2. The profiles emergent from the spherically symmetric atmosphere are deeper. Although the difference is not drastic – only about 2% of the continuum flux in the core of $Pi_\alpha$ line and about 1% in the core of $P_\alpha$ line – plane-parallel atmosphere should not be assumed without any discussion, since the systematic error is always present.

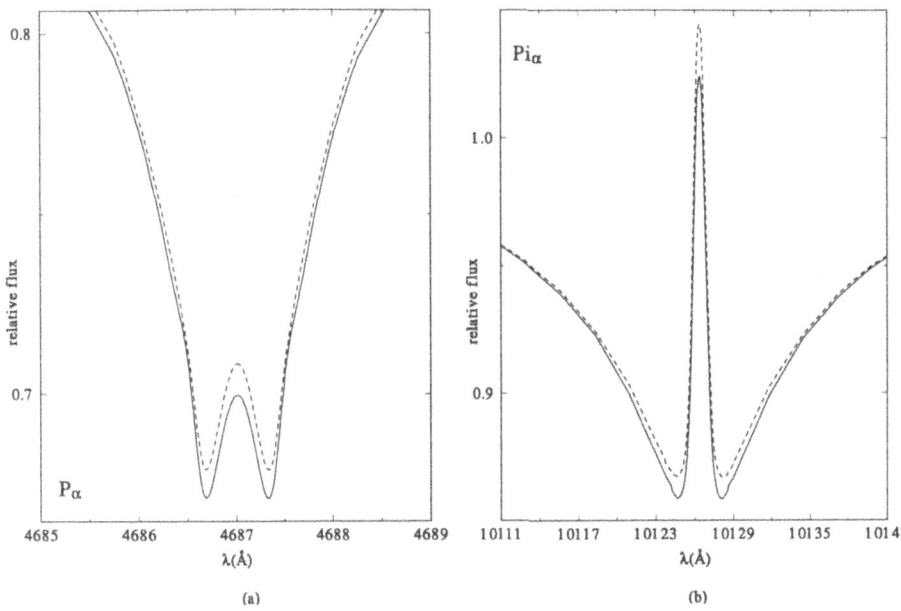

*Figure 2.* The He II $P_\alpha$ (a) and $Pi_\alpha$ (b) lines emerging from the pure helium NLTE model atmosphere $T_{eff} = 100000$K, $\log g = 6.0$, and $M = 0.6 M_\odot$. Full lines are spherically symmetric models, dashed lines denote the plane-parallel models.

## 5. Conclusions

The sphericity effects in the atmospheres of white dwarfs are not drastic, but they are not absent. They are more pronounced for stars with lower gravity. Neglecting of them introduces a *systematic error* into the resulting model atmosphere, and, consequently, into the theoretical emergent spectrum. However, for weak (photospheric) lines (that are forming at continuum optical depths of 1) the plane-parallel approximation is good. Care must be taken for lines forming above the photosphere (hydrogen $H_\alpha$, He II $H_\alpha$, $P_\alpha$, $Pi_\alpha$), since these lines may affect the estimation of the hydrogen-helium ratio.

### Acknowledgements

This work was enabled by continuous support from my family. It was also supported by grants of the Grant Agency of the Czech Republic (GA ČR) 205/94/0025 and 205/96/1198, by internal grants of the Academy of Sciences of the Czech Republic C3003601 and 303401, and by projects K1-003-601/4 and K1-043-601.

## References

Dreizler S., Werner K., 1993, A&A, 278, 199
Dreizler S., Werner K., 1994, Space Sci. Rev., 66, 147
Hubeny I., Lanz T., 1995, in White Dwarfs, D.Koester & K.Werner eds., Lecture Notes in Physics, Vol.443, Springer Verlag Berlin, p.98
Hubeny I., Hummer D.G., Lanz T., 1994, A&A, 282, 151
Hummer D.G., Mihalas D., 1988, ApJ, 331, 794
Kubát J., 1994, A&A, 287, 179
Kubát J., 1995a, in White Dwarfs, D.Koester & K.Werner eds., Lecture Notes in Physics, Vol.443, Springer Verlag Berlin, p.133
Kubát J., 1995b, A&A, 299, 803
Kubát J., 1996a, A&A, 305, 255
Kubát J., 1996b, in Hydrogen Deficient Stars, C.S.Jeffery & U.Heber eds., ASP Conf. Ser., Vol.96, p.266
Kubát J., 1996c, A&A, submitted
Lanz T., Hubeny I., 1995, ApJ, 439, 905
Mihalas D., 1978, Stellar Atmospheres, 2nd ed., W.H.Freeman & Comp., San Francisco
Mihalas D., 1985, J. Comput. Phys., 57, 1
Napiwotzki R., 1992, in The Atmospheres of Early-Type Stars, C.S.Jeffery & U.Heber eds., Lecture Notes in Physics, Vol.401, Springer-Verlag Berlin, p.310
Napiwotzki R., 1993, Ph.D. thesis, Universität Kiel
Werner K., 1996a, in Hydrogen Deficient Stars, C.S.Jeffery & U.Heber eds., ASP Conf. Ser., Vol.96, p.265
Werner K., 1996b, ApJ, 457, L39

## Discussion

*Y. Wu*: Is your 2% difference for some of the line cores consistent with the scale height - radius ratio in the white dwarf?

*J. Kubát*: The answer is not so simple as is the question. Simple geometrical considerations lead to a well known conclusion that for spherical atmospheres the radiation field is forward-peaked, i.e. that the angle cosine increases when we go along a ray outwards. This directly affects the radiation field.

Since the problem of a construction of a self-consistent NLTE model atmosphere is nonlinear, changes in the radiation field cause changes in population numbers and vice versa, so the sphericity effect can be amplified with respect to the LTE case.

*I. Iben*: Do the 1–2% effects you find translate into similar uncertainties in the abundance estimates which rely on the lines with such uncertainties?

*J. Kubát*: The effects are present only in lines which form in the outer layers of the atmosphere, well above the photosphere. If one uses only photospheric lines for abundance estimates, the results will not be influenced by the change of the geometry from plane-parallel to a spherically symmetric one. On the other hand, using the above-photospheric lines may cause some uncertainty. Unfortunately, it is not possible to make a reasonable estimate before the models are calculated, since the interrelations are too complicated. From the analysis of hydrogen and helium atmospheres I suggest to avoid using of $\alpha$ lines for abundance determinations.

# HEAVY ELEMENTS IN WHITE DWARF ENVELOPES

M.A. BARSTOW
*Department of Physics & Astronomy,*
*University of Leicester, UK*

J.B. HOLBERG
*Lunar and Planetary Laboratory,*
*University of Arizona, Tucson, AZ, USA*

I. HUBENY
*USRA, NASA/GSFC,*
*Greenbelt, MD, USA*

AND

T. LANZ
*Astronomical Institute,*
*Utrecht University, The Netherlands*

## 1. Introduction

Since the first soft X-ray and EUV studies of hot DA white dwarfs it has been apparent that the fluxes of white dwarfs in these wavelength ranges are significantly lower than predicted (e.g. Kahn *et al.* 1984; Petre, Shipman & Canizares 1986; Jordan *et al.* 1987; Paerels & Heise 1989; Koester 1989). All this work made the simplifying assumption that helium was the sole photospheric opacity source, although the He in these stars was never identified spectroscopically. Moreover, the shape of the EXOSAT spectrum of Feige 24 could only be matched by models including trace abundances of elements heavier than H or He (Vennes *et al.* 1989). More recent studies combining the soft X-ray and EUV data available from the *ROSAT* sky survey show that He alone cannot be responsible for absorption in these bands and that most of the hottest stars ($T_{\text{eff}} > 40,000$K) must have heavier elements in their atmospheres to explain the observed fluxes (Barstow *et al.* 1993; Wolff *et al;* 1996, Marsh *et al.* 1995). In parallel, observational evidence has been acquired by the International Ultraviolet Explorer (*IUE*) revealing high ionization features attributed to C, N, and Si (e.g. Bruhweiler & Kondo 1981 & 1983). Subsequently, O, Fe and Ni were also directly de-

*I. Isern et al. (eds.), White Dwarfs,* 237–243.

tected in the stars G191−B2B (e.g. Vennes *et al.* 1992; Holberg *et al.* 1994), Feige 24 (Vennes *et al.* 1992), REJ2214−912 (Holberg *et al.* 1993; Holberg *et al.* 1994) and REJ0623−377 (Holberg *et al.* 1993). From this work, the most obvious conclusion to be drawn is that the elements detected, in particular Fe and Ni which have large numbers of transitions in the EUV, are responsible for the observed EUV and soft X-ray opacity.

The availability of spectral data from *EUVE* presents us with an important opportunity for testing the hypothesis by matching the predicted EUV spectrum, based on abundances measured in the UV, with that observed. However, reconciling the shape and flux level of the EUV spectra of the most extreme high-opacity stars like G191−B2B, Feige 24, REJ2214−491 and REJ0623−371 has proved to be extremely difficult (e.g. Barstow *et al.* 1996a and references therein). Indeed, only in the case of the cooler Si-rich DA GD394 ($T_{\mathrm{eff}} = 38,500$ cf $T_{\mathrm{eff}} \approx 55,000 - 65,000$K) has it been possible to compute a model which gives complete agreement between optical, UV and EUV data (Barstow *et al.* 1996b). We present here the results of a new analyses of the optical UV and EUV spectra of the spectroscopic 'twin' stars G191−B2B and REJ0457−281, using improved non-LTE model atmosphere calculations in which the detail in the model atoms of Fe and Ni included is considerably enhanced. For G191−B2B excellent agreement can be achieved between all the data, the first time in a very hot, heavy element-rich star. Both the shape and flux level of the EUV spectrum can be reproduced but an unusually high HeII absorbing column is needed to yield a good match below the HeII 228Å edge. However, in REJ0457−281, while a good match to most of the EUV spectrum can be achieved the predicted flux remains too high at the shortest wavelengths and optical and far UV temperature determinations disagree.

## 2.  Observations and non-LTE model atmospheres

As one of the brightest DA white dwarfs, G191−B2B is the subject of intensive study and has been observed by many instruments, including *IUE*, *HST*, *ROSAT* and *EUVE*. Discovered as a result of the *ROSAT* survey, REJ0457−281 appears to be a spectroscopic twin of G191−B2B and, as a consequence, has been similarly well-studied. Recently, both objects have been observed using the far-UV *Orfeus* spectrometer (Vennes *et al.* 1996). In this paper we deal with the joint analysis of the *EUVE* spectra, the coadded *IUE* echelle spectra (described in detail by Holberg *et al.* 1994 for G191−B2B), the *Orfeus* archival data and optical Balmer line spectra.

We have constructed a small grid of non-LTE, metal line-blanketed model atmospheres, taking into account a total of 26 ions of H, He, C, N, O, Si, Fe, and Ni. The models were calculated using the program TLUSTY (e.g. Hubeny & Lanz 1995). For iron and nickel, all the levels predicted by Kurucz (1988) are included in the models, totaling over 70,000 individual energy levels grouped into 235 non-LTE superlevels. The effect of over 9.4 million iron and nickel lines is accounted for, representing a significant improvement over our previous non-LTE line-blanketed models for hot white dwarfs (Lanz & Hubeny 1995), where only some 11,000 iron lines were taken into account. The full details are discussed by Lanz et al. (1996).

## 3. Spectral analysis

The technique of determining $T_{eff}$ and log $g$ for a white dwarf by fitting the observed Balmer line profiles to the predictions of synthetic spectra is now well established. Our particular version of this method is documented in earlier papers (e.g. Barstow et al. 1994) and we adapted this to deal with the Lyman series lines present in the Orfeus data. Table 1 summarises the values of $T_{eff}$ and log g obtained independently from the Balmer and Lyman lines for each star, utilising the grid of heavy element line-blanketed models. While these separate analyses are in good agreement for G191−B2B, the temperature obtained from the Lyman lines for REJ0457−281 is substantially greater than that measured with the Balmer fits.

TABLE 1. Non-LTE $T_{eff}$, log g and abundance measurements for G191−B2B and REJ0457−281

| Parameter | G191−B2B | REJ0457−281 |
|---|---|---|
| $T_{eff}$ (Balmer) (K) | $58900 \pm 1150$ | $57850 \pm 1800$ |
| $T_{eff}$ (Lyman) (K) | $56600 \pm 600$ | $66500 \pm 1500$ |
| $T_{eff}$ (EUV cont.) (K) | 56000 | 62000 |
| log g (Balmer) | $7.36 \pm 0.07$ | $7.80 \pm 0.10$ |
| log g (Lyman) | $7.30 \pm 0.04$ | $7.42 \pm 0.08$ |
| C/H | $2.0 \times 10^{-6}$ | $1.0 \times 10^{-7}$ |
| N/H | $1.6 \times 10^{-7}$ | $1.0 \times 10^{-7}$ |
| O/H | $1.0 \times 10^{-6}$ | $1.0 \times 10^{-6}$ |
| Si/H | $3.0 \times 10^{-7}$ | $3.0 \times 10^{-7}$ |
| Fe/H | $1.0 \times 10^{-5}$ | $3.0 \times 10^{-6}$ |
| Ni/H | $2.0 \times 10^{-6}$ | $1.0 \times 10^{-6}$ |

We used the programme XSPEC to compare the EUVE data to EUV spectra generated from the model calculations, taking into account the contributions from overlapping higher spectral orders in the instrument effective area, as described in our earlier work (e.g. Barstow, Holberg and Koester 1995). However, since we were working with only few models span-

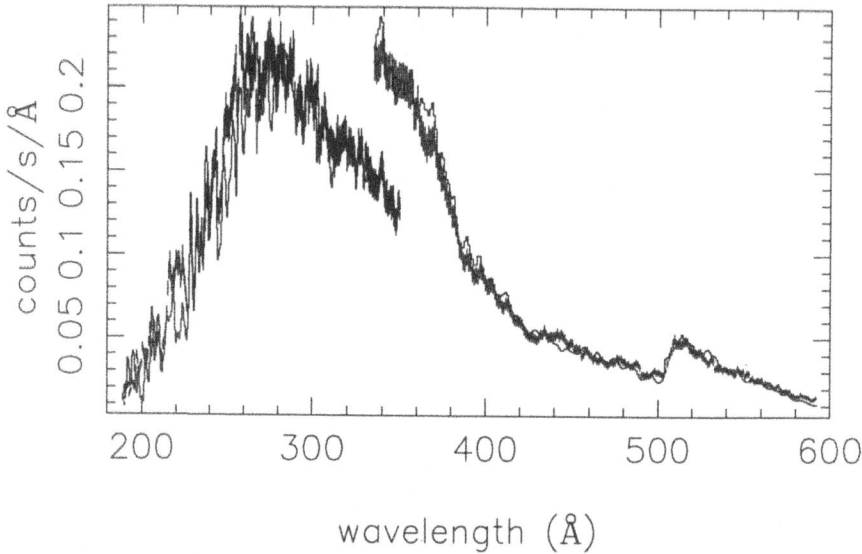

*Figure 1.* EUV count spectrum of G191−B2B covering the wavelength range 180–600Å. The data points (error bars) are compared with the predictions of non-LTE model (see table 1 for abundances) including the effects of interstellar absorption. (HI, HeI and HeII column densities are $2.1 \times 10^{18}$, $1.8 \times 10^{17}$ and $7.9 \times 10^{17}$ cm$^2$ respectively. The discontinuity near 320Å, where the MW and LW ranges overlap arises from differing spectrometer effective area, for which these data are not corrected.

ning a narrow range of abundances a detailed fit was not carried out except that the interstellar HI, HeI and HeII columns were allowed to vary freely to give the best possible agreement between model and data. As before, we used the interstellar model of Rumph, Bowyer & Vennes (1994) but modified to take account of the converging series of He lines near HeI and HeII edges in a manner similar to that described by Dupuis *et al.* (1995). In both G191−B2B and REJ0457−281 the models yield good agreement with the data (see table 1) but only when a significant quantity of HeII is included to suppress the predicted flux below 228Å (figures 1 and 2 respectively). The quality of the fit is very sensitive to the amount of Fe included. Indeed, a degradation in the quality of the agreement is seen at the 90% level if the ratio of Fe/H is increased or lowered by only 20% from the optimum values (see table 1). It is important to note that Fe and Ni are not directly detected in the UV spectrum of REJ0457−281 and that the abundance values quoted are those required to match the EUV continuum while remaining consistent with the upper limits imposed by *IUE*. In the case of REJ0457−281, reproduction of the EUV continuum flux at the longest wavelengths requires a stellar temperature of 62000K, lying in between the Balmer and Lyman results but which is similar to the estimate of Dupuis *et al.* (1995).

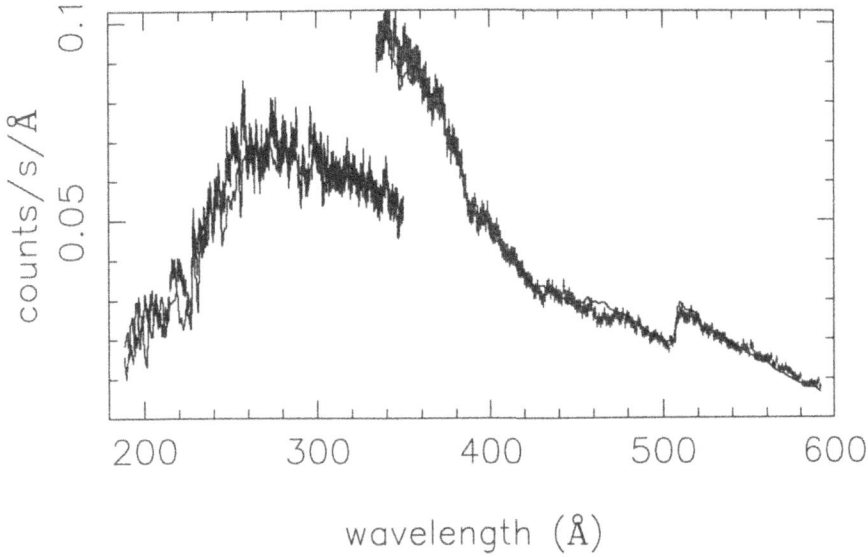

*Figure 2.* EUV count spectrum of REJ0457−281, as described in figure 1 for G191−B2B. (HI, HeI and HeII column densities are $1.1 \times 10^{18}$, $8.4 \times 10^{16}$ and $6.2 \times 10^{17}$ cm$^2$ respectively.

## 4. Discussion

In the EUV analyses we initially assumed that the necessary HeII opacity is completely interstellar, the best fit values for the HI, HeI and HeII column densities are then as given in the figure captions. If we assume that the amount of HeIII is negligible, the implied ionization fractions of He (HeII/[HeI+HeII] column densities) are 80% and 90% for G191−B2B and REJ0457−281 respectively, far greater than anything reliably reported elsewhere (e.g. Holberg *et al.* 1995 give 36% for GD659). Taking a cosmic value (10:1) for the total ratio of H:He in the ISM, the ionization fractions of H would be equally large. In the absence of any measurement against which to compare these figures, it is difficult to assess their plausibility. Such large ionization fractions could be avoided if the HeII were assumed instead to be circumstellar. We have no direct evidence for such a component in the far UV spectrum of G191−B2B, but there does appear to be circumstellar material in REJ0457−281 (see Holberg *et al.* these proceedings).

The HeII contribution need not be entirely in the line of sight material and could reside in the photosphere of these stars. Indeed, the many lines of heavy elements present in the EUV spectrum might easily be blended with members of the HeII Lyman series, masking the signature of photospheric He. Equally good fits to the EUV spectra can be obtained with no interstellar HeII and He abundances of $\approx 5.5 \times 10^{-5}$. However, at this abundance, the predicted strength of the HeII 1640Å (267mÅ) is above the limits of detection in the *IUE* spectra (140mÅ, He/H$\approx 2 \times 10^{-5}$). This problem could be avoided if the photospheric H and He were stratified (see

Vennes *et al.* 1988).

We have shown that recent progress in computing stellar atmosphere models with an unprecedented degree of realism enables us to reproduce the observed spectra in the EUV, far-UV and optical ranges almost fully consistently. Despite this, there are still some remaining problems. The next step will be to relax the simplifying assumption that the helium and heavier elements are homogeneously mixed.

## References

Barstow, M. A. *et al.* 1993, *MNRAS*, 264, 16

Barstow, M. A. *et al.* 1994, *MNRAS*, 267, 653

Barstow, M. A., Holberg, J. B. & Koester, D. 1995, *MNRAS*, 274, L31

Barstow, M. A., Hubeny, I., Lanz, T., Holberg, J. B., & Sion, E.M. 1996a, in *Astrophysics in the Extreme Ultraviolet* eds. S. Bowyer & R. F. Malina, p203, Kluwer

Barstow, M. A., Holberg, J. B., Hubeny, I., Lanz, T., Bruhweiler, F. C., & Tweedy, R. W. 1996b, *MNRAS*, 279, 1120.

Bruhweiler, F. C., & Kondo, Y. 1981, *ApJ*, 248, L123

Bruhweiler, F. C., & Kondo, Y. 1983, *ApJ*, 269, 657

Dupuis, J., Vennes, S., Bowyer, S., Pradhan, A. K., & Thejll, P. 1995, *ApJ*, 455, 574

Holberg, J. B. *et al.* 1993, *ApJ*, 416, 803

Holberg, J. B., Hubeny, I., Barstow, M. A., Lanz, T., Sion, E. M., & Tweedy, R. W., 1994, *ApJ*, 425, L105

Holberg, J.B., Barstow, M. A., Bruhweiler, F. C., & Sion, E. M., 1995, *ApJ*, 453, 313

Hubeny, I. & Lanz, T. 1995, *ApJ*, 439, 875

Jordan, S., Koester, D., Wulf-Mathies, C., & Brunner, H. 1987, *A& A*, 185, 253

Kahn, S. M., Wesemael,. F., Liebert, J., Raymond, J. C., Steiner, J. E., & Shipman, H. L. 1984, *ApJ*, 278, 255

Koester, D. 1989, *ApJ*, 342, 999

Kurucz, R. L. 1988, in IAU Trans., ed. M. McNally, Vol. XXB, (Kluwer: Dordrecht), 168

Lanz, T., Barstow, M.A., Hubeny, I., & Holberg, J.B., 1996, *ApJ*, in press

Lanz, T. & Hubeny, T. 1995, *ApJ*, 439, 905

Marsh, M.C. *et al.* 1995, In *Proceedings of the 9th European Worshop*, (ed. D. Koester & K. Werner). Lecture Notes in Physics, p 328, Springer.

Paerels, F. B. S., & Heise, J. 1989, *ApJ*, 339, 1000

Petre, R., Shipman, H. L., & Canizares, C. R. 1986, *ApJ*, 304, 356

Rumph, T., Bowyer, S. & Vennes S. 1994, *AJ*, 107, 2108

Vennes, S., Chayer, P., Fontaine, G., & Wesemael, F. 1988, *ApJ*, 336, L25

Vennes, S., Chayer, P., Hurwitz M., & Bowyer S. 1996, *ApJ*, in press.

Vennes, S., Chayer, P., Thorstensen, J. R., Bowyer, S., & Shipman, H. L. 1992, *ApJ*, 392, L27

Vennes, S., Pelletier, C., Fontaine, G., & Wesemael, F. 1988, *ApJ*, 331, 876

Wolff, B., Jordan, S., & Koester,D. 1995, *A& A*, 307, 149.

## Discussion

*P. Bradley*: I have contact with people doing line broadening work that might help. Do you already have contacts? Maybe we should get together and see if we can find someone doing Lyman line broadening.

*M. Barstow*: We already get data from the Munich group, but it would certainly be helpful to have contact with others.

*M. van Kerkwijk*: Could the relative strengths of the Balmer lines in REJ0457 be due to binarity (ie. another hottish white dwarf)? In this case the continuum flux distribution should show this.

*M. Barstow*: It is possible, but there is no evidence in the red for a companion. Perhaps we should look a bit harder, but the temperatures would not be too different to those we measure and it ought to be clear in the data. The CIV line doublet features (see Holberg *et al.* these proceedings) are not reproduced with other heavy element lines.

*J. Dupuis*: If you use the metals mixture derived from your non-LTE fit to the G191−B2B *EUVE* spectrum, would you get a fit with an LTE model?

*M. Barstow*: I have not tried it, but it seems that from the paper of Burkhard Wolff (these proceedings) that the temperatures become inconsistent between different wavebands. The 'measured' abundances would also appear to be rather different.

# LYMAN SPECTRA OF HOT DA WHITE DWARFS

DAVID S. FINLEY
*Eureka Scientific, Inc.*

DETLEV KOESTER
*Universität Kiel*

JEFFREY W. KRUK
*Johns Hopkins University*

RANDY A. KIMBLE
*Goddard Space Flight Center*

AND

NICOLE F. ALLARD
*Observatoire de Paris-Meudon*

**Abstract.** During the extremely successful Astro-2 mission flown in March 1995, high-quality spectra were obtained for 9 DA white dwarfs hotter than about 20,000 K with the Hopkins Ultraviolet Telescope. The spectra covered the range of 820–1840 Å and were thus ideal for our purpose of investigating Stark broadening in the Lyman series. The objects observed included GD 50, one of the most massive known DA; RE 0512-004, the least massive known DA in this temperature range; RE 1738+665, the hottest known field DA; the HST standards G191-B2B, GD 153, GD 71, and HZ 43; and GD 394 and Wolf 1346. Our major conclusion is that the wings of the Lyman lines are in reality significantly weaker than predicted using the standard (Vidal-Cooper-Smith) Stark broadening theory. We also discovered satellite features in the Lyman $\beta$ profile of Wolf 1346; these have now been successfully modeled using the same techniques that were previously applied to Ly $\alpha$ satellites.

## 1. Introduction

A correct treatment of Stark broadening is required for the accurate calculation of the opacities of stellar atmospheres. This is especially true for

*I. Isern et al. (eds.), White Dwarfs, 245–251.*
© *1997 Kluwer Academic Publishers.*

degenerate objects. Furthermore, since the hydrogen line profiles provide the best means for determining temperatures and gravities for DA white dwarfs, errors in Stark broadening theory will result in systematic errors in the values obtained from hydrogen line profile fitting. It was pointed out a few years ago (Bergeron, 1993) that there is some discrepancy between laboratory data and observed spectra of white dwarfs and predictions based on the current Stark broadening theory (Vidal, Cooper, & Smith, 1973). In fitting the Balmer lines of white dwarfs hotter than about 13,000 K, Bergeron found systematic differences in the parameters obtained from fitting the different Balmer lines. That was because the theory predicted that the absorption in the wings of the overlapping Balmer lines was stronger than observed, with the maximum effect in observed spectra being of the order of a few percent. Lacking a Stark theory that provided consistent fits, Bergeron obtained consistency by introducing a free parameter that increased the strength of the critical ionizing field in the Hummer-Mihalas occupation probability formalism (Hummer & Mihalas, 1988). Bergeron found that the most consistent results for fitting the Balmer line profiles were obtained by increasing the critical field strength by a factor of 2.

The effect of the error in the Stark opacities in the line wings is small in the Balmer lines of DA. However, at any given temperature the Lyman lines are much stronger than the Balmer lines; hence, errors of similar magnitude in the line opacities would result in much larger errors in calculated fluxes in the Lyman line region. Previous observations of the Lyman series in DA white dwarfs have been unsuitable for critical evaluations of Stark broadening. Two were observed with HUT on Astro-1: G191-B2B (Kimble et al. , 1993a) and HZ 43 (Kimble et al. , 1993b). G191-B2B and MCT 0455-2812 were observed with the ORFEUS spectrometer (Hurwitz & Bowyer, 1995; Vennes et al. , 1996). Given that they were all very hot DA, with $T_{eff} > 50,000$ K, the differences in the line profiles due to temperature variations were not large enough to serve well for testing model effects, and both G191-B2B and MCT 0455-2812 had the added complication of having significant photospheric abundances of heavy elements. Also, in that temperature range, models with and without Bergeron's correction differed by at most several percent in the Lyman lines. However, at 20,000 K, while the models with and without the correction differed by only 2% in the Balmer lines, the differences were nearly 40% in the Lyman lines, simply due to the much greater strength of the Lyman lines.

Recently, the opportunity arose to observe the entire Lyman line series in a number of DA, with the announcement of the reflight of the Hopkins Ultraviolet Telescope (HUT, Davidsen et al. , 1992) as part of the Astro-2 mission (Kruk et al. , 1995). Given that the effect of Stark profile errors was predicted to be quite large for the Lyman lines, it was clearly desirable to obtain high-quality FUV spectra of a number of hot white dwarfs over a wide range of temperatures and gravities in order to evaluate the accuracy of the current Stark theory for the hydrogen Lyman series.

## 2. Observations

Nine DA were observed during the course of the Astro-2 mission aboard the space shuttle *Endeavour*, which took place 2-18 March, 1995. The white dwarf observations were collaboratively obtained as part of an Astro-2 Guest Investigator program (Finley, Kimble, & Koester) and as part of the in-flight HUT instrument calibration effort. The HUT instrument had a peak effective area of just under 25 cm$^2$, and the resolution through the Lyman series ranged from 2 to 4.5 Å FWHM. Exposure times usually ranged from about 800 to 1300 seconds. The peak per-pixel signal-to-noise (S/N) typically exceeded 30. Some targets were observed more than once, some with different instrument configurations as part of the calibration effort. The primary observations of all the DA white dwarfs except G191-B2B were obtained in the fully-open door state. Due to its brightness, G191-B2B was only observed in partially-open door states.

## 3. Calibration

The spectra presented here were fluxed using the preliminary in-flight calibration of HUT described in Kruk *et al.* (1995). The primary calibration, for the instrument with fully-open front doors, was based on the model for HZ 43 using the optically-determined $T_{eff}$ and log $g$ with $f_\beta = 2$, scaled to the V magnitude measured with the HST FOS (Bohlin, Colina, & Finley, 1995). The G191-B2B observation was not suitable for direct comparison with the other spectra, because the G191-B2B model was used to calibrate the instrument in the state in which G191-B2B was observed.

## 4. Results

The calibrated spectra and models for HZ 43 and GD 71 are shown in Fig. 1. This demonstrates that over a 20,000 K temperature range, the models with $f_\beta = 2$ consistently fit the Balmer line profiles and the FUV flux in the Lyman series. For HZ 43, GD 71, GD 153, GD 50, RE 0512-004, the observed spectra and models based on Balmer line profile fits (Finley, Koester, & Basri, 1997) all agreed to within about 5%; the spectra for the latter three are shown in Fig. 2. Fits to the Lyman line profiles of these objects using Ly $\beta$ through Ly $\epsilon$ yielded temperatures and gravities that were consistent with the Balmer fit results at a 2 $\sigma$ (or better) level.

A detailed analysis of the line profiles will be performed after the final calibration is complete.

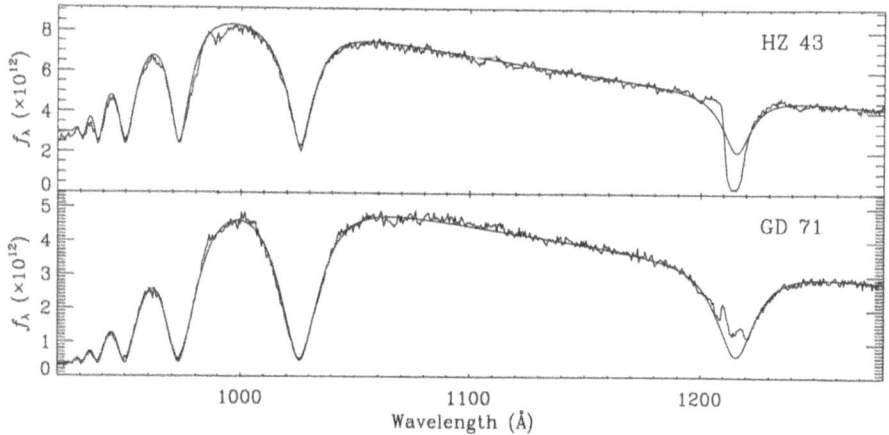

*Figure 1.* The top panel shows the Lyman series from the observed spectrum for HZ 43 and the model calculated with $T_{eff}$/log $g$/V = 50,000 K/8.0/12.914. The bottom panel includes the observed spectrum for GD 71 and the model (32,750 K/7.68/13.032). Both models were calculated with $f_\beta = 2$.

Fitting the observed optical spectra using a model grid with $f_\beta = 1$ yielded slightly different temperatures and gravities. Comparing the HUT spectra to the best-fit models using $f_\beta = 1$ (see Fig. 3), we found a discrepancy for HZ 43 of as much as 7%. The maximum difference for GD 71, however, was about 35%. For both, the models with $f_\beta = 1$ included too much opacity where the line wings overlapped. Also, given the differential effect with $T_{eff}$, fluxing the spectra using the HZ 43 model with $f_\beta = 1$ could not produce consistent fits to the other spectra.

The other DA white dwarfs that were observed include G191-B2B, which is metal-rich, and has not been fully analyzed yet. In the case of GD 394, we could not consistently fit both the Balmer and Lyman lines and FUV continuum fluxes, presumably due to the presence of anomalously high trace element abundances in this star (Shipman & Finley, 1994; Barstow *et al.*, 1996). RE 1738+665 has a significant $H_2$ column density, complicating its analysis. Wolf 1346, at 20,000 K, had previously unobserved satellite features of Ly $\beta$ that have since been successfully modeled (Koester *et al.*, 1996).

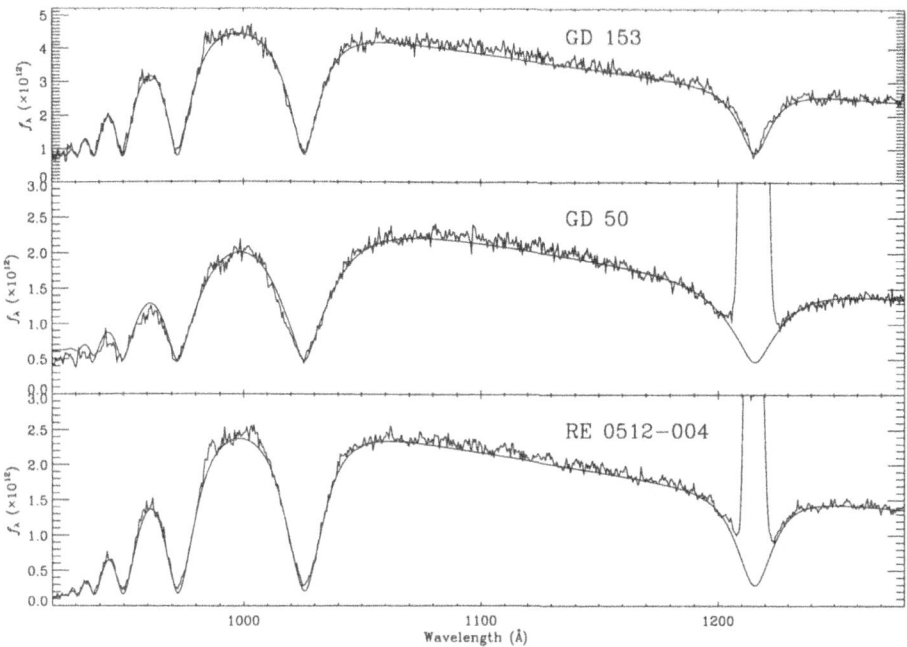

*Figure 2.* Observed spectra and models for GD 153 (38,700 K/7.66/13.346), GD 50 (42,300 K/9.1/14.04), and RE 0512-004 (31,900 K/7.3/13.78). All models were calculated with $f_\beta = 2$. The spectra of GD 50 and RE 0512-008 have not been background subtracted.

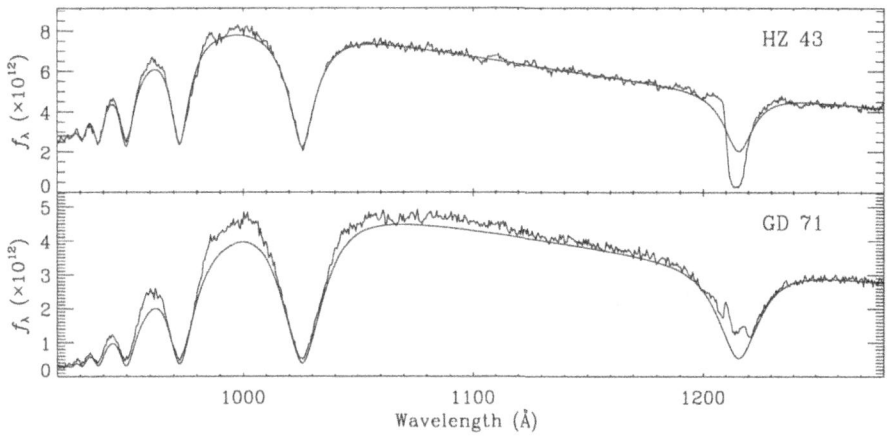

*Figure 3.* Same as Fig. 1, except that the models were calculated with $f_\beta = 1$, using parameters obtained by fitting the Balmer lines using a model grid with $f_\beta = 1$ (49,000 K/8.0/12.914 and 31,900 K/7.7/13.032).

## 5. Conclusions

Our fundamental result to date is that the use of models with $f_\beta = 2$ provides consistent fits to the Balmer line profiles, the Lyman line profiles, and the FUV fluxes over a very wide range of effective temperatures and gravities. On the other hand, the "standard" models predict fluxes between the Lyman lines that are substantially lower than observed, indicating that the VCS opacities are too great in the line wings.

## References

Barstow, M. A., Holberg, J. B., Hubeny, I., Lanz, T., Bruhweiler, F. C., & Tweedy, R. 1996, MNRAS, 279, 1120

Bergeron P. 1993, in White Dwarfs: Advances in Observation and Theory, ed. M. Barstow (Kluwer: 1993), p. 267

Bohlin, R. C., Colina, L., & Finley, D. S. 1995, AJ, 110, 1316

Davidsen, A. F. et al. 1992, ApJ, 392 264

Finley, D. S., Koester, D., & Basri, G. 1997, ApJ, submitted

Hummer D. G., & Mihalas D. 1988, ApJ, 331, 794

Hurwitz, M., & Bowyer, S. 1995, ApJ, 446, 812

Kimble, R. A. et al. 1993a, ApJ, 404, 663

Kimble, R. A., Davidsen, A. F., Long, K. S., Feldman, Paul D. 1993b, ApJ, 408, L41

Koester, D., Finley, D. S., Allard, N. F., Kruk, J. W., & Kimble, R. A. 1996, ApJ, 463, L93

Kruk, J. W., Durrance, S. T., Kriss, G. A., Davidsen, A. F., Blair, W. P., Espey, B. R., Finley, D. S. 1995, ApJ, 454, L1

Shipman, H. & Finley, D. S. 1994, BAAS, 184, 07.08

Vennes, S., Chayer, P., Hurwitz, M., & Bowyer, S. 1996, ApJ, 468, 898

Vidal, C. R., Cooper, J., & Smith, E. W. 1973, ApJS, 25, 37

## Discussion

*J. Kubat*: Do you include also the photoionization opacity below the ionization edge in your Hummer-Mihalas occupation probability formalism?

*D. Finley*: Yes.

*H. Shipman*: How much molecular hydrogen is needed to fit the data [for RE 1738+665] – and is this consistent with what we know about molecular hydrogen in the ISM?

*D. Finley*: So far, I've only determined that it is $H_2$ that is responsible for the extra absorption features we see. We will do a quantitative analysis in the future.

*M. A. Barstow*: RE 1738 appears to have a PN associated with it but the connection between star and gas cloud is rather tenuous. The detection of molecular hydrogen may just provide that evidence.

# DIFFUSION OF SILICON IN THE ENVELOPES OF HOT DA WHITE DWARFS

P. CHAYER
*Center for EUV Astrophysics*
*2150 Kittredge St., University of California, Berkeley, CA 94720-5030, USA*

G. FONTAINE
*Département de Physique, Université de Montréal*
*C.P. 6128, Succ. Centre-Ville, Montréal,*
*Québec, Canada, H3C 3J7*

AND

C. PELLETIER
*Bell-Northern Research*
*16 Place du Commerce, Verdun, Québec, Canada, H3C 1H6*

## 1. Introduction

White dwarf stars are expected to have peculiar elemental abundances in their superficial layers because of their high gravities and thin atmospheres. The gravitational settling, if left unimpeded, is so efficient in these stars that it very rapidly depletes the atmosphere of its heavy element content, leaving only the lightest elements: hydrogen or helium. The detection of traces of heavy elements in the atmospheres of white dwarfs, however, clearly implies that some mechanisms must be at work to counteract the effects of downward settling.

Vauclair, Vauclair, & Greenstein (1979) showed that radiative levitation is an efficient mechanism to keep heavy elements in the atmospheres of hot white dwarfs. The line spectra of heavy elements, under favorable conditions, can generally gain enough momentum through bound-bound absorption from the radiation field to establish an equilibrium between the radiative and gravitational accelerations. Chayer, Fontaine, & Wesemael (1995a) and Chayer et al. (1995b) extended and significantly improved the radiative levitation theory by including, among others, the atomic data computed in

*I. Isern et al. (eds.), White Dwarfs, 253–258.*

the Opacity Project for all cosmically abundant elements (Seaton et al. 1992). Equilibrium abundances were then computed with a large set of DA and DB/DO white dwarf models in order to compare the predicted abundances and those observed in a handful of white dwarfs. Chayer et al.'s results, though, demonstrate that the equilibrium radiative levitation theory is unable to explain quantitatively the observed abundances.

The observed silicon abundances, in particular, are challenging the predictions of radiative levitation theory. For example, in a few DA stars where $T_{eff} > 50,000$ K, the observed silicon abundances exceed the predicted abundances by at least one order of magnitude. This result is surprising because, at these temperatures, the silicon is mainly in its noble gas configuration (Si V), which has excitation energies much greater than the energy corresponding to the maximum of the Eddington flux. Under this condition, radiative levitation can support only small abundances of silicon, while, in contrast, relatively large amounts of silicon are observed. Moreover, for stars having $T_{eff} < 50,000$ K, the silicon seems to have disappeared from the atmosphere even though the radiative levitation theory predicts observable abundances.

Chayer et al. (1995a) pointed out in their concluding remarks that the equilibrium radiative levitation theory was used in its simplest form by equalizing the radiative and gravitational accelerations. Other mechanisms such as mass loss and accretion may have substantial effects on the abundances of heavy elements in the atmospheres of white dwarfs. Consequently, we examine in this paper the effects of mass loss and accretion on the silicon abundance by carrying out time-dependent diffusion calculations in the presence of radiative levitation including either mass loss or accretion in envelopes of hot DA white dwarf models.

## 2. Computations

We consider the problem of H-Si chemical separation in the presence of radiative levitation and mass loss (or accretion) in a model envelope of a typical hydrogen-rich white dwarf with $M/M_\odot = 0.5$, $\log g = 7.36$, and $T_{eff} = 60,000$ K. The evolution of the silicon distribution through the stellar envelope is determined by solving the following transport equation,

$$\frac{\partial}{\partial t}(\rho X_2) = -\frac{1}{r^2}\frac{\partial}{\partial r}[r^2 \rho X_2(w_2 + w_m)], \tag{1}$$

assuming a stellar plasma made of two ionic species of average charge $Z_1$ (hydrogen) and $Z_2$ (silicon). In this expression, $\rho$ is the density, $X_2$ is the mass fraction of silicon, $t$ is the time coordinate, $r$ is the radial coordinate, $w_2$ is the microscopic diffusion velocity of silicon with respect to the star's center of mass, and $w_m$ is a global macroscopic velocity due to the presence of mass loss. The mass loss rates of interest in the present context are sufficiently small to have negligible effects on the structure of the stellar envelope. The appropriate velocity law for a given mass loss rate, $\dot{M}$, is independent of the trace element and, thus, expresses the mass conservation of the dominant element (hydrogen in the present case),

$$w_m = \frac{\dot{M}}{4\pi r^2 \rho}. \tag{2}$$

The microscopic diffusion velocity of silicon in presence the of radiative levitation is given by (see Vennes et al. 1988),

$$\begin{aligned}
w_2 &= \frac{D_{12}(1+\gamma)}{1+(\gamma A_2)/A_1}\left[\frac{-\partial \ln c_2}{\partial r} + \left(\frac{A_1 Z_2 - A_2 Z_1}{Z_1 + \gamma Z_2}\right)\frac{m_p g}{kT}\right.\\
&\left. + \left(\frac{Z_2 - Z_1}{Z_1 + \gamma Z_2}\right)\frac{\partial \ln p_i}{\partial r} + \left(\frac{A_1 A_2}{A_1 + \gamma A_2}\right)\frac{m_p g_{\mathrm{rad}}(\gamma)}{kT}\right], \tag{3}
\end{aligned}$$

where $D_{12}$ is the diffusion coefficient, $\gamma \equiv n_2/n_1$ is the silicon-to-hydrogen number density ratio, $c_2 \equiv n_2/(n_1 + n_2)$, $Z_i$ $(A_i)$ is the average charge (atomic weight) of ions of species $i$ [$= 1$ (H), 2 (Si)], $m_p$ is the proton mass, $g$ is the local gravity, $k$ is the Boltzmann constant, $T$ is the temperature, $p_i = p_1 + p_2$ is the ionic pressure, and $g_{\mathrm{rad}}(\gamma)$ is the upward selective radiative acceleration exerted on silicon in the plasma and function of $\gamma$. The radiative acceleration is computed following the work of Chayer et al. (1995a,b)

We must specify the boundary and initial conditions to resolve equation (1) because this equation represents an initial-boundary-value problem in which the unknown $X_2$ (the silicon abundance) is a function of both time and space. We fix the boundary conditions at the top and bottom of the envelope as all velocity flows are in the radial direction. The bottom of the envelope is chosen to be sufficiently deep that the silicon abundance can be regarded as constant, while at the top of the envelope the rate of change of the surface abundance of silicon is directly proportional to the diffusion velocity of silicon evaluated at the Rosseland photosphere. For our accretion models, we add a source term to the surface boundary condition, as was done by Dupuis et al. (1993), and take $w_m = 0$ in equation (1). The initial silicon abundance through the envelope is assumed to be solar, $\log \mathrm{Si/H} = -4.5$, except in layers where the solar abundance is smaller than the radiative equilibrium abundance. In this case we take the abundance predicted by the radiative levitation theory at equilibrium (see, e.g., the upper and lower left panel of Fig. 1). Finally, we use a Galerkin finite-element method to solve equation (1) (see Pelletier, Fontaine, & Wesemael 1989).

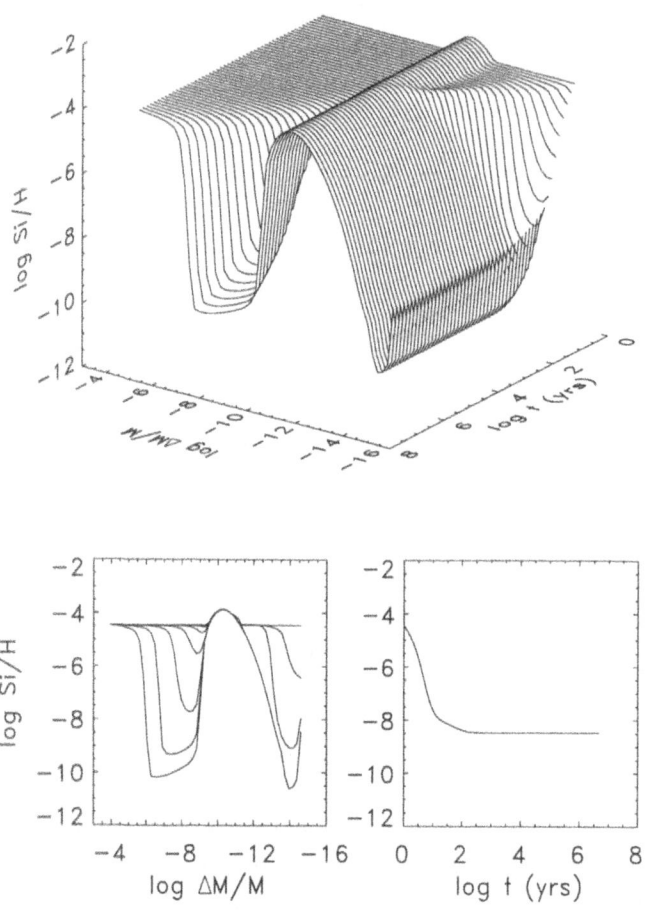

*Figure 1.* Evolution of the silicon distribution (upper and lower left panel) and surface abundance (lower right panel) in presence of radiative levitation in an envelope model with $M/M_\odot = 0.5$, log $g = 7.36$, and $T_{eff} = 60,000$ K.

## 3. Results

We show that the equilibrium radiative levitation approach of Chayer et al. (1995a) is appropriate for estimating the abundances of heavy elements in white dwarfs if no other mechanisms exist to counterbalance the gravitational settling. The upper panel of Figure 1 illustrates the evolution of the silicon distribution in our envelope model in the presence of radiative levitation only. At the beginning of our numerical simulation, we can see that the silicon distribution decreases very rapidly in the outermost layers because of the high settling velocity and reaches a state of diffusive equilibrium in which settling is counterbalanced by radiative levitation.

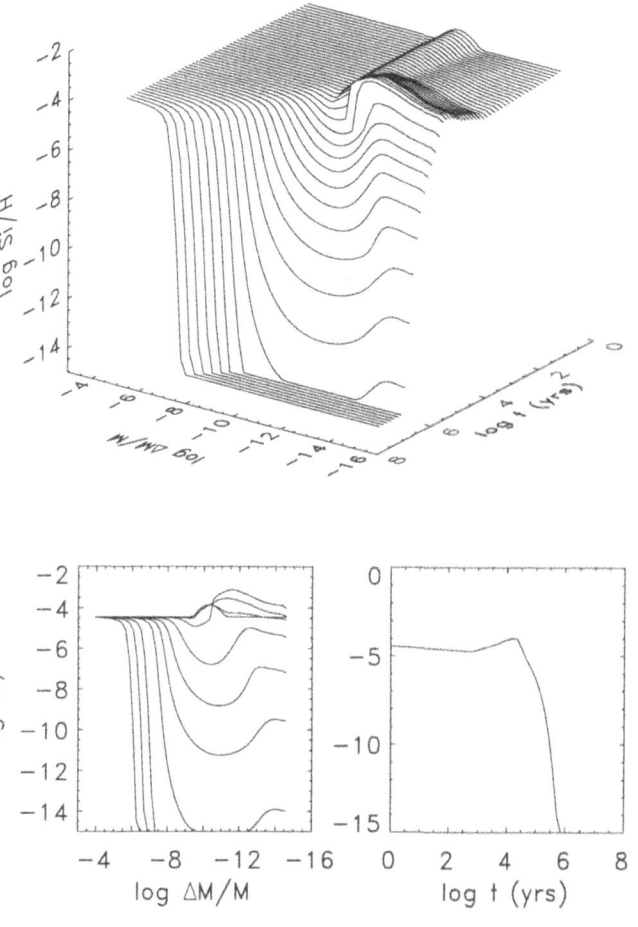

*Figure 2.* Same as in Fig. 1, but with the addition of a mass loss rate of $\dot{M} = 10^{-14}$ M$_\odot$ yr$^{-1}$.

After the silicon abundance has reached that state, it does not change any more as illustrated in the lower right panel of Figure 1, which shows the surface abundance of silicon as a function of time. We note the same trend in the interior where the radiative support is important in the region $-13 < \log \Delta M/M < -9$. In this region a reservoir of silicon builds up with time and reflects the equilibrium radiative support. In fact, the shape of the silicon reservoir is essentially the same as that obtained on the basis of equilibrium radiative levitation theory in which one imposes $w_2 = 0$, *a priori*. Finally, in deeper layers, where radiative levitation becomes negligible, settling goes on unimpeded. The lower left panel of Figure 1 illustrates the evolution of the silicon distribution and the formation of the reservoir in the interior. Each curve represents the silicon distribution at a given time. The last profile in the upper and lower left panel of Figure 1 corresponds to the age of our model, $4.67 \times 10^6$ yrs.

Figure 2 shows that a mass loss of $\dot{M} = 10^{-14}$ $M_\odot$ yr$^{-1}$ completely changes the evolution of the silicon distribution in the presence of radiative levitation. Unlike the case without mass loss, the abundance practically does not vary during the first fifty thousand years and plummets to very small values everywhere in the envelope after that time. The velocity field, generated by the mass loss and given by equation (2), exceeds the settling velocity near the surface of the star and then carries the silicon from the interior toward the surface. This temporary steady state is not maintained because a depth exists (where $g_{\mathrm{rad}} < g$) below which silicon can only settle toward the center of the star. When the amount of silicon is nearly spent above that point, large underabundances appear throughout the envelope and at the surface as shown by each panel of Figure 2.

We carried out numerical simulations with other mass loss rates, and with accretion flows as well. Because of a lack of space, we simply summarize our results in what follows. For a mass loss rate $\dot{M} \leq 10^{-18}$ $M_\odot$ yr$^{-1}$, the silicon abundance profile is essentially given by the equilibrium radiative levitation theory after $4.67 \times 10^6$ yrs of evolution (the age of the model). For $10^{-16} \leq \dot{M} \leq 10^{-14}$ $M_\odot$ yr$^{-1}$, however, the outermost silicon reservoir predicted by radiative levitation is completely depleted. In contrast, a mass loss rate of $\dot{M} = 5 \times 10^{-13}$ $M_\odot$ yr$^{-1}$ is sufficient to maintain relatively large values of the silicon abundance ($\log N(\mathrm{Si})/N(\mathrm{H}) > -5.0$) throughout the envelope as is actually oberved. Similarly, accretion in solar proportion at rates ranging from $10^{-17}$ to $5 \times 10^{-16}$ $M_\odot$ yr$^{-1}$ can maintain silicon abundance of $-6.5 \leq \log N(\mathrm{Si})/N(\mathrm{H}) \leq -5.0$ in the atmosphere of our white dwarf models. Thus, the observed pattern of silicon abundances in these objects can be explained through either mass loss or accretion.

This work is supported by NASA contract NAS5-29298. We also acknowledge financial support from the Natural Sciences and Engineering Research Council of Canada and from the fund FCAR (Québec).

## References

Chayer, P., Fontaine, G., & Wesemael, F. 1995a, ApJS, 99, 189

Chayer, P., Vennes, S., Pradhan, A. K., Thejll, P., Beauchamp, A., Fontaine, G., & Wesemael, F. 1995b, ApJ, 454, 429

Dupuis, J., Pelletier, C., Fontaine, G., & Wesemael, F. 1993, ApJS, 84, 73

Pelletier, C., Fontaine, G., & Wesemael, F. 1989, in IAU Colloq. 114, White Dwarfs, ed. G. Wegner (New York: Springer), 249

Seaton, M.J., Zeipper, C.J., Tully, J.A., Pradhan, A.K., Mendoza, C., Hibbert, A., & Berrington, K.A. 1992, Rev. Mexicana Astron. Af., 23, 19

Vauclair, G., Vauclair, S., & Greenstein, J.L. 1979, A&A, 80, 79

Vennes, S., Pelletier, C., Fontaine, G., & Wesemael, F. 1988, ApJ, 331, 876

# TIME-DEPENDENT DIFFUSION CALCULATIONS FOR HOT PRE-WHITE DWARFS

KLAUS UNGLAUB AND IRMELA BUES

*Astronomisches Institut der Universität Erlangen-Nürnberg*
*Dr.Remeis Sternwarte,*
*Sternwartstr. 7, 96049 Bamberg, Germany*

**Abstract.** The evolution of the chemical composition in the outer regions (with surface layer masses $\leq 10^{-8} M_*$) of hot pre-white dwarfs with time is investigated, because the results of previous works do not sufficiently clarify the influence of gravitational settling and the selective radiative forces in the helium- carbon- and oxygen-rich atmospheres of PG 1159 stars with effective temperatures around 100000 K.

## 1. Introduction

The PG 1159 stars are probably in an intermediate evolutionary stage between the central stars of planetary nebulae and the white dwarfs of spectral type DO. Their atmospheres are helium-rich with no detectable amounts of hydrogen. Typical number fractions of carbon and oxygen are C/He $\approx 0.5$, O/He $\approx 0.1$. The effective temperatures are in the range $65000K \leq T_{eff} \leq 170000K$, the surface gravities in the range $6.0 \leq \log g \leq 8.0$. For a review about these objects see Werner et al. (1996). To explain the surface chemistry, Werner et al. (1991) and Kawaler & Bradley (1994) proposed the "born-again AGB star" scenario: a post-AGB star suffers a late He shell flash, which poses it back onto the AGB. If the mass loss rate during the second superwind phase and subsequent phases is high enough, deep carbon- and oxygen-rich intershell layers become visible at the stellar surface.

In a recent paper (Unglaub & Bues, 1996; Paper I) we started the investigation of an alternative scenario, according to which the effects of gravitational settling and selective radiative forces are responsible for the strange surface composition of PG 1159 stars. According to the results, the feedback effects between chemical composition and radiation flux and the concentration gradients, which have been neglected in previous works, un-

*I. Isern et al. (eds.), White Dwarfs, 259–263.*

der certain circumstances may lead to large abundances of heavy elements in the outer regions with surface layer mass fractions of about $10^{-8} M_*$. For $T_{\text{eff}} = 140000$K, $\log g = 6.0$ the time-independent diffusion calculations of Paper I predicted a thin metal-rich region floating ontop of a helium-rich mantle. According to theoretical evolutionary tracks for post-AGB stars of Wood & Faulkner (1986) the objects PG 1159-035 and PG 1520+525 (both with $T_{\text{eff}} = 140000$K, $\log g = 7.0$; Werner et al., 1991) should have passed through a region in the HRD with these model parameters within a time scale of about 10000 y. If the selective radiative forces were able to transform an originally helium-rich atmosphere with only traces of heavy elements into a metal-rich one within this time scale, diffusion could be an important mechanism in explaining the surface composition. To clarify the influence of gravitational settling and selective radiative, time-dependent diffusion calculations are necessary. For a complete solution the diffusion calculations should be incorporated into a stellar evolution code. As a first step, in our diffusion calculations we assume fixed values of $T_{\text{eff}}$ and $\log g$, which are typical for theoretical post-AGB evolutionary tracks or observed objects. In spite of this simplification the influence of gravitational settling and selective radiative forces in the outer stellar regions with surface layer masses $\leq 10^{-10} M_*$ can be understood sufficiently well. The results are published in Unglaub & Bues (1997; Paper II). Here some additional results will be shown with slightly varied initial conditions.

## 2. Physical assumptions and numerical method

We take into account the elements He, C, N, O and Ne, assuming a plane-parallel, one dimensional statification with no mass-loss. The radiative transfer equation is solved by use of the diffusion approximation. For each element $l$ one momentum equation is used according to Burgers (1969):

$$\frac{dp_l}{dz} - n_l m_l g + n_l \overline{Z}_l eE + n_l \overline{F}_{l,\text{rad}} = \sum_t K_{l,t} (w_t - w_l) \tag{1}$$

The mean radiative force $\overline{F}_{l,\text{rad}}$ acting on the particles of element $l$ is obtained as described in Paper I. The resistance coefficients $K_{l,t}$ are calculated according to Paquette et al. (1986), the electric field $E$ is obtained from the momentum equation of the free electrons, as described in Paper II. Because one of the momentum equations is redundant, it has to be replaced by the condition of zero mass flow. The set of linear algebraic equations can be solved for the diffusion velocities $w_l$ of the various elements.

To solve the time-dependent diffusion problem, an Eulerian differencing scheme is used, which is based on a finite difference approximation for the

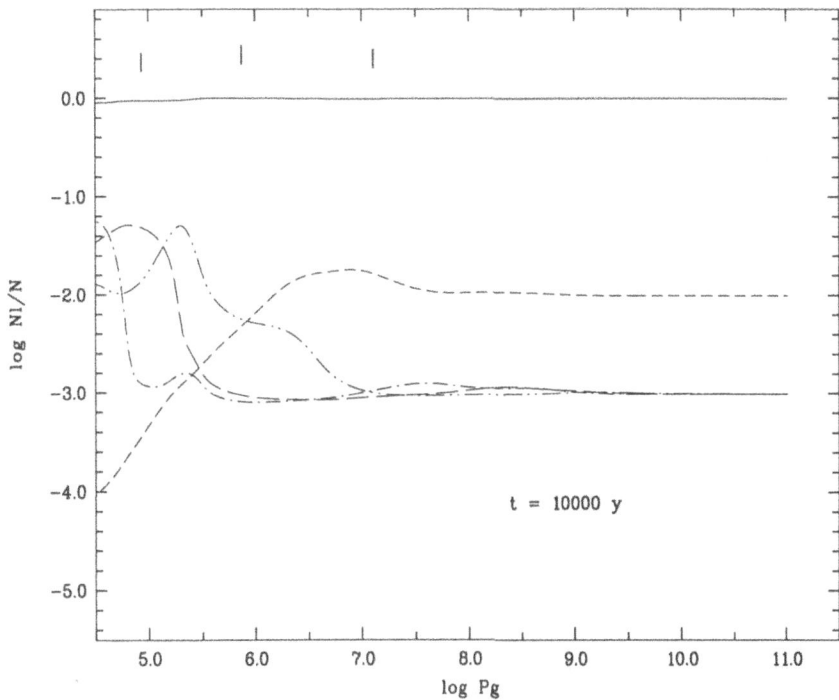

*Figure 1.*    Number fractions of the elements He (fat line), C (- - - -), N (- · - · -), O (–
– – – -), Ne (- ·· - ·· -) as a function of the gas pressure for $T_{eff} = 140000$K, $\log g = 6.0$
after 10000 y. The number fractions are on a logarithmic scale and are defined as the
ratio of all particles of an element to all heavy particles. The gas pressure is in SI-units!
The three tick marks in the upper part of each figure show where the Rosseland mean
optical depth is $\bar{\tau} = 1, 10, 100$, respectively.

spatial part of the differential equations. The details are described in Paper
II.

## 3.  Results

The results shown in Figs. 1, 2 and 3 refer to $T_{eff} = 140000$K, $\log g = 6.0$. As
already mentioned, these model parameters are fixed during the diffusion
calculations.

At a time $t = 0$ we assume the following number fractions for the heavy
elements: C/He=$10^{-2}$, N/He=O/He=Ne/He=$10^{-3}$. The lower boundary
condition is such that the composition remains constant there, so that for
long times the abundance distribution should converge to the stationary
case, for which we predicted an atmosphere dominated by heavy elements
with traces of helium only (see Fig. 4a in Paper I). Figure 1 shows the situ-

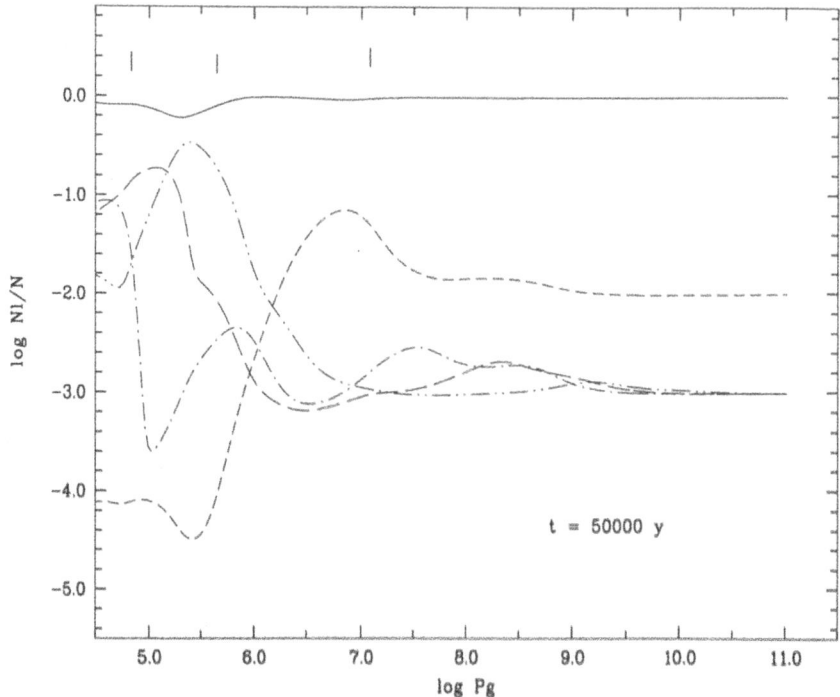

*Figure 2.* The same as Fig.1 after 50000 y.

ation after 10000 y. Near $\bar{\tau} = 1$ the number fraction of oxygen is about 5%, which is only slightly lower than the observed value. However, the results for carbon are disillusioning. In the whole computation domain, its number fraction never exceeds 2%. Near $\bar{\tau} = 1$ the carbon abundance has even decreased. Carbon is mainly $C^{4+}$ in these regions, therefore the radiative forces acting on it are not effective. It is a common feature of all diffusion calculations done up to now, that for effective temperatures around 100000 K carbon is always predicted to be a trace element. Therefore diffusion processes alone cannot explain the observed abundances.

Fig. 2 shows the situation after 50000 y. This time step is already longer than the time scales of stellar evolution of post-AGB stars in the corresponding region of the HRD (according to the evolutionary track in Fig. 1d of Wood & Faulkner, 1986, for $M_* = 0.6M_\odot$). The atmosphere is still far from a steady state, helium is still the most abundant element in all depths. The maximal number fraction of carbon is 7% only near $\bar{\tau} = 100$. So it becomes evident that the transformation of a helium-rich into a metal-rich atmosphere is not possible in time scales which are comparable to the time scales of stellar evolution. This would take at least 200000 y, as can be seen from Paper II.

## 4.  Conclusions

According to the time-independent diffusion calculations of Paper I it seemed possible that originally helium-rich atmospheres of post-AGB stars are transformed into metal-rich ones by diffusion. The present results show that for $T_{eff} = 140000$K, $\log g = 6.0$ equilibrium calculations are inadequate. In time scales comparable to those of stellar evolution, diffusion processes may modify significantly the surface composition. However, they cannot explain the carbon abundances deduced from the observations. These results suggest that the input physics is not yet complete. Therefore in future work it is necessary to investigate the influence of other processes like convective mixing or mass loss on the surface composition in addition.

We expected the heavy elements to be enriched more rapidly in the early stages of post-AGB evolution at lower gravity and to sink back only slowly at the later stages at $\log g \approx 7$. If this were true, the simplified diffusion scenario could be an alternative explanation to the "born again star" scenario mentioned in Sect.1. However, the results published in Paper II show that quite the opposite is true. For cases with $\log g = 7.0$, due to gravitational settling the atmosphere should be depleted of carbon and oxygen within 1000 y. This result again shows that another mechanism not yet taken into account must exist, which prevents carbon and oxygen from sinking.

## References

Burgers J.M., 1969, *Flow Equations for Composite Gases*, Academic Press, New York
Kawaler S.D., Bradley P.A., 1994, *ApJ* 427, 415
Paquette C., Pelletier C., Fontaine G., Michaud G., 1986, *ApJS* 61, 177
Unglaub K., Bues I., 1996, *A&A* 306, 843 (Paper I)
Unglaub K., Bues I., 1997, *A&A*, in press (Paper II)
Werner K., Heber U., Hunger K., 1991, *A&A* 244, 437
Werner K., Dreizler S., Heber U., Rauch T., 1996, in it Hydrogen-Deficient Stars, eds.
    U. Heber and C.S. Jeffery, ASPCS 96, 267
Wood R.P., Faulkner D.J., 1986, *ApJ* 307, 659

# EVOLUTIONARY CALCULATIONS OF CARBON DREDGE-UP IN HELIUM ENVELOPE WHITE DWARFS

JAMES MACDONALD

*Department of Physics and Astronomy, University of Delaware, Newark, DE 19716, U.S.A*

MARGARITA HERNANZ

*Institut d'Estudis Espacials de Catalunya, (CSIC Research Unit), Edifici Nexus-104, C/Gran Capità 2-4, E-08034 Barcelona, SPAIN*

AND

JORDI JOSÉ

*Departament de Física i Enginyeria Nuclear (UPC), Avda. Víctor Balaguer, s/n, E-08800 Vilanova i la Geltrú, SPAIN*

## 1. Introduction

It has been proposed that the presence of carbon in the mainly helium atmospheres of the DQ white dwarfs is due to convective dredge-up of carbon diffusing outwards from a CO core. The first quantitative study of this process is that of Pelletier at al (1986, hereafter P86), who found that agreement with the observationally determined photospheric C/He ratios required that the mass of the helium-rich layer in the white dwarf progenitor be an order of magnitude less than determined from theoretical calculations of the pre-white dwarf evolution.

## 2. Results

Here we present the preliminary results of stellar evolution calculations of carbon dredge-up in cooling white dwarfs that have CO cores and helium envelopes. Our calculations differ from those of P86 in that:

1) we consistently follow the evolution of the thermal structure of the white dwarf together with the evolution of the composition due to diffusion processes, convective mixing and nuclear reactions,

*I. Isern et al. (eds.), White Dwarfs, 265–267.*

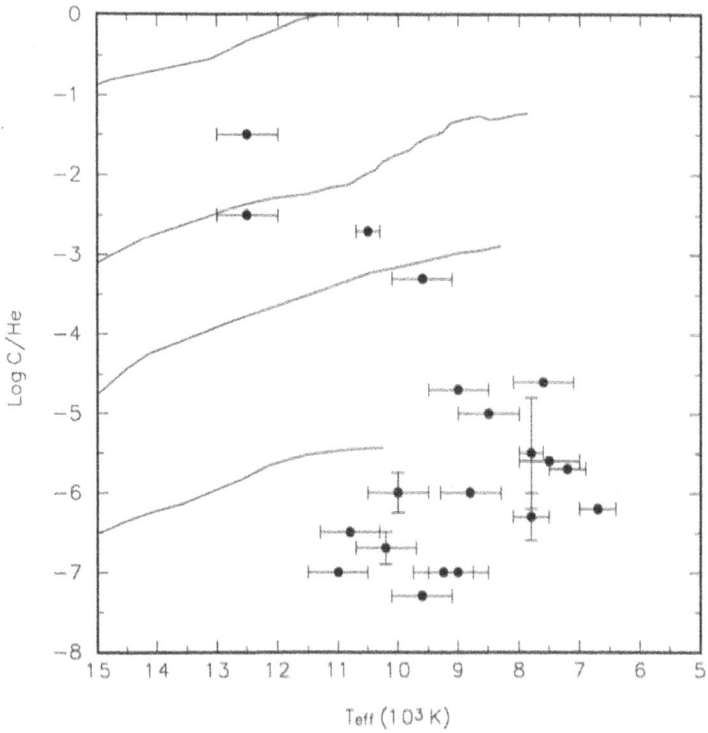

*Figure 1.* Photospheric C/He number density ratios. The solid lines are our theoretically predicted values for a 0.6 $M_\odot$ CO core with helium envelope masses (from top to bottom) of $10^{-6}$, $10^{-4}$, $10^{-3}$, and $10^{-2}$ $M_\odot$. The filled circles with error bars are the observed values.

2) the core composition is 50% C and 50% O by mass, and

3) we use the OPAL92 opacities (Iglesias & Rogers 1993).

Our main result is shown in figure 1, in which the photospheric C/He number density ratio is plotted against effective temperature for helium layer masses of $10^{-6}$, $10^{-4}$, $10^{-3}$, and $10^{-2}$ $M_\odot$ and a core mass of 0.6 $M_\odot$. Also shown are the observed photospheric ratios for DQ white dwarfs taken from the compilation of Weidemann & Koester (1995). Error bars are taken from the original papers, when possible. Our main conclusion is that for a canonical mass of 0.6 $M_\odot$, the majority of DQ white dwarfs have helium layer masses near $10^{-2}$ $M_\odot$, which is an order of magnitude greater than found by P86 and is consistent with current models for the evolution of white dwarf progenitors.

We attribute our finding of significantly larger helium layer masses than those found by P86 to our use of more recent opacity tables. Larger opacity in the helium envelope leads to deeper convection zones and dredge-up from

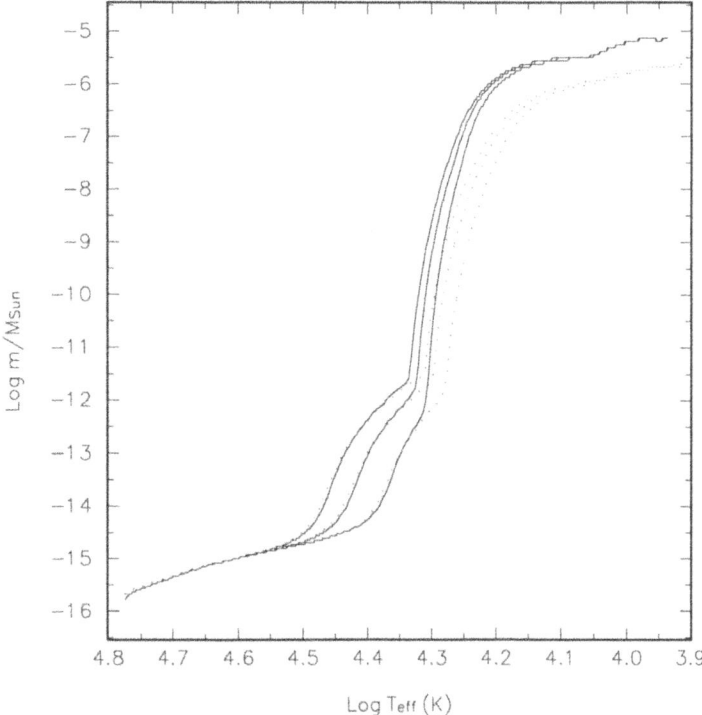

*Figure 2.* The depth in mass units of the inner boundary of the convection zone for envelopes of uniform composition and mass fractions $X_{He} = 0.999$, $X_C = 0.001$. The solid lines are for OPAL opacities and the dashed lines are for the Los Alamos opacities. For each set of three lines, from bottom to top, the mixing length ratio is 1.0, 1.5, 2.0.

regions of higher carbon abundance. To confirm this hypothesis, we have calculated models without diffusion processes for the OPAL92 opacities and the older Los Alamos opacities. The depth of the convection zone for a envelope of uniform composition, $X_{He} = 0.999$, $X_C = 0.001$, is shown in figure 2 for 3 values of mixing length parameter. It can clearly be seen that use of the OPAL opacities gives convection zones that are an order of magnitude deeper than those for the Los Alamos Opacities.

**Acknowledgements** One of us (M.H.) thanks the CIRIT for a grant to do a stage in the University of Delaware, and the DGICYT Project PB94-0111.

### References

Iglesias, C.A. & Rogers, F.J. (1993), *Astrophysical Journal*, **412**, 752

Pelletier, C., Fontaine, G., Wesemael, F., Michaud, G. & Wegner, G. (1986), *Astrophysical Journal*, **307**, 242

Weidemann, V. & Koester, D. (1995), *Astronomy and Astrophysics*, **297**, 216

## References

# ANALYSIS OF OPTICAL AND ULTRAVIOLET SPECTROSCOPY OF THE DAZ WHITE DWARF G74-7

M. BILLÈRES, F. WESEMAEL
*Département de Physique, Université de Montréal*

A. BEAUCHAMP
*CAE Électronique Ltée*

AND

P. BERGERON
*Lockheed Martin Electronic Systems Canada*

## 1. Introduction

White dwarfs with spectral lines of more than one chemical element afford us the possibility of studying the competition between different physical mechanisms in determining the chemical composition of the stellar envelope. Among cool white dwarfs of the DA, or hydrogen-line variety, only the prototype DAZ star G74-7 (Lacombe *et al.* 1983) and the ZZ Ceti star G29-38 (Koester & Provencal 1997) are known to show metal lines. Because of the presence of the Ca II H & K lines in the optical, G74-7 ($V \sim 14.5$; $T_{\rm eff} = 7200$ K) was observed with the *HST* Faint Object Spectrograph by the white dwarf consortium of Shipman *et al.* (1995). The 1700–3200 Å spectrum of this object revealed lines of Mg I, Mg II, Fe I, and Fe II. By providing additional relative abundance ratios, these lines can contribute in a useful way to our understanding of the diffusion process in white dwarfs.

An additional, intriguing feature of G74-7 is the suggestion made by Hammond *et al.* (1991, 1993) that the atmosphere of this classical, hydrogen-line white dwarf could be significantly enriched in helium, with a helium abundance $N({\rm He})/N({\rm H}) \sim 25$, a factor $\sim 2000$ larger than that derived, e.g., by Bergeron *et al.* (1990). We do not address this controversy here, and focus instead on providing an homogeneous analysis of the abundance of heavy elements observed both in the optical and in the ultraviolet spectrum of G74-7 under the assumption of a hydrogen-rich composition for that object.

269

*I. Isern et al. (eds.), White Dwarfs, 269–272.*
© *1997 Kluwer Academic Publishers.*

## 2. Observational Material and Analysis

A new optical spectrum of G74-7 was obtained at the Steward Observatory
2.3 m telescope, and serves as the basis for this reanalysis. $BVRIJHK$

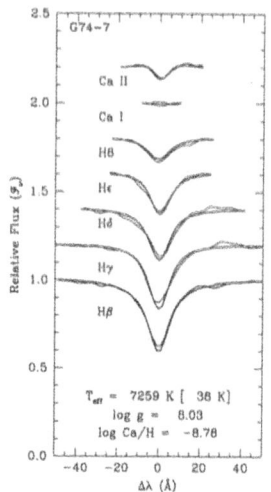

*Figure 1.*     Optimal fit to the optical spectrum of G74-7 based on a model with a
pure-hydrogen composition in which Ca is treated as a trace element. The Ca I λ 4227
and Ca II H and K lines are included in the spectrum calculation. The synthetic spectra
are degraded to a 7 Å resolution.

photometry was also secured in the course of a program of optical and
infrared photometry of cool white dwarfs (Bergeron, Ruiz, & Leggett 1997).
In addition, the two FOS spectra of G74-7 described by Shipman *et al.*
(1995) cover the range 1700-2400 Å at a resolution of 6.6 Å, and the range
2225-3293 Å at a resolution of 2.1 Å. The S/N ratio is, typically, greater
than 10. Attempts at using the absolute FOS fluxes of G74-7 showed that
the flux level was considerably lower than that expected from the available
optical photometry, a result associated with the poor centering of the object
in the 1 arcsec aperture, which led to considerable light loss. Unfortunately,
there are no IUE fluxes which can be compared directly to the FOS data,
and provide a scale factor. Thus the FOS observations cannot contribute
objectively to the construction of a complete energy distribution of that
object, and is usable only for its spectroscopic properties.

     Our analysis of the optical and IR energy distribution of G74-7, using
a pure hydrogen composition, yields the following, optimal combination of
parameters consistent with the measured parallax: $T_{\rm eff} = 7320$ K ($\sigma_{T_{\rm eff}} =$
180 K), log $g = 8.02$ ($\sigma_{\log g} = 0.09$). In parallel, we matched the optical
spectrum by allowing the effective temperature, the surface gravity, and the
Ca/H ratio to vary. This analysis (see Fig. 1) yields $T_{\rm eff} = 7260\pm40$ K, log $g$

$= 8.03 \pm 0.07$, and $\log (\mathrm{Ca/H}) = -8.8 \pm 0.1$, or $\mathrm{Ca/H} = 1.6 \pm 0.4 \times 10^{-9}$. The quoted errors are $1\sigma$ errors obtained from the fitting technique procedure.

Given the internal consistency between the photometric and spectroscopic data, we assume, for the rest of our analysis, that G74-7 is a normal gravity ($\log g = 8.0$), hydrogen-atmosphere ($N(\mathrm{He})/N(\mathrm{H}) = 0$) white dwarf at $T_{\mathrm{eff}} = 7320$ K.

The two FOS spectra of G74-7 were analyzed by Shipman $et\ al.$ (1995), who visually identified lines of Mg I, Mg II, Fe I, and Fe II in smoothed versions of these data. For Mg, a visual match to the data as well as more formal fits suggest an abundance of the order of $\log (\mathrm{Mg/H}) \sim -8.0 \pm 0.3$ as our optimal value for the Mg abundance in G74-7.

Our constraints on the Fe abundance rely on a comparison of the FOS data with synthetic spectra in five spectral windows. The results of this comparison is shown in Fig. 2 and 3 for iron abundances, $\log (\mathrm{Fe/H})$, ranging from $-4.5$ to $-6.0$. On the basis of a visual comparison, the data in all five windows is consistent with $\log (\mathrm{Fe/H}) \lesssim -5$. It seems difficult to be more specific at this stage, given the limited S/N ratio of the data, especially in the 1700-2400 Å range of the G160L grating. GHRS observations, at good S/N ratio and in well-targeted wavelength windows, should allow one to improve considerably upon the preliminary limits set here.

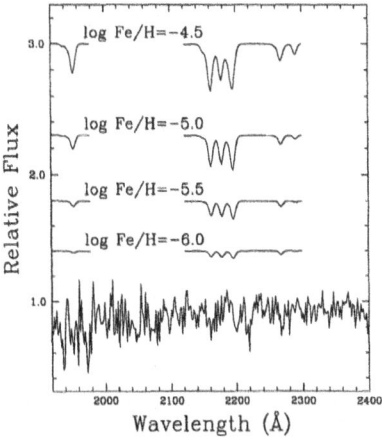

$Figure\ 2.$    Theoretical match to three selected regions of the FOS data where Fe I features are expected: the 1935-1960 Å, 2140-2200 Å, and 2260-2300 Å regions. The synthetic spectra are calculated for $T_{\mathrm{eff}} = 7320$ K, $\log g = 8.0$, $N(\mathrm{He})/N(\mathrm{H}) = 0$. They are degraded to a spectral resolution of 6.6 Å.

## 3.  Concluding Remarks

With this analysis, we have provided new determinations of the abundances of Ca and Mg, as well as limits on the abundance of Fe, in the cool

DA star G74-7: $\log(Ca/H) = -8.8 \pm 0.2$, $\log(Mg/H) = -8.0 \pm 0.3$, and $\log(Fe/H) \lesssim -5$. In terms of deficiencies with respect to the solar abundance ratios, these results can be rewritten in the form $[Ca/H] = -3.1$, $[Mg/H] = -3.6$, $[Fe/H] \lesssim -0.5$. G74-7 clearly shares the general deficiency in heavy elements observed in the more numerous helium-rich white dwarfs of type DBZ or DZ.

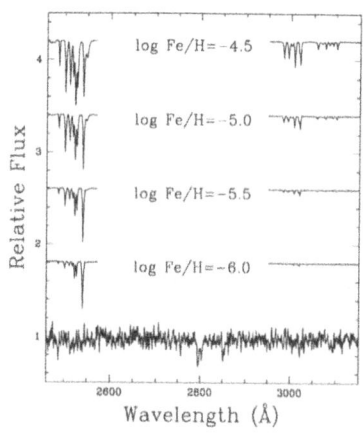

*Figure 3.* Same as Fig. 2, but for the the 2450-2560 Å and 2975-3110 Å regions, where Fe II features are expected. The spectra are degraded to a 2.1 Å spectral resolution. The Mg II λλ 2795.53, 2801.70, and Mg I λ 2852.13 features are seen near the middle of the plot.

This work was supported in part by the NSERC Canada, by the Fund FCAR (Québec), and by a travel grant from the Centre de Coopération Interuniversitaire Franco-Quebécois.

## References

Bergeron, P., Ruiz, M.-T., & Leggett, S.K. 1997, *Ap.J. (Suppl.)*, in press
Bergeron, P., Wesemael, F, Fontaine, G., & Liebert, J. 1990, *Ap.J. (Letters)*, **351**, L21
Hammond, G.L., Sion, E.M., Kenyon, S.J., & Aannestad, P.A. 1991, in *White Dwarfs*, ed. G. Vauclair & E.M. Sion, NATO ASI Series Vol. 336 (Dordrecht: Kluwer), p. 317
Hammond, G.L., Sion, E.M., Aannestad, P.A., & Kenyon, S.J. 1993, in *White Dwarfs: Advances in Observation and Theory*, ed. M.A. Barstow, NATO ASI Series Vol. 403 (Dordrecht: Kluwer), p. 253
Koester, D. & Provencal, J. 1997, these Proceedings
Lacombe, P., Liebert, J., Wesemael, F, & Fontaine, G. 1983, *Ap.J.*, **272**, 660
Shipman, H.L., Barnhill, M., Provencal, J., Roby, S., Bues, I., Cordova, F., Hammond, G., Hintzen, P., Koester, D., Liebert, J., Oswalt, T., Starrfield, S., Wegner, G., & Weidemann, V. 1995, *A.J.*, **109** 1231

# MODELING EXTREME ULTRAVIOLET SPECTRA OF WHITE DWARFS WITH HIGH-METALLICITY LTE MODELS

P. CHAYER, S. VENNES AND J. DUPUIS
*Center for EUV Astrophysics, 2150 Kittredge St.,*
*University of California, Berkeley, CA 94720-5030, USA*

P. THEJLL
*Niels Bohr Institute, Blegdamsvej 17, DK-2100,*
*København Ø, Denmark*

AND

A. K. PRADHAN
*Department of Astronomy, Ohio State University,*
*Columbus, OH 43210-1106*

## 1. Introduction

The presence of heavy elements in the hot hydrogen- and helium-rich atmospheres of DA and DO white dwarfs is evident in high-dispersion, far-ultraviolet (FUV) spectroscopy (see Holberg 1995; Werner, Dreizler, & Wolff 1995). The measured abundances are sometimes greatly in excess of the solar composition, and sometimes much below; the abundance pattern in young degenerate stars remains unexplained (Chayer, Fontaine, & Wesemael 1995). Theory shows that surface abundances rapidly reach diffusive equilibrium, and large departures from model predictions indicate that mass-loss or accretion may determine the atmospheric composition of hot white dwarf stars. Vennes et al. (1996a) also suggest that silicon overabundances observed in several hot DA white dwarfs may result from accretion of grains or larger particles.

The *ROSAT* WFC and the *Extreme Ultraviolet Explorer* (*EUVE*) recently surveyed the population of hot white dwarf stars in the extreme ultraviolet (EUV) spectral range; the presence of heavy elements in a large segment of the EUV-selected population is established in broad-band photometric studies (Barstow et al. 1993; Wolf, Jordan, & Koester 1996; Vennes et al. 1996b). Heavy element opacities were first noted in the *EXOSAT*

273

*I. Isern et al. (eds.), White Dwarfs, 273–276.*
© 1997 *Kluwer Academic Publishers.*

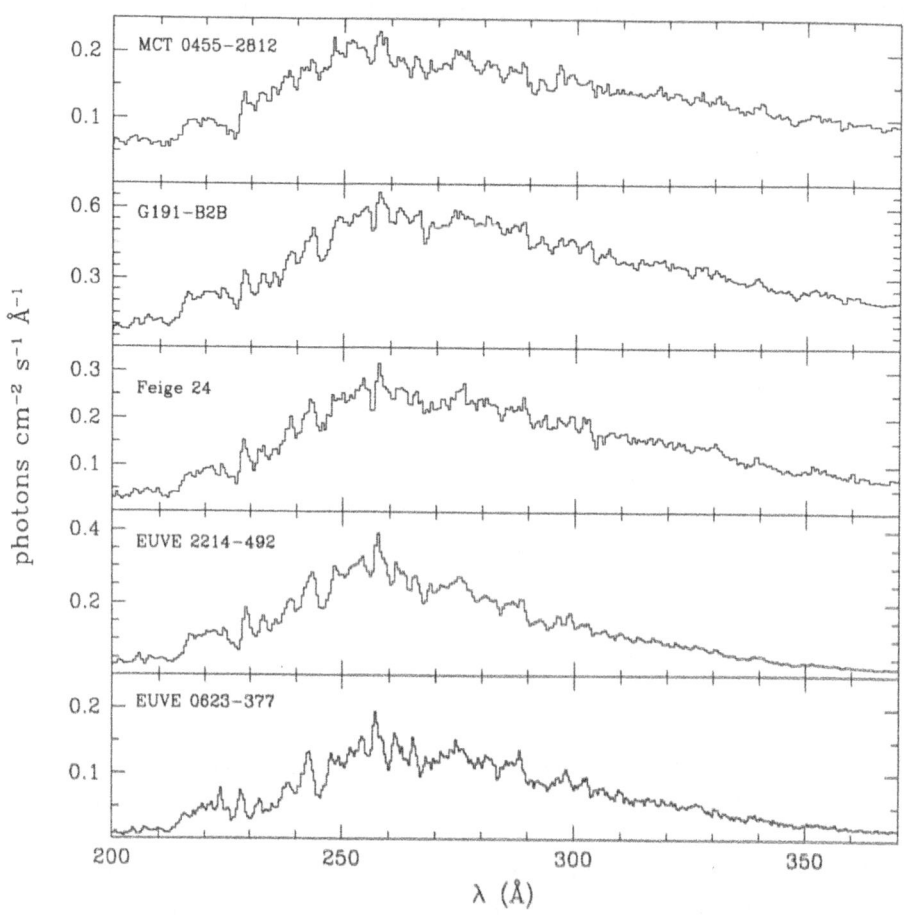

*Figure 1.* *EUVE* spectra of five hot DA white dwarfs.

spectrum of the hot DA star Feige 24 (Vennes et al. 1989); recent medium-resolution *EUVE* spectra confirm the Vennes et al. (1989) proposition and reveal intricate details of EUV heavy element opacities (see Dupuis et al. 1995). We examine some of these spectra and we present spectral synthesis depicting the effects of He, C, N, O, Si, Fe, and Ni opacities.

## 2. EUV Spectra of Hot White Dwarfs

Figure 1 shows the EUV spectra of five hot DA white dwarfs. The spectra are arranged from top to bottom in a sequence with apparently increasing heavy element opacities. All spectra are characterized (1) by continuum emission ($\lambda \geq 260$ Å) gradually depressed by interstellar medium attenu-

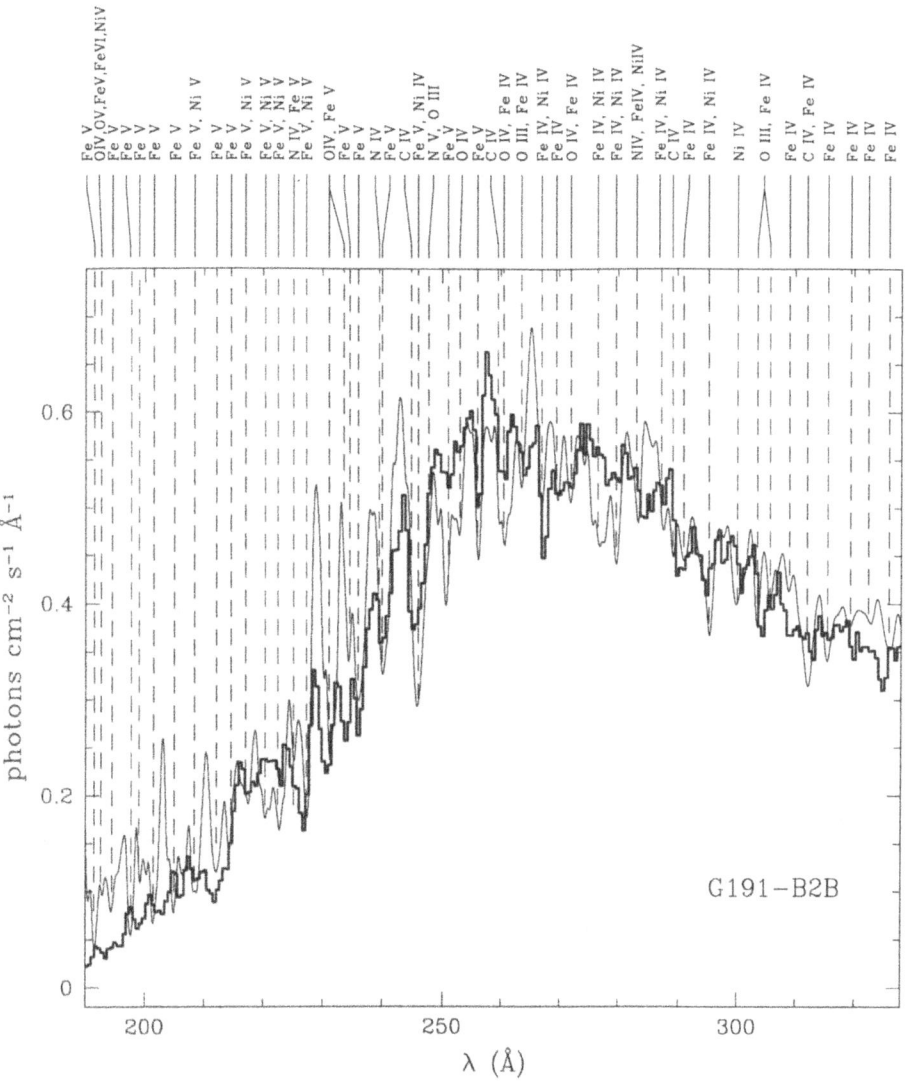

*Figure 2.* Possible model atmosphere representation of the high-metallicity white dwarf G191–B2B (see text and Table 1).

ation and (2) by increasingly dominant heavy element opacities at shorter wavelengths ($\lambda \leq 310$ Å). Four objects in the sample show FUV lines of C IV, N IV–V, O IV–V, Si IV, Fe V–VI, and Ni V–VI; based on this information we have computed a series of EUV spectra using detailed line profiles (Kurucz 1992) and cross sections (Cunto et al. 1993; Verner & Yakovlev 1995). The spectra are computed using models in radiative equilibrium with heavy-element opacities based on Opacity Project data, and using a single set of abundances (see Dupuis et al. 1995). Although potentially inconsistent, this procedure provides greater flexibility in the search for an optimum fit; after establishing a reasonable abundance pattern, we will initiate detailed calculations. Figure 2 shows a comparison between the *EUVE* spectrum of the hot DA white dwarf G191–B2B and a model including heavy element opacities; the most important line features are identified.

TABLE 1. G191–B2B

| $T_{\rm eff} = 56,000$ K | $\log {\rm He/H} = -7.0$ |
|---|---|
| $\log g = 7.5$ | $\log {\rm C/H} = -6.0$ |
| $V = 11.78$ | $\log {\rm N/H} = -6.8$ |
| $N({\rm H\,I}) = 20.7 \times 10^{17}$ cm$^{-2}$ | $\log {\rm O/H} = -6.6$ |
| $N({\rm He\,I})/N({\rm H\,I}) = 0.072$ | $\log {\rm Si/H} = -6.5$ |
| $N({\rm He\,II})/N({\rm H\,I}) = 0.22$ | $\log {\rm Fe/H} = -5.4$ |
| | $\log {\rm Ni/H} = -5.7$ |

## 3. Conclusions

Our spectral synthesis displays numerous correspondences with the observed EUV spectrum of G191–B2B, but also an equal number of discrepancies; Lanz et al. (1997) achieved a similar success using NLTE model atmospheres. However, these efforts underline basic flaws in the modeling of EUV spectra of DA white dwarfs: missing elements and the effect of unaccounted opacities, and large uncertainties introduced by inaccurate atomic cross sections. This work is supported by NASA contract NAS5-30180 and grant NAG5-2636.

## References

Barstow, M.A., et al. 1993, MNRAS, 264, 16

Chayer, P., Fontaine, G., & Wesemael, F. 1995, ApJS, 99, 189

Cunto, W., Mendoza, C., Ochsenbein, F., & Zeippen, C. J. 1993, A&A, 275, L5

Dupuis, J., Vennes, S., Bowyer, S., Pradhan, A.K., & Thejll, P. 1995, ApJ, 455, 574

Holberg, J. 1995, in White Dwarfs, eds. D. Koester & K. Werner (Berlin: Springer), 138

Kurucz, R. L. 1992, Rev. Mexicana Astron. Af., 23, 45

Lanz, T., Barstow, M.A., Hubeny, I., & Holberg, J.B. 1997, ApJ, in press

Vennes, S., Chayer, P., Fontaine, G., & Wesemael, F. 1989, ApJ, 336, L25

Vennes, S., Chayer, P., Hurwitz, M., & Bowyer, S. 1996a, ApJ, 468, 898

Vennes, S., Thejll, P., Wickramasinghe, D.T., & Bessell, M. 1996b, ApJ, 467, 782

Verner, D.A., & Yakovlev, D.G. 1995, A&AS, 109, 125

Werner, K., Dreizler, S., & Wolff, B. 1995, A&A, 298, 567

Wolff, B., Jordan, S., & Koester, D. 1996, A&A, 307, 149

# EXTREME ULTRAVIOLET SPECTROSCOPY OF EUVE J0720–317 AND EUVE J0723–277

JEAN DUPUIS, STÉPHANE VENNES AND STUART BOWYER

*Center for EUV Astrophysics, 2150 Kittredge St.,*
*University of California, Berkeley, CA 94720-5030, USA*

**Abstract.** We present *Extreme Ultraviolet Explorer* (*EUVE*) and optical spectra of the DAO member of the close binary EUVE J0720–317 and of the hot DA EUVE J0723–277. A comparison of density measurements of the local interstellar medium (ISM) toward EUVE J0720–317 and EUVE J0723–277 (only 4° apart) reveals large variations in ISM ionization properties. We speculate that the line of sight toward the close binary EUVE J0720–317 is contaminated with circumbinary material or by unsuspected structures in the local ISM.

## 1. Introduction

Extreme ultraviolet (EUV) spectroscopy provides us with new means for a study of peculiar white dwarfs, such as members of close binary systems. Several new binaries have been identified in EUV all-sky surveys (*EUVE*, *ROSAT* WFC): EUVE J0720–317 (RE 0720–318) is a precataclysmic binary discovered by Vennes & Thorstensen (1994) and described as a DAO white dwarf plus dM0–2 pair with an orbital period of 1.3 days. Barstow et al. (1995) report visual variations on the same period, which Vennes & Thorstensen (1996) attribute to a reflection effect. Vennes & Thorstensen (1996) also present a long-term orbital ephemeris and definitive fundamental parameters of the DAO. EUVE J0723–277 is an apparently normal DA white dwarf also discovered in EUV all-sky surveys (Vennes et al. 1997). EUVE J0723–277 is located at a small angular separation from the DAO EUVE J0720–317, and a differential study of the local interstellar medium (ISM) in this line of sight becomes possible.

*I. Isern et al. (eds.), White Dwarfs, 277–280.*

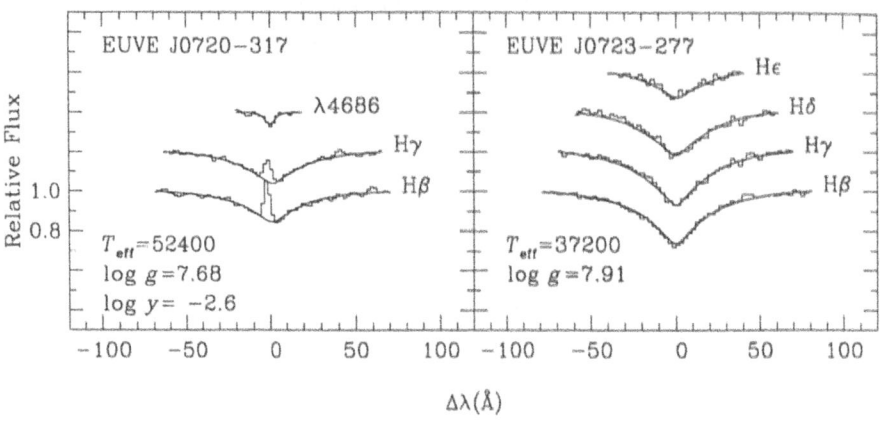

*Figure 1.*  Balmer line spectroscopy for EUVE J0720–317 and EUVE J0723–277.

## 2. Analysis of Optical and EUV Spectroscopy

We observed EUVE J0720–317 with the CTIO 4m telescope on 1994 January 25 and 26, and EUVE J0723–277 with the MDM 2.4m telescope on 1996 January 8. Figure 1 (*left*) shows Balmer and He II $\lambda$4686 line profiles for EUVE J0720–317; Figure 1 (*right*) shows Balmer line profiles for EUVE J0723–277. The spectral resolution is $\sim$ 5 Å. The spectra are analyzed with detailed synthetic line profiles computed from a grid of homogeneous LTE line-blanketed hydrogen/helium models (see Vennes & Fontaine 1992). A fit to H$\beta$, H$\gamma$ and He II $\lambda$4686 implies $T_{\text{eff}}$ = 52400 K and $\log g$ = 7.68 for EUVE J0720–317 (Vennes & Thorstensen 1996). The Balmer line spectrum of EUVE J0723–277 is typical of a normal hot DA white dwarf, and a fit to H$\beta$, H$\gamma$, H$\delta$, and H$\epsilon$ gives $T_{\text{eff}}$ = 37200 K and $\log g$ = 7.91 (Vennes et al. 1997).

Figure 2 (*bottom*) shows the EUV spectrum of EUVE J0720–317 obtained with *EUVE*. The observation began on 1995 December 16 for a total exposure time of 120,000 s. The spectrum is very peculiar and shows strong absorption edges of He I (504 Å) and He II (228 Å); the shape of the continuum also shows evidence of photospheric heavy element opacities. Adopting the effective temperature and surface gravity but *lowering* the helium abundance measured with optical data, we obtained a reasonable fit to the entire EUV range (70 – 600 Å) by adding heavy element opacities computed by A.K. Pradhan. The inferred composition is: He/H = $3.2 \times 10^{-5}$, C/H = $2 \times 10^{-6}$, N/H = $5 \times 10^{-7}$, O/H = $2 \times 10^{-6}$, Si/H = $2 \times 10^{-7}$, S/H = $10^{-7}$, and Fe/H = $2 \times 10^{-7}$. The metallicity corresponds to one hundredth of the solar mixture. The measured ISM column densities of H I, He I, and He II are $2.1 \times 10^{18}$, $1.3 \times 10^{18}$, and $\leq 0.5 \times 10^{18}$ cm$^{-2}$,

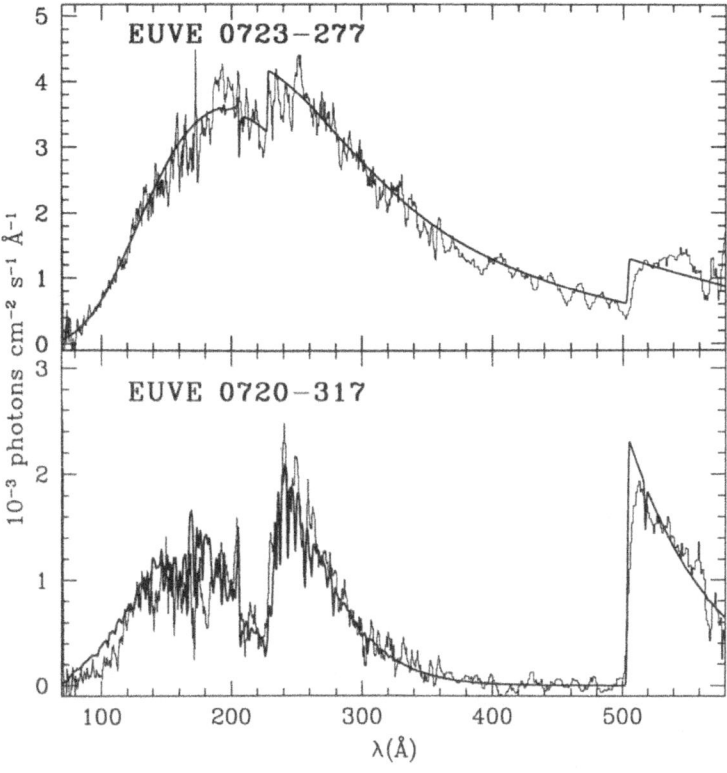

*Figure 2.* Extreme ultraviolet spectra of the DAO white dwarf EUVE J0720–317 and the DA white dwarf EUVE J0723-277.

respectively.

Figure 2 (*top*) shows the EUV spectrum of EUVE J0723–277 observed with *EUVE*, starting on 1995 February 9 for a total exposure of 100,000 s. The spectrum is typical of a hot DA with a pure hydrogen composition, verifying our earlier assumption; the star is detected above the He I edge, suggesting a low hydrogen column in the line of sight. We fit the EUV spectrum with a pure hydrogen model at $T_{\rm eff} = 36,000$ K (assuming the optical surface gravity) in agreement with optical values. As expected, we measure a very low H I ISM column of $9.0 \times 10^{17}$ cm$^{-2}$, and we also determine He I and He II columns of $1.0 \times 10^{17}$ and $1.6 \times 10^{17}$cm$^{-2}$, respectively.

## 3. Discussion

We find abnormal column density ratios of He I/H I, and He II/H I toward EUVE J0720–317 and EUVE J0723–277. The ratio of He I to H I measured in several lines of sight toward hot white dwarfs is normally between 0.05

and 0.08 (Vennes et al. 1993; Dupuis et al. 1995). We determine a value of 0.6 toward EUVE J0720–317. We estimate the ionization fraction of hydrogen by predicting the total hydrogen density (H I + H II), and using the measured helium density (He I + He II) and a cosmic abundance ratio of H-to-He of 10. We obtain an ionization fraction of 88% for hydrogen and 28% for helium.

The ISM column densities toward EUVE J0723–277 are strikingly dissimilar despite the small angular separation between the two objects (4°). The ionization fractions of hydrogen and helium are 65% and 62%, respectively, and are significantly different from the fractions measured toward EUVE J0720–317. The dissimilarity suggests either that the ionization state of the local ISM can vary substantially over small angular scales or that the peculiar ionization fractions measured in the direction of EUVE J0720–317 are due in part to circumbinary material. EUVE J0723–277 is at a distance of 100 pc compared to 168 pc for EUVE J0720–317; the dissimilarity may imply the existence of an ISM structure toward EUVE J0720–317, beyond EUVE J0723–277.

We conclude that the peculiar ISM abundances observed in the direction of EUVE J0720–317 either indicate the presence of a component associated with circumbinary material ejected during a recent ($t \sim 2$ Myr) common envelope phase or reveal the existence of an ionized cloud encompassing the binary EUVE J0720–317. Both white dwarfs reside in the Galactic plane, but in a region often dubbed the "Local Void" (see a review by Bruhweiler 1994). Structures known as the Loop I and the Gum Nebula lie away from the two white dwarfs and probably do not interfere with the the line of sight. The study of EUVE J0720–317's line of sight suggests the presence of a previously unsuspected structure that determines the hydrogen and helium ionization fractions.

This work is supported by NASA contract NAS5-30180 and grant NAG5-2636.

## References

Barstow, M.A., O'Donoghue, D., Kilkenny, D., Burleigh, M.R., & Fleming, T.A. 1995, MNRAS, 273, 711

Bruhweiler, F. 1994, in Frontiers of Space and Ground-Based Astronomy, eds. W. Wamsteker, M.S. Longair & Y. Kondo (Dordrecht: Kluwer), 289

Dupuis, J., Vennes, S., Bowyer, S., Pradhan, A.K., & Thejll, P.A. 1995, ApJ, 455, 574

Vennes, S., Dupuis, J., Rumph, T., Drake, J., Bowyer, S., Chayer, P., & Fontaine, G. 1993, ApJ, 410, L119

Vennes, S., & Fontaine, G. 1992, ApJ, 401, 288

Vennes, S., Thejll, P.A., Génova Galvan, R., & Dupuis, J. 1997, ApJ, in press

Vennes, S., & Thorstensen, J.R. 1994, ApJ, 433, L29

Vennes, S., & Thorstensen, J.R. 1996, AJ, 112, 284

# A NEW DITHERED EUVE SPECTRUM OF PG 1234+482

S. JORDAN AND D. KOESTER

*Institut für Astronomie und Astrophysik*
*D-24098 Kiel, Germany, jordan@astrophysik.uni-kiel.de*

AND

D. FINLEY

*Center for EUV Astrophysics, Berkeley, CA 94720, USA*

## 1. Introduction

With the exception of the hot DO white dwarfs RE J 0503-289 (Barstow et al. 1994, Vennes et al. 1994a,b), PG 1034+001 (Werner et al. 1995), and HD 149499 B (Jordan et al. 1996a) all white dwarfs from which photospheric emission was detected in the EUV or X-ray are of type DA. At 100-300 Å a hydrogen atmosphere becomes very transparent and radiation from rather deep and hot layers can reach the surface of the star; therefore white dwarfs with pure hydrogen atmospheres emit much more flux in this spectral region than black bodies at the same temperature.

However, if traces of heavier elements are present in the atmosphere, the EUV/X-ray flux can be strongly reduced, even if they cannot be detected in the optical. Measurements with the EINSTEIN, EXOSAT, and ROSAT satellites have shown (Kahn et al. 1984, Petre et al. 1986, Jordan et al. 1987, Paerels & Heise 1989, Barstow et al. 1993a,b, Jordan et al. 1994, Wolff et al. 1996) that the observed X-ray flux is often significantly lower than predicted by pure hydrogen atmospheres. Thus it was concluded that helium or metals must be present in the outer layers.

## 2. Metals in hot DA white dwarfs

Traces of metals can survive in the atmospheres of hot white dwarfs due to selective radiation pressure, while helium is predicted to sink down into deeper layers (Vennes 1988). The first studies of the radiative levitation of

*I. Isern et al. (eds.), White Dwarfs, 281–284.*

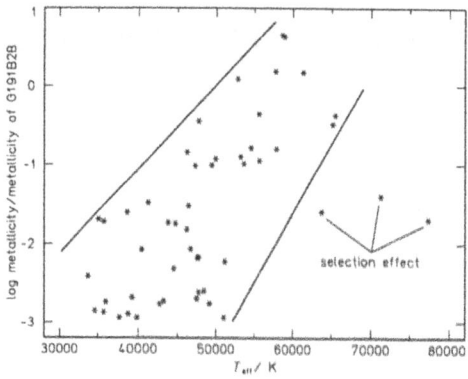

*Figure 1.* Metallicity (relative to G 191-B2B) vs. effective temperature for 50 DA white dwarfs with $T_{\rm eff} > 34000$ K. Below about 40 000 K the results are compatible with pure hydrogen atmospheres within the error limits. See Jordan et al. (1996b) for details

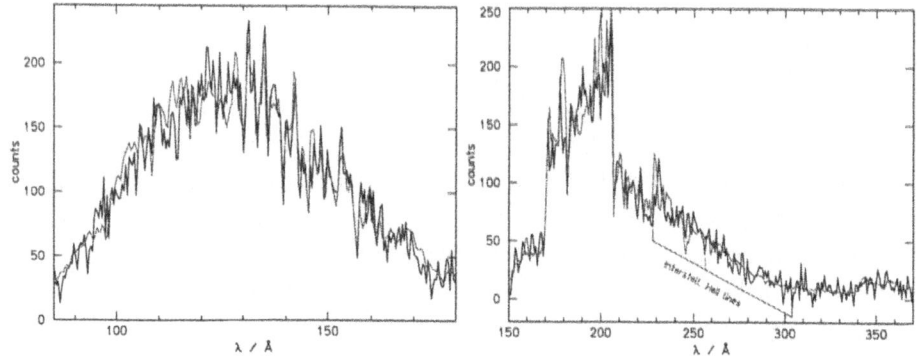

*Figure 2.* EUVE spectra of PG 1234+482 (left: SW spectrum, right: MW spectrum) compared to the prediction from a DA model atmosphere with 55 600 K, $\log g = 7.5$, and the element abundances given in the text. The flux level is slightly adjusted

metals have been performed by Vauclair et al. (1979), and later by Morvan et al. (1986), Vauclair (1989), and Chayer et al. (1989, 1991, 1995).

Since it was not possible to derive a unique composition of metal absorbers in white dwarf atmospheres from the EXOSAT, ROSAT, and EUVE photometry, Jordan et al. (1996b) used a procedure similar to Finley (1995) to quantify the metal content: With continuum opacities for several metals the overall energy distribution in the EUVE spectrum of G 191-B2B was fitted with Koester model atmospheres. Subsequently, it was assumed that the resulting number ratios for the individual metals in all DA white dwarfs are the same as in G 191-B2B if multiplied by a factor that we have called "metallicity". In Fig. 1 this metallicity is plotted as a function of the effective temperature for 50 DA stars above 34 000 K. The diagram shows the trend that the metallicity increases with effective temperature, as predicted by diffusion theory, but the scatter at a given temperature is rather large. We also tested the influence of the gravity on the metallicity by comparing the results for stars with about the same temperature (45-55 kK). In contrast to the predictions from diffusion theory no clear dependence on $\log g$ could be inferred.

## 3. The EUVE spectrum of PG 1234+482

Spectroscopy with the EUVE satellite offers the unique opportunity to identify the heavy elements directly. At the last WD workshop in Kiel Jordan et al. (1995) have presented an analysis of the EUVE spectrum of PG 1234+482 (a star that has a "metallicity" of about 0.25), and have shown that the strongest features in the short (SW) and medium (MW) wavelength spectrum can be approximately reproduced by model atmospheres containing Fe and some additional metal absorbers. However, since the spectrum (exposure time 100 ksec) was not taken in dithered mode it could be suspected that the spectrum was contaminated by fixed pattern noise.

In April 1996 we have obtained new dithered 200 ksec spectra and it turned out that the the strongest features remained unchanged. Their strength can be explained by a model containing hydrogen plus iron with $Fe/H = 2.5 \cdot 10^{-7}$. However, the flux below $200\,\text{Å}$ is up to a factor of three too high, if only Fe+H are taken into account.

In a second step we have calculated other bi-elemental model atmospheres in order to find out which other elements may be present besides hydrogen and what their maximum abundances are. We have calculated several mixtures and have adjusted the abundances until a relatively good fit was obtained. Although it is still extremely difficult to determine abundance without further information from the UV (c.f. paper by Wolff et al. in these proceedings), we are now relatively sure that Fe, C, N, O, Ni, and Ne are present in the atmosphere of PG 1234+482. For the other elements in our calculations we used upper limits for the abundances so that the spectrum is still compatible with the observations.

Fig. 2 shows the best fit of our analysis of the new EUVE spectrum. We used the following number ratios: $Fe/H = 6 \cdot 10^{-7}$ ($1.5 \cdot 10^{-5}$), $Ni/H = 2 \cdot 10^{-8}$ ($2.0 \cdot 10^{-6}$), $Ca/H = 5 \cdot 10^{-9}$ ($2.0 \cdot 10^{-6}$), $C/H = 8 \cdot 10^{-6}$ ($1.8 \cdot 10^{-4}$), $N/H = 5 \cdot 10^{-8}$ ($2.5 \cdot 10^{-4}$), $O/H = 2 \cdot 10^{-8}$ ($1.3 \cdot 10^{-4}$), $Si/H = 1 \cdot 10^{-8}$ ($< 10^{-8}$), $Ne/H = 2 \cdot 10^{-7}$ ($< 3.2 \cdot 10^{-7}$), $Ar/H = 2 \cdot 10^{-6}$ ($10^{-5}$). Many elements are significantly less abundant compared to the values predicted by the current theory of radiative levitation (Chayer et al. 1995), which are given in parenthesis.

Although not every absorption feature is fitted in every detail, many absorption lines that canot be explained by iron alone are now approximately reproduced. In fact, to our knowledge PG 1234+482 is the only DA white dwarf in which iron and other metal features could be well reproduced in detail by synthetic spectra. Other analyses (see e.g. Wolff et al. in these proceedings) are mostly based on fits to the overall energy distribution. The flux level predicted by our model now differs by less than 10-20% from the model predictions.

This small discrepancy may arise from the fact that we have still used an effective temperature determined by Balmer line fitting, without taking into account the effect of metal blanketing on the optical temperature determination; this could lead to a temperature decrease of about 1000-2000 K.

Acknowledgements. We thank the CEA for carrying out the EUVE observations and the DARA (project 50 OR 9409 1 and 50 OR 9617 3) for financial support.

## References

Barstow M.A., Fleming T.A., Finley D.S., Koester D., Diamond C.J. 1993a, MNRAS 260, 631

Barstow M.A., Fleming T.A., Diamond C.J. et al., 1993b, MNRAS 264, 16

Barstow M.A., Holberg J.B., Werner K., Buckley D.A.H., Stobie R.S., 1994, MNRAS 267, 653

Chayer P., Fontaine G., Wesemael F., 1989. In: White Dwarfs, ed. G. Wegner, Springer, p. 176

Chayer P., Fontaine G., Wesemael F., 1991. In: White Dwarfs, eds. G. Vauclair, E. Sion, Kluwer, Dordrecht, p. 249

Chayer P., Fontaine G., Wesemael F., 1995, ApJS 99, 189

Finley D., 1995, in Proc. IAU Colloq. 152, Astrophysics in the Extreme Ultraviolet, ed. S. Bowyer and R.F. Malina, (Dordrecht: Kluwer), p. 223

Jordan S., Koester D., Wulf-Mathies C., Brunner H., 1987, A&A 185, 253

Jordan S., Wolff B., Koester D., Napiwotzki R., 1994, A&A 290, 834

Jordan S., Koester D., Werner K., Finley D.S., Dreizler S., 1995, Lecture Notes in Physics Vol. 443, Springer, Berlin, p. 332

Jordan S., Napiwotzki R., Koester D., Rauch T., 1996a, A&A, in press

Jordan S., Finley D., Koester D., Wolff B., 1996b, in Röntgenstrahlung from the Universe, MPE Report 263, eds. H.U. Zimmermann, J.E. Trümper, and H. Yorke, p. 5

Kahn S.M., Wesemael F., Liebert J., et al., 1984, ApJ 278, 255

Morvan E., Vauclair G., Vauclair S., 1986, A&A 163, 145

Paerels F.B.S., Heise J., 1989, ApJ 339, 1000

Petre R., Shipman H.L., Canizares C.R., 1986, ApJ 304, 356

Vennes S., Dupuis J., Bowyer S., Fontaine G., Wiercigroch A., Jelinski P., Wesemael F., Malina R., 1994a, ApJ 421, L35

Vauclair G., Vauclair S., Greenstein J.L., 1979, A&A 80, 79

Vauclair G., 1989. In: White Dwarfs, ed. G. Wegner, Springer, p. 176

Vennes S., Pelletier C., Fontaine G., Wesemael F., 1988, ApJ 331, 876

Vennes S., Pradhan A.K., Fontaine G.,Dupuis J., Wesemael F., 1994b, Bul. AAS 26, 868

Werner K., Dreizler S., Wolff B, 1995, A&A 298, 567

Wolff B., Jordan S., Koester D., 1996, A&A 307, 149

# LINE OF SIGHT H AND HE IONIZATION FRACTIONS IN THE LOCAL INTERSTELLAR MEDIUM

P.D. DOBBIE AND M.A. BARSTOW

*Department of Physics & Astronomy,*
*University of Leicester, UK*

## 1. Introduction

Hot DA white dwarfs are well suited for studies of the local interstellar medium (LISM). Their EUV flux is very sensitive to hydrogen, helium and heavy elements along the lines of sight. The LISM is regarded as the dust and gas contained within a region of space extending several hundred parsecs from the Sun. It is believed the Sun lies in a diffuse cloud, the Local Cloud, on the perimeter of the Loop I radio shell. It is thought that this cloud extends several parsecs from the Sun, with T$\sim$7000K and n$\sim$0.1cm$^{-3}$.

Recent observations of hot DA white dwarfs with the EUVE show substantial He ionization along lines of sight to GD246 (Vennes *et al.* 1993) and GD659 (Holberg *et al.* 1995). From these studies values for the ionization of H and He are obtained which appear to agree with Lyu & Bruhweiler's (1996) time dependent ionization calculations for the Local Cloud.

We present here the results from our detailed study of 13 hot DA white dwarf spectra taken from both our own EUVE GO programme and from the data archive. We have analysed the entire usuable spectrum of each object with both homogeneous and stratified H+He model atmospheres (Koester *et al.* 1991) and also heavy element blanketed versions. For simplicity we will only consider the homogeneous H+He fits here. The majority of stars in our sample show no evidence of photospheric heavy elements, the exceptions being PG1123 and GD246. For these two cases, heavy element blanketed models developed for the analysis of G191-B2B (Lanz *et al.* 1996) are used.

## 2. Columns and Ionization Fractions

The atmospheric models are used in conjunction with the ISM model of Rumph, Bowyer & Vennes (1994). This is modified to take into account the converging lyman series of HeI and HeII. On fitting, the parameters

*I. Isern et al. (eds.), White Dwarfs, 285–288.*

TABLE 1. The ionization fractions of helium and hydrogen in
the interstellar medium, calculated from the line-of-sight column density measurements using homogeneous H+He photospheric models (Barstow *et al.* 1996).

| Star | $f_H$ | | | $f_{He}$ | | |
|------|-------|------|------|----------|------|------|
|      | Value | $-1\sigma$ | $+1\sigma$ | Value | $-1\sigma$ | $+1\sigma$ |
| GD659 | 0.16 | 0.00 | 0.50 | 0.42 | 0.29 | 0.58 |
| GD71 | 0.00 | 0.00 | 0.28 | 0.04 | 0.00 | 0.32 |
| REJ0715−705 | 0.00 | 0.00 | 0.79 | 1.00 | 0.00 | 1.00 |
| REJ1032+535 | 0.46 | 0.30 | 0.56 | 0.27 | 0.20 | 0.34 |
| GD153 | 0.00 | 0.00 | 0.28 | 0.00 | 0.00 | 0.36 |
| HZ43 | 0.19 | 0.00 | 0.34 | 0.40 | 0.29 | 0.50 |
| CoD−38 10980 | 0.00 | 0.00 | 1.00 | 0.00 | 0.00 | 1.00 |
| BPM93487 | 0.17 | 0.00 | 1.00 | 1.00 | 0.00 | 1.00 |
| REJ2009−605 | 0.00 | 0.00 | 0.63 | 0.04 | 0.00 | 1.00 |
| REJ2156−546 | 0.34 | 0.00 | 0.57 | 0.39 | 0.28 | 0.46 |
| REJ2324−547 | 0.57 | 0.00 | 0.74 | 0.13 | 0.00 | 0.35 |

HI, HeI, HeII and log H/He are allowed to vary freely. The atmospheric
temperature and log g are constrained by the $1\sigma$ limits of their optical
values (Marsh *et al.* 1996). The best fit is found using a $\chi^2$ minimization
technique described by Barstow *et al.* (1996).

To calculate the ionization fractions from the columns we use the equations shown below and assume the cosmic abundance ratio (He/H) has its
value 0.1.

$$f_{He} = \frac{HeII}{HeI+HeII}$$

$$f_H = 1 - \frac{HI}{10(HeI+HeII)}$$

## 3. Results

It is possible to split our sample into 3 groups. We have detected the HeII
228Å lyman edge along 5 lines of sight. For objects GD659, REJ1032+53,
REJ2156-56, GD246 and PG1123 we can measure directly the HeII column
density. Although the spectra of HZ43, GD153 and GD71 extend beyond
700Å we are unable to detect a HeII edge, possibly an instrumental effect.
The remaining spectra are either severely attenuated by HI or are insuffi-
ciently exposed to allow us to resolve features longward of 200Å. For these
latter two groups of objects we can only infer upper limits to the HeII column. None of the stars in our sample show signs of photospheric helium,

TABLE 2. The ionization fractions of helium and hydrogen in the interstellar medium, calculated from the line-of-sight column density measurements towards PG1123+189 and GD246 using (top) homogeneous H+He photospheric models and (bottom) models incorporating heavy elements (Barstow *et al.* 1996).

| Star | $f_H$ | | | $f_{He}$ | | |
|------|-------|-----|-----|----------|-----|-----|
| | Value | $-1\sigma$ | $+1\sigma$ | Value | $-1\sigma$ | $+1\sigma$ |
| PG1123+189 | 0.00 | 0.00 | 0.16 | 0.28 | 0.20 | 0.38 |
| PG1123+189 | 0.43 | 0.13 | 0.65 | 0.20 | 0.15 | 0.26 |
| GD246 | 0.00 | 0.00 | 0.19 | 0.29 | 0.20 | 0.36 |
| GD246 | 0.20 | 0.04 | 0.45 | 0.23 | 0.17 | 0.30 |

characterised by a converging series of broad absorption lines which gives an apparent edge at 235Å as opposed to 228Å.

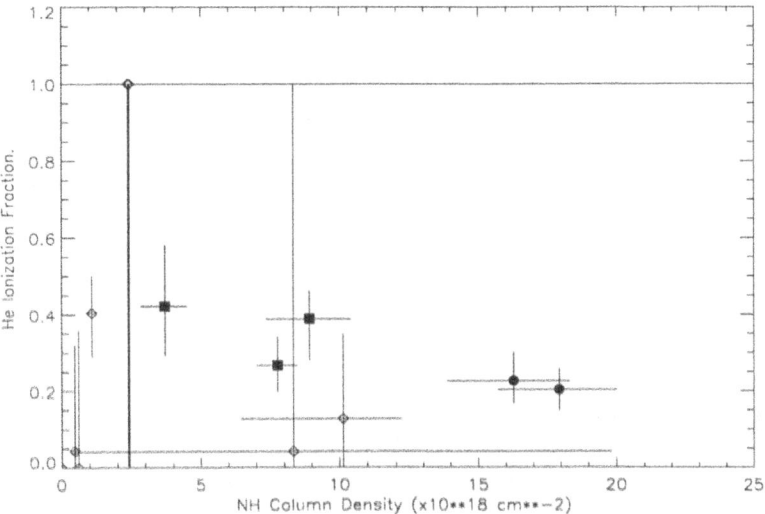

*Figure 1.* The observed fractional ionization of He for each star in the sample as a function of the total column density of hydrogen along the line-of-sight. The shaded shapes indicate those stars where He II is detected directly, squares corresponding to objects analysed with H+He models (GD659, REJ1032, REJ2156) and circles objects analysed with heavy element models (PG1123 and GD246). The open diamonds represent those stars where He II was not detected and only broad limits on the ionization fraction obtained (Barstow *et al.* 1996).

Our calculated values for H and He ionization fractions are shown in tables 1 and 2.

## 4. Conclusion

We have directly measured the amount of HeII along 5 lines of sight in the LISM. The ionization fractions show no correlation with direction, distance or line of sight hydrogen column and volume densities. We note no relationship with stellar temperature, indicating that this material is unlikely to be circumstellar. For the 5 directions, we calculate ionization fractions which indicate that the LISM maybe uniformly ionized. The limits on the fractions along the remaining lines of sight do not contradict this conclusion. Our average values for the ionization fractions along these 5 lines of sight of $0.35\pm0.1$ for H and $0.27\pm0.04$ for He agree with the work of Lyu & Bruhweiler (1996). However, assuming that all the neutral HI in the line of sight has $n\sim0.1\text{cm}^{-3}$, we calculate excessively large distances to the edge of the Local Cloud. Therefore, we conclude that significant material must lie beyond the Local Cloud and agreement with Lyu & Bruhweiler (1996) is possibly just fortuitous.

## References

Barstow, M. A. *et al.* 1993, *MNRAS*, 264, 16

Barstow, M. A., Holberg, J. B., Hubeny, I., Lanz, T., Bruhweiler, F.C. & Tweedy, R. W. 1996, *MNRAS*, 279, 1120

Barstow, M. A., Dobbie, P.D., Holberg, J. B., Hubeny, I. & Lanz, T. 1996, *MNRAS*, accepted

Bruhweiler, F. C. 1996, in *Astrophysics in the Extreme Ultra Violet*, eds Bowyer, S. and Malina, R. F., Kluwer, 261

Gry, C., Lemonon, L., Vidal-Madjar, A., Lemione, M. & Ferlet, R. 1995, *A & A*, 302, 497

Holberg, J.B., Barstow, M. A., Bruhweiler, F. C., & Sion, E. M., 1995, *ApJ*, 453, 313

Lyu, C.-H. & Bruhweiler, F., C. 1996, *ApJ*, 459, 216

Marsh, M. C.,*et al* 1996, *MNRAS*, in press

Koester, D. 1991, in *Michaud G., Tutukov A.*, eds, Proc. IAU Symp. 145, Evolution of Stars: The Photospheric Abundance Connection. Kluwer, Dordrecht, 435

Rumph, T., Bowyer, S. & Vennes S. 1994, *AJ*, 107, 2108

Vennes, S., Dupuis, J., Rumph, T., Drake, J. & Bowyer, S. 1993, *ApJ*, 410, L119

# AN INSTRUMENT FOR OBSERVATIONS OF EUV OPACITY SOURCES IN G191-B2B

KURT GUNDERSON, JAMES GREEN AND RYAN MCLEAN

*Astrophysical Research Laboratories*
*1255 38th Street*
*Boulder, CO 80303*

**Abstract.** We present a sounding rocket payload design to perform high resolution EUV spectroscopy of G191-B2B. Previous rocket-borne and EUVE observations show that the EUV spectra of several DA white dwarfs roll off at longer wavelengths than pure H model or H-He stratified model atmospheres predict. These data, however, lack the spectral resolution to identify the opacity sources unambiguously. Our instrument's capability of 0.1 Å resolution is a factor of ten improvement over the previous observations (EUVE $\sim$ 1-2 Å) in the bandpass of 210-290 Å, where G191-B2B is most affected by the elusive EUV opacity sources. Therefore, this observation could lead to their identification. Knowledge of the chemical composition and associated ionization states responsible for the EUV absorption will constrain models of the atmospheric structure and provide clues to explain the details of how the constituents are supported against downward diffusion. The instrument is scheduled for launch in late 1997.

## 1. Background

Despite observations of G191-B2B by several instruments, EUV opacity sources remain unidentified. EXOSAT observations using the Al/P (170-400 Å) and the LX3000 (44-300 Å) thin foil survey photometers first suggested that one or more short wavelength opacity sources existed in the photosphere of G191-B2B (Jordan, *et al*, 1987; Paerels and Heise, 1989). Originally, HeII was considered the likely candidate, as the high surface gravity was thought to preclude the presence of any higher Z metals. Then the EUVS sounding rocket experiment showed that HeII was not the opacity source, but was unable to identify the actual opacity sources (Wilkin-

*I. Isern et al. (eds.), White Dwarfs, 289–292.*
© 1997 *Kluwer Academic Publishers.*

son, Green, and Cash, 1992). Theoretical work at that time indicated that radiative levitation would support metals against downward diffusion, particularly iron and nickel, but did not preclude the abundant low Z metals such as C, N and O (Vennes, *et al*, 1992; Dreizler and Werner, 1992; Becker and Butler, 1992). It was hoped that the better signal-to-noise spectrum obtained with EUVE would resolve the issue, but the EUVE spectrum does not identify the source of the opacity unambiguously. However, model spectra by Vennes, *et al.*, (1992) of an H-Fe atmosphere degraded to 0.1 Å resolution show resolvable lines in the 520–560 Å wavelength region. Photon counting statistics will not support a similar $\lambda/\Delta\lambda$ in a 210–290 Å bandpass for a sounding-rocket observation, so we have set a resolution goal of $\lambda/\Delta\lambda$ of 2500.

## 2. The Instrument

Achieving 0.1 Å resolution with acceptable signal-to-noise requires improvements in optics, rockets, and instrument design. The instrument presented here minimizes slit and reflective losses and maximizes imaging capabilities by implementing a normal incidence Wadsworth system. A Wadsworth system suffers reflective losses from only one surface, and making that surface a spherically curved grating eliminates the need for a slit which will obstruct starlight if alignment is imperfect. Although normal incidence reflections maximize imaging resolution, few materials offer significant EUV reflectivities except at grazing incidence. However, applying a multi-layer

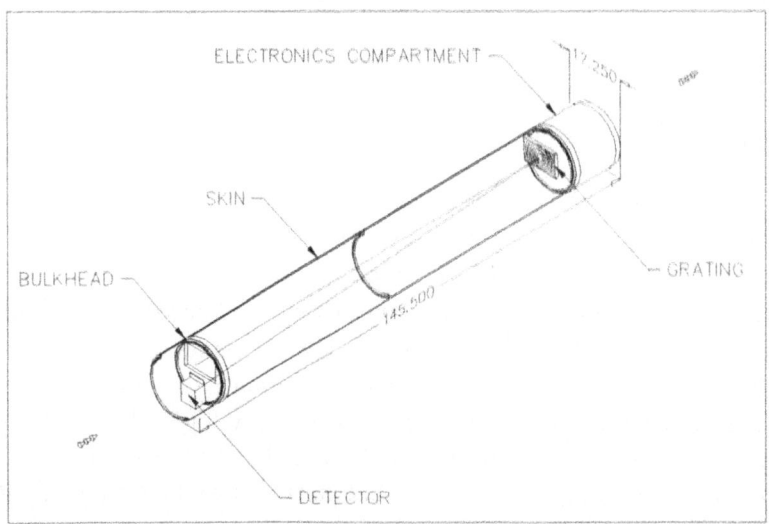

*Figure 1.* Preliminary design of the instrument. Dimensions are given in inches.

coating can produce a grating with the necessary reflectivity over the broad, 210–290 Å bandpass despite the coatings' typical uses as narrow band, high efficiency, reflective filters. Launching the instrument with the larger Mark 70 booster will allow a longer exposure time at higher altitude to minimize atmospheric attenuation and maximize the total number of incident photons.

Figure 1 shows a conceptual layout of the instrument encased in seventeen inch (inner diameter) skin sections. Axial arrows indicate the direction of the rocket's nose during flight. The shutter door opens from the bottom (left hand side of the Figure 1), exposing a holographically ruled (2485 1 mm$^{-1}$), multilayer coated grating through an 18 × 27 cm entrance aperture. The spherically curved ($R_c = 600$ cm) surface of the grating focuses the spectrum onto the detector. Figure 2 shows ray trace models of the instrument (including the detector's point spread function) for two sets of two point sources. The upper set shows spots at 210.0 and 210.1 Å, and the lower shows spots at 290.0 and 290.1 Å. At both wavelength limits, the

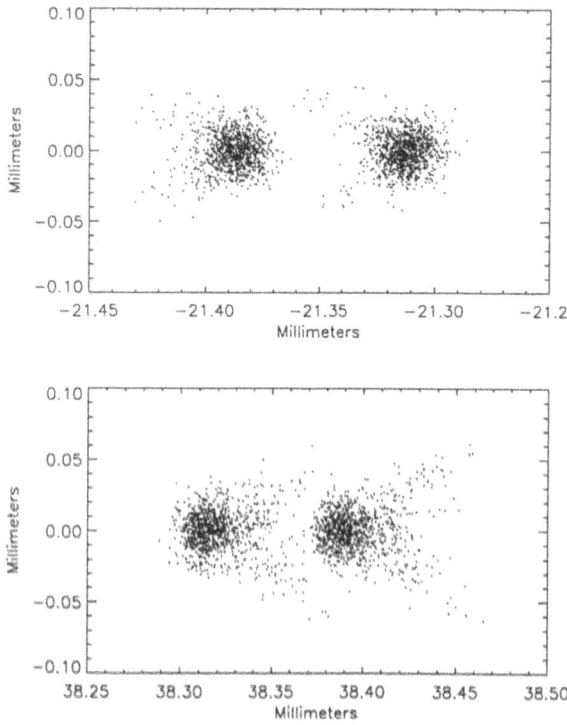

*Figure 2.* Ray traces of point sources of wavelengths 210.0 & 210.1 Å (*upper*) and 290.0 & 290.1 Å (*lower*). Coordinates represent positions on a 10×60 mm detector face.

spots are clearly resolved. A time delay microchannel plate detector with 60 mm of active area in the spectral direction and 20 $\mu$m resolution will support the spectral resolution.

The payload is scheduled for launch from White Sands Missile Range in late 1997.

## References

Becker, S. R. and Butler, K. 1992, *Astr. Ap.*, **265**, 647

Dreizler, S., and Werner, K. 1992, *8th Eur. Conf. on White Dwarfs*, ed. M. A. Barstow (Kluwer), p. 205

Jordan, S., Koester, D., Wolf-Mathies, C., and Brunner, H. 1987, *Astr. Ap.*, **185**, 253

Lanz, T., and Hubeny, I. 1995 *ApJ*, **439**, 905

Paerels, F. B. S., and Heise, J. 1989, *ApJ*, **339**, 1000

Vennes, S., Chayer, P., Thorstensen, J. R., Bowyer, S., and Shipman, H. L. 1992, *ApJ Let*, **392**, L27

Wilkinson, E., Green, J. C., and Cash, W. 1992, *ApJ Let*, **397**, L51

# METALS IN THE VARIABLE DA G29-38

D. KOESTER

*Institut für Astronomie und Astrophysik*
*Universität Kiel, 24098 Kiel, Germany*

AND

J. PROVENCAL AND H.L. SHIPMAN

*University of Delaware*
*Newark, Delaware 19716, USA*

## 1. Introduction

G29-38 (WD2326+049) is one of the brighter members of the class of variable DA or ZZ Ceti stars. It has one of the largest amplitudes in its variations and shows the typical complex variable lightcurves of other large-amplitude DAVs. Zuckerman and Becklin (1987) found an infrared excess in the stars spectrum above 2 $\mu$m. Two different models have been proposed to explain this excess: a brown dwarf companion or a near-by dust cloud. The Whole Earth Telescope observed this star in 1988 and Winget et al. (1990) reported a large phase change in the largest amplitude oscillation, indicating the presence of a massive companion. On the other hand, radial velocity measurements showed that there is no massive companion in the system, although a *low-mass* orbiting object is still possible (Barnbaum and Zuckerman 1992). However, recent photometric observations over a five year time span (Kleinman et al. 1994) did not show any evidence of orbital motion in the system.

## 2. Observations

A UV spectrum of G29-38 was obtained with the FOS on the Hubble Space Telescope, covering the range 1150 to 3000 Å with a resolution of 6 Å. To support the analysis, we have also obtained two optical spectra at the DSAZ (Calar Alto) with very high S/N at a resolution of 3.8 Å. All spectra are dominated by the hydrogen lines and the quasi-molecular satellites of Ly $\alpha$

*I. Isern et al. (eds.), White Dwarfs, 293–296.*
© *1997 Kluwer Academic Publishers.*

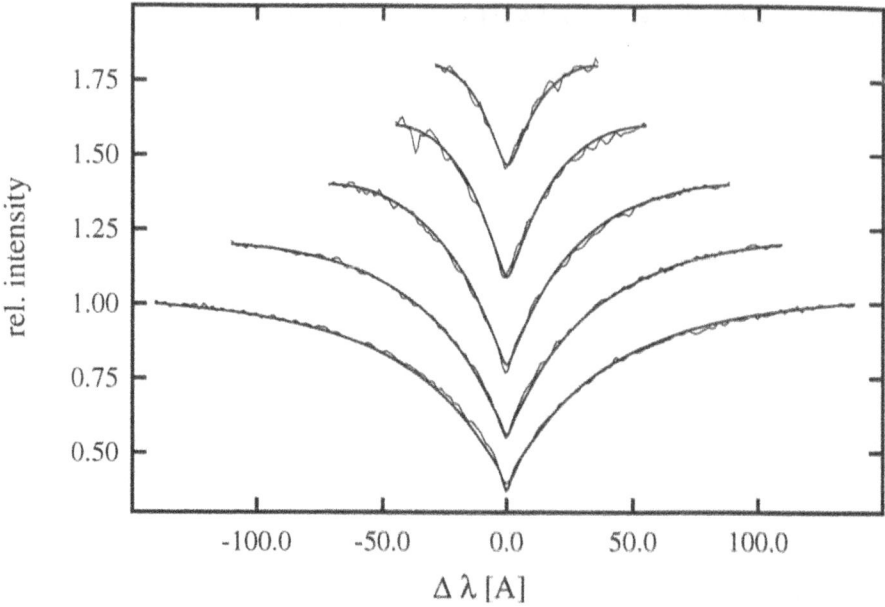

*Figure 1.* Model fit to Balmer lines H$\beta$ to H8 from bottom at T$_{eff}$ = 11600 K, log g = 8.05

in the UV. We have analyzed the observations using an extensive grid of DA model atmospheres with convection included at medium efficiency as found appropriate for the DAV by Koester et al. (1994) and Bergeron et al. (1995). Fig. 1 shows the fit to the Balmer lines at the parameters adopted for the analysis: T$_{eff}$ = 11600 K, log g = 8.05.

Inspection of the blue wing of H$\epsilon$ in Fig. 1 shows an absorption feature, which is present in both optical spectra and can easily be identified as the H line of the CaII resonance doublet. The red part of the FOS spectrum also clearly shows metal lines: the MgII resonance lines and several strong features due to numerous FeII lines can be identified. The features are weaker, but generally similar to those observed e.g. in L745-46A (Koester and Allard 1996); many of the FeII lines originate from exited levels and not the ground state, we therefore assume for our analysis that these absorption features are photospheric and not caused by circumstellar material. This would make G29-38 a DAZ spectroscopically, to our knowledge the only one in the typical DA range between 10000 and 25000 K. Since accretion from material near the star is the only plausible explanation for the metals this also supports the explanation of the infrared excess as originating from circumstellar dust.

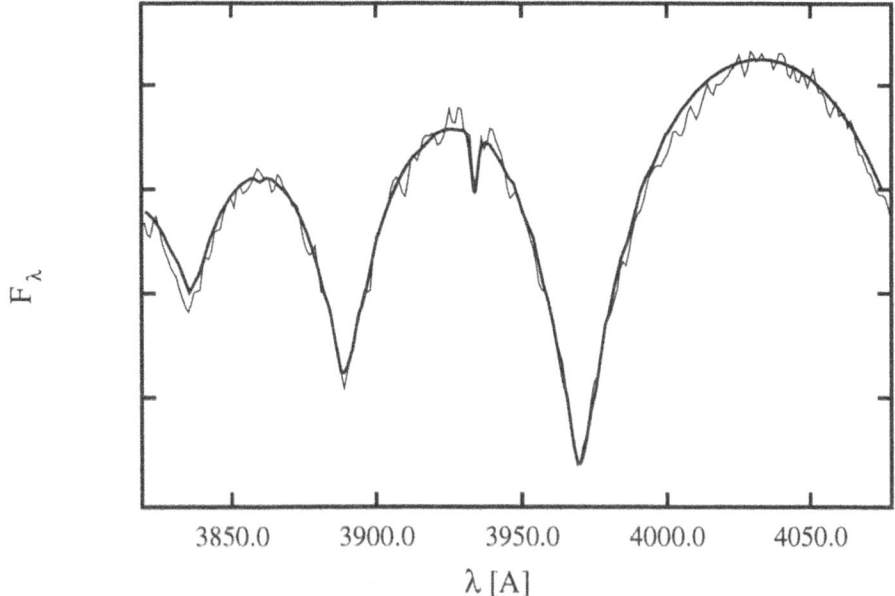

*Figure 2.* CaII H resonance line in G29-38

## 3. Results

Under the photospheric assumption we have determined the abundances of Ca, Mg, and Fe from detailed comparisons with synthetic spectra. Fig. 2 shows the fit to the CaII line, Fig. 3 displays the MgII resonance lines and the region toward shorter wavelengths, where numerous FeII lines are visible.

The line list used in our calculation is obviously incomplete and the fit to the observations is far from perfect. However, using only stronger features, we believe that the abundances of the 3 elements are fairly well determined. The resulting number abundances (relative to H) are:

$$Mg/H = 5.5 \; 10^{-7}, \; Ca/H = 7.0 \; 10^{-8}, \; Fe/H = 6.0 \; 10^{-7}.$$

In their study of the accretion/diffusion scenario for metals in cool white dwarfs Dupuis et al. (1993) give only one result for Ca in DA. Extrapolating their Fig. 5 slightly, our value lies exactly on their result for *high* accretion rates. Given the obvious presence of dusty material close to the star, the accretion process is perhaps still ongoing.

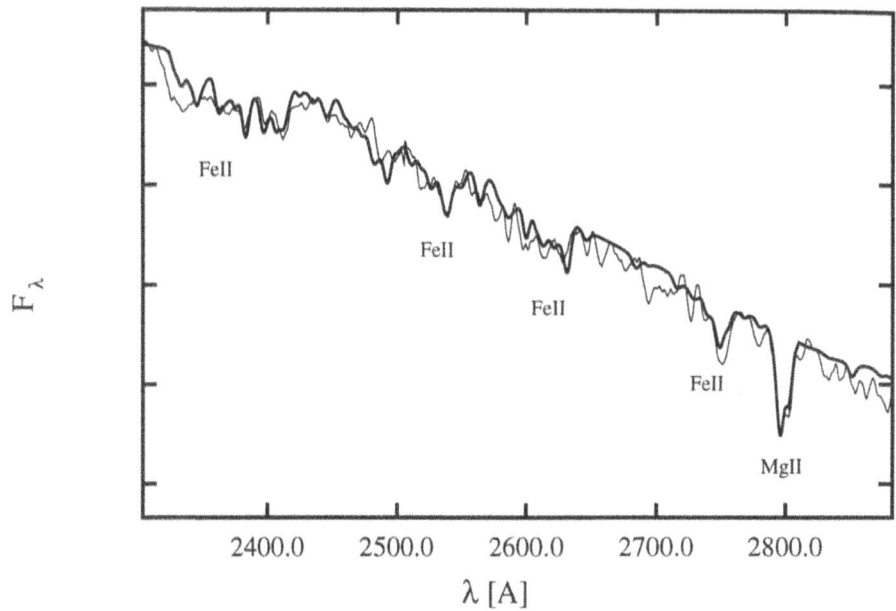

$\lambda$ [A]

*Figure 3.* MgII resonance lines and FeII lines in G29-38

## References

Barnbaum C., Zuckerman B. 1992, ApJ 396, L31
Bergeron P. et al. 1995, ApJ 449, 258
Dupuis J., Fontaine G., Wesemael F. 1993, ApJS 87, 345
Kleinman S.J. et al. 1994, ApJ 436,875
Koester D., Allard N.F., Vauclair G. 1994, A&A 291, L9
Koester D., Allard N.F. 1996, ASP Conf.Ser. 96, 324
Winget D.E. et al. 1990, ApJ 357, 630

# Section IV:
# White Dwarfs in Binaries

# IN SEARCH OF THE DOUBLE DEGENERATE
# PROGENITORS OF TYPE IA SUPERNOVAE

REX A. SAFFER
*Department of Astronomy & Astrophysics*
*Villanova University*
*800 Lancaster Ave.*
*Villanova, PA 19085*
*rsaffer@ucis.vill.edu*

AND

MARIO LIVIO
*ST ScI*
*3700 San Martin Dr.*
*Baltimore, MD 21218*
*mlivio@stsci.edu*

**Abstract.** We describe radial velocity observations of a large sample of apparently single white dwarfs in an effort to discover close, double-degenerate (DD) pairs which might comprise viable Type Ia Supernova progenitors. We have identified a number of candidate variables, including a new short-period subdwarf B/white dwarf pair. The observations appear to be in general agreement with the predictions of theory. Further observations are planned to solve for the orbital parameters of the candidates, from which it will be possible to place meaningful observational constraints both on DDs as SN Ia progenitors and, more generally, on the theory of binary star evolution.

## 1. Introduction

Both theoretical and observational investigations into the origins of Type Ia Supernovae (SN Ia) have focused on binary star evolution. Within this framework, scenarios are divided into "single degenerate" (Whelan & Iben 1973) and "double degenerate" (Webbink 1984; Iben & Tutukov 1984) models. In the former models, the white dwarf (WD) accretes from a nondegenerate companion, while in the latter, two WDs with a total mass exceeding

*I. Isern et al. (eds.), White Dwarfs, 299–305.*
© *1997 Kluwer Academic Publishers.*

the Chandrasekhar limit ($M_{ch}$) are assumed to coalesce. The latter scenario predicts that there should exist a population of close, short-period ($P \lesssim 10^h$) binary WD systems, or double degenerates (DD), with total system masses in excess of $M_{ch}$. They are brought to short-period orbits by common envelope (CE) evolution, with the energy necessary to expel the envelope taken from orbital shrinkage. Subsequently, further loss of orbital angular momentum through emission of gravitational wave radiation provides the mechanism through which the merger can take place.

Saffer et al. (1988) discovered that L 870–2 (EG 11) is a DD with $P \sim 1^d.56$. However, the period is too long for the system to merge on an astrophysically interesting time scale. When the merger does take place, a SN Ia probably will not result since the total system mass is well below $M_{ch}$ (Bergeron et al. 1989). In recent years, three systematic searches for DD have been performed. Two (Robinson & Shafter 1987; Foss et al. 1991) failed to discover any candidates at all. The third (Bragaglia et al. 1990) did discover one DD (and 4 other candidates), but the period ($P \sim 1^d.18$) and very low primary mass ($M = 0.16 \, M_\odot$) of the confirmed DD render it an unlikely SN Ia progenitor. Even more recently, observations (Marsh et al. 1995; Marsh 1995) targeting the low-mass tail of the WD mass distribution of Bergeron et al. (1992) spectacularly confirmed the prediction that (at least some) WDs with masses less than the canonical core-helium ignition mass should be post-common envelope objects residing in short-period binary systems. However, their low masses again make them unlikely SN Ia progenitors.

These negative results have cast doubt on the merging DD scenario as the favored mechanism for producing SN Ia. All DD were assumed to have been born into longer-period orbits, then to have evolved to short-period orbits where they could more easily be detected. In fact, an improved theoretical treatment of the evolution of the Galaxy's binary population (Yungelson et al. 1994) shows that a significant number of DD are born directly into short-period orbits, but that evolution to merger takes place so rapidly that previous searches had sample sizes too small to detect them. On the other hand, it is clear that the DD that *were* discovered (Saffer et al. 1988; Bragaglia et al. 1990; Marsh et al. 1995; Marsh 1995) fall within the peak of the expected DD number distribution (cf. Figure 1 of Yungelson et al. 1994). Yungelson et al. (1994) suggested that further searches with modestly increased sample sizes are likely to reveal the existence of the short-period, massive DD population. In this paper, we report such observations of an enlarged sample in order to place meaningful observational constraints both on DDs as SN Ia progenitors and, more generally, on the theory of binary star evolution.

## 2.  Observations and Radial Velocity Measurements

Our sample comprises ~100 apparently single DA (hydrogen line) and DB (helium line) WDs, drawn from the catalog of McCook & Sion (1987) and which are (largely) not in common with the previous three systematic surveys. Over two observing seasons, in a total of 21 nights in 4 observing runs in 1994 and 1995, we have obtained radial velocity spectroscopy at the Kitt Peak National Observatory's 2.1-m reflector equipped with GOLDCAM. This instrumentation provided 3Å FWHM spectral resolution over a spectral range of ~700Å centered on Hα and including HeI λ6678. Exposures were chosen to achieve a signal-to-noise ratio of ~30.

The advantages of the technique are: 1) In all but the hottest DA WDs, Hα has a sharp (FWHM ~1Å) non-LTE core, making possible radial velocity measurements with 20–30 km s$^{-1}$ errors despite the extreme width of the Stark-broadened absorption line. The HeI λ6678 line in DB WDs provides similar precision, 30–40 km s$^{-1}$. The close, short-period DD of interest have peak-to-peak radial velocity amplitudes nearly an order of magnitude larger than these detection limits, and any existing systems would have a high probabilty of detection. 2) The method detects objects with similar temperatures, like the prototype L 870–2, that have double-lined spectra. The technique of differential photometry (Robinson & Shafter 1987) in the wings of the absorption line does not detect this type of system. 3) Spectroscopic observations near Hα offer the maximum velocity resolution for any optical features observable from the ground at intermediate resolution. 4) The large throughput at intermediate spectral resolution makes it possible to observe large, relatively faint samples with modest telescope apertures.

Our targets were observed twice on any given night with the observations separated by 1–2 hr, followed by a third observation a day or two later. Short-period systems are easily detected simply by comparing spectra by eye (a.k.a. "astrophysics with a ruler", H. Bond, private communication.) In Figure 1 we show one such comparison, where the cores of one object's Hα line are separated by ~150 km s$^{-1}$ in two exposures on successive nights. The laboratory wavelength of Hα is marked with a dashed line.

We have found a total of 12 such candidates, for which followup observations will be required to solve for the orbital parameters. In particular, the periods and velocity semiamplitudes will determine the mass functions and provide lower bounds on the WD masses. In addition to the DD candidates, we also have detected two systems with narrow, variable Hα emission in the core of the WD absorption line, betraying the probable presence of low-mass main sequence (MS) companions. Moreover, the subdwarf B (sdB) stars Feige 36 and Ton 245, pre-white dwarfs, have recently been found to be in close binary systems with periods of 8$^h$5 and 2$^d$5, respectively (Saffer

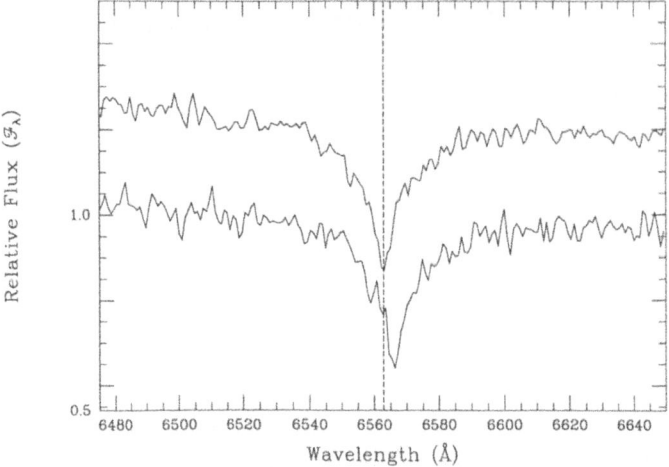

*Figure 1.* Radial velocity variability at Hα. The laboratory wavelength of Hα is marked with a dashed line.

et al. 1997, in preparation). Low-mass MS companions are ruled out by large mass functions and the non-detection of infrared excesses. The companions are thus inferred to be compact, probably white dwarfs. Finally, Holberg et al. (1995) recently discovered that the DAO WD Feige 55 is binary with a period of $1^d\!.49$ and a compact companion.

## 3. Comparisons with Theoretical Predictions

In Table 1, we present all known WDs in close (short-period) binary systems having measured orbital periods, including the two sdB–WD pairs Feige 36 and Ton 245. We exclude possible short-period carbon-oxygen (CO) + helium (He) mass-transfer systems such as GP Com, which might well prove not to be viable SN Ia progenitors. Successive table columns give the object, period in days, logarithm of the period in hours, an estimate or lower bound on the mass of the primary, total system mass when known, and the source of the data. We now wish to compare the results of the observations with the predictions of theory.

The calculations of Yungelson et al. (1994) make predictions about the total numbers of close DD among WDs, as well as the number and mass distributions vs. orbital period. We restrict the comparison to two samples, the one described here and that of Bragaglia et al. (1990), as these have similar numbers of objects and are sensitive to the same range of orbital periods ($\sim$3 hr to a few days). Three comparisons are possible:

1. In Table 2 of Yungelson et al. (1994), where a flat mass ratio distribution and a value of $\alpha_{CE} = 1$ have been assumed, the fraction of

| Object | P(d) | log P(h) | $M_1$ | $(M_{tot})$ | Reference |
|--------|------|----------|-------|-------------|-----------|
| L 870–2 | 1.56 | 1.57 | 0.47 | (0.99) | Saffer et al. (1988) |
| 0957−666 | 1.15 | 1.44 | 0.16 | | Bragaglia et al. (1990) |
| 1101+364 | 0.15 | 0.54 | 0.31 | (0.58) | Marsh et al. (1995) |
| 1241−010 | 3.35 | 1.91 | 0.31 | | Marsh (1995) |
| 1317+453 | 4.80 | 2.06 | 0.33 | | |
| 1713+332 | 1.12 | 1.43 | 0.35 | | |
| 2331+290 | 0.17 | 0.60 | 0.39 | | |
| Feige 55 | 1.49 | 1.55 | 0.40 | | Holberg et al. (1995) |
| Feige 36 | 0.35 | 0.93 | 0.50 | | Saffer et al. 1997, |
| Ton 245 | 2.50 | 1.78 | 0.50 | | in preparation |

TABLE 1. Derived parameters of known close DD.

WD+MS pairs and DD pairs in the two samples is predicted to be 1/21 and 1/10, respectively, whereas in the Bragaglia et al. (1990) sample and the one presented here the observed fractions are 2/54 and 5/54, and 2/100 and 13/100, respectively. Here we include both confirmed and candidate close binary systems. These fractions are in good agreement with the predicted fractions.

2. In Figure 1 of Yungelson et al. (1994), the expected number distribution of close DD is given. We reproduce a portion of that figure with permission (see Figure 2). The predicted DD distribution is the solid histogram, the predicted WD+MS distribution is the dashed histogram, and the dot-dashed histogram is the observed distribution of the DD in Table 1 of this paper. The observed distribution (arbitrarily normalized) is generally in agreement with the predicted distribution, albeit possibly offset to longer periods. We note that *not one* convincing SN Ia progenitor system, i.e., a pair of CO WDs with a total system mass $M_{tot} > M_{ch}$ and a period $P < 10$ hr, has been observed. Is this a cause for alarm? Possibly not, as the next comparison shows.

3. In Figure 2 of Yungelson et al. (1994), the expected total system mass distribution of close DD is given. There are three distinct peaks at 0.5, 0.9, and 1.3 $M_\odot$, corresponding to He+He, He+CO, and CO+CO pairs, and there is a small (but not insignificant) tail extending to

masses in excess of $M_{ch}$. The fraction of systems with $M_{tot} > M_{ch}$ is predicted to be ~1/25. To date, only 10 systems have measured orbital periods; the sample of confirmed short-period DDs is still too small to draw definitive conclusions, although we note that systems with the lowest masses tend to have the shortest orbital periods. However, the 12 candidate DDs from the survey presented here will increase the number of confirmed short-period DDs to 22 systems, nearly the number predicted to include a super-$M_{ch}$ system.

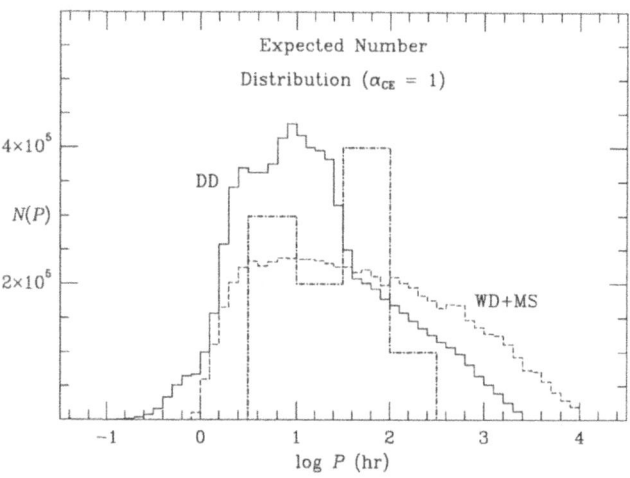

*Figure 2.* Comparison of the predicted and observed DD number distributions vs. log P(hr). The heavy dot-dashed histogram is the observed distribution.

## 4. Conclusions

We have performed a radial velocity survey of nearly 100 field WDs not previously known to be binary. Among the sample, we have discovered 12 new DD candidates, 2 WD+MS pairs, one new short-period DD (Feige 55, Holberg et al. 1995), and one new short-period sdB+WD pair (Feige 36). Our conclusions are:

1. The population of close, short-period DD systems that the theory of binary star evolution predicts to exist *does in fact exist.* The observed number fraction of all confirmed short-period DD, as well as their observed number and mass distributions with respect to orbital period, *are not in contradiction with the predictions of theory* (Yungelson et al. 1994). In particular, the number distribution vs. orbital period of the observed DD systems falls very nearly in the peak of the predicted dis-

tribution, although possibly shifted slightly to longer periods. We note that this small difference could be made to vanish by a small increase in the value of $\alpha_{CE}$, a free parameter in the theory ("the parameter of ignorance", I. Iben).

2. The very low-mass systems observed by Marsh et al. (1995) and Marsh (1995) tend to have the shortest orbital periods, however, their total system masses are probably too small to make them viable SN Ia progenitors. Further, the primary stars almost certainly have He cores, while CO white dwarfs are thought to be the best candidates to undergo the carbon deflagration required to produce a SN Ia explosion.

3. No "loaded gun" has yet been found, i.e., a super-$M_{ch}$ CO–CO system with $P \lesssim 10$ hr. On the other hand, the sample of confirmed short-period DD is still smaller than the number predicted to contain such a system. The 12 new DD candidates discovered in our survey will increase the fraction of confirmed short-period DDs among all WDs nearly to that ($\sim 1/25$) which theory predicts will include a massive, super-$M_{ch}$ system.

We plan followup observations in the coming year to solve for the orbital parameters of the 12 new DD candidates found in our survey. *Even if no convincing SN Ia progenitor is found, a meaningful test of the predictions of the theory of close binary evolution now seems within reach.*

Acknowledgements: This work has been supported by NASA Grant NAGW-2678 and by the Director's Discretionary Research Fund at the Space Telescope Science Institute.

# References

Bergeron, P., Wesemael, F., Fontaine, G., & Liebert, J. 1989, ApJ, 345, L1
Bergeron, P., Saffer, R.A., & Liebert, J. 1992, ApJ, 394, 228
Bragaglia, A., Greggio, L., Renzini, A., & D'Odorico, S. 1990, ApJ, 365, L13
Foss, D., Wade, R.A., & Green, R.F. 1991, ApJ, 374, 281
Holberg, J.B., Saffer, R.A., Tweedy, R.W., & Barstow, M.A. 1995, ApJ, 452, L133
Iben, I. Jr., & Tutukov, A.V. 1984, ApJ, 282, 615
Marsh, T.R., Dhillon, V.S., & Duck, S.R. 1995, MNRAS, 275, 828
Marsh, T.R. 1995, MNRAS, 275, L1
McCook, G.P & Sion, E.M. 1987, ApJS, 65, 603
Robinson, E.L. & Shafter, A.W. 1987, ApJ, 322, 296
Saffer, R.A., Liebert, J., & Olszewski, E.W. 1988, ApJ, 334, 947
Webbink, R.F. 1984, ApJ, 277, 355
Whelan, J. & Iben, I. Jr. 1973, ApJ, 194, 657
Yungelson, R.L., Livio, M., Tutukov, A.V., & Saffer, R.A. 1994, ApJ, 420, 336

# THE DOUBLE DEGENERATE POPULATION

M. HERNANZ, J. ISERN & M. SALARIS

*Institut d'Estudis Espacials de Catalunya (CSIC Research Unit)*
*Edifici Nexus-104, C/Gran Capità 2-4, 08034 Barcelona, Spain*

**Abstract.** The expected population of double degenerate (DD) systems in the solar neighborhood is computed, taking into account the scale height of these systems over the galactic plane, as well as the white dwarf's cooling times. The predictions of our theoretical models are in agreement with the scarcity of DD detections resulting from the systematic searches carried up to now (Robinson & Shafter, 1987, Foss *et al.*, 1991 and Bragaglia *et al.*, 1991).

## 1. Scenario code and method

The aim of population synthesis codes is to reproduce the observed characteristics of a given population from an estimate of the primordial population and using theoretical evolutionary models. Predictions obtained in this way can be used to guide the observational search.

The number of double degenerate systems which are born per unit time and separation interval with an initial separation $a$ at instant $t$ can be written as (Isern *et al.*, 1997):

$$b(a,t) = \int_{M_1} \int_{M_2} \phi(M_1, M_2)\Psi(t - \tau_b)H_0(A_0)\frac{dA_0}{da}\, dM_1 dM_2$$

where $M_1$, $M_2$ and $A_0$ are the masses and the separation at the zero age main sequence, $a(M_1, M_2, A_0)$ is the present separation, $\tau_b(M_1, M_2, A_0)$ is the time necessary to form a double degenerate system, $\phi(M_1, M_2)$ is the initial mass function of the binary, $H_0(A_0)$ is the distribution of separations and $\Psi(t)$ is the star formation rate per unit volume. If we prefer the birthrate as a function of the period $P$ instead of $a$, we can write $b(P,t)dP = b(a,t)da$.

*I. Isern et al. (eds.), White Dwarfs, 307–314.*
© *1997 Kluwer Academic Publishers.*

The birthrate of DDs is obtained from the analysis of all the evolutionary channels leading to this type of binary system, mainly those situations in which Roche lobe overflow occurs when the envelope of the star is convective. An homogeneous set of stellar models has been constructed using the FRANEC code (Chieffi and Straniero 1989), covering the mass range from 0.8 to 8 $M_\odot$. Details concerning the evolutionary paths leading from primordial to final systems can be found in Tornambé (1995) and are summarized in Table 1. The maximum radii on the main sequence, red giant branch and asymptotic giant branch phases, as well as the final masses of the resulting He and CO degenerate cores are shown in Figure 1. The steps to be performed to know the nature (He or CO) and the separation of the white dwarfs present in the binary system are the following: for a given primary mass, $M_1$, and separation, $A$, we estimate if the common envelope will occur. If this is the case, we estimate the mass of the remnant of the primary, $M_{1r}$, and the new separation $A'$. Then we determine whether a second common envelope for the secondary will occur. If this is the case, we estimate the mass of the remnant of the secondary, $M_{2r}$ and the final separation $a$.

TABLE 1. Final systems after common envelope evolution

| $M_1$ | RLP1 | $M_2$ | RLP2 | Intermediate system | Final system |
|-------|------|-------|------|---------------------|--------------|
| 0.7–2.3 | RGB | 0.7–2.3 | RGB | | DHe+DHe |
| 2.3–8 | RGB | 2.3–8 | RGB | He star + He star | DCO+DCO |
| 0.7–8 | AGB | 0.7–2.3 | RGB | | DCO+DHe |
| 2.3–8 | AGB | 2.3–8 | RGB | DCO + He star | DCO+DHe |
| 0.7–8 | AGB | 0.7–8 | AGB | | DCO+DCO |

RLP1: Primary's Roche lobe overflow phase. RLP2: Secondary's Roche lobe overflow phase. DHe: He white dwarf. DCO: Carbon–oxygen white dwarf

The other ingredients necessary to compute b(a,t) are the input population properties: the initial mass function, the separation or period distribution and the star formation rate. The initial mass distribution can be written as the mass distribution of the primary times the mass ratio distribution: $\phi(M_1, M_2)dM_1 dM_2 = \phi(M_1)f(q)dM_1 dq$. For the mass distribution of the primary we have adopted Salpeter's law. The mass ratio distribution is taken to be $f(q) \propto q^\alpha$ where $q = M_2/M_1$. Here we adopted $\alpha = 1$. This distribution is one of the most critical inputs. The constants appearing in those expressions are obtained from the normalization condition (Popova et al., 1992, Mazeh et al., 1992). The distribution of separations adopted

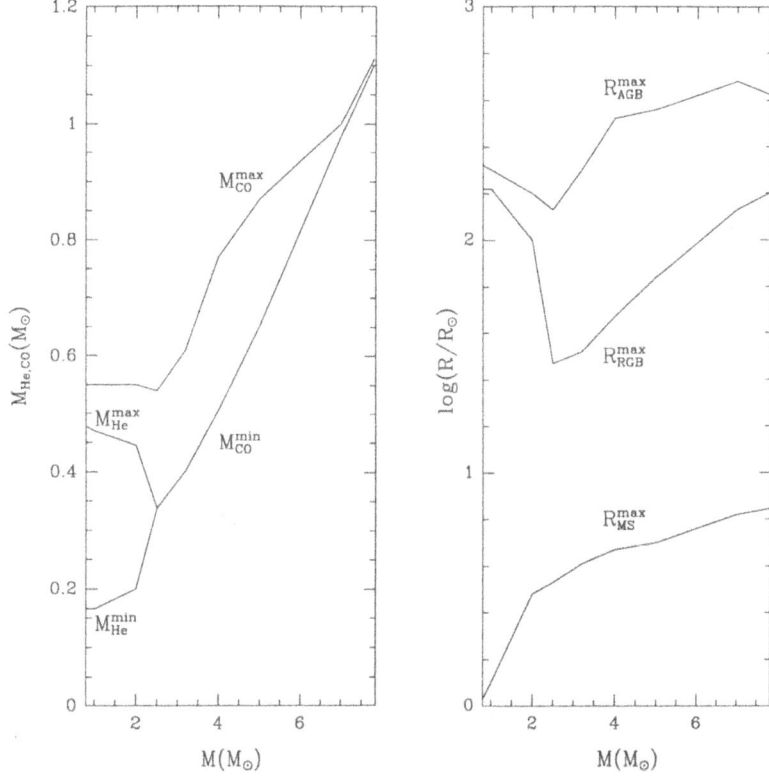

*Figure 1.* Final He and CO degenerate masses (left panel) and maximum radii on the main sequence, the red giant branch and the asymptotic giant branch phases (right panel) as a function of the progenitor's mass (all in solar units). He (CO) white dwarfs have masses in the range $M_{He}^{min}$-$M_{He}^{max}$ ($M_{CO}^{min}$-$M_{CO}^{min}$)

is $H(A_0) = 1/(5ln10)/(A_0/R_\odot)$ (Yungelson and Tutukov 1993), where the constant comes from normalization within the limits $10 \leq A_0/R_\odot \leq 10^6$.

Concerning the star formation rate. It is worthwhile to remember that (as it was pointed out by Iben and Tutukov 1991) the distribution of white dwarfs (and hence of DDs) is a local function while the supernova rate is an average over the galactic disk. Both properties can be connected through a model of chemical evolution and adopting a scale height for the white dwarfs. A reasonable star formation rate per unit of galactic disk surface can be obtained with the following approximations (Abia Canal and Isern, 1991; Bravo *et al.*, 1993): the star formation rate $\Sigma(t)$ is related to the gas surface mass density via $\Sigma(t) = \alpha\sigma_g^n(t)$, with $\alpha = 0.6$ Gyr$^{-1}$ and n=1. In order to avoid the G-dwarf problem, an exponentially decreasing unenriched infall with an e–folding time of 5 Gyr has been included in such a way that the present accretion rate would be $f \leq 1.0$ M$_\odot$pc$^{-2}$Gyr$^{-1}$ (Mirabel and Morras 1984), and the total present surface mass density, $\sigma_T \simeq 70$ M$_\odot$pc$^{-2}$ (Bahcall and Soneira 1984). The star formation rate par unit volume can be obtained dividing by the scale height of white dwarfs: $\Psi(t) = \Sigma(t)/h_{wd}$. This value is estimated to be in the range of 200–250 pc, although it is quite ucertain (Case 1). The scale height, however, has not remained constant during time. It is well known that old objects have higher scale heights over the galactic plane than young ones. This can be due either to a different scale height at birth or to a diffusion out of the galactic plane due to collisions with massive objects. We have determined the scale height that, together with $\Sigma(t)$, better fits the observed luminosity function of white dwarfs (Isern et al., 1995a,b). This is an effective rate that includes not only the star formation but also the diffusion of stars in/out the solar neighborhood (Case 2). Both approximations give the same present star formation rate per unit volume or surface at the present time.

The velocity at which the DD system shrinks due to the emission of gravitational waves is $da/dt = -K/a^3$, with $K = m_1m_2(m_1+m_2)/0.6$ and masses in solar units, separation in solar radii and time in Gyr. For the periods, we have $dP/dt = -(3/2)Q/P^{5/3}$, with $Q = \beta^{8/3}m_1m_2/(0.6(m_1 + m_2)^{1/3})$, being $\beta=8760$ if the period is in hours ($P_h = \beta a^{3/2}/(m_1+m_2)^{1/2}$). Therefore, the DDs having a period $P_D$ at the instant $t_D$, at which they are observed, are those born at the instant $t$ with an initial period $P = [4Q(t_D - t) + P_D^{8/3}]^{3/8}$, where it is assumed that the origin of time corresponds to the moment at which the galaxy was formed. Since $t_D \geq t$, $P \geq P_D$. The equation that governs the density of DDs (systems per pc$^3$) as a function of the period is (see Isern et al., 1997, for the same formulation but as a function of the separation):

$$\frac{\partial n}{\partial t} + \frac{\partial(n\dot{P})}{\partial P} = b(P,t)$$

which has as a solution:

$$n(P_D, t_D) = \frac{1}{\dot{P}(t_D)} \int_0^{t_D} b[P(t), t]\dot{P}(t)\, dt$$

where it is assumed that $n(P,0) \equiv 0$. Notice that when $P_d \to 0$, $n(P_D, t_D) \to 0$ and that when $P_0 \to \infty$ (in which case the emission of gravitational waves is negligible)

$$n(P_D, t_D) \to \int_0^{t_D} b[P(t), t]dt$$

which is just the number of DDs born with a period $P_D$ during the life of the Galaxy.

Since white dwarfs cannot be detected below a given luminosity, it is necessary to correct this equation. If, in order to simplify, it is assumed that all white dwarfs are born with the same luminosity and that they cool

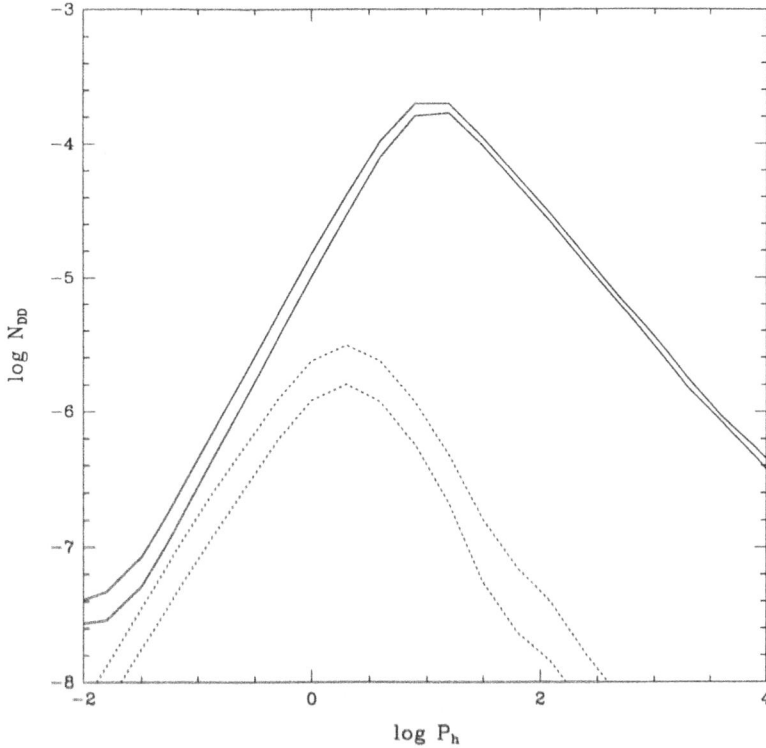

*Figure 2.* Density (in pc$^{-3}$) of all DD systems (solid lines) and of DD systems younger than 0.3 Gyr (dotted lines) versus the period (in hours). For each set of curves, the upper one refers to the stellar formation rate compatible with the models of chemical evolution of the Galaxy, whereas the lower one refers to the star formation rate compatible with the white dwarf's luminosity function.

at the same rate, the number of DDs with both components brighter than $L_0$ is:

$$n(P_D, t_D) = \frac{1}{\dot{P}_D} \int_{t_D - \delta t}^{t_D} b(P, t)\dot{P}dt$$

where $\delta t$ is the time necessary to cool down to $L_0$ ($\delta t \sim 0.3$ Gyr, for $L_0 \sim 10^{-2.5}L_\odot$, which is the threshold luminosity for the Bragaglia *et al.* (1990) survey, for instance). Notice that here it is also assumed that both white dwarfs have started to cool down simultaneously. This is not strictly correct, as explained below.

## 2. Results and discussion

In Figure 2 we show the density of DD systems obtained from our models, adopting the two different prescriptions for the star formation rate explained above and with (lower curves) or without (upper curves) the selection effect related to the brightness of the white dwarfs taken into account. In these last cases, we have rejected all the DD systems with a secondary older, in terms of cooling times, than $3 \times 10^8$ years. This produces a drastic reduction of the number of observable objects by a factor larger than 100. This is, of course, an oversimplification. The bias is due to the luminosity of the white dwarfs which, in turn, depends on the amount of cooling of the primary before the second Roche lobe overflow, the reheating produced during this phase and, finally, the simultaneous cooling of both components that depends on their mass and chemical composition. Assuming that all stars are born as binaries, our models predict densities $2.6 \times 10^{-1}$ pc$^{-3}$ and $2.1 \times 10^{-1}$ pc$^{-3}$ when all systems are taken into account, whereas these densities decrease to $7 \times 10^{-4}$ pc$^{-3}$ and $3 \times 10^{-4}$ pc$^{-3}$ when only white dwarfs younger than $3 \times 10^8$ years are considered (for cases 1 and 2 respectively). These numbers have to be compared with the estimation of $3.3 \times 10^{-3}$ pc$^{-3}$ for all white dwarfs (Liebert *et al.*, 1988, 1989) and $8 \times 10^{-4}$ pc$^{-3}$ (Fleming *et al.*, 1986) for the bright ones. Both quantities are estimated to be uncertain by a factor of two at least. The discrepancy between the observations and the theoretical predictions is not very large when only bright white dwarfs are considered, whereas theory predicts more white dwarfs than observed if all of them are included. One has to take into account that the theoretical values are uncertain due to many unknown inputs related to the birthrate of binaries, such as the function $f(q)$, for instance. However, the ratios between densities for different intervals of periods are more confident. In Table 2 are shown the densities of DD systems in the intervals explored spectroscopically as well as their ratio versus the single white dwarf density. Our results agree fairly well with those of Yungelson *et al.* (1994) (see the last column in their Table 2) and indicate that the negative results of the systematic searches carried up to now are in accordance with the theoretical expectations.

## 3.  Conclusions

We have computed the expected population of double degenerate binary systems in the solar neighborhood taking into account the cooling of white dwarfs and their scale height over the galactic plane. We show that the predicted number of double degenerates in the observationally explored period ranges is compatible with the scarce results obtained up to now by the observations (in agreement with the results of Yungelson *et al.*, 1994). Further observations will help to establish better constraints on the double degenerate scenario as progenitor of type Ia supernovae and on the theory concerning the evolution of binary stars (see Saffer *et al.*, this same volume).

**Acknowledgements** This work has been supported by the DGICYT grants PB94-0111, by the CIRIT grant GRQ94-8001, the AIHI94-082A, the grant

TABLE 2.

| Predicted densities of DD systems (pc$^{-3}$) | | |
| --- | --- | --- |
| | RS | FWG | BGRD |
| Case 1 | $2 \times 10^{-5}$ | $2 \times 10^{-5}$ | $6 \times 10^{-5}$ |
| Case 2 | $8 \times 10^{-6}$ | $1 \times 10^{-6}$ | $3 \times 10^{-5}$ |
| DD systems/single WD | | |
| Observed | 0/44 | 0/25 | 5/54 |
| Case 1 | 1/44 | 1/25 | 4/54 |
| Case 2 | 1/44 | 1/25 | 4/54 |

RS: Robinson &Shafter, 1987 FWG: Foss *et al.*, 1991 BGRD: Bragaglia *et al.*, 1991

ERBCHGECT920009 of the program "Human Capital and Mobility" and by the CESCA.

## References

Abia, C., Canal, R. and Isern, J. (1991) *ApJ* **366**, 198.

Bahcall, J.N. and Soneira, R.M. (1984) *ApJS* **55**,67.

Bragaglia, A., Greggio, L., Renzini, A. and D'Odorico, S. (1990) *ApJ* **365**, L13.

Bravo, E., Isern, J. and Canal, R. (1993) *A&A* **270**, 288.

Chieffi, S. and Straniero, O. (1989) *ApJS* **71**, 47.

Fleming, T.A., Liebert, J., Green, R.F. (1986) *ApJ* **308**, 176.

Foss, D, Wade, R.A., Green, R.F. (1991) *ApJ* **374**, 281.

Iben, I. and Tutukov, A.V. (1991) in *Frontiers in Stellar Evolution*, ed. D.L. Lambert, ASP Conference Series 20, p.403.

Isern, J., García–Berro, E., Hernanz, M., Mochkovitch, R., Burkert, A. (1995a) in *White Dwarfs*, eds. D. Koester and K. Werner, Springer, p.15.

Isern, J., Hernanz, M., Mochkovitch, R., Burkert, A. and García–Berro, E. (1995b) in *The formation of the Milky way*, eds. E.J. Alfaro and A.J. Delgado, CUP, p.179.

Isern, J., Hernanz, M., Salaris, M., Bravo, E., García–Berro, E. and Tornambé, A. (1997) in *Thermonuclear Supernovae*, eds. P. Ruiz–Lapuente, R. Canal and J. Isern, Kluwer, p.127.

Liebert, J., Dahn, C.C., Monet, D.F. (1988) *ApJ* **332**, 891.

Liebert, J., Dahn, C.C., Monet, D.F. (1989), in *IAU 114*, p.15.

Mazeh, T., Goldberg, D., Duquennoy, A. and Mayor, M. (1992) *ApJ* **401**, 265.

Mirabel, J.F. and Morras, R. (1984) *ApJ* **279**, 86.

Popova, E.I., Tutukov, A.V. and Yungelson, L.R. (1982) *Astrophys. Space Sci.* **88**, 55.

Robinson, E.L. and Shafter, A.W. (1987) *ApJ* **322**, 296.

Tornambé, A. (1995) in *Menorca School of Astrophysics: Supernovae* eds. E.Bravo, R.Canal, J.M. Ibáñez and J.Isern (Fundació Catalana per a la Recerca. Barcelona) p.59.

Yungelson, L., Livio, M., Tutukov, A.V. and Saffer, R.A. (1994) *ApJ,* **420**, 336.

Yungelson, L.R. and Tutukov, A.V. 1993 in *New Frontiers in Binary Star Research*, eds. J.C. Lenng and J.S. Nha, ASP Conference Series 38, p.21.

## Discussion

*M.H. van Kerkwijk*: What mass ratio distribution dif you use? How would the results change if you assume that the secondary mass is drawn from the IMF as well (or use Duquenoy & Mayor empirical f(q))

*M. Hernanz*: We have adopted $f(q) = 2q$. This function has a big influence in the final numbers (or densities) of systems, but not so much in the fractions of systems in the different period ranges.

# WHITE DWARF+M DWARF SYSTEMS: A NEW INFRARED SEARCH

ALESSANDRA G. GEMMO

*Osservatorio Astronomico di Padova – Progetto Galileo*
*Riviera Tiso da Camposampiero, 35100 Padova (Italy)*

## 1. Background

The knowledge of the fraction of white dwarfs (WDs) with an M star companion is of fundamental importance in various fields of astrophysics, such as cosmology and the theory of stellar evolution.

In particular, after it was first reported that $\sim$30%–60% of the mass in the solar neighborhood could not be attributed to known matter, some astronomers proposed that the presence of many faint, previously undetected M dwarfs might explain the discrepancy. If not, there might be a sufficient, additional number of brown dwarfs (BDs, objects with masses < 0.08 $M_\odot$, which are unable to spark hydrogen burning in their interiors) to comprise the remainder of the missing mass. Mature BDs are too dim to be observed directly by telescope, although there have been searches for them via gravitational microlensing (e.g. Aubourg *et al.*, 1993 and Alcock *et al.*, 1995). But BDs and M dwarfs may be formed by the same physical process. Thus, by examining the mass function of M dwarfs, it may be possible to extrapolate below the hydrogen–burning limit and study the contribution of BDs to the Galactic disk and halo.

Until recently, it was quite difficult to detect isolated very low mass (VLM) stars because of their extreme faintness and very red colour that made then inaccessible to most ground–based instrumentation. The growth and steady improvement of infrared astronomical technology has provided the opportunity to perform systematic searches. In this context, VLM stars that are companions to WDs offer higher chances of detection because of the unique properties of WDs, which are hot but very small, so it becomes possible to infer the existence of a secondary solely from the infrared colours of the combined system.

*I. Isern et al. (eds.), White Dwarfs, 315–320.*

Two of the most extended searches for low–luminosity objects orbiting WDs have been those of Probst (1983) and Zuckerman and Becklin
(ZB, 1987, 1992). Both these surveys have concluded that the fraction of
WDs showing an infrared excess is no more than 10%. ZB discovered excess infrared radiation from the DA4 WD GD 165 (Becklin & Zuckerman,
1988). Follow–up infrared imaging verified the presence of a faint companion, designated as GD 165B. Because of its extremely red colours and
low–luminosity, GD 165B has been regarded as one of the most promising
BD candidates, until the discovery by D. A. Golimovski and B. R. Oppenheimer of the coolest BD ever found, Gl 229B, which is ten times fainter
than previously known BD candidates and is orbiting an M1 type dwarf
only 5.7 parsecs from the Earth (Nakajima *et al.*, 1995).

M dwarfs and BDs however, are probably about to be ruled out as the
most plausible candidates for baryonic dark matter, especially after Bahcall
*et al.* (1994) have published the results of a *Hubble Space Telescope* search
for M dwarfs. They have placed a conservative upper bound of 6% on the
mass of the halo stars with (V−I) colours between 2 and 3, and have also
stated that M dwarfs contribute at most 15% to the mass of the disk.
Moreover, Graff & Freese (1996) have remarked that likely extrapolations
of the mass function for VLM stars to the BD regime imply that the total
mass of BDs in the halo is less than about 3% of the local mass density.

From the point of view of stellar evolution, WD+M dwarf systems serve
at once as a description of the outcome of common–envelope (CE) evolution
and as a population of direct progenitors to the mass–transfer cataclysmic
variables (CVs). Pre–CVs can become CVs if angular momentum loss can
drive the two stars together sufficiently rapidly. The fraction of WDs with
a cool secondary and the distribution of orbital periods of such systems
is a useful constraint upon models of the CV population. Furthermore, in
these pre–CV binaries it is possible to study each component separately,
addressing questions such as how CE evolution has affected the WD when
compared with isolated examples. It might be that individual objects have
unusual chemical composition, ot that the whole sample has a different
mass distribution to the isolated stars.

In this paper, I describe the preliminary results of a new infrared search
for VLM companions to WDs. This new search has been performed on
83 WDs identified by the *Homogeneous Bright QSO Survey* (Cristiani *et
al.*, 1995, hereafter HBQS) in six $25^o \times 25^o$ areas of the sky: the ESO/SERC
fields F290, F295, F468, F470, F479, located around the South Galactic Pole
($b < -60^o$) and the Selected Area 94 at $b \sim -49^o$. The sample is statistically
complete in the magnitude range $14 < B < 17.5$. Previous studies of this
kind relied mostly on catalogue WDs, chosen for being relatively young
and nearby. The old samples, however, were in large majority the product

of blue star surveys that usually selected against objects whose B and V colours were contaminated by the presence of a red companion. Now a much more reliable approach is possible because in the selection of the HBQS samples, particular care has been taken not to neglect all those objects that in a (U−B), (B−V) diagram distinctly fall below the main sequence and, relatively to isolated WDs with the same (U−B), appear to have abnormally "red" (B−V) and (V−I) color indexes.

## 2. Infrared Observations and Data Reduction

The observations were performed during 11 nights belonging to three different runs (October 1993: 3 nights, October 1994: 4 nights and October 1995: 4 nights) at the ESO–La Silla Observatory, Chile. I used the 2.2m ESO/MPI telescope together with the IRAC2 camera, equipped with a NICMOS III array detector ($40\times40$ $\mu m^2$ pixel size). All the observations were obtained with lens C, which gives an image scale of 0.49 arcsec/pix and a field of view of $125\times125$ arcsec$^2$. At first, it was my intention to take only K frames for as many WDs as possible, and to observe also in the J and H passbands only the objects that showed an infrared excess in K. However, most of the objects seemed to show at least a slight infrared excess, which was almost invariably coupled to red (B−V) and (V−I) colour indexes, so I ended up with doing J, H and K photometry for about 80% of the program WDs. Typically, I employed exposure times of 10, 15 and 30 min for the J, H and K passbands, respectively. These long exposure times have been reached by adding and averaging a large number of very short integrations. This way it has also been possible to obtain signal–to–noise ratios at least equal to 3 and in most cases higher. A raw sky subtraction was performed with the on–line software at the telescope in order to check the quality of the data and to look for a possible infrared excess. Groups of infrared standard stars were observed at least three times during each night. A standard data reduction was later performed with the *IRAC2 CONTEXT* of the *MIDAS* software. The estimated accuracy of the measured magnitudes is about 0.05 mag for $14 < K < 15$ and 0.1 mag for $K > 17$.

## 3. White Dwarfs with Infrared Excess

Table 1 lists the names, together with the B and K apparent magnitudes and the U−B, B−V, B−R, V−I, J−H and H−K colour indexes of the 83 program WDs that showed infrared excess. Note that, for brevity reasons, only the WDs with positive detection in the infrared are listed here. The complete list of all the objects searched by this project will be included in a forthcoming paper (Gemmo, 1997). The names of the objects are indicative of the area of the sky where each WD belongs. The U−B, B−V, B−R and

V−I colour indexes were measured by the above mentioned HBQS, while the J, H and K magnitudes have been obtained by the author as described in Section 2.

TABLE 1. HBQS White Dwarfs with IR Excess

| Name | B | U−B | B−V | B−R | V−I | K | J−H | H−K |
|------|------|-------|-------|-------|-------|-------|-------|-------|
| F290/3 | 16.99 | −0.56 | +0.14 | +0.02 | +0.45 | 17.00 | −0.42 | +0.48 |
| F290/8 | 17.19 | −0.53 | +0.22 | +0.62 | +1.53 | 13.71 | +0.61 | +0.29 |
| F290/9 | 14.20 | −0.39 | +0.27 | +0.32 | +0.25 | 13.38 | +0.15 | +0.04 |
| F290/12 | 16.27 | −1.15 | −0.67 | −0.76 | +0.23 | 16.54 | * | * |
| F290/13 | 17.14 | −1.21 | −0.38 | −0.53 | +0.09 | 17.56 | * | * |
| F295/2 | 17.47 | −1.18 | +0.14 | +0.47 | * | 14.38 | +0.59 | +0.15 |
| F295/12 | 17.26 | −0.79 | +0.35 | +0.48 | * | 17.05 | −0.09 | −0.03 |
| F295/13 | 15.77 | −0.25 | +0.30 | +0.35 | * | 15.01 | +0.29 | −0.11 |
| F295/22 | 17.07 | −1.03 | −0.34 | +0.22 | * | 16.78 | * | * |
| F295/24 | 15.67 | * | * | * | * | 11.73 | −0.08 | +0.93 |
| F295/26 | 16.81 | * | * | * | * | 14.89 | +0.53 | +0.33 |
| F468/1 | 15.68 | −0.26 | +0.34 | +0.53 | +0.40 | 14.48 | +0.30 | +0.05 |
| F468/7 | 16.69 | −0.42 | +0.27 | +0.74 | +0.26 | 15.11 | +0.28 | +0.07 |
| F468/10 | 16.72 | −0.63 | +0.07 | +0.36 | * | 14.95 | +0.59 | +0.85 |
| F468/12 | 16.06 | −1.31 | +0.20 | +0.50 | +0.41 | 13.81 | +0.34 | +0.44 |
| F468/14 | 14.67 | −0.05 | −0.03 | +0.06 | * | 14.67 | +0.08 | −0.05 |
| F470/1 | 14.87 | −0.68 | +0.17 | +0.27 | +0.35 | 14.06 | +0.19 | +0.09 |
| F470/2 | 16.69 | −0.94 | −0.35 | +0.25 | +0.47 | 15.56 | +0.46 | +0.37 |
| F470/3 | 16.94 | −1.13 | −0.48 | −0.19 | +0.58 | 15.55 | +0.51 | +0.44 |
| F470/9 | 17.18 | −1.21 | −0.14 | 0.00 | +0.22 | 15.61 | +0.68 | +0.22 |
| F470/11 | 17.28 | −0.47 | +0.48 | +0.74 | +0.38 | 15.58 | +0.53 | +0.03 |
| F479/1 | 15.37 | −0.30 | +0.64 | +1.08 | +1.28 | 11.61 | +0.40 | +0.46 |
| F479/4 | 17.25 | −0.34 | +0.49 | +0.98 | +0.39 | 16.34 | +0.55 | +0.57 |
| F479/9 | 16.79 | −0.38 | +0.25 | +0.56 | +0.20 | 16.14 | −0.11 | +0.41 |
| F479/12 | 16.72 | * | * | * | * | 13.00 | +0.24 | +0.14 |
| SA94/3 | 17.11 | −1.09 | 0.00 | +0.24 | * | 16.87 | +0.40 | −0.15 |
| SA94/4 | 17.43 | −0.21 | +0.16 | +0.32 | * | 16.78 | +0.40 | −0.30 |
| SA94/5 | 15.76 | −0.52 | +0.38 | +0.61 | * | 14.61 | +0.29 | −0.03 |
| SA94/8 | 16.72 | −1.17 | +0.34 | +0.76 | * | 14.24 | +0.78 | −0.10 |
| SA94/10 | 17.46 | −1.06 | −0.14 | −0.01 | * | 15.71 | * | * |
| SA94/14 | 14.62 | −0.43 | +0.47 | +0.53 | * | 13.91 | +0.24 | −0.12 |
| SA94/16 | 17.00 | −1.55 | −0.14 | −0.16 | * | 16.30 | +0.54 | +0.28 |
| SA94/17 | 15.18 | * | * | * | * | 14.09 | +0.27 | +0.01 |
| SA94/19 | 16.80 | −0.96 | +0.04 | +0.01 | * | 16.52 | * | * |

## 4. Discussion and Conclusions

The first conclusion we can draw regarding the results of this search for infrared companions to WDs is of purely statistical nature: 42% (that is 35 out of 83) of the program WDs show an infrared excess which can probably be attributed to an unresolved VLM companion. However, if we calculate the fraction of WD+M dwarf systems for each of the six different areas of the sky that have been considered in this project, we note an interesting thing: while the fraction of binary WDs is on the average lower than 40% for the high Galactic latitude fields, it rises to 69% in the case of SA 94, which is situated about halfway between the Galactic Equator and the South Galactic Pole. A possible explanation could be that the WD+M dwarf systems identified in the SA94 belong to a younger and metal–richer stellar population, such that the late–type companions are usually redder and hence much easier to detect in the infrared. On the contrary, many of the VLM companions to WDs observed around the South Galactic Pole may be part of the older and metal–poorer disk population which, as remarked e.g. by Allard & Hauschildt (1995) is bluer because the $H_2$ opacities centered on 2 $\mu$m depress the continuum and the flux emerges only in the passabands bluer than H. Metal–poor and very cool companions to WDs can be detected only in the favorable case of hotter WDs where most of the flux escapes in the U and B passabands and almost no radiation is emitted at longer wavelengths. WDs of this kind are detectable even at distances of a few hundred parsecs and thus they might be objects located high on the Galactic plane.

## References

Alcock, C., et al. (1996) ApJ. 461, 84.

Allard, F. and Hauschildt, P.H. (1995) ESO Workshop on The Bottom of the Main Sequence and Beyond, ed. C.G. Tinney (Heidelberg: Springer), 32.

Aubourg, E. et al. (1993) Nature 365, 623.

Bahcall, J.N., Flynn, C., Gould, A., and Kirhakos, S. (1994) ApJ. 435, L51.

Becklin, E.E. and Zuckerman, B. (1988) Nature 336, 656.

Cristiani, S., La Franca, F., Andreani, P., Gemmo, A., Goldschmidt, P., Miller, L., Vio, R., Barbieri, C., Bodini, L., Iovino, A., Lazzarin, M., Clowes, R., MacGillivray, H., Gouiffes, Ch., Lissandrini, C., and Savage, A. (1995) Astron. Astrophys. Suppl. Ser. 112, 347.

Gemmo, A.G. (1997) In preparation.

Graff, D.S. and Freese, K. (1996) ApJ. 467, L65.

Nakajima, T., Oppenheimer, B.R., Kulkarni, S.R., Golimovski, D.A., Matthews, K., and Durrance, S.T. (1995) Nature 378, 463.

Probst, R.G. (1983) ApJ. Suppl. 53, 335.

Zuckerman, B. and Becklin, E.E. (1987) ApJ. 319, L99.

Zuckerman, B. and Becklin, E.E. (1992) ApJ. 386, 260.

## Discussion

*M. van Kerkwijk:* Just a comment: the fact that you did not find any "good" BD candidate may be due to the fact that you selected whether to observe after taking the K band picture, since BDs peak at J (see the spectrum of Gl 229B).

*A. Gemmo:* Yes, I am aware of that, but the HBQ survey determined the U, B, V, R, I colours of all the program WDs so, whenever the B−R and V−I colour indexes of an object were positive, I almost always detected an excess of radiation in the K band. As a consequence, to be on the safe side, I observed about 80% of the targets in J, H and K. After a careful data reduction, some of these "suspect" IR excesses simply disappeared, but many of them were confirmed.

# THE EC14026 STARS: A NEW CLASS OF PULSATING STAR

D.E. O'DONOGHUE
*Dept. of Astronomy*
*University of Cape Town*
*Rondebosch 7700*
*South Africa*

C. KOEN, D. KILKENNY AND R.S. STOBIE
*South African Astronomical Observatory*
*P.O. Box 9*
*Observatory 7935*
*South Africa*

A.E. LYNAS-GRAY
*Dept. of Astrophysics*
*University of Oxford*
*1 Keble Rd*
*Oxford*
*England*

AND

S.D. KAWALER
*Dept. of Physics and Astronomy*
*Iowa State University*
*Ames*
*Iowa*
*USA*

## 1. Introduction

Hot subdwarfs comprise typically 50 per cent of faint blue star surveys at high galactic latitude (e.g. the Palomar-Green (PG) survey: Green, Schmidt & Liebert 1986; the Edinburgh-Cape (EC) survey: Stobie et al. 1996a, Kilkenny et al. 1996a). In contrast to the PG survey, the EC survey has found that a large fraction ($\sim$ 15 per cent) of its hot subdwarf sample is

*I. Isern et al. (eds.), White Dwarfs, 321–327.*

TABLE 1.   Oscillation Periods Of The
EC14026 Stars

| Star | $P_1$ (s) | $P_2$ (s) | $P_3$ (s) |
|---|---|---|---|
| EC 14026–2647 | 144.3 | 133.6 | |
| EC 10228–0905 | 152.3 | 147.5 | 139.8 |
| EC 20117–4014 | 157.4 | 142.0 | 137.3 |
| PB 8783 | 134.4 | 123.6 | 122.7 |

comprised of sdB stars showing the Ca II K line. This implies that the hot subdwarf has a relatively luminous companion of spectral type F or G.

It took some time before the nature of these composite objects was recognized during which, as described in Kilkenny et al. (1996b), the star EC14026–2647 was suspected to be a cool white dwarf of spectral type DAZ. This, in combination with its $B - V$ colour (which was typical of a DAV pulsating white dwarf), caused it to be monitored photometrically for rapid variability. Oscillations with a period of 144 s and an amplitude of 1–2 per cent were discovered. Only subsequently was the true nature of the spectrum of EC14026–2647 understood as comprising the sum of an sdB star and a G star. As emphasised by Kilkenny et al. (1996b), it is ironic that the star would never have been monitored photometrically had it been realised that it was not a white dwarf.

As a result of this serendipitous discovery, a search was instigated for oscillations in stars similar to EC14026–2647. This resulted in the discovery of rapid oscillations in EC10228-0905 (Stobie et al. 1996b), EC20117–4014 (O'Donoghue et al. 1996) and, widening the search to composite subdwarfs outside the EC survey, PB 8783 (Koen et al. 1996). In parallel to the photometric studies, moderate resolution optical spectra of these four stars was obtained leading to atmospheric parameters of the components (O'Donoghue et al. 1996). It is the purpose of this brief review to summarize the properties of the first four members of this new class of pulsating star for which the name "EC14026 stars" has been suggested.

## 2.  Photometric Properties

Fig. 1 shows a selection of Fourier amplitude spectra of a selection of light curves of EC20117–4014. It is clear that, although a single period at 137 s is dominant, two other periods are seen. These three periods, along with those derived from amplitude spectra of the other three stars, are listed in Table 1.

TABLE 2. Pulsation Periods Of A ZAEHB
sdB Star

| Radial Index | Degree $\ell$ | | |
|:---:|:---:|:---:|:---:|
| n | 0 | 1 | 2 |
| 0 | 149.65 | | 162.69 |
| 1 | 132.62 | 135.27 | 132.89 |
| 2 | 103.95 | 118.39 | 106.45 |

From these data it is clear that the oscillations in the EC14026 stars
are multiperiodic with periods occupying a relatively narrow range. This
is highly suggestive that the cause is stellar pulsation - as discussed in
Kilkenny et al. (1996b), no other plausible mechanism can be devised to
explain the phenomenon observed. From multicolour observations, Koen
et al. (1996) were able to show that the blue component in PB 8783 is
responsible for the oscillations.

Using atmospheric parameters for the four stars derived by O'Donoghue
et al. (1996) (see below), pulsation models appropriate to sdB stars on the
Zero Age Extended Horizontal Branch (ZAEHB) were reported in Stobie
et al. (1996). The periods of low order (n), low degree ($\ell$), radial and non-
radial p-modes for a model with a mass of 0.485 $M_\odot$, effective temperature
of 34400 K and log g of 5.98 are listed in Table 2. A comparison of Tables
1 and 2 shows that the model pulsation periods are, in general, similar to
those observed in the EC14026 stars, thus confirming that stellar pulsation is the cause of the oscillations. No detailed agreement could, however,
be found between the periods of any individual EC14026 star and any of
the limited range of models calculated. A comprehensive grid of models is
clearly required.

## 3. Spectroscopic Properties

High signal-to-noise spectra of moderate resolution ($\sim 1 - 3$ Å) were acquired for the stars listed in Table 1. The appearance of these spectra is
typical of the spectra of sdB stars: broad Balmer lines up to H10 on a
blue continuum. In addition, it was obvious from close inspection that the
presence of a cool companion of spectral type F or G was required to explain the Ca II K line and the numerous weak metallic absorption features:
for example, in EC14026–2647 and EC10228–0905 the G-band was clearly
present.

The spectra of the EC14026 stars were interpreted with the aid of spec-

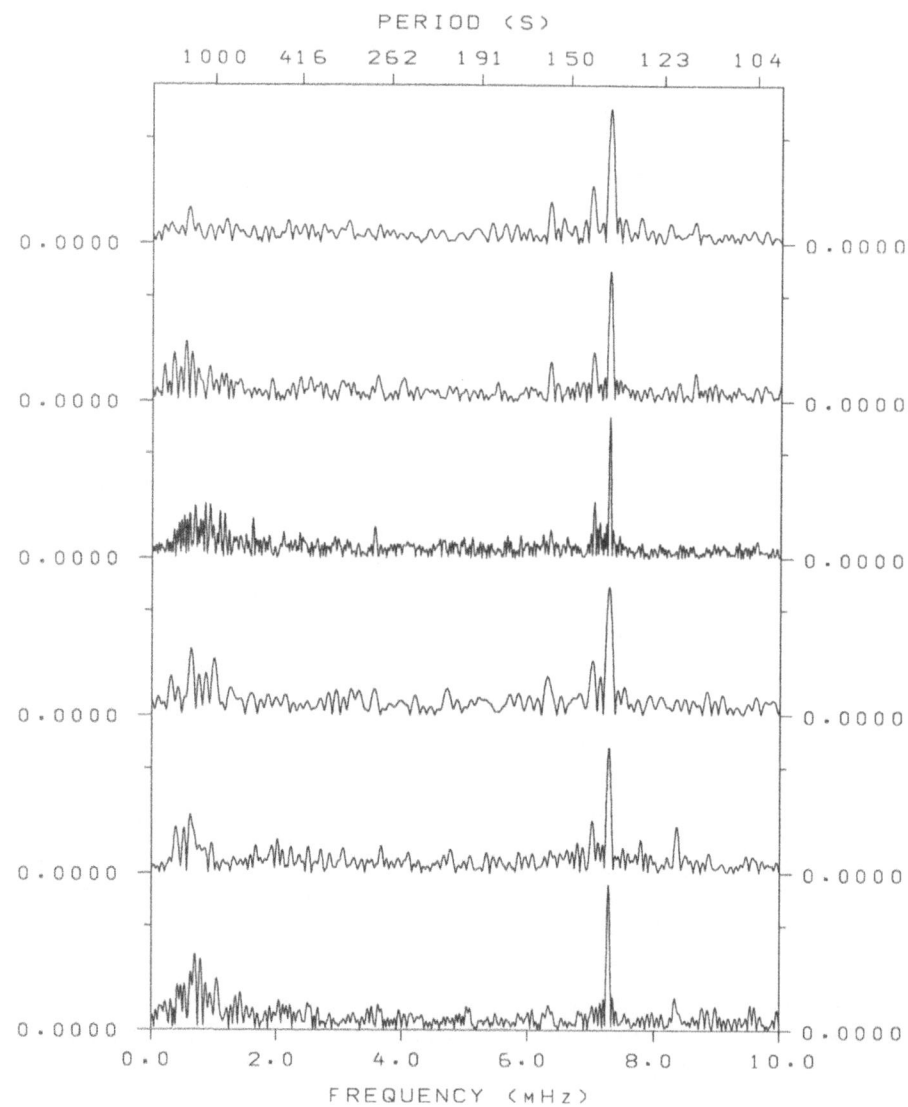

*Figure 1.*   Fourier amplitude spectra of EC20114–4014.

tra of a range of F and G main sequence stars published by Jacoby, Hunter & Christian (1984), higher resolution observations of similar stars obtained by us, and a grid of model atmospheres appropriate for high gravity sdB stars computed using ATLAS 9. Accurate Balmer line profiles up to H10 for the sdB star were calculated using the theory of Vidal, Cooper & Smith (1973). The relative contributions of the sdB star and the cool companion were determined by a least squares fit to the observed spectrum. The sdB star component was isolated by subtracting the contribution of the cool companion from the observed spectrum and the effective temperature and gravity of the sdB star were determined from fits of the model sdB Balmer line profiles (e.g. Saffer et al. 1994, Saffer & Liebert 1995). Full details of the models and these procedures can be found in O'Donoghue et al. (1996).

The results show that the effective temperatures and gravities of the sdB components of all four stars are consistent with each other: $\sim$ 35000 K and $\sim$ 5.9 respectively. The spectral types of the cool companions were found to be early G for EC14026–2647 and EC10228–0905, and early F for EC20117–4014 and PB 8783. The luminosities of the cool companions were found to be consistent with those of main sequence stars. The atmospheric parameters of the sdB stars place them at the blue end of the ZAEHB, close to the He main sequence. This implies that the masses of their H envelopes are $\sim$ 0.1 per cent of the mass of the star or smaller (see fig. 8 of O'Donoghue et al. 1996).

## 4. Exploitation

The pulsations of the EC14026 stars can be used to obtain insight into these stars on two fronts. The first is the traditional hope for multiperiodic pulsating stars: asteroseismology. Detailed comparison between pulsation models for sdB stars and the observed periods (as hinted at by Tables 1 and 2) will undoubtedly provide constraints on the internal structure of sdB stars (assuming that the pulsators are typical of their spectral class).

The second avenue of attack involves using the pulsations to probe the binary properties of these systems: the variable time delay in the arrival time of the pulsations due to the motion of the sdB star around the system barycentre will permit the binary period to be measured. Using this along with reasonable estimates for the total system mass provides the orbital separation (independent of needing to know the inclination). This is the key parameter in ascertaining whether these systems have been the product of close binary evolution as discussed in more detail in the next section.

## 5. The Binary Nature Of sdB Stars

In view of the remarkably thin hydrogen envelopes possessed by sdB stars, it has long been suggested that they may be the result of close binary evolution (Mengel, Norris & Gross 1976). Iben (1990 and references therein) have also proposed that close binary evolution is responsible, although in this scheme the merger of a pair of low mass white dwarfs is involved. Saffer et al. (1994) find a very low dispersion in the effective gravities of a large sample of field sdB stars from which they infer that the Iben (1990) mechanism must not be relevant (as a much larger dispersion in the masses and hence gravities would be expected). In contrast, the binary model has been dismissed by Heber et al. (1984) who suggest that normal single star mass loss accounts for the thin hydrogen envelope.

The present authors believe that the binary model of Mengel et al. (1976) for sdB stars is the most likely scenario: it is the most natural way to explain the extremely thin H envelope because Roche lobe overflow by a red giant will terminate when the H envelope is almost completely lost. Observational evidence in favour of the binary model is accumulating (e.g. Allard et al. 1996; see O'Donoghue et al. 1996 for a complete list of references). It is opportune to advance the hypothesis that *all* sdB stars are the result of close binary evolution in the hope that counter examples may be found and the hypothesis discredited. A consequence of the hypothesis is that every sdB star must have a companion at an orbital separation sufficiently close that close binary evolution has taken place in the past. The EC14026 stars discussed in this paper will provide a crucial test of the hypothesis. More generally, a large program of red/infrared photometry will be needed to check for cool companions to sdB stars. Companions as early as those seen in these first four EC14026 stars will be easy to discover; M dwarfs will be very difficult, but not impossible, to detect.

## References

Allard F., Wesemael F., Fontaine G., Bergeron P. and Lamontagne R. (1994) *AJ*, 107, 1565

Green R.F., Schmidt M. and Liebert J. (1986) *ApJS*, 61, 305

Heber U., Hunger K., Jonas G. and Kudritzki R.P. (1984) *A&A*, 130, 119

Iben I. (1990) *ApJ*, 353, 215

Jacoby G.H., Hunter D.A. and Christian C.A. (1984) *ApJS*, 56, 257

Kilkenny D., O'Donoghue D., Koen C., Stobie R.S. and Chen A. (1996a) *MNRAS*, in press

Kilkenny D., Koen C., O'Donoghue D. and Stobie R.S. (1996b) *MNRAS*, in press

Koen C., Kilkenny D., O'Donoghue D., van Wyk F. and Stobie R.S. (1996) *MNRAS*, in press

Mengel J.G., Norris J. and Gross P.G. (1976) *ApJ*, 204, 488

O'Donoghue D., Lynas-Gray A.E., Kilkenny D., Stobie R.S. and Koen C. (1996) *MNRAS*, in press

Saffer R.A., Bergeron J., Koester D. and Liebert J. (1994) *ApJ*, 432, 351

Saffer R.A. and Liebert J. (1995) in Koester D., Werner K. (eds), *White Dwarfs, Lect. Not. in Phys.*, p.221, Springer-Verlag, Berlin

Stobie R.S. et al. (1996a) *MNRAS*, in press

Stobie R.S., Kawaler S.D., Kilkenny D., O'Donoghue D. and Koen C. (1996b) *MNRAS*, in press

Vidal C.R., Cooper J. and Smith E.W. (1973) *ApJS*, 25, 37

## Discussion

*I. King*: With regard to the extremely blue horizontal branch stars in NGC 6752, these stars must have masses $\leq 0.8\ M_\odot$ (the turn-off mass); otherwise they would not be found in the outer field that gave rise to the color-magnitude diagram that you showed. I they were binaries such as you describe, with a total mass of 2 $M_\odot$, they would be confined almost totally to the dense central core of this highly concentrated cluster, and thus, unobservable from the ground.

*D. O'Donoghue*: All I say is that all SdBs are binaries. The companions could be much lower mass in globular clusters.

*H. Shipman*: Have you searched any brightest subdwarf B stars in search of variability?

*D. O'Donoghue*: We have searched all subdwarfs known to us brighter than $14^{\mathrm{mag}}$ and $\delta < +15°$.

*Y. Wu*: What's the pulsation amplitude of the SdB stars after correcting for the F/G components?

*D. O'Donoghue*: Probably $\sim 0.02^{\mathrm{mag}}$.

# HOT WHITE DWARFS IN NON-INTERACTING BINARIES FROM THE ROSAT EUV SURVEY

MATTHEW BURLEIGH AND MARTIN BARSTOW

*XRA Astronomy Group, University of Leicester, UK*

## 1. Introduction

More than 120 hot white dwarfs have been identified in the extreme ultraviolet (EUV) survey of the ROSAT Wide Field Camera (WFC, Pye et al., 1995). The majority of these are isolated stars, but, in addition to those found in interacting cataclysmic variables, over 30 are now known to lie in detached binary systems. These include wide, resolved binaries like Sirius, and close WD+dM pairs that are identified from composite optical spectra. ROSAT has also detected a previously unidentified sample of hot white dwarfs in unresolved pairs with much more luminous main sequence stars. These degenerates are masked by their companions at optical wavelengths, but follow-up spectroscopic observations in the far-UV with the International Ultraviolet Explorer (IUE) have enabled us to confirm the identifications.

Studying the white dwarfs in detached binaries allows us to place constraints on binary evolution models (e.g. de Kool and Ritter, 1993). In addition, white dwarfs in binaries may have evolved in vastly different circumstances to isolated stars. For example, in close systems (P~few days) they may be the product of common envelope (CE) evolution. Thus by studying these objects we can not only add to our overall picture of white dwarf evolution, but we can investigate the similarities and differences between the isolated and binary populations.

White dwarfs are usually identified in blue colour or proper motion surveys, and so the vast majority of the ~2000 known white dwarfs are single, isolated objects, even though it is generally assumed that over half of all stars should be in binary or multiple systems. Clearly, there is a missing population of white dwarfs hidden in detached, but unresolved, binaries. This has profound implications for our knowledge of the white

*I. Isern et al. (eds.), White Dwarfs, 329–335.*

TABLE 1. New ROSAT WFC white dwarf binaries

| Name | SpT | D(pc) | $T_{WD}$ | log g | M (M$_\odot$) | $V_{WD}$ |
|------|-----|-------|------|-------|--------|------|
| HD2133[a] | F7−F8V | ~125 | 28,700 | 8.25 | 0.79 | 16.3 |
| RE J0357+28[a,b] | K2V | >107 | 31,000 | 7.9 | 0.60 | 15.2 |
| BD+27 1888[a] | A8−F2V | ~200 | 34,000 | 7.25 | 0.39 | 15.0 |
| RE J1027+32[a,c] | G0−G4V | ~450 | 32,500 | 7.5 | 0.44 | 17.3 |
| RE J0500+36 | G | ~400 | 42,000 | 7.5 | 0.47 | 18.0 |
|  | K | ~700 | 50,000 | 8.5 | 0.97 |  |

Refs: a) Burleigh, Barstow and Fleming, 1996 b) Jeffries, Burleigh and Robb, 1996 c) Genova et al., 1995

dwarf luminosity function, formation rate and space density, which is largely determined from observations of the isolated stars (e.g. Fleming, Liebert and Green, 1986).

## 2.  A search for hot white dwarfs in unresolved binaries

Any companion of spectral type K or earlier will completely dominate the optical spectrum of a white dwarf. Although a number of unresolved WD+MS binaries have been detected serendipitously in the past, it has been the ROSAT and EUVE surveys (Bowyer et al., 1996), together with follow-up observations in the far-UV with IUE, which has greatly increased our knowledge of this population.

The initial discovery of a white dwarf companion to $\beta$ Crateris (Fleming et al., 1991) was followed by nine others, (Barstow et al., 1994). These white dwarfs are all relatively bright EUV sources. Comparisons with the count rates of known white dwarfs in the WFC catalogue left little doubt as to their true nature, and observations with IUE were needed merely to confirm the identifications. However, it is obvious from even a cursory glance at the WFC catalogue that there are many faint hot white dwarfs with rather unremarkable EUV colours and low count rates. There are still a significant number of unidentified EUV sources in the WFC catalogue, and there are also a large number of late type stars in the catalogue that were not previously known to be chromospherically active. Indeed, in some cases follow-up optical studies (e.g. Mason et al., 1995) failed to find any evidence of activity. Determining which EUV source is due to an active star, and which may be due to a hidden hot white dwarf companion is not an easy task. In recent episodes of IUE, therefore, we have had a continuing programme to search for further non-interacting WD+MS binaries.

The results of the search are summarised in Table 1. In 1994/95 we found four more binaries from a total of 10 targets, but in 1995/96 we were less successful, identifying only one further system from 13 targets. The degenerate in this system, RE J0500−36, is, at V≈18, one of the faintest white dwarfs to be detected by the ROSAT WFC. This result suggests that there are few further WD+MS binaries remaining to be identified in the EUV catalogues. The most interesting object in Table 1 is RE J0357+28. The primary in this system is a rapidly rotating K2V star. Jeffries and Stevens (1996) suggest that it has been spun-up through accretion from the wind of the companion during its red giant phase, and they have christened it a WIRRing star (Wind-accretion Induced Rapid Rotator).

## 3.  The EUV spectrum of the DAO+dM pre-CV RE J0720−31

DAO white dwarfs are a rather disparate collection of objects with a variety of possible origins (Bergeron et al., 1994). They have hybrid optical spectra showing both hydrogen and helium lines, including a distinctive HeII feature at 4686Å. Three DAO stars, all in pre-catalcysmic variable binaries, have been detected by ROSAT. Previous studies (e.g. Tweedy et al., 1993) have speculated that the helium in these stars may result from accretion of the M dwarf companion's stellar wind.

We present here an analysis of the EUVE spectrum of one of these binaries, RE J0720−31, observed by the satellite in December 1995. The entire spectrum is shown in Figure 1. The most striking feature is the saturated HeI absorption edge at 504Å. Nothing like this has ever been seen in any other EUVE white dwarf spectra (e.g. Dupuis et al., 1995). We have fitted a set of fully line blanketed H+He stratified and homogeneous models to the EUV data, using the V magnitude as a normalisation point. Both models assume LTE conditions. The stratified model assumes a thin H layer ($10^{-16} > M_{H\odot} > 10^{-11}$) overlying a helium atmosphere; the second model assumes a homogeneous distribution of H+He.

In contrast to the optical spectrum, which can only be matched by a homogeneous model, we find the EUV spectrum can only be fitted with a stratified structure. The best fit model is shown in Figure 1 and its parameters summarized in Table 2. All the features are accurately reproduced, in particular the HeII series of absorption dips converging on 228Å, the 206Å feature, and the HeI 504Å edge. This result has important implications for our understanding of the white dwarf's structure and evolution, and for the binary itself.

The helium present in the optical spectrum could reside on the surface of the white dwarf in a thin mixed layer, perhaps having being accreted from the stellar wind of the dM companion, and may not be intrinsic to

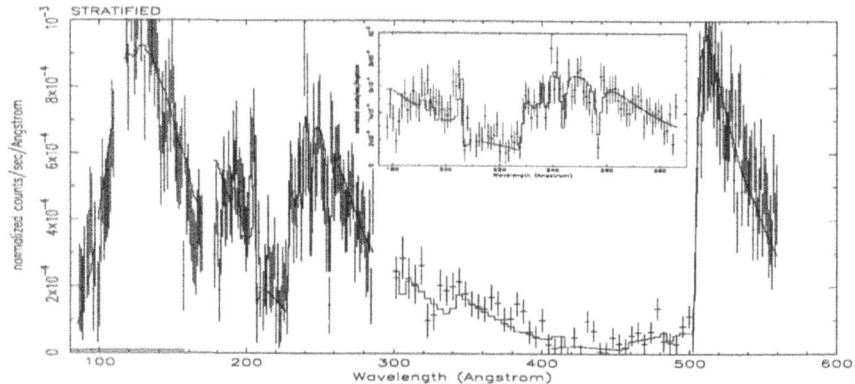

*Figure 1.* EUVE spectrum of RE J0720−318 with the best fit stratified model. Inset: Best fit to the medium waveband.

the star itself. The hydrogen layer mass is the lowest found for any white dwarf from an EUV analysis. As this is a close binary system, the thin H layer may well be the result of mass loss from the white dwarf progenitor during an earlier common envelope phase. Thereafter, any moderate mass loss may bring underlying helium to the surface. However, careful analysis of the HeII series converging on 228Å suggests that while the lines are photospheric in origin, the edge itself is due to less dense interstellar or even circumstellar material.

TABLE 2. Best fit parameters from the stratified model

| Parameter | Value | $1\sigma$ limits |
|-----------|-------|------------------|
| nHI | $2.30\times10^{18}$ | $2.25-2.36\times10^{18}$ |
| nHeI | $1.44\times10^{18}$ | $1.41-1.47\times10^{18}$ |
| nHeII | $9.78\times10^{17}$ | $8.58-10.03\times10^{18}$ |
| $M_H$ | $3.07\times10^{-14}M_\odot$ | $2.93-3.21\times10^{-14}M_\odot$ |

The size of the HeI 504Å edge suggests that there is an unusually high line-of-sight helium column density to this system. If this is assumed to be interstellar, the implied hydrogen ionization fraction is 90%. This is the highest measured in any direction in the sky from EUVE data (Barstow et al., 1997), and therefore we suggest instead that the helium resides in the system itself in the form of a circumstellar gas. Recent theoretical studies (e.g. Terman and Taam, 1996) suggest that circumbinary disks are likely to form in post-CE binaries. The inclination of this system may be as high as

85° (Barstow et al., 1995), and thus it is possible we are seeing the system through an optically thin disk.

This work is presented in more detail by Burleigh, Barstow and Dobbie (1996); the system is also discussed by Jean Dupuis in these proceedings.

## 4. The ROSAT binary mass distribution

The mass distribution of the ROSAT WFC binary white dwarfs is shown in Figure 2, and is compared to the Marsh et al. (1996) sample of 89 mainly isolated WFC white dwarfs, and also with the optically selected sample of Bergeron, Saffer and Liebert (BSL, 1992). The Marsh et al. and BSL samples have been normalised to the size of the binary sample.

The most striking feature of the distribution is the excess of high mass objects at $\sim 1 M_\odot$ in the two EUV-selected samples, when compared to BSL. This implies that the most massive white dwarfs have the largest EUV fluxes, which is surprising since they should have the smallest radii and consequently the lowest luminosity for a given T and composition. It would appear that the decrease in flux as a result of a smaller radius is more than offset by an increase in the intrinsic luminosity and, since these stars have higher gravities, any heavy elements in their photospheres will sink out to leave them with lower EUV opacities. In addition, higher gravity stars will cool far more slowly than lower mass objects, hence they are more likely to be detected in the hot regime of the EUV. It should also be noted that it is difficult to get a white dwarf with $M_{WD} > 0.9 M_\odot$ from single star evolution. The isolated high mass objects could therefore be the result of double degenerate mergers, and the high mass white dwarfs in binaries may represent evidence of past interaction.

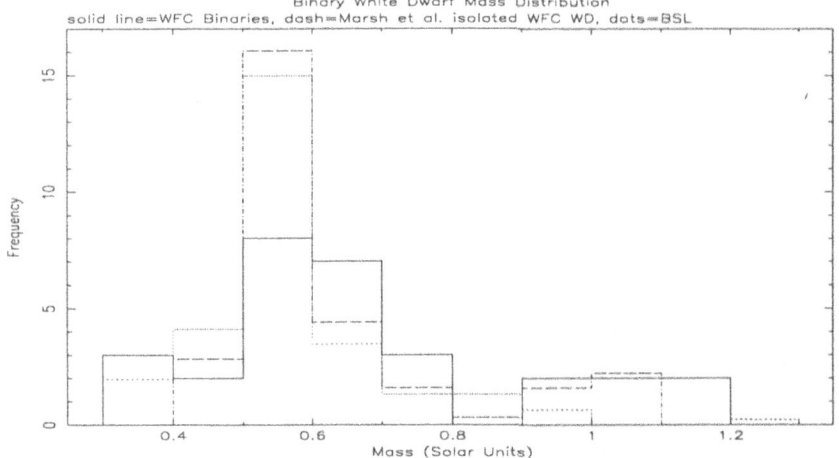

*Figure 2.*   ROSAT WFC white dwarf binary mass distribution

## 5. Conclusions - are there any more white dwarf binaries in the EUV surveys?

Approximately 25% of the ROSAT white dwarfs are in detached binary systems. Given that we expect >50% of all stars to be in binaries, we must ask if there are any more awaiting discovery in the WFC catalogue. As was discussed in section 2, our searches with IUE have been fairly exhaustive and we do not expect to discover many more 'hidden' white dwarfs. A number of B stars are detected in the EUV with count rates similar to known isolated white dwarfs. We would not normally expect these stars to be luminous at these wavelengths, and we suggest that, since the B star will still dominate the spectrum in the far-UV, EUVE observations are urgently needed to determine if the source is a hidden hot white dwarf. At the other end of the spectrum, there may be a significant population of WD+faint, low mass dM binaries awaiting discovery in the survey. With the current observational approach (looking for strong Balmer emission due to reprocessing of EUV radiation from the white dwarf on the surface of the red dwarf), companions with masses <dM6 are almost undetectable (Vennes and Thorstensen, 1994).

## References

Barstow, M.A., et al., 1994, *MNRAS*, **270**, 499
Barstow, M.A., et al., 1995, *MNRAS*, **273**, 711
Barstow, M.A., et al., 1997, *MNRAS*, **submitted**
Bergeron, P., Saffer, R.A. and Liebert, J., 1992, *ApJ*, **394**, 228
Bergeron, P., et al., 1994, *ApJ*, **432**, 305
Burleigh, M.R., Barstow, M.A. and Dobbie, P.D., 1996, *A&A*, **submitted**
Burleigh, M.R., Barstow, M.A. and Fleming, T.A., 1996, *MNRAS*, **submitted**
Bowyer, S., et al., 1996, *ApJS*, **102**, 129
de Kool, M., and Ritter, H., 1993, *A&A*, **267**, 397
Fleming, T.A., Liebert, J. and Green, R.G., 1986, *ApJ*, **308**, 176
Fleming, T.A., et al., 1991, *A&A*, **246**, L47
Genova, R., et al., 1995, *AJ*, **110**, 788
Jeffries, R.D., Burleigh, M.R. and Robb, R.M., 1996, *A&A*, **305**, L45
Jeffries, R.D. and Stevens, I.R., 1996, *MNRAS*, **279**, 180
Marsh, M.C., et al., 1996, *MNRAS*, **submitted**
Mason, K.O., et al., 1995, *MNRAS*, **274**, 1194
Pye, J.P., et al., 1995, *MNRAS*, **274**, 1165
Terman, J.L. and Taam, R.E., 1996, *ApJ*, **458**, 692
Tweedy, R.W., et al., 1993, *AJ*, **105**, 1938
Vennes, S. and Thorstensen, J.R., 1994, *ApJ*, **433**, L29
Vennes, S., et al., 1995, *A&A*, **299**, L29

## Discussion

*J. Liebert*: Aren't there two classes of DAO stars — those of low mass which are not EUV sources, and this small group of pre-cataclysmic variables, where the primary star is detected by ROSAT and EUVE (Bergeron et al. 1994)? Hence, a different origin of the helium seems plausible to me.

*M. Burleigh*: Yes. In fact at the Kiel conference I presented a paper in which we attempted to explain the non-detection of isolated DAOs with ROSAT, and we concluded that in many cases they must have large quantities of heavy elements in their photospheres blocking the EUV flux. In this case the helium is either being accreted onto the surface of the white dwarf, or this is in fact a plain DA star and all the He is material in the system. There are other DAOs where the origin of the He need to be explained. GD50 was recently classified as DAO from its EUVE spectrum, yet it is very massive and probably the result (Vennes et al. 1996) of a merger. The isolated low mass DAO Feige 55 turns out to be a double degenerate (Holberg et al. 1995) — what is the origin of the He in these cases? What is the role of DAOs in the white dwarf cooling sequences? Are they just a group of peculiar, disparate objects with a whole bunch of origins?

# HST OBSERVATIONS OF AM CANUM VENATICORUM

J.-E. SOLHEIM
*Auroral Observatory,*
*University of Tromsø, N-9037 Tromsø,Norway*

J. L. PROVENCAL
*Department of Physics,*
*University of Delaware, Newark, DE 19716, USA*

AND

E. M. SION
*Department of Astronomy and Astrophysics,*
*Villanova University, Villanova, PA 19085, USA*

**Abstract.** For the ultra-short period binary system AM CVn we obtained time-resolved UV spectroscopy, spanning $\lambda$ 1260 Å to $\lambda$ 1550 Å , with the HST Faint Object Spectrograph. Strong, asymetric features of NV and CIV gave direct evidence of mass loss from the system via a stellar wind. AM CVn's UV spectrum is dominated by features of silicon and maybe iron. The spectra have been binned giving 125 flux points spanning approximately 7 hours. The strongest UV variation corresponds to a modulation period of $1014 \pm 2$ s, and has an amplitude much larger than the 1011.5 s modulation observed in the visible part of the spectrum. This strong UV modulation indicates a different origin for this modulation than for the harmonics of the 1051 s period observed in white light photometry, and maybe we for the first time observe pulsations on the accreting primary star.

## 1. Introduction

AM CVn stars, ultrashort period interacting binary white dwarf systems, have blue colours like DB white dwarfs, but the spectra are different. Their optical spectra show wide asymetric He I lines indicating a disk origin, and no traces of Hydrogen lines. (Provencal et al. 1995)

337

*I. Isern et al. (eds.), White Dwarfs, 337–343.*

A helium disk model, assuming reasonable masses for the primary and secondary stars, demonstrates a disk temperature from 13 to 43 000 K, with a mean of 25 000 K, just in the DBV range (Solheim et al, 1997). The fit of a simple Planckian disk model indicates that a central star with an effective temperature of 180 000 K is necessary to explain the continuum spectrum in the UV (Nymark, 1997).

We expect the central star to be much easily detected in the UV than in the optical spectral regions. We planned our HST observations with the goal of detecting the primary object, which should reveal its presence trough spectral features, line modulations, or intensity variations in the UV region.

Before the HST observations are discussed, we should mention that the system shows low amplitude light variations in the optical region with periods 1028,1011,525,350,262,and 175 s. The 4 last periods are all harmonically related to a fundamental period of 1051 s which is not observed in the optical range. A WET campaign on this object in 1990 showed that the peaks at 525, 350 and 262 s were muliplets with a frequency splitting of 21 $\mu$Hz (Solheim et al, 1991, Provencal et al, 1995). The 21 $\mu$Hz splitting corresponds to a period of 13.2 hours, which is also observed as spectral line modulations by Patterson et al (1993). All the above mentioned periods and modulations, except 1011 s, fit well with a model containing a small excentric disk with tidal bulges, if the binary period is 1051 s and the primary mass is about 1 solar mass (Solheim et al, 1997).

The 1011 s modulation is unexplained by this model. This variation behaves quite differently than the other periodicities, demonstrating large amplitude changes on secular, and even shorter, timescales (Provencal et al, 1995). Simultaneous multicolour optical photometry for AM CVn shows the 1011 s variation to have a larger amplitude in the U band, as compared to B, V, R and I (Massacand and Solheim, 1995). This variation may be related to the hot central object, and therefore be stronger in the far UV, while the amplitudes of the other periods should decrease in the far UV if they originate in the cool outer part of the disk.

## 2.  The HST observations

We obtained time-resolved UV spectroscopy, spanning $\lambda$ 1260 Å to $\lambda$ 1550 Å, with the HST Faint Object Spectrograph. Strong, asymetric features of SiIV, SiIII and CIV give direct evidence of mass loss from the system via a stellar wind. The spectrum (figure 1) is dominated by features of silicon. We also find narrow interstellar lines from CII and SiII. A CII line at $\lambda$ 1335.7 Å as expected for a DB white dwarf atmosphere is not observed.

In order to determine intensity modulation periods, we binned the spec-

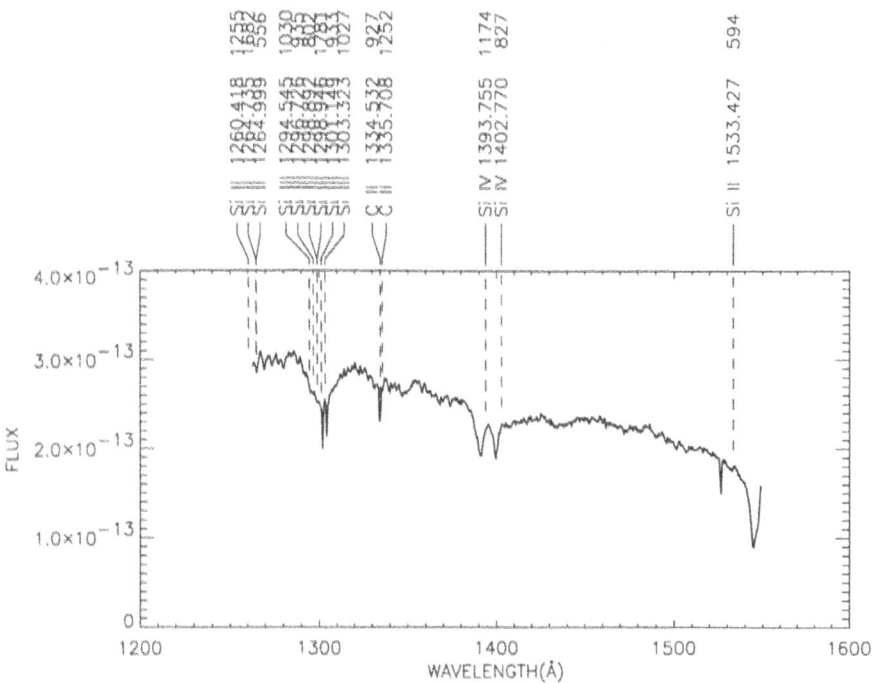

*Figure 1.* HST average spectrum for the region λ 1260 Å to λ 1550 Å. The vertical lines refer to spectral lines expected for a 25 000 K DB star.

tra in the region from about λ 1440 Å to λ 1485 Å giving 125 flux points spanning approximately 7 hours. Each data point represents about 50 s integrations. The coverage is very "unique". We have three points separated by about 50 seconds, followed by an approximately 230 second gap (flux calibrations and reading), with three more points separated by 50 seconds, another 200 second gap etc, until the end of the orbit where there is a large gap, and then the process repeats for the next orbit. This gap pattern results in a peculiar window function as seen in figure 2.

In figure 3 we present a discrete Fourier transform (DFT) of the "light curve". The dominant peak is at 1014 ± 2 s with an amplitude of 30 ± 3 mma. If we prewhite the light curve with this peak and its window function, we find the next strongest peak at 1051 s with amplitude 14 mma. The pattern in this frequency region, after prewhitening with the peak at 1014 s, is almost identical with the window function in figure 2, indicating that the 1051 s peak is single. When we prewhiten with the 1051 s modulation, we find the third strongest peak at 526.4 s with an amplitude of ≤10 mma, but with a pattern different from the window function, indicating that this region may contain more than one peak (as in the optical region) or that we have artifacts.

Studies of simulated light curves, demonstrate that the 1014 and 1051 s

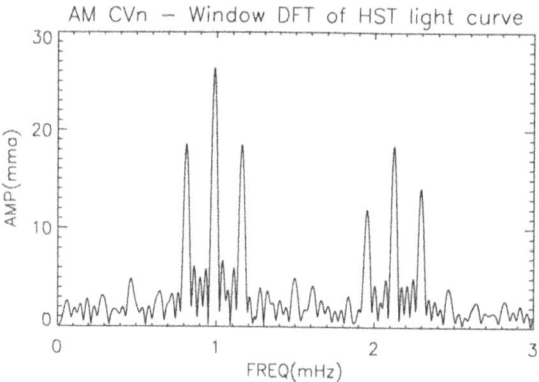

*Figure 2.* Window function for the light curve constructed from spectra in the region λ 1440 Å to λ 1485 Å.

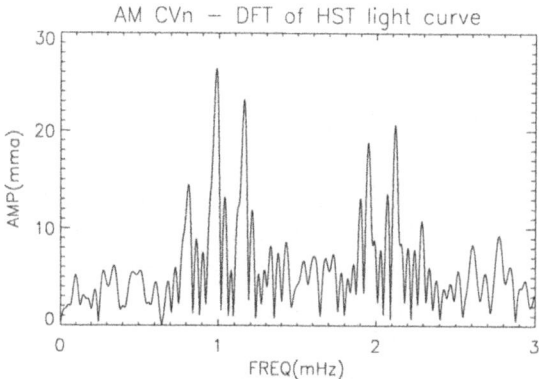

*Figure 3.* Discrete Fourier Transform of the light curve constructed from spectra in the region λ 1440 Å to λ 1485 Å.

modulations can reproduce the light curve quite well. If we instead simulate with the 1014 and 525 s modulations, we also get an acceptable fit. Because of our unfortunate pattern of gaps every 200 s, we cannot determine with certainty which of the two frequencies 525 or 1051 s are most likely present. However, based on the prewhitening and the window function, we favour the conclusion that the 1051 s modulation is detected with an amplitude of 14 ± 3 mma.

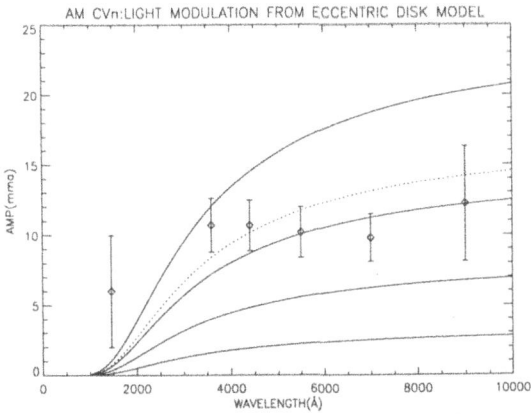

*Figure 4.* The precessing disk model amplitude predictions for the amplitudes of the 525 s modulation. The dotted and solid lines refer to various parameter choices. The UBVRI observations with error bars are from Massacand et al (1995), and the far UV point from the HST data.

## 3. Discussion

One model for explanation of the modulations, is the precessing disk model proposed by Patterson et al (1993) and worked out in more detail by Solheim et al (1997). In this model the 525 s modulation is explainded by a precessing disk with tidal bulges. A simple version of this model predicts that the amplitude of the 525 s modulation should decrease significantly in the far UV as seen in figure 4.

The detection of a period at 1051 s in the far UV modulation is an important result for this system, since this period is proposed as the orbital period, but not detected in the optical lightcurve (Solheim et al 1997). The origin of the far UV modulation with this period must be from an area which has to be small and hot, in order to be undetected in the visible region of the spectrum. An accreting area on the mass receiving star, which then has to corotate with the secondary, is one possibility. Another possibility is a hot spot on the disk, where the accretion mass stream hits the disk, and maybe creates a shock. This explanation is less likely, since we then also would observe modulations at longer wavelengths. Work on detailed modelling of the disk shape, and the resulting modulations, is in progress and will be reported elsewhere.

Another unexpected result is the large amplitude of the 1014 s modulation, which is probably the same as the 1011.4 s modulation which has been observed with strongly secular variable amplitude during the years (Provencal et al, 1995). The maximun amplitude observed in the visible range is about 15 mma, and we find almost twice that value in the far UV.

One possible explanation for this beaviour is that we observe a pulsation on the central white dwarf. This period is of the right order as predicted for H–deficient pre–white dwafs, and observed for PG 1159 stars (Saio, 1996). Since the pulsation period is close, but not exactly equal to the orbital period, we may expect parametric resonances which may explain the large secular amplitude variations.

We have in progress a study of line profile variations, and have also detected a large number of iron lines, possibly from a circumbinary cloud.

## References

Massacand C. and Solheim, J.-E. (1995),*Baltic Astronomy* 4, 378

Nymark, T. K. (1997), *This proceedings*, p 00

Patterson, J, Halpern, J., and Shambrook, A.,(1993), *ApJ* 419, 803

Provencal, J.L., Winget, D.E., Nather, R.E., Robinson, E.L., Solheim, J.-E. Clemens, J.C., Bradley, P.A., Kleinman, S.J., Kanaan, A., Claver, C.F., Hansen, C.J., Marar, T.M.K., Seetha, S., Ashoka, B.N., Leibowitz, E.M., Meištas, E.G., Bruvold, A, Vauclair, G., Dolez, N., Chevreton, M., Barstow, M.A., Sansom, A.E., Tweedy, R.W., Fontaine, G., Bergeron, P., Kepler, S.O., Wood, M.A., and Grauer, A.D, (1995), *ApJ* 445, 927

Saio, H., (1996), *in C.S. Jeffery and U. Heber (eds) Hydrogen-Deficient Stars, ASP Conference Series* 96, p 361

Solheim, J.-E., Emanuelsen, P.-I., Vauclair, G., Dolez, N., Chevreton, M., Barstow, M.A., Sansom, A.E., Tweedy, R.W., Kepler, S.O., Kanaan, A., Fontaine, G, Bergeron, P., Grauer, A.D.,Provencal, J.L., Winget, D.E, Nather, R.E., Bradley, P.A., Claver, C.F., Clemens, J.C., Kleinman, S.J., Hine, B.P., Marar, T.M:K., Seetha, S, Ashoka, B.N., Leibowitz, E.M., and Mazeh, T., (1991) *in G. Vauclair and E. Sion (eds.) White Dwarfs, Kluwer*, p 431

Solheim, J.-E., Provencal, J.L., Bradley, P.A., Vauclair, G., Barstow, M.A., Kepler, S.O., Fontaine, G., Grauer, A.D.,Winget, D.E., Nather, R.E., Marar, T.M.K., Leibowitz, E.M., Emanuelsen, P.-I., Chevreton, M., Dolez, N., Kanaan, A., Henry, G.A., Bergeron, P., Claver, C.F., Clemens, J.C., Kleinman, S.J., Hine, B.P., Seetha, S., Ashoka, B.N., Mazeh, T., Sansom, A.E., Tweedy, R.W., Meištas, E.G., Bruvold, A., and Massacand, C.M., (1997) *A&A*, submitted

## Discussion

*H. Shipman*: I'd be cautious about drawing too strong conclusions about observing — or failing to observe — the CII 1335 line. We've only seen it in one DB white dwarf — GD 358.

*J-E. Solheim*: Thanks. This strengthens our belief that it must be hotter than a DB. May be the central object is a DD or will emerge as a DD one day.

*M. Wood*: Two questions: 1) Have you phase-binned the spectra to look for radial velocity variations? 2) Your elliptical disk model, if correct, should be something that could be confirmed by observations of other $q \approx 0.1$ cataclysmic variables with known orbital periods. Have you made this comparison, and if not, do you plan to do so?

*J-E. Solheim*: 1) We have not done this yet. 2) The model we finally arrived at has two tidal bulges — just as we observe lunar tides on the earth. An elliptical shape with the star in the center is a good analytical representation. We have not looked at CVs with small $q$'s to check this model, and do not plan to do so in the near future.

# EFFECTS OF IRRADIATION ON THE SECONDARY STAR IN A TIGHT BINARY SYSTEM, EXEMPLIFIED BY AM CANUM VENATICORUM

T.K.NYMARK

*Auroral Observatory, University of Tromsø*

**Abstract.** Aiming at understanding the nature of the secondary of AM CVn, we investigate the effects of irradiation in a tight binary system. Observations of the spectrum place an upper boundary on the temperature of the secondary, while the observed light modulations limit the temperature difference between the illuminated and the unilluminated side. The model atmosphere program TLUSTY is used to calculate models at various temperatures. The resulting spectral features will be searched for in the observed spectrum of AM CVn.

## 1. Introduction

The effect of irradiation on low mass helium white dwarfs is a poorly investigated subject, yet it may be of great importance for a series of objects. In tight binary systems with a hot primary, like the AM CVn systems, the secondary is exposed to a high irradiative flux. This must strongly affect the net flux from the secondary, and thereby the spectrum of the system.

## 2. The continuum spectrum of AM CVn

The source of the photometric variations of AM CVn is an object of much debate. The lightcurve variations have been suggested to originate in a pulsating white dwarf, or in aspect variations in a semi-detached binary system. The shape of the continuum spectrum may give important clues to the answer to this problem. If AM CVn is a pulsating white dwarf, the spectrum should resemble that of a single white dwarf, whereas the spectrum of a binary system should show signs of a second star, as well

345

*I. Isern et al. (eds.), White Dwarfs*, 345–351.

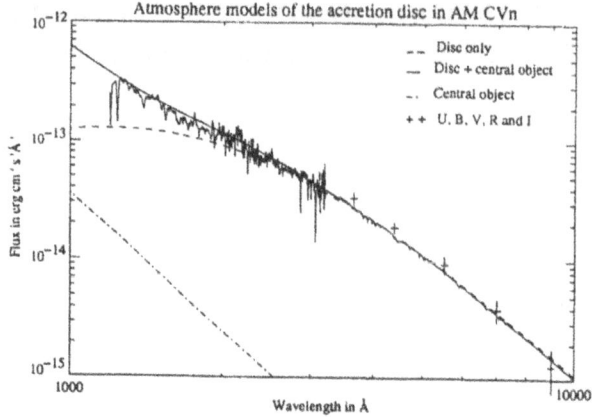

*Figure 1.* The observed spectrum of AM CVn with a synthetic spectrum of the accretion disc and primary. From (Bard, 95).

as an accretion disc. The latter problem was addressed by Bard (Bard,95), who calculated a synthetic spectrum of a hot central star with an accretion disc. Using rings of **ATLAS** atmospheres for the accretion disc and a black body for the central star, he obtained a better fit to the observed spectrum than could be achieved with a single white dwarf. As figure 1 shows, the fit is very good for wavelengths shorter than approximately 3000 Å. At higher wavelengths, however, the fit is not as good. The discrepancy seems small, since the plot is logarithmic in the flux, but if we subtract the synthetic flux from the observed flux in the U, B, V, R, I and I bands, we see that the difference is comparable to a black body of 12,000 K (figure 2).

This implies that another object is present, which contributes to the spectrum. The aim now is to describe this object and its spectrum.

## 3. The secondary star

Evolutionary calculations by Savonije, de Kool and van den Heuvel (Savonije et al., 1986) show that the secondary star of a tight binary system with a degenerate primary should become semi-degenerate when the mass is reduced below 0.3 $M_\odot$, and remain semi-degenerate for the rest of its lifetime. Calculations based on the observed photometric period of AM CVn give a mass of 0.09 $M_\odot$ for the secondary (Solheim et al., 1996), which places it in the semi-degenerate regime. Savonije et al. also find that for masses lower than 0.36 $M_\odot$ the mass transfer drives the star out of thermal equilibrium, and it becomes subluminous (figure 3).

Their calculations stop at a mass of 0.117 $M_\odot$, for which the effective temperature is reduced to approximately 400 K. Since the mass of the

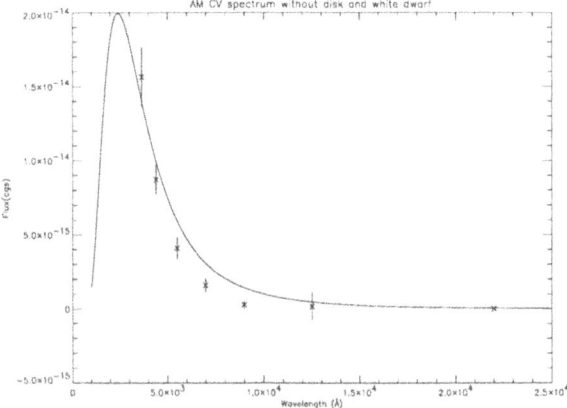

*Figure 2.* The difference between the observed flux and the flux from the combined disc and primary in the wavelengths of the photometric bands. The solid line is the flux of a black body of 12,000 K.

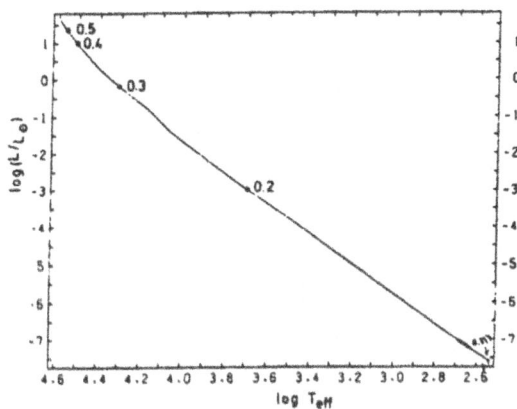

*Figure 3.* The path of the secondary in the HR-diagram. From (Savonije et al., 86)

secondary of AM CVn is even smaller, we have assumed an effective temperature of 400 K also for this object. The luminosity is then so low that it should not give any noticeable contribution to the total spectrum of the system. However, as figure 2 shows, the presence of the secondary is indeed detectable in the spectrum. This can only be explained if the secondary receives energy from the primary through irradiation.

## 4.  Model Atmosphere Calculations

The spectrum of the irradiated star is computed by using the model stellar atmosphere package TLUSTY by I. Hubeny, which in its most recent version (TLUSTY193) allows irradiation to be included. The basic input parameters are the effective temperature of the undisturbed atmosphere, the temperature of the irradiating star and the distance between them. The latter is included through the dilution factor, which is defined by:

$$w_{dil} = \frac{R_1^2}{r^2}$$

where $R_1$ is the radius of the primary and $r$ the distance between the two stars. Since the spectrum of AM CVn shows no signs of hydrogen, we assume a pure helium composition. The temperature of the primary, $T_1$, is kept as a free parameter, as well as the position of the system relative to the Earth, $d$. For the temperature of the secondary we use the lowest temperature for which we were able to compute a model atmosphere, $T_2{=}900$ K. When the temperature is so low, we would expect the spectrum to be completely dominated by the energy added to the secondary through irradiation from its extremely hot companion. Therefore we do not expect the difference between a temperature of 900 K and the previously estimated temperature of 400 K to be of any great importance.

When we have calculated the spectrum of the secondary, we add this to the spectrum of the primary and the accretion disc, and compare with the observed spectrum of AM CVn. For the primary we use a non-irradiated TLUSTY model atmosphere composed only of HeI, and the disc spectrum is calculated from a Planckian disc.

## 5.  The dilution factor

The results depend upon how we choose to calculate the dilution factor. If the radius of the irradiated star is small compared to the distance between the stars ($R_2 \ll r$), the dilution factor can be assumed to be constant across the surface of the star. If, however, the stars are very close, the distance to the source from a given point on the surface of the secondary will vary significantly, and the dilution factor must therefore be modified. In order to account for this effect, we have divided the illuminated surface into rings with different dilution factors. We then calculated a model atmosphere for each of these rings, and computed an average spectrum for the star.

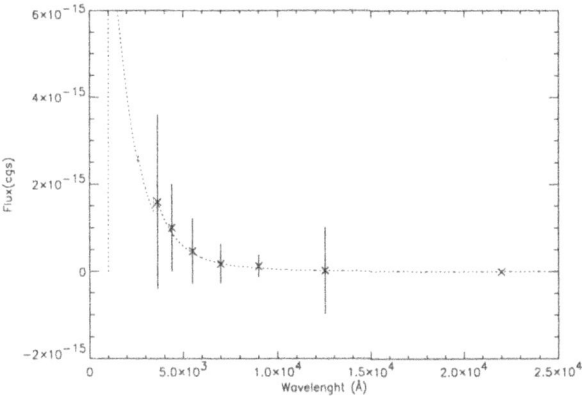

*Figure 4.*   Synthetic spectrum of an illuminated atmosphere (dotted line). The crosses denote the difference between the observed spectrum and a synthetic spectrum for the disc and primary, similar to figure 2

| $T_1$ | $T_2$ | distance | $q$ | $M_1$ | $M_2$ |
|---|---|---|---|---|---|
| $1.8 \cdot 10^5$ K | 900 K | 125 pc | 0.088 | 0.99 | 0.087 |

TABLE 1.   Parameters for AM CVn.

## 6.  Results

In the case of a single dilution factor, the best fit to the observed spectrum was found for $w_{dil} = 1.3 \cdot 10^{-3}$, $T_1 =$180.000 K and a distance of 130 parsecs. The discrepancy between the observed and synthetic spectra of the primary and disc is now greatly reduced compared to figure ??, because we now use a better model for the primary, and the fit of the secondary spectrum to the difference is considerably improved. When we use rings of different dilution factors, the best fit is found for $T_1$=180.000 K and $d$=125 parsecs, as is shown in figure 4.

Thus this improvement does not change the temperature of the source, but it does imply that the distance of the system from the Earth is smaller than what was found by the other method. Dividing the surface into rings in this way does not completely account for the effect of varying dilution factor, but it is closer to the true case than the model with a single value for $w_{dil}$. We therefore adopt the parameters found by the latter fit (table 1).

The total spectrum of the system is shown in figure 5, with the observed spectrum of AM CVn, similar to figure 1.

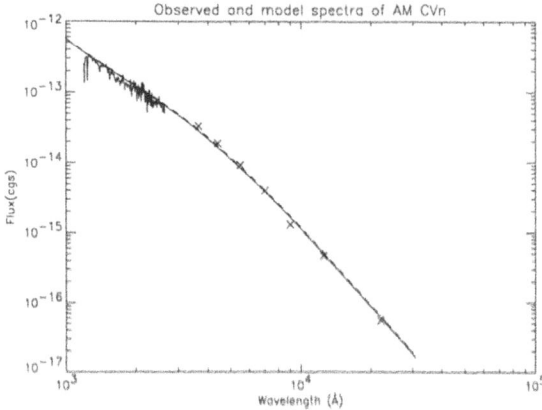

*Figure 5.* The total synthetic spectrum for AM CVn. The temperature of the primary is $T_1$ =180,000 K, the temperature of the secondary before irradiation is $T_2$ =900 K and the distance to the system is 125 parsecs.

## 7. Conclusion

By including the contribution of the secondary, we have obtained a better fit to the spectrum of AM CVn than what was previously achieved. This suggests that there is a secondary star, and that the effect of irradiation is important for its spectrum. However, our models leave out a number of possibly significant effects. The most important is the assumption that only HeI is present in the primary. Even with a pure helium composition, this must necessarily fail at the high temperature in question. The next step would therefore be to include HeII in the models for the primary. Further improvement might be achieved by including other elements in the composition of the models, taking into account the effects of the accretion disc, and calculating line spectra instead of continuum spectra for all components of the system.

## References

Bard,S., Master thesis, University of Tromsø,1995

Savonije, de Kool & van den Heuvel, *A&A*, 1986, **155**, 51-57

Solheim,J.-E., Provencal,J.L., Bradley, P.A., Vauclair, G., Barstow, M., Kepler, S.O., Fontaine, G., Grauer, A.D., Winget, D.E., Nather, R.E., Marar, T.M.K., Leibowitz, E.M., Emanuelsen, P.-I., Chevreton, M., Dolez, N., Kanaan, A., Henry, G.A., Bergeron, P., Claver, C.F., Clemens, J.C., Kleinman, S.J., Hine, B.P., Seetha, S., Ashoka, B.N., Mazeh, T., Meištas,E.G., Bruvold, A. and Massacand, C.M.,*A&A*, 1996, in prep.

## Discussion

*I. Iben*: Did you take into account the shadowing of the primary by the disk?

*T. Nymark*: No, I have assumed the primary to be a point source, and ignored all effects of the accretion disc. Only a small part of the secondary will be in the shadow of the disc, but in the future I plan to include this effect.

*J. Kubát*: What are the differences between irradiated and non-irradiated model atmospheres in temperature structure?

*T. Nymark*: That depends upon the initial tempereature of the two stars, but in general the temperature will be higher in the irradiated atmosphere than in the non-irradiated atmosphere. If the irradiation is strong compared to the intrinsic flux of the star, the outer layers are strongly heated, but also the deeper layers are affected. If the irradiation is smaller than the flux of the irradiated star, the temperature structure is approximately the same in the two cases for $\tau \leq 1$. In the deeper layers the temperature is higher in the irradiated atmosphere.

# WHITE DWARFS IN AM HERCULIS SYSTEMS

BORIS T. GÄNSICKE AND KLAUS BEUERMANN
*Universitäts-Sternwarte, Geismarlandstr. 11, 37083 Göttingen, Germany*

DOMITILLA DE MARTINO
*Osservatorio Astronomico di Capodimonte, Via Moiariello 16, 80131 Naples, Italy*

AND

STEFAN JORDAN
*Institut für Astronomie und Astrophysik, Universität Kiel, Olsenhausenstr. 40, 24098 Kiel, Germany*

**Abstract.** We have determined the photospheric temperatures of the white dwarfs in AM Herculis systems in a uniform way, using all available IUE/HST observations. We find that the white dwarf can unequivocally be identified through its photospheric Lyα absorption profile in seven systems only: BY Cam, V834 Cen, AM Her, DP Leo, AR UMa, QQ Vul, and the new, ROSAT-discovered system RX J1313-32. All systems contain a moderately hot white dwarf ($T_{wd} = 15\,000 - 25\,000$ K,). Most observations indicate that a large ($f \sim 0.015 - 0.1$) fraction of the white dwarf is heated to $\sim 25\,000 - 35\,000$ K.

## 1. Introduction

In AM Herculis systems (*polars*), a synchronously rotating, strongly magnetized white dwarf accretes matter from the Roche Lobe filling late-type secondary. The magnetic field funnels the accretion flow onto one or both poles of the white dwarf, resulting in locally confined heating and enrichment with heavy elements of the white dwarf atmosphere. Even though the number of identified polars increased dramatically over the last years to ~ 60 (see Beuermann et al. 1995 & 1996), few details are known about the thermal response of the white dwarf to this asymmetric accretion. Mainly two effects are expected: (a) The photospheric temperature of the white dwarf depends on the *long term accretion-induced heating* which coun-

*I. Isern et al. (eds.), White Dwarfs, 353–358.*
© *1997 Kluwer Academic Publishers.*

teracts the secular core cooling to some extent (with additional heating occuring during nova outbursts). Sion (1986, 1991) found from observed photospheric temperatures of accreting white dwarfs that the white-dwarf temperature decreases towards shorter periods. This can be understood in the general picture of cataclysmic variable evolution where the systems evolve towards shorter periods; hence, the orbital period can be considered a clue to the *age* of the system. Sion (1991) found also a weak evidence for lower temperatures in polars. However, the number of polars with reliable temperature determinations is too small for definite conclusions. (b) The infalling matter is magnetically channeled to a small accretion spot near one or both magnetic poles where it passes through a shock. Depending on the accretion geometry, a rather large area on the white dwarf can be heated through irradiation with cyclotron radiation and bremsstrahlung from the post-shock region. Gänsicke et al. (1995) showed that the UV flux emitted by the hot spot in AM Herculis itself is a crucial ingredient for the overall energy balance of the accretion process.

We report here first results of our project to determine reliable effective temperatures of the white dwarfs in AM Her stars and, if possible, to derive the temperature distribution over their surface.

## 2. The data

The most promising wavelength region to derive the effective temperature of the accretion-heated white dwarf in polars is the UV. In the optical, the white-dwarf emission is diluted in most cases by cyclotron radiation and by emission from the secondary star and from the accretion stream. During the 19th (and last) episode of IUE, we obtained low-resolution SWP spectra of AM Her, QQ Vul, AR UMa and RX J1313-32. These spectra were complemented with all archive IUE and HST data of polars available at the end of 1995. The SWP spectra used for our analysis have been reprocessed with the IUE Final Archive procedures (NEWSIPS), yielding a better flux calibration and a better S/N, especially for underexposed spectra. Our complete dataset contains UV observations for 28 polars. All observations are average spectra over the binary orbit, except for AM Her, QQ Vul and DP Leo, where phase-resolved spectra exist.

## 3. Analysis and results

All spectra were screened for any evidence of photospheric Lyα absorption from the white dwarf, which was detected in seven out of 28 systems: BY Cam, V834 Cen, AM Her, DP Leo, AR UMa, QQ Vul and RX J1313-32. For these systems, we checked the possible contribution to the Lyα profile from intrinsic and interstellar neutral hydrogen absorption using column densities derived from X-ray data and find that it is negligible in all cases. Spectral fitting with white-dwarf model spectra from Gänsicke et al. (1995) is described below for the individual systems, see also fig. 1

## 3.1. BY CAM

BY Cam ($P = 3.3\,\mathrm{hrs}$) is a peculiar member of the polar family as it is slightly asynchronous (Silber et al. 1992) and shows peculiar emission line ratios with extremely strong $\mathrm{N\,V}\,\lambda\,1240$ emission (Bonnet-Bidaud & Mouchet 1987), both possible hints of a "recent" (within the last $\sim 1000\,\mathrm{yrs}$) nova explosion. The only IUE low-state spectrum was analysed by Szkody et al. (1990) who derived a lower limit of the white-dwarf temperature of $70\,000\,\mathrm{K}$. We find, however, that the NEWSIPS-processed spectrum shows clearly Ly$\alpha$ absorption which can be fitted with a $22\,000\,\mathrm{K}$ white dwarf. Using $E_{B-V} = 0.05$ (Bonnet-Bidaud & Mouchet 1987), the white dwarf is expected to have $V \approx 17.8$, which is consistent with $V = 17.5$ observed simultaneously to the IUE spectrum. The distance implied by the scaling factor of the model spectrum, assuming a white-dwarf radius of $8 \times 10^8\,\mathrm{cm}$, is $250\,\mathrm{pc}$, agreeing with the distance estimate made by Szkody et al. (1990) from the $J$ magnitude of BY Cam.

## 3.2.  V834 CEN & RX J1313-32

Despite the different orbital periods of V834 Cen ($P = 1.68\,\mathrm{hrs}$) and RX J1313-32 ($P = 4.25\,\mathrm{hrs}$), the UV spectra of both systems are very similar. In both systems, it is not possible to fit the observed spectrum with a single-temperature white-dwarf model spectrum. However, we can describe the observed spectra with a two-temperature model (white dwarf plus heated polecap), yielding almost identical parameters for both systems: a white-dwarf of $T_{\mathrm{wd}} \approx 15\,000\,\mathrm{K}$, a hot spot of $T_{\mathrm{HS}} \approx 30\,000\,\mathrm{K}$ covering $\sim 2\%$ of the white dwarf and an implied distance of $d \approx 130-140\,\mathrm{pc}$ for a white-dwarf radius of $R_{\mathrm{wd}} = 8 \times 10^8\,\mathrm{cm}$. The $V$-magnitudes of the white dwarfs agree with the observed low-state brightness of the systems. The white-dwarf and spot temperatures of V834 Cen are in agreement with the estimates obtained by Schwope (1990) from optical Zeeman and cyclotron spectroscopy.

## 3.3.  AM HER

An extensive analysis of phase-resolved IUE spectroscopy of AM Her in high and low state ($P = 3.1\,\mathrm{hrs}$) has been presented in Gänsicke et al. (1995), showing that a large fraction ($f \sim 0.1$) of the white dwarf ($T_{\mathrm{wd}} = 20\,000\,\mathrm{K}$) is heated to $\sim 24\,000\,\mathrm{K}$ and $37\,000\,\mathrm{K}$ in the low and high state, respectively. Here, we analysed additional IUE spectra taken during the decline into a low

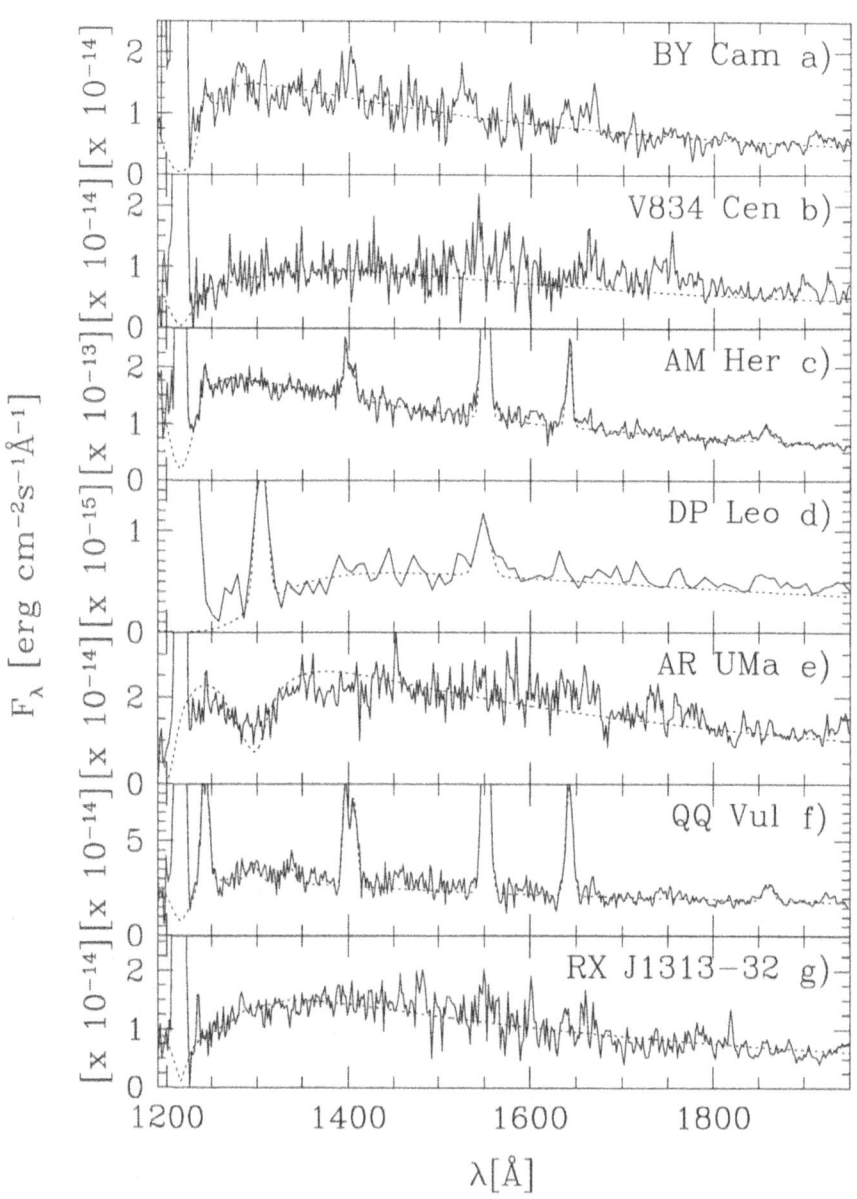

*Figure 1.* UV spectra of polars reveiling Lyα absorption from the white-dwarf photosphere. Full lines: observed spectra. Dashed lines: best fit. From top to bottom: (a)
orbital average of BY Cam in low state; (b) orbital average of V834 Cen in low state; (c)
bright phase of AM Her in an intermediate state; (d) faint phase of DP Leo in low state
(the emission at 1306 is geocoronal O I); (e) orbital average of AR UMa in low state; (f)
bright phase of QQ Vul in high state; (g) orbital average of RX J1313-32 in low state.
See text for details.

state, when the system was in an intermediate state of accretion ($V \approx 13.8$), yielding a temperature of $\sim 27\,000$ K for the hot spot. This shows that the localized heating of the white-dwarf surface is strongly correlated to the level of accretion. However, sofar no time series covering the transition between different accretion states exists, hence, no timescales of the white-dwarf cooling/heating can be derived.

### 3.4. DP LEO

Stockman et al. (1994) fitted a two-temperature blackbody model to phase-resolved HST/FOS low-state spectroscopy of DP Leo ($P = 1.5$ hrs) and derived a white-dwarf of $T_{\mathrm{wd}} = 16\,000$ K and a hot spot of $T_{\mathrm{HS}} \approx 50\,000$ K, covering a small ($f \sim 0.006$) fraction of the white dwarf. We repeated this analysis using more realistic white-dwarf model spectra and find $T_{\mathrm{wd}} \approx 14\,000$ K. However, the hot spot in our model is much cooler, $T_{\mathrm{HS}} \approx 27\,000$ and larger ($f \sim 0.03$) than in that of Stockman et al. (1994). The distance derived from our fit, again assuming $R_{\mathrm{wd}} = 8 \times 10^8$ cm, $d = 390$ pc, is in good agreement with the distance estimate by Biermann et al. (1985). The $V$-magnitude of the white dwarf is consistent with the observed low-state magnitude of DP Leo.

### 3.5. AR UMA

AR UMa ($P = 1.93$ hrs) is the long-sought example of a polar with a high-field magnetic white dwarf (Schmidt et al. 1996). Hitherto, all analysed polars contained white dwarfs with $B < 100$ MG, a puzzling fact as single magnetic white dwarfs with $B$ up to $1\,000$ MG are known (Schmidt & Smith 1995). Fig. 1 shows a fit of a magnetic white-dwarf model spectrum to the IUE data, yielding $B = 190$ MG and $T_{\mathrm{wd}} = 25\,000$ K. However, these values should be regarded as preliminary as the IUE spectrum is an orbital average, smearing out any information about the field geometry.

### 3.6. QQ VUL

Our phase-resolved IUE high-state spectroscopy of QQ Vul ($P = 3.7$ hrs) allows to discern the Ly$\alpha$ absorption of the white dwarf/hot spot. However, the contribution of the accretion stream to the UV flux is much stronger than in the other systems, hampering a clear separation of the contributions from the white dwarf, the hot spot and the stream. We applied, similar to our previous analysis of AM Her (Gänsicke et al. 1995), a two-component model of a white-dwarf model spectrum and a blackbody to account for the white dwarf and the stream emission, respectively, yielding a preliminary value of $T_{HS} \sim 30\,000$ K. A more sophisticated analysis of the complete dataset of QQ Vul is under way and will be reported elsewhere.

## 4. Conclusions

We have determined the temperatures of the white dwarfs in AM Herculis systems, covering a large range of orbital periods ($P = 1.5 - 4.25$ hrs). We find, in agreement with Sion (1991), a tendency for lower temperatures at shorter periods. However, RX J1313-32, being a long-period system is found to have a remarkably low temperature. All analysed systems seem to have rather large, moderately heated UV emitting regions, in contrast to the usually very small EUV/soft X-ray spots. This may indicate rather high standing accretion shocks in most polars.

## Acknowledgements

We thank Cathy Imhoff for her assistance with the NDADS IUE archive. This work was supported in part by the DARA under project number 50 OR 9210 1. BTG would like to thank the Astronomische Gesellschaft for financial travel support.

## References

Beuermann, H., Burwitz, V., K., Reinsch, Schwope, A.D., Thomas, H.-C., 1995, in *Cataclysmic Variables*, eds. A. Bianchini et al., p. 381 (Dordrecht: Kluwer)

Beuermann, H., Burwitz, V., K., Reinsch, Schwope, A.D., Thomas, H.-C., 1996, in *Röntgenstrahlung from the Universe*, MPE Report 263, eds. H.U. Zimmermann et al., 107

Biermann, P., Schmidt, G.D., Liebert, J., Stockman, H.S., Tapia, S., Kühr, H., Strittmaier, P.A., West, S., Lamb, D.Q., 1985, ApJ 293, 303

Bonnet-Bidaud, J.-M., Mouchet, M., 1987, A&A 188, 89

Gänsicke, B.T., Beuermann, K., de Martino, D., 1995, A&A 303, 127

Schmidt, G.D., Smith, P.S., 1995, ApJ 448, 305

Schmidt, G.D., Szkody, P., Smith, P.S., Silber, A., Tovmassian, G., Hoard, D.W., Gänsicke, B.T., de Martino, D., 1996, ApJ *in press*

Schwope, A.D., 1990, Reviews in Modern Astronomy 3, 44

Silber, A.D., Bradt, H.V., Ishida, M., Ohashi, T., Remillard, R.A., 1992, ApJ 389, 704

Sion, E.M., 1986, PASP 98, 821

Sion, E.M., 1991, AJ 102, 295

Stockman, H.S., Schmidt, G., Liebert, J., Holberg, J.B., 1994, ApJ 430, 323

Szkody, P., Downes, R., Mateo, M., 1990, PASP 102, 1310

# HST SYNTHETIC SPECTRAL ANALYSIS OF U GEM IN EARLY AND LATE QUIESCENCE: A HEATED WHITE DWARF AND ACCRETION BELT

F.H. CHENG, E.M. SION

*Villanova University, Villanova, PA 19085. fhcheng@ucis.vill.edu, emsion@ucis.vill.edu & (FHC) Center for Astrophysics, Univ. of Science and Technology of China, Hefei, Anhui, People's Republic of China.*

K. HORNE

*University of St. Andrews, School of Physics and Astronomy, North Haugh, St. Andrews, Fife KY16 9SS, Scotland. kdh1@st-and.ac.uk*

I. HUBENY

*Code 681, Goddard Space Flight Center, Greenbelt, MD 20771. hubeny@stars.gsfc.nasa.gov*

M. HUANG

*Villanova University, Villanova, PA 19085. huang@ucis.vill.edu*

AND

S.D. VRTILEK

*Harvard-Smithsonian Center for Astrophysics, 60 Garden Street, Cambridge, MA 02138. saku@head-cfa.harvard.edu.*

**Abstract.** We have re-examined two archival *HST* FOS G130H spectra of the prototype dwarf nova U Geminorum obtained during its quiescence 13 days and 70 days after a *wide* outburst. Using synthetic spectral fitting with two flux-emitting components, a slowly rotating white dwarf photosphere and a rapidly spinning accretion belt ($V_{rot} \sin i = 3200$ km s$^{-1}$) we found significantly improved spectral fits at the 99.73% confidence level. We found clear evidence for the cooling of the white dwarf, confirming earlier results, and evidence for the cooling of the accretion belt as well. For the white dwarf and accretion belt respectively, 13 days post-outburst, we find $T_{wd}$ = 37,000K ± 400K and $T_{belt}$ = 45,000K ± 2,500K while at 70 days post outburst, we find $T_{wd}$ = 33,500K ± 700K and $T_{belt}$ = 37,500K ± 4,000K.

*I. Isern et al. (eds.), White Dwarfs, 359–366.*
© *1997 Kluwer Academic Publishers.*

The percentage emitting area and flux contribution of the accretion belt decreased from 14.9% and 24.5% respectively, (at 13 days post outburst), to 9.4% and 15.0% respectively (at 70 days post outburst). These results are compared with our *HST* GHRS G160M observations obtained 13 days and 61 days after a *narrow* outburst of U Gem.[1]

## 1. Introduction

On purely theoretical grounds an accretion belt is plausibly expected when matter with angular momentum shears onto the surface of a white dwarf in a dwarf nova during outburst (Kippenhahn and Thomas 1978; Sparks *et al.* 1993). Since matter accretes tangentially at the equator and does so every several weeks, an equilibrium state may be established such that the equatorial region spins rapidly while the higher latitudes and polar regions do not. The controlling parameter for this equilibrium state is the Richardson number (Kippenhahn and Thomas 1978), which implies a differential rotation going both inward into the white dwarf and poleward along the white dwarf surface. The spectroscopic detection of such accretion belts would represent an important milestone in our understanding of accretion physics.

In this paper, we investigate these questions for the well-observed case of U Geminorum during dwarf nova quiescence. U Geminorum is the prototype for those CVs known as dwarf novae. It has a 4.25 hours orbital period with a ∼ 1 $M_{\odot}$ white dwarf and an M4.5 secondary star (Wade, 1981). About every 120 days, U Geminorum undergoes a dwarf nova outburst, where the V magnitude of the system rises from 14 to 9, and the system is above 14 mag for 7–14 days (Szkody and Mattei, 1984). If the outburst lasts longer, say ∼ 2 weeks, the light curve during the outburst looks wider, and it is called *wide* outburst; otherwise, if it lasts shorter, say 1 week or less, the light curve during the outburst looks narrower, and it is called a *narrow* outburst. In §2 we review previous observations and spectral fitting work on U Geminorum; our synthetic spectral analysis and fitting techniques are described in §3; and finally we present our discussion and conclusions in §4.

[1]Based on observations with the NASA/ESA *Hubble Space Telescope* obtained at the Space Telescope Science Institute, which is operated by the Association of University Research in Astronomy, Incorporated, under NASA contract NAS5–26555.

## 2.  HST FOS Spectra of U Geminorum

Two 360 second *HST* FOS G130H spectra were obtained by Long, Sion, Huang, and Szkody (1993, hereafter LSHS) 13 days and 70 days into the quiescence interval following a *wide* outburst of U Geminorum beginning on 1992 August 29. The mid-time of the exposures occurred at orbital phases 0.11 and 0.51 respectively. The spectra cover the wavelength range 1150-1610 Å. The flux from U Gem at 1400 Å dropped by ∼28% between the first and second observations. There is a "hot" diode near 1508 Å in the second observation. The observed absorption lines show a system redshift around 0.3 - 0.4 Å (LSHS).

Single temperature white dwarf synthetic spectral fits to the FOS data reveal that the white dwarf cools by 7000K during quiescence, with $T_{eff}$ ∼ 39,000K 13 days POB and $T_{eff}$ ∼32,000K 70 days POB (LSHS). However, starting with Kiplinger, Sion and Szkody (1991), evidence for two component temperature fits to the white dwarf has been mounting from *IUE*, *HUT* and *HST* data (Kiplinger, Sion and Szkody 1991; Long *et al.* 1993, 1994, 1995; Huang *et al.* 1996b). This notion is reinforced by the presence of He II (1640) and N V (1240) absorption features which cannot be associated in their observed strength with a ∼35,000K photosphere.

It is apparent that in the two FOS spectra of U Gem in quiescence, the presence of continuum curvature is considerably less pronounced than in VW Hydri. Nonetheless visual inspection reveals the continuum curvature characteristic of a Keplerian-broadened component, most obvious in the first FOS observation (see upper panel in Figure 1). We constructed special white dwarf plus accretion belt model grids to fit these two FOS spectra in the present paper.

## 3.  White Dwarf Plus Accretion Belt Models

The model atmosphere (TLUSTY, Hubeny 1988), and spectrum synthesis (SYNSPEC, Hubeny, Lanz, and Jeffery 1994) codes and details of our $\chi^2$ minimization fitting procedure can be found in Sion *et al.* (1995 a,b) and Huang *et al.* (1996a, b). We calculated a grid of white dwarf models (30,000 K < $T_{wd}$ < 40,000K) with log $g$ = 8.0 and solar abundance. Since the white dwarf in U Gem has very low rotational velocity (Sion *et al.* 1994), we took $V_{rot} \sin i$ = 0.0 for the white dwarf models. We also included the Lyα satellite absorption opacity (Allard and Kielkopf 1991 Allard and Koester 1992, Allard, private communication) in calculating the synthetic spectra of white dwarf (the Lyα red wing was slightly improved in the temperature range considered here). All model fluxes were normalized to a solar radius at a distance 1 kpc, in unit of mJy.

We also calculated a grid of belt models ($30,000 < T_{belt} < 60,000$ K) with $\log g = 7.0$ and $V_{rot} \sin i = 3200$ km s$^{-1}$. The chemical abundances in the belt are the same as that used in Huang *et al.* (1996a), and Sion *et al.* (1996) except Si, *i.e.* C=20 solar, Si=100 solar, others=solar. The enhanced abundances of C and Si may be due to the dredge-up of the metal beneath the white dwarf atmosphere caused by the strong shearing near the equator. During the spectral fitting we masked the Ly$\alpha$ region to avoid the effect of airglow emission and emission from the disk of U Gem. The N V and C IV regions were excluded in the spectral fitting because these features do not arise from the white dwarf photosphere (LSHS). We also excluded the "hot" diode in the second FOS spectrum. The observed spectra have systematic redshifts, we blue-shifted the first FOS spectrum by 0.65 Å and the second FOS spectrum by 0.3 Å in order to bring the observed and synthetic Si IV absorption profiles into precise wavelength coincidence. Since U Gem has very low absorption in its sightline (LSHS), we took a color excess E(B-V)=0.0 in our spectral fitting.

For a pair of white dwarf and belt temperatures, we can find a best fit $\chi^2$ value by adjusting the belt emitting area. We calculate a grid of $\chi^2$ values for different pairs of white dwarf and belt temperatures, and determine the best white dwarf and belt temperatures according to the least $\chi^2$ value. To construct a 1-3 $\sigma$ contour diagram in the white dwarf temperature – accretion belt temperature plane we take $\Delta\chi^2 = 4.72$, 9.70, and 16.3 (for 4 free parameters) respectively (Lampton, Margon, & Bowyer 1976). This exercise yielded the uncertainty estimates for the white dwarf and the belt temperatures.

When synthetic spectral fits combining a white dwarf photosphere and a rapidly spinning accretion belt were carried out, the least $\chi^2$ values dropped significantly relative to the single temperature "pure" white dwarf fit. The least $\chi^2$ values for the 1st and the 2nd FOS spectra decreased to 2309 (DOF=1640) and 2287 (DOF=1620). In our fitting experiments, there were four free parameters: scale factors and temperatures for both the white dwarf and accretion belt. The $\chi^2$ values are significantly improved at the 99.73% confidence level.

We noticed that the absorption depths of the Si IV ($\lambda\lambda$1393, 1402) lines in the white dwarf plus belt models are slightly shallower than the observations. The belt spins near Keplerian velocity (we fixed the belt velocity $V_{rot} \sin i$ 3200 km s$^{-1}$, the spectrum from the belt was smoothed, the Si IV absorption depth from the white dwarf's contribution is not enough to match that of the observations because the white dwarf models with solar abundance contributed $\sim$ 85% and 91% of the emitting area for the 1st and 2nd observations respectively. We therefore enhanced the silicon abundance of the white dwarf by factors 1.5, 2.0 and 2.5, we found the $\chi^2$ reduced to

2302 and 2281 for 1st and 2nd FOS fits when the silicon in the white dwarf
has 1.5 solar abundance.

The percentage emitting areas of the belt and the white dwarf, the radii
of the white dwarf for the 1st and the 2nd FOS observations are listed in
Table 1. Table 1 also lists the parameters used and the results obtained
from the best synthetic spectral fits. We see from Table 1 that cooling was
indicated not only for the white dwarf, but also for the accretion belt during
quiescence. We also find a decreased belt area at 70 days POB, consistent
with Figure 1 in which the curvature between 1380 Å and 1550 Å was
less pronounced than is seen 13 days POB. Note also that the curvature
hallmark is not at all evident in the Astro 2 spectrum of Long *et al.* (1995)
that was observed 185 days POB. The best fitting models to the two FOS
spectra are displayed in Fig. 1 with the top panel showing the belt plus
white dwarf fit to the 1st FOS spectrum obtained 13 days POB and the
bottom panel showing the best fit to the 2nd FOS spectrum 70 days POB.
We also plot the contribution from the belt in each panel for comparison
(see dotted smooth curves in Fig. 1).

## 4. Discussions and Conclusions

From the white dwarf plus belt model fitting, we obtained white dwarf radii
of $3.85 \times 10^8$ cm and $3.99 \times 10^8$ cm for the two observations respectively (see
Table 1). This is equal to a $\sim 7\%$ difference of solid angle of the white dwarf
between the 1st and 2nd FOS observations. If the solid angle did not greatly
change between outbursts, the greatly reduced solid angle difference in our
white dwarf plus belt model looks more reasonable than in the single white
dwarf model.

The belt temperature and the belt emitting area decreased during quiescence, and may disappear entirely at the very late quiescent stage, *e.g.* in
the Astro 2 observations 185 days POB (Long *et al.* 1995), as evidenced by
the disappearance of the hallmark continuum curvature. The belt temperature derived for the earlier FOS spectrum (13 days POB) is of the same
order as the residual temperature (53,000K) of the hot source in U Gem
found by Long *et al.* (1995) when they subtracted the *HUT* ASTRO 2 spectrum of U Gem obtained 185 days POB from the *HUT* ASTRO1 spectrum
obtained early in the quiescence of a different outburst. This source had a
hot continuum but showed C IV absorption which could not be associated
with a rapidly spinning belt. It is reasonable that this feature originates
from the same source which produces the N V absorption.

The N V $\lambda 1240$ absorption features in both FOS spectra cannot be
accounted for with either single temperature white dwarf photospheric fits
(LSHS) or with our white dwarf plus accretion belt fits. These features

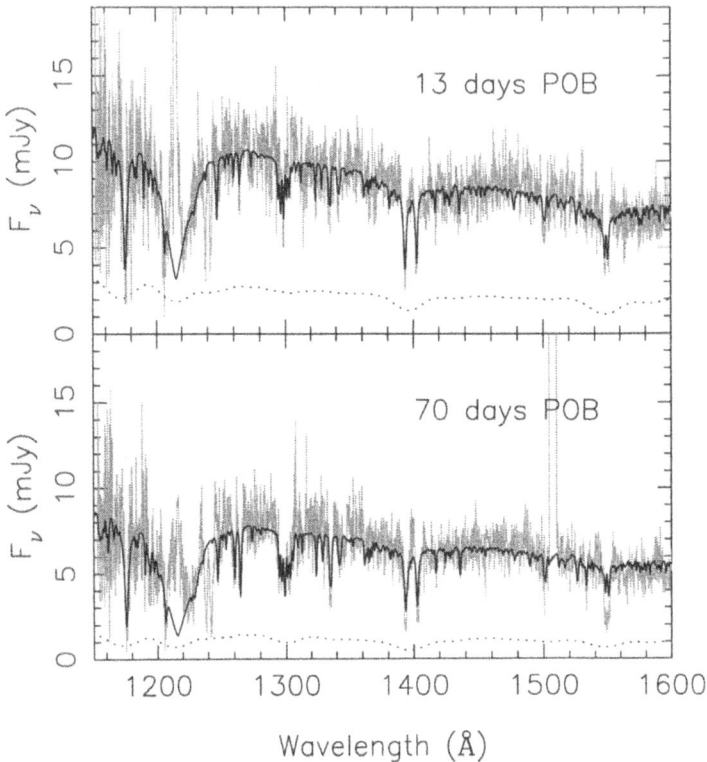

*Figure 1.* Best white dwarf plus belt models in spectral fits. The upper panel is the fit for the FOS observations 13 days post outburst; the lower panel is the fit for that 70 days post outburst. In the panels the dotted smooth curves are the contributions from the accretion belt.

require a higher temperature and are simply too deep and sharp to arise from the white dwarf or belt. They probably arise from hot gas in front of the white dwarf above the disk plane. Our current studies on FOS G130H spectra reveal that the N V velocities are not associated with the Einstein-redshifted white dwarf photospheric frame. Curiously, the N V absorption (and also the C IV absorption) is stronger in the later (70 days POB) FOS spectrum than in the early (13 days POB) spectrum. This strengthening is not expected for a cooling photosphere/belt during quiescence. Moreover, this behavior is contrary to the marked weakening of N V absorption in the very late quiescence (185 days POB) *HUT* spectrum of Long *et al.* (1995). While our solar composition synthetic spectral fits do provide significant C IV absorption, they fail to fully account for the observed C IV strengths. When the C abundance is increased until agreement with the observed depth of C IV is achieved (a C abundance of 50 times solar), other C II and C III features appear which are not present in the FOS data. Thus, it seems likely that part of the C IV absorption is due to the same non-photospheric

TABLE 1. The Parameters Used And the Fitting Results White Dwarf + Accretion Belt Fits

| Parameter | FOS1 (13d POB) | | FOS2 (70d POB) | |
|---|---|---|---|---|
| | WD | BELT | WD | BELT |
| $\log g$ | 8.0 | 7.0 | 8.0 | 7.0 |
| $V_{rot} \sin i$ (km/s) | 0 | 3200 | 0 | 3200 |
| Ly$\alpha$ satellite | yes | yes | yes | yes |
| Abundance* | Si:1.5 | C:20, Si:100 | Si:1.5 | C:20, Si:100 |
| | others:1.0 | others:1.0 | others:1.0 | others:1.0 |
| T/1000 (K) | 37.0±0.4 | 45.0±2.5 | 33.5±0.7 | 37.5±4.0 |
| Scale factor($10^{-3}$) | 3.22 | 0.56 | 3.67 | 0.38 |
| Emitting area (%) | 85.1 | 14.9 | 90.6 | 9.4 |
| Flux Contribution (%) | 75.5 | 24.5 | 85.0 | 15.0 |
| $\chi^2$ | 2302 (DOF=1640) | | 2281 (DOF=1620) | |
| Radius of WD (cm)** | $3.85 \times 10^8$ | | $3.99 \times 10^8$ | |

* in units of solar abundance.

** by assuming a distance 90 pc

hot source as the N V.

The accretion belt temperatures (45,000K to 37,500K) for U Gem are higher by 9,000K to 17,000K than the accretion belt temperatures we derived for VW Hydri (Sion *et al.* 1996) which are in the range 31,000K to 28,000K. This result is consistent with a longer outburst interval and probably higher accretion rate for U Gem versus VW Hydri, a shorter period system below the period gap. Note however that the relative elevation of the belt temperature over the $T_{eff}$ of the white dwarf's higher latitudes appear similar for the two systems. For example, $T_{belt}$ - $T_{eff} \sim$ 8000K to 4,000K for U Gem 13 days and 70 days POB respectively whereas for VW Hydri $T_{belt}$ - $T_{eff} \sim$ 4,000K to 8,000K 1 day post normal outbourst and 10 days post superoutburst respectively. This could imply similar meridional temperature gradients resulting from tangential accretion onto the two white dwarfs.

Finally these results viewed comparatively with our recent analysis of *HST* GHRS G160M observations obtained 13 days and 61 days after a *narrow* outburst of U Gem (Cheng *et al.* 1997) showed the accretion belt is far less prominent, and the C and Si abundances in the belt are solar, indicating empirically for the first time the difference in heating, angular momentum, chemical abundance and mass accretion into the white dwarf, between a *wide* and a *narrow* outburst of U Gem. The white dwarf temperature is lower in both quiescent spectra and the belt emitting area is only 3%.

This work is supported by NASA through grant AR-5817 and GO-6019 from the Space Telescope Science Institute, which is operated by the Association of Universities for Research in Astronomy Inc., under NASA Contract NAS5-26555. We thank Nicole Allard for providing us the Ly$\alpha$ satellite absorption opacity.

# References

Allard, N.F., & Kielkopf, J.F. 1991, A&A, 242, 133
Allard, N.F., & Koester, D. 1992, A&A, 258, 464
Gansicke, B., & Beuermann, K. 1996, A&A, 309, L47
Huang, M., Sion, E.M., Hubeny, I., Cheng, F.H.,& Szkody, P. 1996a, ApJ, 458, 355
Huang, M., Sion, E.M., Hubeny, I., Cheng, F.H., & Szkody, P., 1996b, AJ, 111, 2386
Hubeny, I., 1988, Comput. Phys. Comm., 52, 103
Hubeny, I., & Lanz, T., 1995,
Hubeny, I., Lanz, T., & Jeffery, S. 1994, Newsletters on Analysis of Astronomical Spectra
    (St. Andrews University), 20, 30
Kiplinger, A.L., Sion, E.M., & Szkody, P. 1991, ApJ, 366, 569
Kippenhahn, R., & Thomas, H.-C. 1978, A&A, 63, 265
Lampton, M., Margon, B., & Bowyer, S. 1976, ApJ, 208, 177
Long, K, S., Blair, W.P., Bowers, C.W., Davidsen, A.F., Kriss, G.A., Sion, E.M., &
    Hubeny, I., 1993, ApJ 405, 327
Long, K.S., Blair, W.P., & Raymond, J.C., 1995, ApJ 454, L39
Long, K.S., Sion, E.M., Huang, M. and Szkody, P., 1994, ApJ, 424, L49 (LSHS)
Cheng, F.H. et al. , 1997, in preparation
Sion, E.M., Cheng, F.H., Huang, M., Hubeny, Ivan, & Szkody, P., 1996, ApJ, 471, L41
Sion, E.M., Long, K.S., Szkody, P., & Huang, M. 1994, ApJ, 430, L53
Sion, E.M., Szkody, P., Cheng, F.H., & Huang, M. 1995a, ApJ, 444, L97
Sion, E.M., Huang, M., Szkody, P., & Cheng, F.H. 1995b, ApJ, 445, L31
Sparks, W.M., Sion, E.M., Starrfield, S.G., & Austin, Scott 1993, in Proceedings of
    Cataclysmic Variables and Related Physics, Eds. Oded Regev and Giora Shaviv, p.96
Szkody, P., & Mattei, J.A. 1984, PASP, 96, 988
Verbunt, F., Hassall, B., Pringle, J., Warner, B., & Marang, F., 1987, MNRAS, 225,113
Wade, R.A. 1981, ApJ, 246, 215

# WFPC2 OBSERVATIONS OF PROCYON B

J. L. PROVENCAL AND H. L. SHIPMAN
*Department of Physics and Astronomy, University of Delaware*
*Newark, DE*

F. WESEMAEL
*Département de Physique, Université de Montréal*
*C.P. 6128, Succ. Centre Ville, Montréal, Québec, Canada H3C*
*3J7*

P. BERGERON
*Lockheed Martin Electronic Systems Canada*
*6111 Avenue Royalmount, Montréal, Québec, Canada H4P 1K6*

H. E. BOND
*Space Telescope Science Institute*
*3700 San Martin Drive, Baltimore, MD 21218*

JAMES LIEBERT
*Department of Astronomy and Steward Observatory*
*University of Arizona, Tucson, AZ 85721*

AND

E. M. SION
*Department of Astronomy and Astrophysics*
*Villanova University, Villanova, PA 19085*

## 1. Introduction

Procyon B represents a critical object for testing the fundamental physics of stellar degeneracy. Although this important star is the third brightest white dwarf known, we know very little about it, as Nature has conspired to place it a maximum of 5 arcsec from one of the brightest stars in the sky, rendering ground-based observations of this object nearly impossible. We have a rough visual magnitude of $m_v \approx 10.7$ (Eggen & Greenstein 1965). The system has a period of $40.82 \pm 0.06$ yrs, and the white dwarf mass is

367

*I. Isern et al. (eds.), White Dwarfs, 367–373.*

0.622±0.023 $M_\odot$ (Girard et al. 1996). Walker et al. (1994) utilize speckle interferometry, and their High Resolution Camera to obtain $m_v = 11.3$, $B - V = +0.26$, and $B - I = +0.62$. Unfortunately, their technique is not designed with photometry in mind, and the authors do not quote error estimates. This brief introduction outlines the extent of our knowledge of Procyon B. We do not know its temperature, chemical composition, or accurate visual magnitude.

Procyon B is one of seven known white dwarfs which are members of visual binaries, and which provide observational support for the validity of stellar degeneracy. The mass-radius relation for a degenerate configuration (Chandrasekhar 1933), assuming carbon cores, is an underlying assumption in most of our techniques of determining white dwarf masses. Visual binaries provide the best method to determine both white dwarf masses and radii without assuming a mass-radius relation. Knowledge of the white dwarf's mass comes from its binary motion, which, when combined with its temperature and distance, reveals the stellar radius. Observational confirmation of stellar degeneracy rests on the three visual binaries for which ground based measurements are sufficient to determine the white dwarf's characteristics: 40 Eri B (Koester & Weidemann 1991), Stein 2051 B (Hebert 1976), and Sirius B (Shipman 1979). HST now gives us access to the remaining white dwarfs in such systems, the best known of which is Procyon B.

With Procyon B currently at its largest separation ($\approx$ 5 arcsec), we obtained a series of 16 WFPC2 images of the Procyon system with a wide variety of filters. In the following, our main goal is accurately determining Procyon B's effective temperature. Combining this information with the well known parallax of the system will give us the white dwarf's radius, and place it on the theoretical mass-radius relation. Finally, using the images acquired in narrow bands, we hope to obtain general information about the white dwarf's chemical composition.

## 2. The Data

The basic reduction process converts the counts found in each image to the WFPC2 magnitude system, and finally to fluxes (Holtzman et al. 1995). The details of the data reduction process can be found in Provencal et al. (1996).

The most direct, and accurate, method to obtain Procyon B's effective temperature is to fit the energy distribution with atmosphere models. Because of the large uncertainties associated with the UV and narrow filters (Holtzman et al. 1995), we emphasize the HST equivalent UBVRI filters (F336W, F439W, F569W, F675W and F791W) in our temperature determination, and use the narrow filters (F487N, F502N, F631N, F656N and

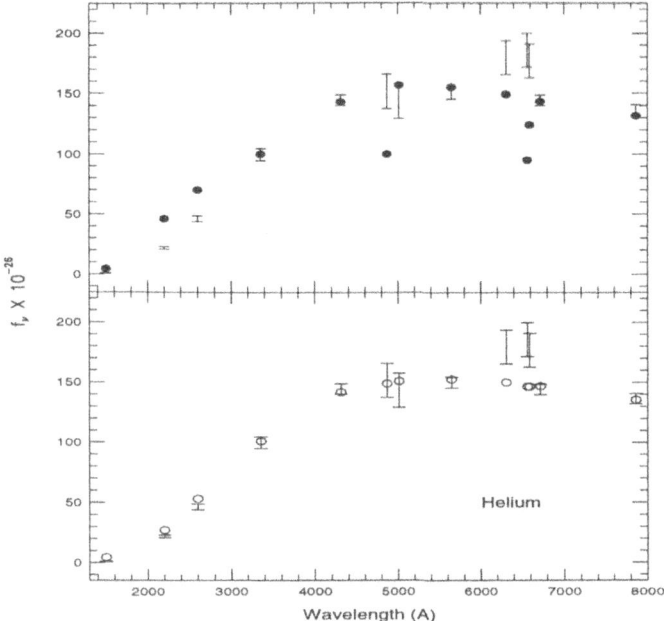

*Figure 1.* The fluxes for Procyon B, including model atmosphere fits to filters F336W, F439W, F569W, F675W, F791W, F487N and F502N. The error bars denote our observations, the hollow circles are helium models, and the filled circles are hydrogen. Both models use log g=8.00.

F658N) to distinguish overall chemical composition. The fits to F336W, F439W, F569W, F657W and F791W for hydrogen and helium models, at log g=8.0, are adequate, and point to rather high effective temperatures of 9184 K (hydrogen) and 8648 K (helium). We added two narrow filters, F487N and F502N, to the fit (Figure 1). F487N, centered on $H_\beta$, and F502N, 30 Å away, contain discriminating information on composition. The hydrogen rich temperature decreases significantly, to 7957 K (log g=8.0), while the temperature based on the helium-rich models is barely affected (8693 K). The $H_\beta$ point is clearly discrepant in the hydrogen fit, and, combined with the self-consistent helium temperatures, argues strongly for a helium-rich composition for Procyon B.

While we believe our temperature of 8650±200 K is the best possible fit for our current data, two problems remain. The flux points from F656N ($H_\alpha$), F631N, and F658N are inconsistent with a helium-rich model containing no Balmer lines, yet the observed flux level in this region is too high to account for the rapidly varying flux level with the Balmer lines of a DA white dwarf. In addition, the measured UV fluxes are inconsistent with either pure hydrogen or helium models. We believe these disagreements

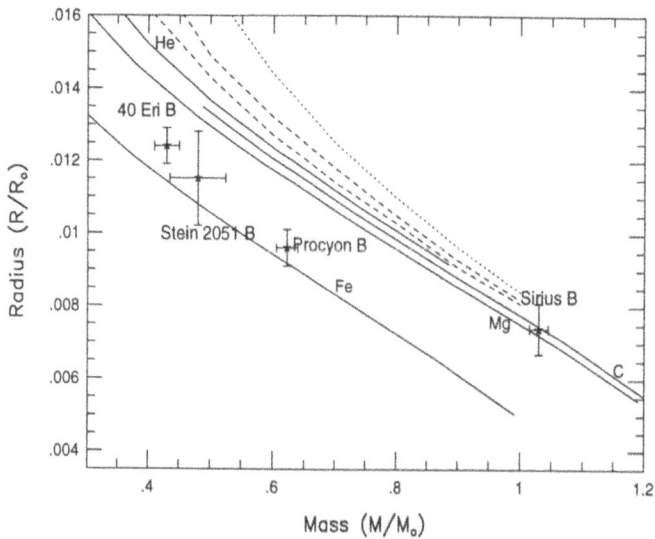

*Figure 2.* The zero-temperature relations of Hamada & Salpeter (1961) (solid lines) and the mass-radius relations of Wood (1996). The positions of Sirius B, Stein 2051 B, 40 Eri B, and our results for Procyon B are given. The error bars indicate $1\sigma$ errors in mass and radius.

have two causes: firstly, our insufficient understanding of HST UV and narrow filter calibrations (Holtzman et al. 1995), and secondly, the inability of current models to accurately describe the far-ultraviolet flux in cool white dwarfs (Koester & Allard 1995).

## 3. The Radius of Procyon B

We now proceed to combine our determination of Procyon B's effective temperature with our knowledge of its apparent magnitude and parallax, to derive the white dwarf's radius. Following the footsteps of Shipman (1979), using

$$f_\nu = 4\pi H_\nu R^2 / D^2$$

where $H_\nu$ is the monochromatic Eddington flux, which we took from the model atmospheres of Bergeron et al. (1995), we arrive at a best radius of $6.27 \pm 0.05 \times 10^8$ cm, or $0.0095 \pm 0.0005 R_\odot$.

Figure 2 displays the zero temperature mass-radius relation of Hamada & Salpeter (1961) for He, C, Mg, and Fe core compositions (solid lines). The dotted line represents a 30,000 K mass-radius relation, and the dashed lines are T=8,000 K and T=15,000 K from Wood (1996). All Wood (1996) relations assume carbon cores, have log q(He) = $-4$, and no hydrogen. Stein 2051 B, Sirius B, and 40 Eri B mark the three objects previously defining the white dwarf mass-radius relation (Shipman & Sass 1980). Our results place Procyon B $3\sigma$ below Hamada & Salpeter and $5\sigma$ below the corresponding Wood (1996) relationship.

We must now consider two distinct possibilities. The first is that the astrometric mass of $0.622 \pm 0.023$ (Girard et al. 1996) for Procyon B is erroneous. But what stellar mass would we need to insure self consistency? Our method of analysis gives us some handle on this problem. Our energy distribution fit, rather than being carried out at log g=8.0, as we have done above, can also be carried out at a surface gravity which is consistent with the fitted solid angle and known distance of Procyon B. If we *assume the validity* of the Wood (1996) mass-radius relationship for carbon, this value of log g comes out to 8.34, corresponding to a stellar mass of 0.78-0.80 $M_\odot$. On the other hand, the astrometric masses determined by Strand (1951), Irwin et al. (1992), and Girard et al. (1996) are too self-consistent to be so far in error.

The second possibility negates the assumption of carbon core composition commonly used for white dwarf stars. Our results for Procyon B are consistent with a heavier core, perhaps as heavy as iron, and hence smaller radius. To narrow the possible core compositions, we would require an improved temperature measurement, which at this point is obtainable only through an actual spectrum of the object. All four objects plotted in Figure 2 are members of binaries, which may or may not affect their individual evolution, perhaps holding a key to a correlation between progenitor mass, white dwarf mass, and core composition.

## References

Bergeron, P., Wesemael, F., & Beauchamp, A. 1995, *PASP*, **107**, 1047

Chandrasekhar, S. 1933, *MNRAS*, **93**, 390

Eggen, O. J., & Greenstein, J. L. 1965, *ApJ*, **141**, 83

Girard, T. M., We, H., Lee, J. T., Dyson, S. E., Horch, E. P., van Altena, W. F., Ftaclas, C., Gilliland, R. L., Schaefer, K. G., & Bond, H. E. 1996, *preprint*

Hamada T., & Salpeter, E. E. 1961, *ApJ*, **405**, 298

Hebert, J. W. 1976, *ApJ*, **210**, 715

Holtzman, J., et al. 1995, *PASP*, **107**, 1065

Irwin, A. W., Fletcher, J. M., Stephenson, L. S. Y., Walker, G. A. H., & Goodenough, C. 1992, *PASP*, **104**, 489

Koester, D., & Allard, N. 1995, *Proceedings of the Ninth European Workshop on White Dwarfs*, ed. D. Koester & K. Werner, (Berlin: Springer-Verlag), 197

Koester, D., & Weidemann, V. 1991, *AJ*, **102**, 1152

Provencal, J. L., Shipman, H. L., Wesemael, F., Bergeron, P., Bond, H. E., Liebert, J., & Sion, E. M. (1996), WFPC2 Photometry of the Bright, Mysterious White Dwarf Procyon B, *ApJ*, submitted

Shipman, H. L. 1979, *ApJ*, **228**, 240

Shipman, H. L., & Sass, C. A. 1980, *ApJ*, **235**, 17
Strand, K. A. 1951, *ApJ*, **113**, 1
Walker, G. A., Walker, A. R., Racine, R., Fletcher, J. M., McClure, R. D. 1994, *PASP*, **106**, 356

## Discussion

*I. Iben*: This is an unfair question - about 40 Eridani B rather than Procyon B. The error bars for the astrometric and spectroscopic estimates of the mass of 40 Eri B do not overlap. Is this understood? Is there any concern that the astrometric estimate of the mass of Procyon B is in error by more than you estimate?

*J. Provencal*: We believe the mass of Procyon B is well known. We have several independent measurements (Girard et al. 1996, Irwin et al. 1992, and Strand 1951) which come up with the same answer. The problem with 40 Eri B is its parallax. There is some question about how well this value is known. Also, the spectroscopic estimates for 40 Eri B assume a mass radius relation, which may or may not be valid.

*H. Shipman*: Most people who have thought about 40 Eri B - many of whom are in this room, have concluded that the best way to fit all the data we have is to assume that the astrometric mass is incorrect, probably because the parallax is incorrect and $M \approx \pi^3$. The parallax of Procyon is much better known than the parallax of 40 Eri B.

*I. King*: The discrepancy in the $H_\beta$ filter disturbs me. I wonder if there is some way to use synthetic photometry, with the known curves of the HST filters, to bootstrap a continuum curve. But what I really want to ask is, why don't you have a spectrum, and when will you get one?

*J. Provencal*: I believe our model atmosphere fits do take into account the HST filter responses. To answer your second question, we are working on it. We recently tried to obtain a spectrum with FOS, but ran into technical difficulties. This is not an easy target, because of the close proximity of Procyon A, and the subsequent danger to FOS. We had to use a very small acquisition aperture, and slew from an offset star. The telescope failed to acquire Procyon B, and probably set up on a diffraction spike. Hopefully we will be successful with STIS.

*S. Kawaler*: Can you comment on how the new astrometric determination of the orbit of Procyon by Girard et al. (1996) will affect your results?

*J. Provencal*: The Girard et al. determination will not affect our results for Procyon B. They find a similar mass for the white dwarf, and don't comment on the parallax of the system. The new mass for Procyon A, however, resolves the long standing conflict between Procyon A's position on the HR diagram and its estimated mass with current evolution theory.

# EXTREME ULTRAVIOLET LIGHT CURVE AND PHASE-RESOLVED SPECTROSCOPY OF THE HOT WHITE DWARF IN V471 TAURI

JEAN DUPUIS, STÉPHANE VENNES AND PIERRE CHAYER
*Center for EUV Astrophysics, 2150 Kittredge St.,*
*University of California, Berkeley, CA 94720-5030, USA*

AND

SCOTT CULLY AND TED RODRIGUEZ-BELL
*Space Sciences Laboratory, University of California, Berkeley,*
*CA 94720-7450, USA*

## 1. Introduction

The close binary V471 Tauri ($P = 12.5$ hr) is one of a few systems predicted to evolve into a cataclysmic variable within a Hubble time (King et al. 1994). The properties of this K0v plus DA white dwarf pair are reviewed in detail in Bois, Lanning, & Mochnacki (1988), Skillman & Patterson (1988), and Ibanoglu et al. (1994). Van Buren, Charles, & Mason (1980) identified V471 Tauri with an X-ray source in the *HEAO-1* survey, and an X-ray flare was possibly detected during an *Einstein* pointed observation (Young et al. 1983). Jensen et al. (1986) observed V471 Tauri with *EXOSAT* and discovered that the extreme ultraviolet (EUV) emission, mostly from the white dwarf, is variable with a period of 555 s. They speculate that the white dwarf is either pulsating or has a "spotted" atmosphere resulting from accretion of a wind from the K dwarf companion. The same period was later detected in optical *U*-band light (Robinson, Clemens, & Hines 1988). A Whole Earth Telescope observation of V471 Tauri (Clemens et al. 1992) revealed an additional period (561 s), explained in terms of re-processed white dwarf EUV radiation in the K dwarf photosphere, which argues against a pulsation origin of the 555 s period.

The "spotted" model postulates that accretion on a magnetic hot DA star leads to an accretion-diffusion steady-state abundance profile of helium and heavier elements over the polar regions of the magnetic field. The

*I. Isern et al. (eds.), White Dwarfs, 375–381.*
© *1997 Kluwer Academic Publishers.*

spots appear darker in the EUV, and the observed variability is explained by rotation of the white dwarf at a period of 555 s. Mullan et al. (1991) report a non-detection of helium and heavy elements in high-dispersion *International Ultraviolet Explorer* (*IUE*) spectra. However, Shipman et al. (1995) report the detection of Si IV in higher quality *Hubble Space Telescope* (*HST*) spectra. We present an analysis of spectroscopic observations of V471 Tauri obtained with the *Extreme Ultraviolet Explorer* (*EUVE*), and we provide definitive evidence in favor of the "spotted" model.

## 2. EUV Spectroscopy and Light Curves

We observed V471 Tauri with *EUVE* between 1994 November 28 and December 2 with an exposure time of 100,000 s. Spectroscopy in three spectral ranges (short wavelength "SW" 70–190 Å, medium wavelength "MW" 140–380 Å, and long wavelength "LW" 280–760 Å) was obtained concurrently with the Deep Survey (DS) broad-band photometry (Lexan/boron band, 60–180 Å). The multichannel plate detectors offer the advantage of retaining the timing information of individual photon events and, therefore, facilitate timing studies and sophisticated data time-filtering. Cully et al. (1996) and a forthcoming publication by Dupuis et al. (1997) describe the observation in greater detail.

Figure 1 shows the SW spectrometer and DS light curves folded on the period of 554.63 s measured by Clemens et al. (1992). The count rate is about 12 times larger in the DS than in the SW, in agreement with the ratio of the effective areas of the two instruments. The pulse shapes of the phased light curves show two unsymmetrical minima separated by half a cycle; a similar shape was observed with *EXOSAT*, and *U*-band optical light curves are in antiphase with EUV light curves. Such properties are intuitively consistent with the type of light curves expected from a rotating star with a spotted atmosphere. A separation of half a cycle between minima suggests the presence of two spots diametrically located on the surface of the white dwarf. We will propose likely locations and sizes for the spots.

Figure 2 shows EUV spectroscopic variations over the 555 s period. The top spectrum is registered between phases 0.9 and 1.0 during the lowest minimum of the light curve; the bottom spectrum combines the 0.2–0.4 and 0.6–0.8 phase intervals (maximum phases). The continuum of the white dwarf short of 230 Å reveals subtle variations consistent with an opacity effect. Note the presence of He II λ304 emission probably originating from the K star. Overall, the spectrum is dominated by the hot white dwarf continuum.

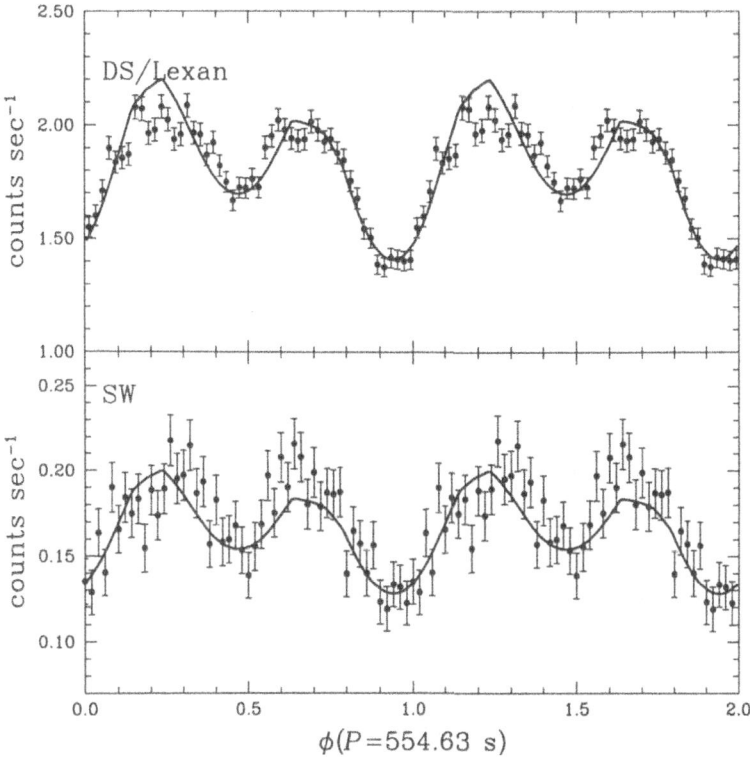

*Figure 1.* Deep Survey and Short Wavelength light curves folded on the period of the white dwarf variation. The dark spots light curves are overplotted (continous lines).

## 3. Light Curve Modeling

As mentionned above, we can extract from the shape of the light curve the location, size, and chemical abundance of the two spots. We adopt an approach used in modeling the light curves of active late-type stars with spotted atmospheres. The formalism of this model is described by Friedemann & Gürtler (1975) and is directly applicable to the case of V471 Tauri. Briefly, the free parameters of the model are the spot size, the location of the spot on the white dwarf surface, the inclination of the rotation axis, and the brightness of the spot, which depends on the composition. A spot appears darker in the EUV for a larger abundance of helium and heavier elements. We assume that the spots share the same effective temperature with the rest of the stellar surface. This assumption is reasonable because the accretion rate must be of the order of $10^{-17}$ $M_\odot$ yr$^{-1}$, and accretion would not produce appreciable heating. The presence of a magnetic field is assumed to channel the accretion flow onto the spots, although there is yet no evidence for a magnetic field. The non-detection of Zeeman splitting in the Lyα profile obtained with *HST* GHRS implies a magnetic field upper limit of about $10^6$ G (Jordan 1996). Mullan et al. (1991) argue that a magnetic field of this intensity would effectively prevent gas accretion from the K-star wind, but that a small, unknown, fraction of the wind material may still reach the surface of the white dwarf.

Figure 1 presents fits to the DS/Lexan and SW light curves (continuous line). We assumed that the rotation axis of the white dwarf is orthogonal to the orbital plane. An acceptable fit is obtained for two circular spots centered on the equator of the rotation axis. The size and brightness of the two spots are significantly different. The spot corresponding to the deepest minimum in the light curve has an angular diameter of 40°, or about 40% of the hemisphere, and has to be almost completely dark in the EUV (spot A). The second spot, with only half the depth of the first one, is slightly more extended with an angular diameter of 50° (about 60% of the hemisphere surface area) and shows less contrast with the pure hydrogen portion of the stellar surface (spot B). We constrain the spot compositions more precisely in the next section.

## 4. Phase-Resolved Spectra and Spot Composition

An unknown ingredient of the dark spot model is the source of opacity. Previous attempts to determine the spot composition were limited by the rather poor quality of the *IUE* high dispersion spectra of V471 Tauri (see Mullan et al. 1991). Even recent *HST* spectra provide little information (Shipman et al. 1995). In the present investigation, we take advantage of the sensitivity of the EUV spectrum of V471 Tauri to the presence of trace amounts of photospheric helium, carbon, nitrogen, and oxygen. We use the spot geometry inferred from the light curve to generate phase-dependent spectra. Linear combinations of pure hydrogen synthetic spectra and hydrogen-dominated synthetic spectra with an admixture of He/C/N/O are computed to model the total flux emerging from the spotted surface. The contribution of the spot spectrum is weighted according to the fraction of the surface area occupied on the hemisphere seen by the observer. Limb darkening is neglected. The phases closest in appearance to a pure hydrogen hemisphere are best modeled with $T_{eff}$ of 32000 K, $\log g$ of 8.5, and ISM hydrogen column density of $1.5 \times 10^{18}$ cm$^{-2}$.

The composition of the accreted material is unknown and one expects a departure from the solar composition because accretion is possibly selective (Mullan et al. 1991). Because we assumed solar proportions, helium dominates the opacity in the spots. The phase-resolved spectra (Fig. 2) are too noisy to allow detections of individual spectral lines, but we clearly observe

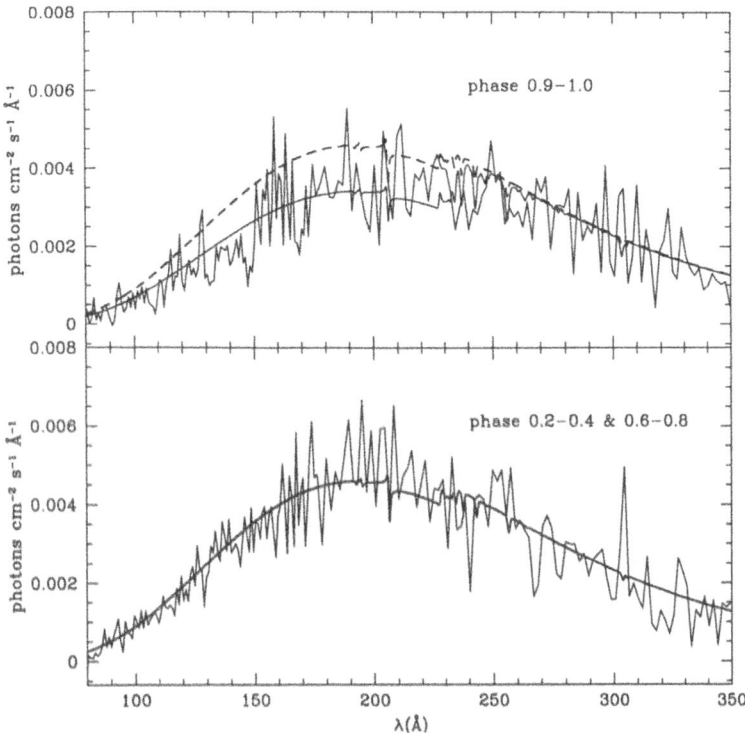

*Figure 2.* Phase-resolved spectra of V471 Tauri. *Top:* flux miminimum. *Bottom:* flux maximum. Synthetic spectra accounting for the presence of spots are shown (continuous lines).

a depression of the continuum below the He II edge at 228 Å, suggesting the presence of helium in the spots. The composition that best matches the phase-resolved spectra is $10^{-3} \times$ (solar abundance of He, C, N, and O) for spot A, and $10^{-4} \times$ (solar abundance of He, C, N, and O) for spot B. A higher abundance cannot be reconciled with the data for the darkest spot (A) because a continuum depression would be observed over the entire EUV range. The continuum level above 228 Å does not vary, and a folded light curve of the LW spectrum shows no variations either (Dupuis et al. 1997). Note also the absence of the He II $\lambda 304$ emission line in the top spectrum of Fig. 2; the emission line may be partially cancelled by the photospheric absorption line, which is predicted to be strongest at this phase.

## 5. Discussion and Conclusions

We may constrain the effective accretion rate on the spots with the help of diffusion calculations. Time-dependent simulations using an approach described in Dupuis et al. (1993; see also Chayer, Fontaine, & Pelletier in these proceedings) demonstrate that a total accretion rate of $10^{-17}$ $M_\odot$ yr$^{-1}$ is sufficient to produce the measured abundance on spot A ($10^{-18}$ $M_\odot$ yr$^{-1}$ for spot B). These very low accretion rates agree with Mullan et al.'s (1991) results. Our measurements suggest that the accretion flow is screened-out considering that the Hoyle-Bondi accretion rate may be as high as $10^{-13}$ $M_\odot$ yr$^{-1}$. Our estimates of the accretion rate are about 5 orders of magnitude smaller than Barstow et al.'s (1992) estimate, which assumed that $U$-band variability originates mainly from accretion heating. We have verified in our model spectra that we can explain the U-band light curve with a flux redistribution effect; an accretion rate of the order of $10^{-13}$ $M_\odot$ yr$^{-1}$ would dramatically alter the chemical composition of the spots, an effect not observed.

*EUVE* observations of V471 Tauri support a "spotted" model of the white dwarf surface that explains the 555 s variability observed in the soft X-ray, EUV, and optical light curves. We have successfully reproduced the bipulsed light curves with a model assuming the presence of circular dark spots on the white dwarf surface coupled with a white dwarf rotation period of 555 s. We find the spots located along the equator of the axis of rotation and covering 40% and 60% of the surface with $10^{-3} - 10^{-4}$ of the solar abundances of He+CNO. We do not understand the observed asymmetry of the spot configuration.

This research is supported by NASA grant NAG5-3424. J.D. travel to the meeting was supported by NASA grant NCC5-138.

## References

Barstow, M. A., et al. 1992, MNRAS, 255, 369

Bois, B., Lanning, H. M., & Mochnacki, S. W. 1988, AJ, 96, 157

Clemens, J. C. et al. 1992, ApJ, 391, 773

Cully, S. L., Dupuis, J., Rodriguez-Bell, T., Basri, G., Siegmund, O. H. W., Lim, J., & White, S. M. 1996, in *Astrophysics in the Extreme Ultraviolet*, eds. S. Bowyer & R. F. Malina, (Kluwer: Dordrecht), 349

Dupuis, J., Cully, S.L, Rodriguez-Bell, T., & Vennes, S. 1997, in preparation

Dupuis, J., Fontaine, G., Pelletier, C., & Wesemael, F. 1993, ApJS, 84, 73

Friedemann, C., & Gürtler, J. 1975, Astron. Nachr. Bd., 296, 125

Ibanoglu, C., Keskin, V., Akon, M. C., Evren, S., & Tunca, Z. 1994, A&A, 281, 811

Jensen, K. A., Swank, J. H., Petre, R., Guinan, E. F., Sion, E. M., & Shipman, H. L. 1986, ApJ, 309, L27

Jordan, S. 1996, private communication

King, A. R., Kolb, U., de Kool, M., & Ritter, H. 1994, MNRAS, 269, 907

Mullan, D., Shipman, H. L., Sion, E. M., & MacDonald, J. 1991, ApJ, 374, 707

Robinson, E. L., Clemens, J. C., & Hine, B. P. 1988, ApJ, 331, L29

Shipman, H. L. et al. 1995, AJ, 109, 1220

Skillman, D. R. & Patterson, J. 1988, AJ, 96, 976

Van Buren, D., Charles, P. A., & Mason, K. O. 1980, ApJ, 242, L105

Young, A., Klimke, A., Africano, J. L., Quigley, R., Radick, R. R., & Van Buren, D., 1983, ApJ, 267, 655

# Discussion

*B. Gänsicke*: Is the accretion rate you derive for the spots ($10^{-17}$ $M_\odot$ yr$^{-1}$) enough to produce the observed X-rays?

*J. Dupuis*: No. The X-ray emission is explained by a combination of coronal emission from the active K dwarf and photospheric emission from the white dwarf.

*M. Barstow*: Have you looked at how the 304 Å emission line varies with orbital phase? Might its disappearance be related to that effect.

*J. Dupuis*: Yes, I have produced a light curve of the He II $\lambda$304 emission line folded on the 12.5 hour orbital period. The line intensity is modulated with the orbital phase in a manner such that it disappears during the eclipse of the white dwarf. It also reaches maximum intensity when the K dwarf is at inferior conjunction. The disappearance of the emission line occurs at a phase of the white dwarf rotation that does not always coincide with the WD eclipse.

*I. King*: I appreciate all this enthusiasm in comparing the optical with the EUV, but you should really first update the ephemeris. Photometric ephemerides are notoriously unreliable when they are extrapolated.

*J. Dupuis*: Concerning the extrapolation, I am using an ephemeris presented at the Phoenix AAS meeting in 1992 by Clemens, Winget, & Kleinman. Their ephemeris is precise enough to be extrapolated to the epoch of the *EUVE* observation in December 1994. Ideally, we would have liked to obtain rapid photometry in the optical contemporaneous to the *EUVE* observation but were unable to do so.

*Y. Wu*: Question: In one of your transparencies, a magnetic field of 10 kG is written down ... Do you include the effect of magnetic pressure on thermal pressure (thus causing a cooler/darker spot as on the Sun)?

*J. Dupuis*: No, we have assumed that the magnetic forces have a negligible effect on the thermal and mechanical structure of the atmosphere.

*D. O'Donoghue* (1) Is the discrepancy of your light curve with respect to the optical light curve an indication of a changing rotation period? (2) Have you compared your data with the *HST*/Optical studies of AE Aqr?

*J. Dupuis*: (1) No, I don't expect the rotation period to vary significantly over only a few years but that is an interesting possibility. (2) I haven't made a detailed comparison yet with the case of the magnetic cataclysmic variable AE Aqr. However, the two systems exhibit interesting similarities and it might be the case that V471 Tauri will eventually evolve in a configuration similar to that of AE Aqr.

# UNIQUE WHITE DWARFS ACCOMPANYING RECYCLED PULSARS

M. H. VAN KERKWIJK
*Palomar Observatory*
*California Institute of Technology, mail stop 105-24*
*Pasadena, CA 91125, USA*

## 1. Introduction and Summary

Knowledge of the properties of the white-dwarf companions of radio pulsars can provide unique constraints on the characteristics and evolution of these binaries, as well as on those of its constituents. Many white-dwarf properties can be determined accurately from their spectra, but for the very faint pulsar companions spectroscopy has only become feasible with the advent of large telescopes like the Keck. Here, I introduce the two classes of pulsar, white-dwarf binaries, and describe for each what we have learned from a specific system, PSR J1012+5307 and PSR B0655+64, respectively, summarising what has been done (Van Kerkwijk & Kulkarni 1995; Van Kerkwijk, Bergeron, & Kulkarni 1996; hereafter Papers I & II), presenting new results, and discussing what the future may hold.

Briefly, for the companion of PSR J1012+5307 we find a DA spectrum, and infer a mass of $\sim 0.16\ M_\odot$, the lowest among all spectroscopically identified white dwarfs. Combined with a radial-velocity orbit, a neutron-star mass between 1.5 and 3.2 $M_\odot$ (95% confidence) is derived. The companion of PSR B0655+64 shows strong Swan $C_2$ bands, i.e., it is a DQ star. Unlike anything reported for other DQs, however, it shows variations in strength of the bands by a factor two. Most likely, the variations are periodic, with a period of $\sim 9.7$ h. This is substantially shorter than the 1 day orbital period, which can likely be understood in terms of its past evolution.

## 2. PSR J1012+5307

PSR J1012+5307 is a member of the largest group of binary pulsars, those with low-mass helium white dwarf companions. These systems have pre-

*I. Isern et al. (eds.), White Dwarfs*, 383–389.
© *1997 Kluwer Academic Publishers.*

sumably descended from the low-mass X-ray binaries, in which mass is transferred onto a neutron star from a less massive companion. The mass transfer in these systems is stable and relatively well understood, and a number of predictions can be made (for a review of binary evolution involving neutron stars, see Bhattacharya & Van den Heuvel 1991). First, the neutron star will likely have accreted a substantial fraction, if not all, of the up to 0.7 $M_\odot$ lost by the companion (e.g., Van den Heuvel & Bitzaraki 1995). Hence, one expects the neutron star to have increased substantially in mass, and to be spun up[1]. For reasons not quite understood, the magnetic field seems to decay at the same time, perhaps by being quite literally buried (Romani 1990). If the neutron stars indeed have masses increased to $\gtrsim 2\,M_\odot$ (assuming they started with the "canonical" 1.4 $M_\odot$; for a recent census, see Van Kerkwijk et al. 1995), this would be very interesting, as it would strongly constrain the equation of state (EOS) at supra-nuclear densities (e.g., Cook, Shapiro, & Teukolsky 1994): for softer EOS, like the one recently proposed by Brown & Bethe (1994), such a massive neutron star would collapse into a black hole. The only system for which a neutron-star mass estimate is available, is PSR B1855+09, for which Kaspi, Taylor, & Ryba (1994) found $M_{\rm NS} = 1.50^{+0.26}_{-0.14}\,M_\odot$ (68% confidence). As yet, the uncertainty is too high to allow one to draw a strong conclusion.

For the white dwarf, one predicts that it will have a helium core, since the companion never reached helium ignition. Furthermore, for the systems with orbital periods $\gtrsim 10$ d, there should be a relation between the orbital period and the white-dwarf mass (as a consequence of the core-mass, radius relation for giants; Refsdal & Weigert 1971; most recently for pulsar binaries, Rappaport et al. 1995), as well as a statistical relation between the orbital period and the eccentricity (Phinney 1992). The latter depends only on the assumption of having a Roche-lobe filling giant with a convective envelope at the end of the evolution, and has been confirmed (ibid.). The orbital-period, mass relation has only been verified at the short-period end, using (again) PSR B1855+09, for which Kaspi et al. (1994) determined an accurate companion mass from Shapiro delay in the pulse arrival times. At the long period end, we hope to obtain an additional constraint using PSR B0820+02, which has a DA companion (Paper I).

PSR J1012+5307 is a recently discovered 5.26 ms pulsar, which is in a 0.60 d orbit with a very low-mass companion (Nicastro et al. 1995). This system was deemed especially interesting as, given its small orbital period, it seemed possible to determine the radial-velocity amplitude and thus the mass ratio. Combined with the white-dwarf mass, this would give the mass of the neutron star.

---

[1]These pulsars are called "recycled" because the spin-up process allows the radio-pulsar mechanism to work again after the mass transfer ceases.

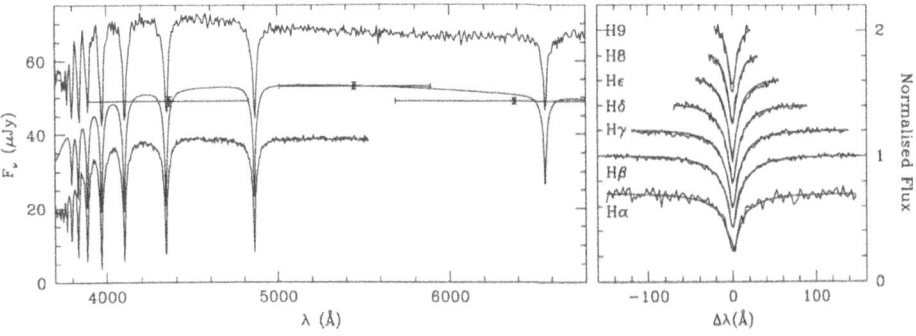

*Figure 1.* The spectrum of the white-dwarf companion of PSR J1012+5307 (taken from Paper II). Shown in the left-hand panel are a 10 Å resolution classification spectrum (top curve; offset by 15 $\mu$Jy) and the average of eight 4 Å resolution spectra (bottom curve; offset by $-15$ $\mu$Jy). Also shown are the broad-band fluxes of Lorimer et al. (1995) and the best-fit pure Hydrogen model spectrum. The latter was derived from a fit to the profiles of H$\beta$ up to H8, and has $T_{\text{eff}} = 8550 \pm 25$ K and $\log g = 6.75 \pm 0.07$. In the right-hand panel, the observed line profiles, including those of H$\alpha$ and H9, are shown with the modeled ones superposed.

Fortunately, the optical counterpart has $V = 19.6$ (Lorimer et al. 1995), making it the brightest pulsar companion currently known. Partly, it is so bright because it is relatively hot, with a colour-temperature of $T_{\text{BB}} \simeq 9400$ K. This indicates a cooling age of only a couple $10^8$ yr, much shorter than the pulsar spin-down age of $7\,10^9$ yr. Since the pulsar presumably started spinning down at the time the white dwarf was formed, this is puzzling. It probably indicates that the pulsar did not start spinning at much shorter periods, as implicitly assumed in calculating the spin-down age, and as would have been thought based on simplistic models for the mass transfer. It could also be, however, that these low-mass white-dwarf have some residual hydrogen burning for quite a while after losing the red-giant envelope, and that the cooling-age estimate is wrong (Alberts et al. 1996).

We found that the companion was a DA star, showing H$\alpha$ up to H12 (Paper II; see Fig. 1). From a model-atmosphere fit (Fig. 1), we find $T_{\text{eff}} = 8550 \pm 25$ K and $\log g = 6.75 \pm 0.07$ (cgs units). To infer the mass, we need a mass-radius relation. Unfortunately, for these very low-mass helium white dwarfs, this is not well known. Using the Hamada-Salpeter zero-temperature relation, with an approximate finite-temperature correction based on models of Wood (1995), we find $M_{\text{WD}} = 0.16 \pm 0.02$ $M_\odot$, the lowest among all spectroscopically identified white dwarfs.

We also measured radial velocities, and found a radial-velocity amplitude $K_{\text{WD}} = 280 \pm 15$ km s$^{-1}$, leading to a mass ratio $M_{\text{NS}}/M_{\text{WD}} = 13.3 \pm 0.7$. Combined with the white-dwarf mass and the pulsar mass function, we infer that with 95% confidence $1.5 < M_{NS}/M_\odot < 3.2$ (Paper II).

This determination is not yet accurate enough to constrain the equation of state, or to test evolutionary theory, but it does show that further study may well prove fruitful. It will be relatively straightforward to improve the accuracy of the radial-velocity amplitude and thus the mass ratio, which might already lead to an interesting constraint on the mass of the neutron star. It will be less easy to improve the estimate of the white-dwarf mass, because of the uncertainties in the mass-radius relation for these very low-mass white dwarfs, as well as the possible presence of helium in the atmosphere. If helium is present, the true surface gravity—and thus the inferred mass—will be lower (Bergeron, Wesemael, & Fontaine 1991; Reid 1996 for an observational indication).

The pulsar is relatively nearby and bright, however, and it may well be possible to derive an accurate distance using radio VLBI or timing. This would allow one to obtain a direct estimate of the radius. If this is the same as the predicted one ($0.028 \pm 0.002\ R_\odot$), it would give confidence in the result. If it is not, one can either assume there is a problem with the mass-radius relation, but not with helium pollution, and infer a mass from the radius in combination with the observed surface gravity, or one can assume that there is helium pollution, but that the mass-radius relation is fine, and use that to derive a mass from the radius. Another possibility is to search carefully for Shapiro delay in the pulse arrival times, which would give a constraint on a combination of the white-dwarf mass and the inclination.

## 3. PSR B0655+64

The companion of PSR B0655+64 has a mass $> 0.67\ M_\odot$ (for a 1.4 $M_\odot$ pulsar), and thus it must have been a relatively massive star. Most likely, the evolution has been similar to that leading to double neutron-star binaries like the Hulse-Taylor pulsar, with the system going through a phase as a wide high-mass X-ray binary, followed by spiral-in during a common-envelope phase, leading to its current 1.03 d orbital period (e.g., Bhattacharya & Van den Heuvel 1991). Presumably, the helium core left was not massive enough to form a second neutron star. Since common-envelope evolution is rather poorly understood, there are few predictions for these systems, except that one expects a carbon-oxygen white dwarf. Most likely, it will have a helium atmosphere, since all the hydrogen left after the spiral-in will probably have disappeared during a second stage of mass transfer when the star became a helium giant (this is expected for low-mass helium stars; Paczynski 1971; Habets 1986).

The companion was identified by Kulkarni (1986). It has $V = 22.2$ and

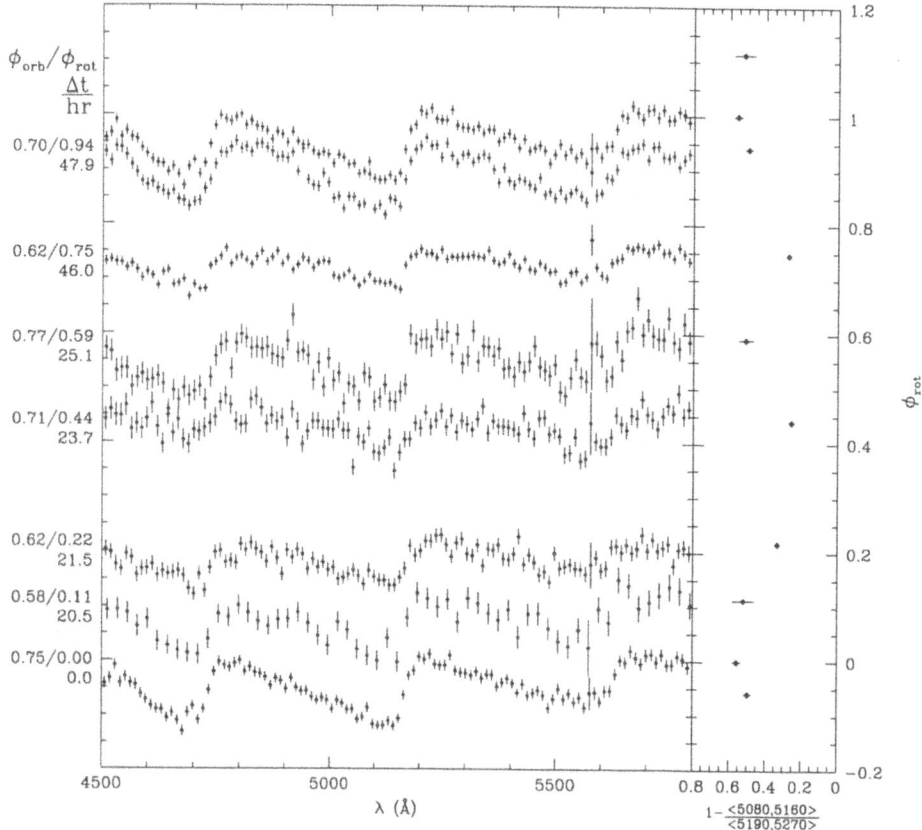

*Figure 2.* Spectra of the companion of PSR B0655+64. In the right-hand panel, a measure of the strength of the Swan bands is given. The numbers on the left indicate the time in hours since the first observation, the orbital phase, and the rotation phase for a period of 9.7 h.

$V - R = 0.1$, indicating a temperature of about 6000 to 9000 K. First Keck spectra showed strong $C_2$ Swan bands (Paper I; see also Fig. 2), i.e., it is a DQ star, with a helium atmosphere sufficiently shallow and convective to allow trace amounts of carbon to be dredged up (Pelletier et al. 1986).

Uniquely among DQ stars, a large variation in the strength of the Swan bands was observed, by about a factor two in less than two hours. In Paper I, this was interpreted as due to brighter and darker spots on the white-dwarf surface, possibly related to the presence of a magnetic field (a locally higher magnetic field strength might lead to a change in gas pressure and thus temperature, or in convective efficiency). Based on the speed and amplitude of the variation, it was shown that if it was periodic, the period had to be $\gtrsim 3$ and $\lesssim 12$ h. Thus, the modulation could not be orbital, but was most likely due to the white dwarf rotation. Spectra obtained in November 1995 confirm this (Fig. 2), and a period of $9.7 \pm 0.1$ h is indicated.

In Paper I, it was noted that if the star was rotating synchronously with the orbit when it was a Roche-lobe filling helium giant, it would have been spun up due to conservation of angular momentum when it shrunk to form a white dwarf. If the spin-up is mostly due to the angular momentum contained in the remaining giant envelope, the envelope mass must have been $\sim 10^{-4.5}$ $M_\odot$, interestingly similar to the typical helium-layer masses inferred for DQ stars (Pelletier et al. 1986; Weidemann & Koester 1995; Dehner & Kawaler 1995).

## Acknowledgements

I thank Yanqin Wu, Shri Kulkarni, and Pierre Bergeron for useful discussions, and acknowledge support from a NASA Hubble Fellowship.

## References

Alberts, F., Savonije, G. J., Pols, O. R., & van den Heuvel, E. P. J. 1996, Nature, 380, 676

Bergeron, P., Wesemael, F., & Fontaine, G. 1991, ApJ, 367, 253

Bhattacharya, D., & van den Heuvel, E. P. J. 1991, Phys. Rep., 203, 1

Brown, G. E., & Bethe, H. A. 1994, ApJ, 423, 659

Cook, G. B., Shapiro, S. L., & Teukolsky, S. A. 1994, ApJ, 424, 823

Dehner, B. T., Kawaler, S. D. 1995, ApJ, 445, L141

Habets, G. M. H. J. 1986, A&A, 165, 95

Hamada, T., & Salpeter, E. E. 1961, ApJ, 134, 683

Kaspi, V. M., Taylor, J. H., & Ryba, M. F. 1994, ApJ, 428, 713

Kulkarni, S. R. 1986, ApJ, 306, L85

Lorimer, D. R., Lyne, A. G., Festin, L., & Nicastro, L. 1995, Nature, 376, 393

Nicastro, L., Lyne, A. G., Lorimer, D. R., Harrison, P. A., Bailes, M., & Skidmore, B. D. 1995, MNRAS, 273, L68

Paczynski, B. 1971, Acta Astron., 21, 1

Pelletier, C., Fontaine, G., Wesemael, F., Michaud, G., & Wegner, G. 1986, ApJ, 307, 242

Phinney, E. S. 1992, Phil. Trans. R. Soc. Lond., 341, 39

Rappaport, S., Podsiadlowski, Ph., Joss, P. C., Di Stefano, R., & Han, Z. 1995, MNRAS, 273, 731

Refsdal, S., & Weigert, A. 1971, A&A, 13, 367

Reid, I. N. 1996, AJ, 111, 2000

Romani, R.W., 1990, Nature, 347, 741

Van den Heuvel, E. P. J., & Bitzaraki, O. 1995, A&A, 297, L41

Van Kerkwijk, M. H., Kulkarni, S. R. 1995, ApJ, 454, L141 (Paper I)

Van Kerkwijk, M. H., Bergeron, P., & Kulkarni, S. R. 1996, ApJ, 467, L89 (Paper II)

Van Kerkwijk, M. H., van Paradijs, J., & Zuiderwijk, E. J. 1995, A&A, 303, 497

Weidemann, V., & Koester D. 1995, A&A, 297, 216

Wood, M. A. 1995, in 9th European Workshop on White Dwarfs, NATO ASI Series, ed. D. Koester & K. Werner (Berlin: Springer), 41

## Discussion

*M.T. Ruiz*: What is the magnitude of the magnetic field that you would need to explain the variability of the $C_2$ Swan bands? Do you expect to detect a blue shift of the bands due to the magnetic field?

*M. Van Kerkwijk*: A magnetic field of order the equipartition value should suffice. This would likely not result in observable shifts of the bands (the data shows no evidence for shifts, although it is hard to derive strong constraints). From Dr. Bues, I understand that in some other DQ stars, the magnetic field is seen to cause shifts, but not changes in strength, so possibly the magnetic field is not directly responsible for the changes in strength (it might still influence the dredge-up process, which could also lead to surface inhomogeneities).

*I. Iben*: You have assumed that, in the case of millisecond pulsars with helium WD companions, all of the envelope of the red giant donor in the precursor system is accreted by the NS. However, accretion of only about 10% of the envelope mass at Keplerian velocities is sufficient to spin the NS up to observed periods. It is quite possible that an intense wind from the red giant, occasioned by absorption of X rays from the accretion disk of the NS and formation of a corona, could cause most of the red giant envelope to be lost from the system rather than accreted (which may be from the wind of the red giant; it need not fill its Roche lobe).

*M. Van Kerkwijk:* I agree that the evolution may well be more complex than I indicated. Indeed, a larger total mass-loss rate from the giant might help to understand the discrepancy between the LMXB and LMBP birth rates. I hope that with good NS masses we will be able to determine observationally how much of the mass in the envelope was accreted.

# MERGING OF DIFFERENT MASS RATIO WHITE DWARFS: FIRST MOMENTS OF THE 0.4-1.2 $M_\odot$ CASE

J. GUERRERO AND J. ISERN
*Centre d'Estudis Avançats de Blanes*
*Camí de Sta. Bàrbara s/n, Blanes (Girona) 17300, Spain*

W. BENZ
*Steward Observatory*
*University of Arizona, Tucson, AZ 85721, USA*

E. GARCIA-BERRO
*Departament de Física Aplicada*
*Mòdul B4/B5, Campus Nord UPC, Barcelona 08034, Spain*

AND

R. MOCHKOVITCH
*Institut d'Astrophysique*
*98 BIS, Boulevard Arago, 75014 Paris, France*

**Abstract.** This is the first of a group of high resolution SPH simulations meant to test the limit mass ratio in double degenerate systems for which the merging, instead of being catastrophic owing to the expansion of the secondary, is stable because of its orbital momentum gain.

## 1. Introduction

There has been some work done in the last years about the result of the merging of two white dwarfs (or two compact objects, in general) (Benz *et al.*, 1989; Cameron and Iben, 1986; Mochkovitch and Livio, 1989; Mochkovitch and Livio, 1990; Mochkovitch *et al.*, 1995). One of the questions yet to be solved is how the accretion depends on the mass ratio of the dwarfs: it is clear that due to the emission of gravitational waves, every binary system will come eventually so close that one of them will overfill its Roche lobe and merging will occur (though only close binary systems can do it in a time smaller than the Hubble time), yet, once the merging

391

*I. Isern et al. (eds.), White Dwarfs, 391–394.*
© 1997 *Kluwer Academic Publishers.*

begins, it is not clear what will happen: on one side, the less massive star will lose mass and therefore will expand, speeding the merging process; on the other hand, the secondary may gain enough angular momentum in the process to move away of the primary and stop, or at least slow down the merging (Benz *et al.*, 1990). The results obtained by Benz et al. (Benz *et al.*, 1990) indicated that there was a critical mass ratio above which the merging is always unstable, but, up to now, there has not been a systematic computation of mergings at different mass ratios to prove it. In this work we present the results obtained for the initial moments of one of a series of those mergings.

## 2.  Technical data

The code used for this computation is a SPH code, with gravity computed through a binary tree (the lagrangian nature of SPH, as well as its simplicity, makes this technique specially suited for hydrodynamical 3-D calculations). The kernel we have used is the well-known spline kernel of Monaghan and Lattanzio (Monaghan and Lattanzio, 1985). Density is calculated through kernel superposition, as usual in SPH, velocities are calculated using the SPH version of the equations of hydrodynamics, and temperature is computed using the first $TdS$ equation. The artificial viscosity used is Balsara's viscosity, and the equation of state includes non zero temperature and coulombian corrections, as well as radiation pressure. Nuclear burning is allowed, though it has not been used in this calculation. The initial model consists of two white dwarfs, one of 0.4 $M_\odot$ and the other one of 1.2 $M_\odot$ (this is the most extreme case expected, though a very unlikely one). Since the primary is much more massive than the secondary we do not expect it to undergo significant changes, and it has been modelled with 4000 SPH particles, while the secondary, likely to be heavily disrupted, is modelled with 20000 SPH particles (so, there is a large difference between the masses of the particles of the two white dwarfs, and one must be careful on interpreting the point density of the plots presented directly as mass density).

## 3.  Discussion

The simulation has been carried to the point in which an accretion disk develops, but is not yet fully formed. At this stage is not possible yet to ascertain the fraction of the secondary mass that will be accreted, but some interesting features can be seen: first of all, we can see that the bulge of the infalling material lays itself in a high velocity rotating protodisk at some distance of the surface of the primary; secondly, we can see some particles in the exterior layers of the secondary which are developing radial outwards velocities, so it is likely that a large, and quickly rotating disk develops. The amount of angular momentum transferred to the primary by this disk will probably determine the outcome of the merging. Pictures show the evolution of the system. Points correspond to single particles.

**Acknowledgements:** This work has been supported by a FI grant from the *Comissionat per a Universitats i Recerca de la Generalitat de Catalunya*, by CICYT Grants PB94-0111, PB94-0827-C02-02, by the AIHF 237-B, and by the $C^4$ Consortium.

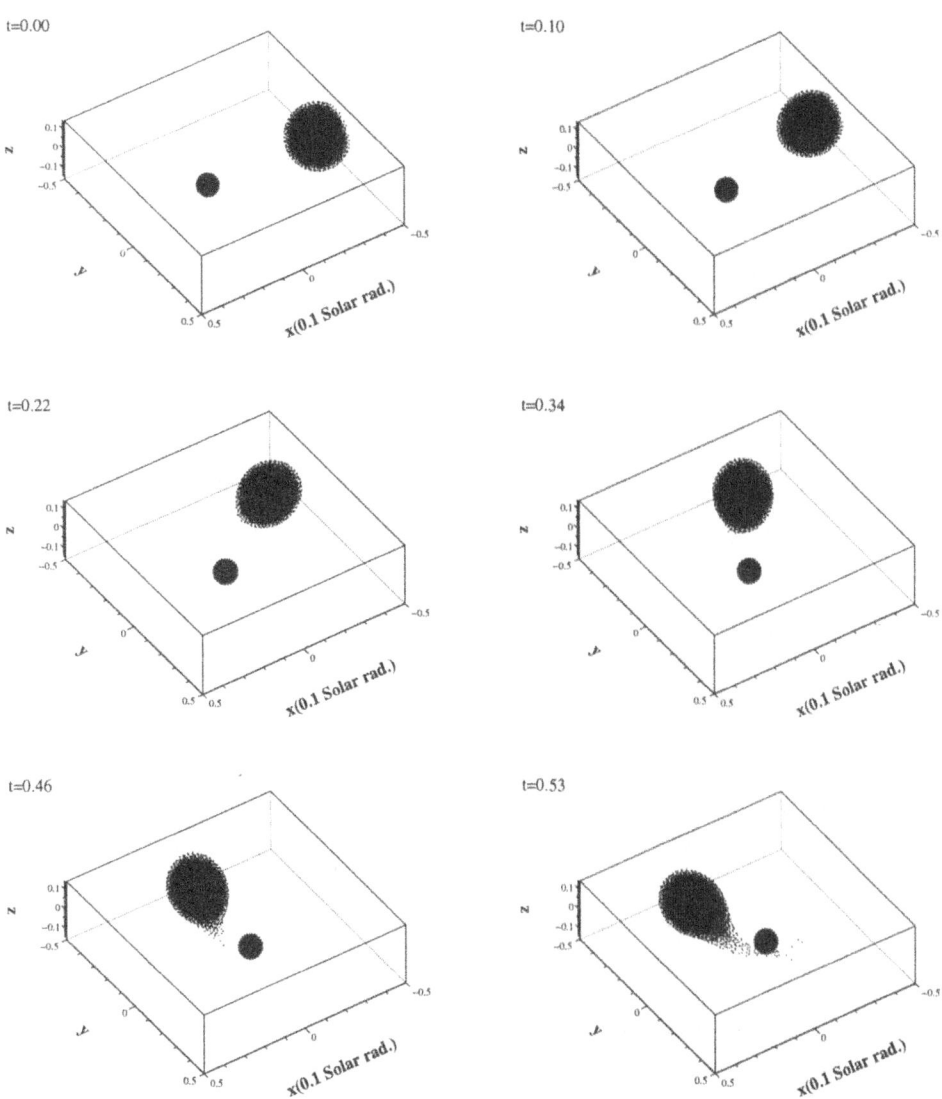

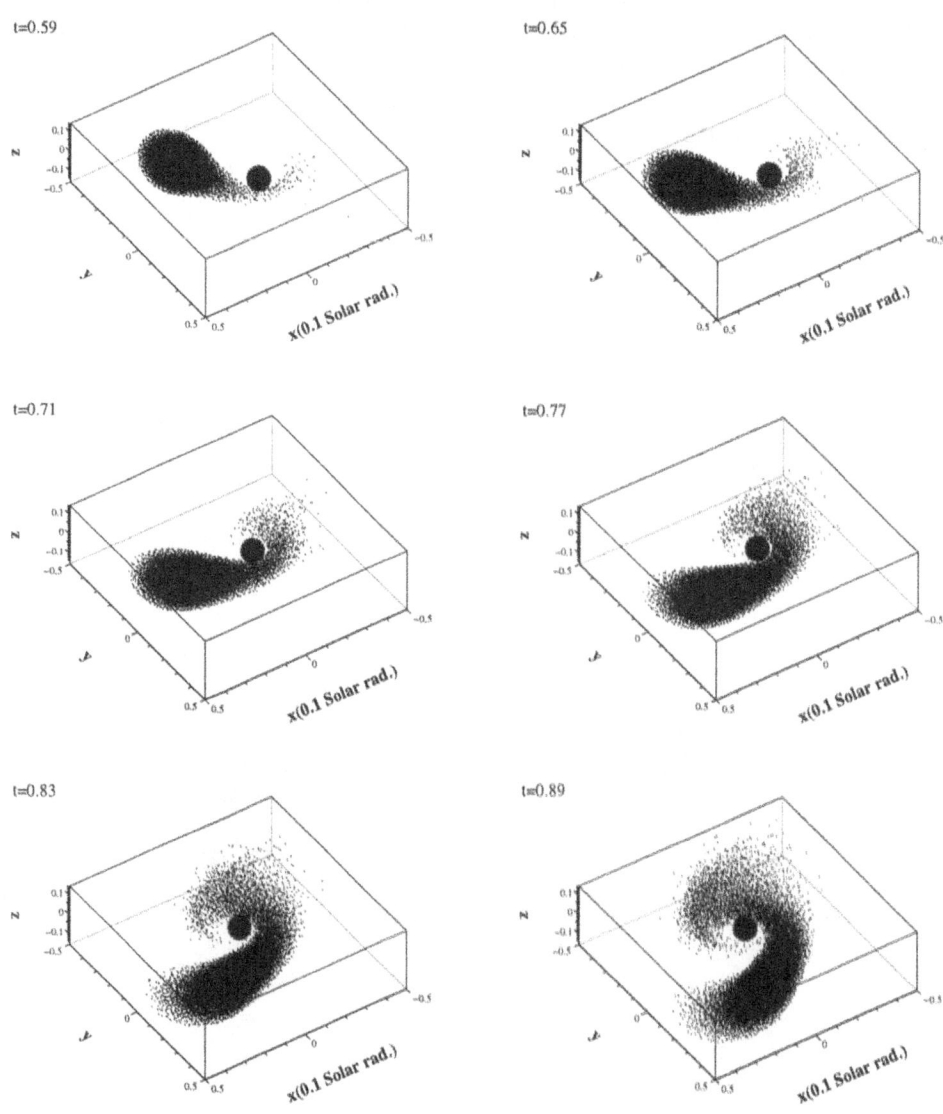

## References

Benz, W., Hills, J.G., Thielemann, F.-K., 1989 *ApJ.* **342** 986

Benz, W., Bowers, R.L., Cameron, A.G.W., Press, W.H., 1990 *ApJ.* **348** 647

Cameron, A.G.W., Iben, I. Jr., 1986 *ApJ* **305** 228

Mochkovitch, R., Livio, M., 1989 *Astr. Ap.* **209** 111

Mochkovitch, R., Livio, M., 1990 *Astr. Ap.* **236** 378

Mochkovitch, R., Guerrero, J., Segretain, L., 1995 *ASI Workshop on Thermonuclear Supernovae*, Aiguablava, in press

J. Monaghan, J., Lattanzio, J.C., 1985 *Astr. Ap.* **149** 135

# Section V:
# Magnetic White Dwarfs

# NEW RESULTS OF MAGNETIC WHITE DWARF SPECTROSCOPY

S. JORDAN

*Institut für Astronomie und Astrophysik*
*D-24098 Kiel, Germany, jordan@astrophysik.uni-kiel.de*

## 1. Introduction

In about 50 (2%) of the 2100 known white dwarfs (McCook & Sion 1996) magnetic fields have been detected with fields ranging from about 40 kG up to 1 GG. Table 1 lists all currently known magnetic white dwarfs and is an update of a similar table with 42 objects published by Schmidt & Smith (1995). Since most surveys that look for magnetic fields in white dwarfs (e.g. Putney 1996, and in these proceedings) are limited to the brighter known white dwarfs and fields below a few MG are not easily recognized in the fainter stars, it is possible that the percentage of magnetic stars is somewhat higher than 2%. However, since the stronger fields are easier to detect and since no rapid increase in number of magnetic white dwarfs with declining field strength has been found, it sounds reasonable to assume that not more than 4% of all white dwarfs are magnetic. This is also supported by the fact that there is also a selection effect in favor of magnetic white dwarfs. In case of the objects discovered by the HE survey, follow-up spectroscopy has been performed because of their peculiar objective prism spectra (e.g. Reimers et al. 1996). Therefore, the number statistics is still consistent with the assumption that Ap stars are the progenitors of magnetic white dwarfs, in which the field strengths are enhanced by magnetic flux conservation during the evolution.

The goal of magnetic white dwarf spectroscopy is to determine the field strength, the detailed geometry of the magnetic field, and the rotational period of the star (which is very difficult to measure in non-magnetic white dwarfs). This is only possible with the help of models for the radiative transfer through a magnetized stellar atmosphere (see e.g. Jordan 1992). The results provide important constraints for the theory of the origin of magnetic white dwarfs.

*I. Isern et al. (eds.), White Dwarfs, 397–403.*
© 1997 *Kluwer Academic Publishers.*

TABLE 1.  Table of the 50 known magnetic white dwarfs

| WD | V | T (K) | Spec. Feat. | Rot. Per. | B/MG | Note/Ref. |
|---|---|---|---|---|---|---|
| G234−4 | 15.09 | 4500 | H | | ≈0.04 | 1 |
| LHS 1038 | 14.36 | 6400 | H | 2–20h | ≈0.09 | 2 |
| LP 907−037 | 14.55 | 9500 | H | | ≈0.1 | 2 |
| LB 8827 | 18.83 | 20000 | He | | ≈ 1 | 1 |
| GD 077 | 14.80 | ≈10000 | H | | 1.2 | 3 |
| PG 0136+251 | 15.83 | 40000 | H | | 1.3? | 4 |
| G 141−2 | 15.91 | 5600 | H | | 2? | 5 |
| PG 1658+440 | 14.9 | 30500 | H | | 2.2 | 6, 7 |
| PG 1220+234 | 15.57 | 27200 | H | | ≈3 | 8 |
| G 99−37 | 14.60 | 6300 | $C_2$, CH | | ≈3.6 | 9, 10 |
| G 256−7 | 16.00 | 5600 | H | | 4.9 | 11 |
| MWD 0159−032 | 17.1 | 26000 | H | | 6 | 12 |
| LHS 1734 | 15.7 | 5300 | H | 16min–1yr | 7.3 | 1, 13 |
| G 62−46 | 17.11 | ≈6050 | H | | 7.4 | 14 |
| HS 1440+7518 | 14.9 | 40000 | H | | ≈8 | 15, 53 |
| HS 1254+3440 | 17 | 10–15000 | H | | ≈9.5 | 16 |
| GD 90 | 15.74 | 11000 | H | | 10 | 17, 18 |
| MWD 0307−428 | 16.3 | 25000 | H | | 10 | 12 |
| PG 1312+098 | 16.37 | 15000 | H | 5.43h | 10 | 19 |
| LHS 2273 | 16.48 | ≈6000 | H | | ≈10 | 20 |
| G 183−35 | 16.4 | ≈7000 | H | 50min–?yr | < 14 | 1, 11 |
| GD 356 | 15.06 | 7500 | H(em) | | ≈14 | 21 |
| KUV 03292+0035 | 16.7 | 19000 | H | | 12 | 22, 23 |
| KPD 0253+5052 | 15.22 | ≈15000 | H | 4.1h | 17 | 12, 19 |
| LHS 1044 | 15.3 | 6000 | H | 4.4h? | 16.7 | 1, 13 |
| G 99−47 | 14.10 | 5600 | H | 1h? | 27 | 10, 24, 25 |
| RE 0616−649 | 18.4 | 35000 | H | | 20 | 26 |
| LBQS 1136−0132 | 18 | 15000 | H | | 24 | 27 |
| ESO 439−162 | 18.77 | 5400 | $C_2$ | | 0–30? | 28, 29 |
| PG 1533−057 | 15.32 | 17000 | H | ≈1d | 31 | 12, 21 |
| HE 1045−0908 | 16.5 | 9000 | H | | 31 | 30 |
| Feige 7 | 14.46 | 20000 | H, He | 2.2h | 35 | 31, 32 |
| BPM 25114 | 15.62 | 20000 | H | 2.8d | 36 | 33, 24 |
| KUV 23162-1230 | 15.38 | 11800 | H | 17.9d | 56 | 12, 19, 25 |
| GD 116 | 15.96 | 16000 | H | | 65 | 34 |
| HE 1211−1707 | 16.9 | 20–25000 | H | 1.75 | 80-? | 35 |
| G 195−19 | 13.79 | 8000 | ? | 1.33d | ≈100 | 36 |
| PG 1015+014 | 16.33 | 14000 | H | 1.65h | 120 | 19, 37 |
| HE 0000-3430 | 15.0 | 7000 | H | | 120 | 35 |
| LP 790−29 | 15.9 | 7500 | $C_2$ | ≳100y | ≈200 | 38, 39, 40 |
| G 227−35 | 15.58 | 7000 | H | ≳100y | 205 | 25, 41 |

TABLE 1. continued

| WD | V | $T\,(K)$ | Spec. Feat. | Rot. Per. | B/MG | Note/Ref. |
|---|---|---|---|---|---|---|
| Grw +70°8247 | 13.19 | 15000 | H | $\gtrsim$100y | 320 | 22, 33, 42, 43 |
| G 240−72 | 14.15 | 6000 | ? | $\gtrsim$100y | ≈200 | 36 |
| G 111−49 | 16.28 | 8400 | H | | ≈220 | 25, 44 |
| HE 0127−3110 | 16.1 | 18000 | H | | 345 | 35 |
| HE 2201−2250 | 16.2 | 18000 | H | | 345 | 35 |
| RE 0317−853 | 14.8 | 50000 | H | | 660 | 45 |
| LB 11146 | 14.32 | 16000 | H+? | | 670 | 46, 47 |
| SBS 1349+5434 | 16.4 | 11000 | H | | 760 | 48 |
| PG 1031+234 | 15.10 | ≈ 15000 | H | 3.4h | 500–1000 | 49, 50 |
| GD 229 | 14.85 | 16000 | ? | ≈100y? | ≈1000? | 51, 52 |

REFERENCES: (1) Putney 1996. (2x) Schmidt & Smith 1994; (3x) Schmidt et al. 1992b; (4) Friedrich et al. 1996; (5) Greenstein 1986 (6x) Liebert et al. 1983; (7x) Schmidt et al. 1992a; (8x) Schmidt & Smith 1995, Bergeron priv. comm.; (9x) Angel 1977; (10x) Bues & Pragal 1989; (11) Putney 1995; (12x) Achilleos et al. 1991; (13x) Bergeron et al. 1992; (14x) Bergeron et al. 1993; (15x) Dreizler et al. 1994, missprint HS1412+6115→HS1440+7518; (16) Jordan 1989; (17x) Angel et al. 1974; (18) Jordan 1993; (19x) Schmidt & Norsworthy 1991; (20x) Schmidt & Smith 1995; (21x) Greenstein & McCarthy 1985; (22) Jordan 1992; (23x) Wegner et al. 1987; (24x) Wickramasinghe & Martin 1979; (25x) Putney & Jordan 1995; (26x) Jordan & Finley, unpubl.; (27x) Foltz et al. 1989; (28) Ruiz et al. 1988; (29) Schmidt et al. 1995; (30x) Reimers et al. 1994; (31x) Achilleos et al. 1992; (32) Martin & Wickramasinghe 1986; (33) Jordan 1988; (34x) Saffer et al. 1989; (35) Reimers et al. 1996; (36x) Angel 1978; (37x) Wickramasinghe & Cropper 1988; (38x) Liebert et al. 1978; (39x) Schmidt et al. 1994; (40) Bues', 1993 (41x) Cohen et al. 1993; (42x) Angel et al. 1985; (43x) Wickramasinghe & Ferrario 1988; (44) Guseinov et al. 1983; (45) Barstow et al. 1995; (46x) Liebert et al. 1993; (47x) Glenn et al. 1994; (48x) Liebert et al. 1994; (49x) Schmidt et al. 1986; (50x) Latter et al. 1987; (51x) Schmidt et al. 1990; (52) Schmidt et al. 1996; (53) This paper. The references which are marked with an 'x' are listed in Schmidt & Smith 1995

## 2. Recent discoveries and analyses

Barstow et al. (1995) have found a wide double degenerate, consisting of a nonmagnetic 16 000 K DA white dwarf (LB 9802) and RE J 0317-853, a 50 000 K magnetic white dwarf. The range of field strengths (200-660 MG) in this hottest known magnetic white dwarf cannot be explained by a centered magnetic dipole. Since the star is rotating with a period of only 725 seconds, only phase resolved spectroscopy will allow to determine the exact field geometry. This is also important in order to solve the question, why the visual brightness of the RE J 0317-853 varies with an amplitude of $\pm 0.^{m}1$ during the rotational period. The most puzzling result is that RE J 0317-853 has a mass of about 1.35 $M_{\odot}$, which is approaching the Chandrasekhar

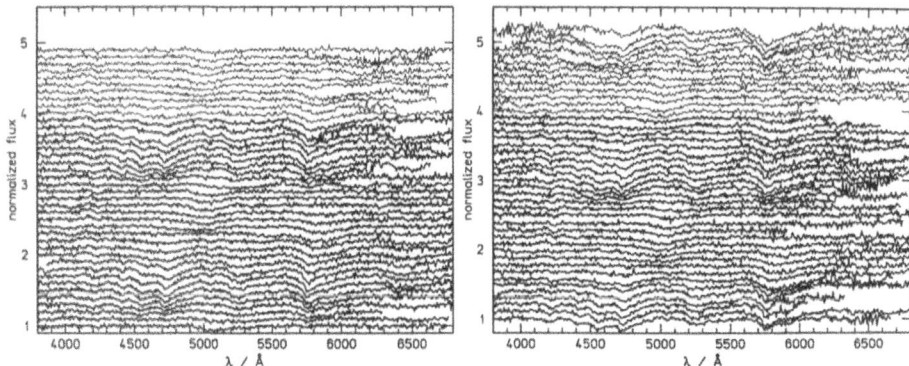

*Figure 1.* Spectra of HE 1211-1707 taken during one night on April 16/17, 1996 (ESO service observations by S. Köhler). The exposure time for each spectrum is 5 minutes. The continua are normalized to unity and the spectra are shifted by 0.1 units in succession from bottom to top. The right part of the figure is the continuation of the left one after a gap of about 20 minutes. The period of spectral variation is 110 minutes corresponding to about 20 spectra

limit and is the highest value ever found in a white dwarf. According to standard evolution theory an object with more than 1.08 $M_\odot$ should be a Mg/Ne white dwarf with a high mass (6-8 $M_\odot$) progenitor (e.g. Becker & Iben 1979), unless fast rotation has increased the limiting mass of C/O white dwarfs (Domínguez et al. 1996). However, single star evolution is rather improbable for RE J 0317-853, since it should now be much cooler than the less massive (0.7 $M_\odot$) companion LB 9802, which is about 600 au away from RE J 0317-853. Therefore, it must be a product of merging which could also explain the fast rotation.

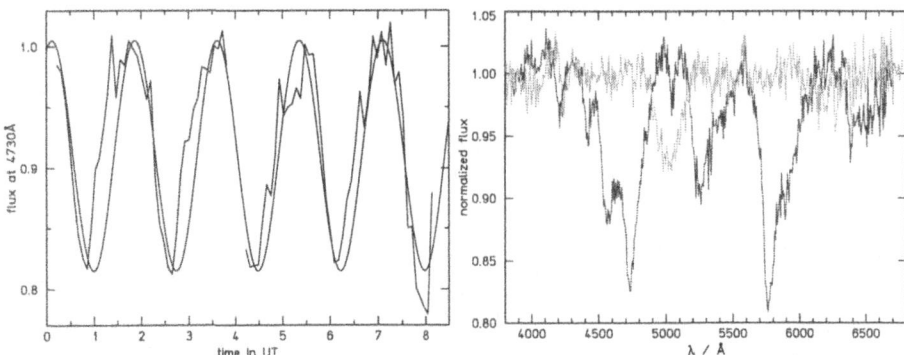

*Figure 2.* Left: Variation of the depth of the strong feature at 4730 Å in the spectrum of HE 1211-1707 with time. For comparison a sinus curve with a period of 110 minutes is shown. Right: Four spectra of HE 1211-1707 taken during the "strong" and the "weak" feature phase, respectively, were added up in order to obtain a better signal-to-noise ratio and are shown for comparison

Putney & Jordan 1995 have analyzed the spectra and polarization of three magnetic white dwarfs (G 99−47, KUV 23162−1230, and G 227−35). They found that the observed data could only be reproduced if off-centered dipoles or dipole+quadrupole configurations were used. This is consistent with the findings of Achilleos & Wickramasinghe (1989) who also needed off-centered magnetic dipoles to account for the observed spectropolarimetry.

Several new magnetic white dwarfs have been discovered by the Hamburg ESO Survey for bright quasars (Reimers et al. 1994, Reimers et al. 1996). While the spectrum of HE 1045-0908 could be explained by a centered magnetic dipole the other four stars possess more complicated field configurations. The observed spectra of HE 0127-3110, HE 2201-2250, and HE 0000-3430 need offcentered dipoles in order to be explained by shifted Balmer lines of hydrogen. The absorption lines in HE 0127-3110 exhibit small variations, probably due to rotation. One star, HE 1211-1707, however, failed to be interpreted in detail and is very peculiar. Fig. 1 shows 83 spectra of HE 1211-1707 with an exposure time of five minutes each during one night with only very small gaps between the exposures. It turned out that the period of spectral variation is about 110 minutes and the spectra can be added up according to there rotational phase (Fig. 2). We were not yet able to reproduce the observations with synthetic spectra, but it seems that some of the features in the optical and the UV agree with the predictions for a magnetic field at about 80 MG. Additionally, there must be parts on the stellar surface with a much higher field, or, alternatively, some of the features may be due to absorption by helium for which no atomic data in the intermediate field rime exist. Currently, we are analyzing a time resolved HST spectrum in the UV, that has been obtained on June 20, 1996. By looking for the position of theLyman $\alpha$ $\sigma_+$ component we hope to obtain a consistent model of HE 1211-1707 in the near future.

## 3. The case of deviations from centered dipoles

Chanmugam & Gabriel (1972) and Fontaine et al. (1973) have calculated the time scale for the decay of magnetic fields of white dwarfs. They showed that the decay times are $\gtrsim 10^{10}$ yr with the higher modes decaying more rapidly than the fundamental. This could lead to the assumption that the magnetic field becomes more dipolar during evolution. However, Muslimov et al. (1995) have shown that a weak quadrupole (or octupole, etc.) component on the surface magnetic field of a white dwarf may survive the dipole component under specific initial conditions: Particularly the evolution of the quadrupole mode is very sensitive (via Hall effect) to the presence of internal toroidal field. For a 0.6 $M_\odot$ white dwarf with a toroidal fossil magnetic field of strength $< 10^9$ G the dipole component declines by a factor of three in $10^9$ yr, while the quadrupole component is practically unaffected. Without an internal toroidal field the dipole component still declines by a factor of three but the quadrupole component is a factor of six smaller after 10 Gyr.

This shows that the detection of higher-order multipoles provides us with information about internal magnetization of white dwarfs and the initial conditions from the pre-white dwarf evolution. Therefore, further investigations of the complex magnetic fields of white dwarfs remain important.

Acknowledgements. We thank Susanne Köhler, who carried out the observations of HE 1211-1707 at ESO, Chile, and the DARA (project 50 OR 9409 1) and DFG (KO 738/7/1) for financial support.

# References

Achilleos N., Wickramasinghe D.T., 1989, ApJ 346, 444

Barstow M.A, Jordan S., O'Donoghue D., Burleigh M.R., Napiwotzki R., Harrop-Allin M.K., 1995, MNRAS 277, 971

Becker S.A., Iben I. Jr., 1979, ApJ 232, 831

Bues I, 1993, in White Dwarfs: Advances in Observation and Theory, ed. M.A. Barstow, Kluwer, p. 213

Chanmugam G., Gabriel M., 1972, A&A 16, 149

Domínguez I., Straniero O., Tornambé A., Isern J., 1996, ApJ, in press

Fontaine G., Thomas J.H., Van Horn H.M., 1973, ApJ 184, 911

Friedrich S., Östreicher R., Schweizer W., 1996, A&A 309, 227

Greenstein J.L., Henry R.J.W., O'Connell R.F., 1985, ApJL 289, L25

Guseinov O.H., Novruzova H.I., Rustamov Yu.S., 1983, ApSS 97, 305

Jordan S., 1988, PhD thesis, University of Kiel

Jordan S., 1989, in White Dwarfs, Lecture Notes in Physics 328, Ed. G. Wegner, Springer, p. 333

Jordan S., 1992, A&A 265, 570

Jordan S., 1993, in White Dwarfs: Advances in Observation and Theory, ed. M.A. Barstow, Kluwer, p. 333

Martin B., Wickramasinghe D.T., 1986, ApJ 301, 177

McCook G.P., Sion E.M., 1987, ApJS 65, 603

McCook G.P., Sion E.M., 1996, priv. comm.

Muslimov A.G., Van Horn H.M., Wood M.A., 1995, ApJ 442, 758

Putney A., 1995, ApJ 451, L67

Putney A., 1996, PASP 108, 638

Putney A., Jordan S., 1995, ApJ 449, 863

Reimers D., Jordan S., Köhler Th., Wisotzki L., 1994, A&A 285, 995

Reimers D., Jordan S., Koester D., Bade N., Köhler Th., Wisotzki L., 1996, A&A 311, 572

Ruiz M.T., Maza M.T., 1988, ApJ 335, L15

Schmidt G.D., Smith P.S., 1995, ApJ 448, 305

Schmidt G.D., Bergeron P., Fegley Jr. B., 1995, ApJ 443, 274

Schmidt G.D., Allen R.G., Smith P.S., Liebert J., 1996, ApJ 463, 320

## Discussion

*M. Van Kerkwijk*: Are there magnetic white dwarfs for which you see temperature effects due to the magnetic field (dark spots, etc.).

*S. Jordan*: The most plausible hypothesis for the sinusoidal variability of the visual magnitude of RE J 0317-853 is, that the surface brightness is not constant so that the apparent luminosity changes over the rotational period. However, I do not have a straightforward explanation, why this variation exists and how it is correlated with the magnetic field. One also has to explain the fact, why the magnitude of other rotating magnetic white dwarfs is practically constant (e.g. HE 1211-1707).

*H. Shipman*: What is the distribution of magnetic field strengths in the stars in the HQS/HE survey.

*S. Jordan*: There are objects with field strengths as low as about 8 MG (HS 1440+7518) and others with more than 300 MG (HE 0127−3110 and HE 2201−2250). In fact, high field magnetic white dwarfs are easier found in the Hamburg ESO survey, since the objective prism spectra (450 Å/mm at H$\gamma$compared to 1390 Å/mm in the HQS) already show that the spectra are rather peculiar.

# 1-D AND 2-D CALCULATIONS OF DYNAMO-GENERATED MAGNETIC FIELDS IN WHITE DWARFS

J. ANDREW MARKIEL, JOHN H. THOMAS AND H. M. VAN HORN

*Department of Physics and Astronomy and*
*C.E.K. Mees Observatory*
*University of Rochester, Rochester, New York 14627 USA*

**Abstract.** We summarize 1-D calculations and present new 2-D calculations of dynamo generated magnetic fields in white dwarfs. We find that kG magnetic fields with oscillation periods of a few years can be generated in the surface layers of these stars. These magnetic fields could be detected through their effect on the pulsations of variable white dwarfs, but care must be taken when analyzing data taken in different observing runs as they may sample different phases of the dynamo.

## 1. Introduction

From an analysis of the *g*-mode oscillation spectrum of the white dwarf GD 358, Winget *et al.* (1994) found evidence for differential rotation and a weak ($\sim 1$ kG), time-varying magnetic field in the outer layers of this star. Motivated by this result, we demonstrated (Thomas, Markiel, & Van Horn 1995, hereafter Paper I) that dynamos producing kG strength fields with oscillations periods of a few years were likely in white dwarfs with well-developed convection zones and some differential rotation. We argued that such a dynamo could be operating in GD 358 to produce the inferred magnetic field. We summarize some of these results in section 3.

The model used in Paper I was radially averaged and considered only individual Fourier modes in latitude. In section 4 we present preliminary results from a 2-D code which solves the dynamo equations in spherical geometry, including the full radial and latitudinal structure. These calculations performed for a white dwarf of lower luminosity than GD 358 show that higher order modes can indeed be excited in these cooler white dwarfs.

*I. Isern et al. (eds.), White Dwarfs, 405–411.*
© *1997 Kluwer Academic Publishers.*

## 2. The Nonlinear Dynamo Model

In our $\alpha$-$\omega$ dynamo model, the magnetic field is amplified through the effects of differential rotation and cyclonic convection. Starting from a weak initial poloidal field, the radial differential rotation shears the poloidal field, producing loops of toroidal field. The strongest toroidal fields are produced just below the convection zone, where the effects of magnetic buoyancy are reduced. As these toroidal flux tubes rise into and through the convection zone, the Coriolis force rotates the flux tubes producing a poloidal component of the field, which diffuses back into the shear layer below the convection zone to complete the cycle (Parker 1955).

We solve our local, nonlinear dynamo equations in spherical coordinates $(r, \theta, \phi)$, in which the mean magnetic field consists of a toroidal component $B\mathbf{e}_\phi$ and a poloidal component $\mathbf{B}_p = \nabla \times (0, 0, A)$, where $A\mathbf{e}_\phi$ is a vector potential. The dynamo equations are

$$\frac{dA}{dt} = \alpha B + \eta \left( \nabla^2 - \frac{1}{r^2\sin^2\theta} \right) A, \qquad (1)$$

$$\frac{dB}{dt} = \nabla\Omega \times \nabla \left( Ar\sin\theta \right) \cdot \mathbf{e}_\phi + \eta \left( \nabla^2 - \frac{1}{r^2\sin^2\theta} \right) B - \frac{1}{r}\frac{d}{dr} \left( ruB \right). \qquad (2)$$

In these equations, $\nabla\Omega$ is the differential rotation which shears poloidal field lines into toroidal loops, and $\alpha$ represents the regeneration of the poloidal field by cyclonic convection. Here $\eta$ is the magnetic diffusivity and $u$ is the rise velocity of a toroidal flux tube.

These equations have wave solutions which represent propagation of the field toward the equator or poles. Non-decaying solutions are found only when the dynamo number $|N_D| > 1$, where $N_D \equiv (\alpha L^4 \nabla\Omega)/\eta^2$ is the product of two magnetic Prandtl numbers. For $N_D$ negative in the northern hemisphere, the field propagates toward the equator (as in the Sun).

The growth of the dynamo is limited by two nonlinear effects. First, the regeneration of the poloidal field by the $\alpha$-effect is exponentially quenched according to the ratio of the energy density of the toroidal field to the typical energy density of the convective motions. This quenching represents the fact that strong magnetic flux tubes will resist the twisting due to convective motions that produce a poloidal field from the toroidal field. Second, toroidal magnetic flux tubes are unstable to magnetic buoyancy and rise to the surface to form active regions (starspots). The last term in equation (2) represents the loss of toroidal flux due to buoyancy, where $u \propto B^2$ is the terminal rise velocity of a flux tube of strength $B$ (Parker, 1979).

These nonlinear effects limit the dynamo field strength to a finite amplitude. To calibrate the free parameters in our model, we compute a solar dynamo model and match the observed 22 year solar period and the field strength $\sim 10^5$ G believed to occur near the base of the solar convection zone (D'Silva & Choudhuri 1993, Moreno-Insertis 1994). We then use these calibrated parameters for our calculations of white-dwarf dynamos.

The white dwarf model parameters required for these calculations were obtained from convective envelope models developed by Fontaine (1973; see also Fontaine & Van Horn 1976). We assume a $M_* = 0.6 M_\odot$ and use the ML2 version of mixing length theory (as in Paper I). For the stellar rotation we use the parameters inferred by Winget *et al.* (1994), i.e. a surface rotation rate of 0.89 d and a rotation rate of the outer core of 1.6 d. In analogy to the solar differential rotation, we assume this shear extends from the middle of the convection zone into the top portion of the underlying radiative zone over a distance comparable to a scale height. We also consider a model where the differential rotation is reduced from that inferred for GD 358.

## 3. 1-D Results

The dynamo equations (1,2) are nonlinear partial differential equations. If we radially average the equations and expand $A$ and $B \sim e^{ikr_c\theta}$, the equations reduce to ordinary differential equations which can be integrated in time with a 4th order Runge-Kutta algorithm to find the maximum field strength and oscillation period for a given $k$ (Markiel, Thomas, & Van Horn 1994; also Paper I). Selected models are listed in Table 1 (Model 3 is at the luminosity of GD 358).

At high luminosities (Models 1-3, which lie within the instability strip for DBs), the convection zone in these stars is extremely thin (as can be seen from the scale height $L$) and the dynamos are quite weak. At lower effective temperatures, the thicker convection zone produces a much more vigorous dynamo, with field strengths of $\sim 10^4$ G for $\log L_*/L_\odot = -1.75$. For this model we also find that a large number of higher order modes can be excited; it is not clear which of these modes would be expected to dominate. The dynamo periods are of order a few years for all the models listed. We find similar results for DA white dwarfs at and below the DA instability strip.

Model 3b has half the differential rotation of Model 3, which shows that we still expect dynamo activity if the differential rotation is less than that inferred by Winget *et al.* (1994). For Model 3c we have used the ML1 formulation of mixing length theory, which indicates that the dynamo field generated is very sensitive to the modelling of the convective envelope.

TABLE 1.  1-D Dynamo Models

| Model | $\log L_*/L_\odot$ | $L$ | $kr_c$ | $|N_D|^{1/2}$ | $B_{max}$ (G) | Period (yr) |
|-------|--------------------|------|--------|---------------|----------------|-------------|
| 1     | $-1.00$            | 0.98 | 2      | 3.1           | 60             | 2.50        |
| 2     | $-1.15$            | 1.60 | 2      | 3.9           | 190            | 2.09        |
| 3*    | $-1.30$            | 2.07 | 2      | 5.8           | 380            | 2.02        |
| 3a    | $-1.30$            | 2.07 | 4      | 2.1           | 310            | 1.08        |
| 4     | $-1.50$            | 2.52 | 2      | 8.8           | 680            | 2.12        |
| 5     | $-1.60$            | 3.85 | 2      | 18.2          | 2480           | 2.17        |
| 6     | $-1.75$            | 9.82 | 2      | 47.9          | 19100          | 1.89        |
| 6a    | $-1.75$            | 9.82 | 4      | 16.9          | 18900          | 0.95        |
| 6b    | $-1.75$            | 9.82 | 6      | 9.2           | 18600          | 0.64        |
| 6c    | $-1.75$            | 9.82 | 8      | 6.0           | 18200          | 0.48        |
| 6d    | $-1.75$            | 9.82 | 10     | 4.3           | 17600          | 0.39        |
| 6e    | $-1.75$            | 9.82 | 12     | 3.3           | 16900          | 0.33        |
| \multicolumn{7}{Reduced Differential Rotation} | | | | | | |
| 3b    | $-1.30$            | 2.07 | 2      | 3.8           | 220            | 3.35        |
| \multicolumn{7}{ML1 mixing length formulation} | | | | | | |
| 3c    | $-1.30$            | 0.73 | 2      | 7.6           | 36             | 3.77        |

## 4.  2-D Results

The equations used in the above calculations were radially averaged, and computed only individual Fourier modes in latitude. Here we describe results obtained from a 2-D calculation which includes the full radial and latitudinal structure of the dynamo. The equations were solved by a finite difference scheme with 40 grid points in radius and 257 grid points in latitude (we found that the presence of higher order modes required very high resolution in latitude). The integration domain extended from pole to pole in latitude, and from the surface down radially to approximately twice the depth of the convection zone. The boundary conditions chosen were: $A = B = 0$ at the poles, $A = B = 0$ at the inner boundary, and $B = 0$ and $A$ matching onto a potential field at the surface.

The luminosity of this model is $\log L_*/L_\odot = -2.0$, which is somewhat fainter than GD 358. We chose this model to investigate the presence of higher order modes expected to occur in these cooler stars. The time series shown extends for several years near the end of the calculation, but due to numerical problems associated with the multiple field reversals we were unable to follow the evolution long enough to be certain the dynamo had completely stabilized. We are currently working to extend the calculation.

The spatial structure of the field is shown in Figure 1. Loops of toroidal field are produced just below the base of the convection zone (the dotted

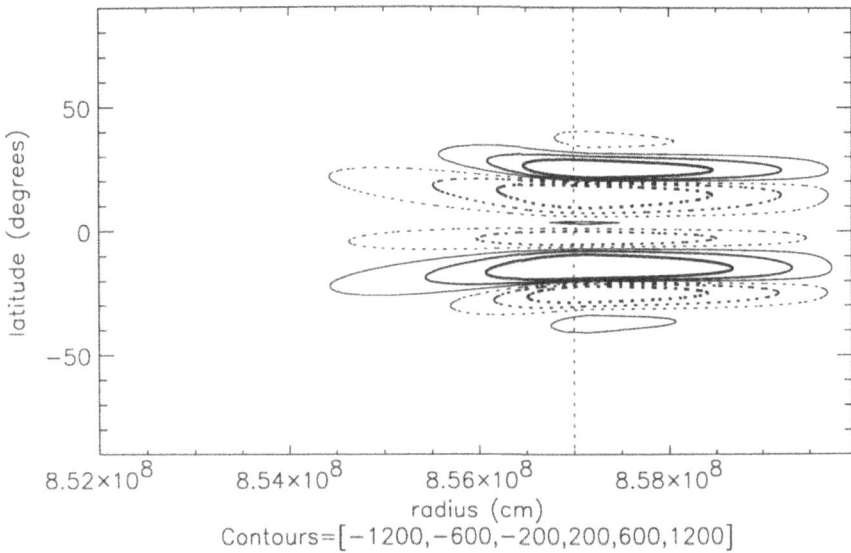

*Figure 1.* Spatial structure of the dynamo. Contours of the toroidal field are plotted, solid lines are positive polarity. The dotted vertical line indicates the base of the convection zone.

vertical line) and rise into the convection zone, where the $\alpha$-effect produces the poloidal field (not shown). The field geometry is clearly not a simple dipole. The field strength is much less than that inferred from the 1-D model. This may be due to the inability to completely follow the calculation until saturation, but we believe this indicates the importance of radial diffusion, which was neglected in the 1-D calculations but is expected to be important for the thin dynamo regions found in white dwarfs.

Figure 2 shows a time series of the toroidal field near the base of the convection zone. We see bands of toroidal field migrating toward the equator, in a manner similar to the solar dynamo.

## 5. Conclusions

We have extended our previous dynamo model for white dwarfs by developing a 2-D code to compute the full radial and latitudinal structure of the generated field. Preliminary results support our argument that dynamos may well be found in many white dwarfs, but more work is needed to extend the calculations.

The most promising possibility for detecting these fields would be through asteroseismological techniques, but this is only possible for white dwarfs within the instability strip. We predict strong dynamo activity at effective

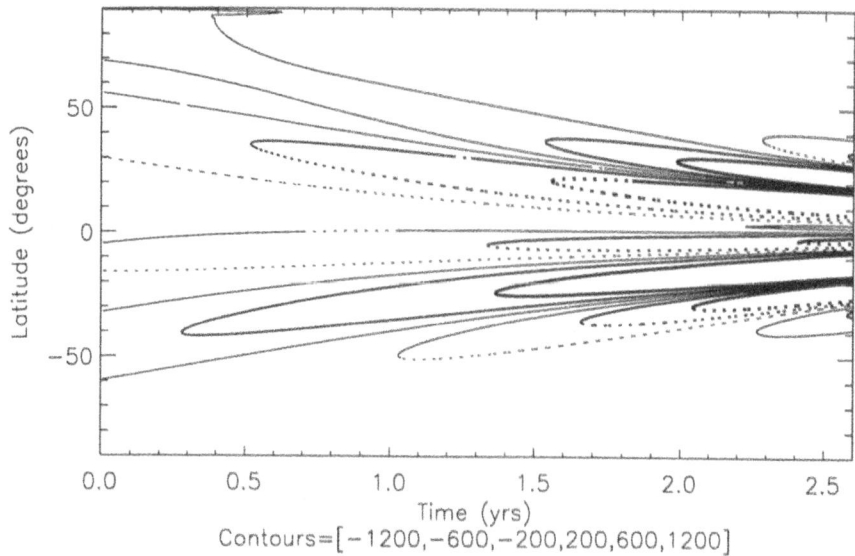

*Figure 2.* Butterfly diagram showing contours of toroidal field near the base of the convection zone as a function of time.

temperatures lower than the instability strip, but this result is subject to uncertainties in the modelling of the convective envelopes as well as the dynamo model. We find typical dynamo periods of a few years, which suggests that separate observations of a single pulsating star could sample different phases of the dynamo and thus have intrinsic differences in the observed pulsation frequencies. The field may also have non-dipole components, which would further complicate analysis of the observations.

## Acknowledgements

We are grateful to the conference organizers for generous financial support for one of us (J.A.M.) to attend this workshop. This research has been supported by NASA grant NAGW-2444 and NSF grant AST-9528398.

## References

D'Silva, S. and Choudhuri, A. R. 1993, *Astron. Astrophys.*, **272**, 621.
Markiel, J. A., Thomas, J. H., and Van Horn, H. M. 1994, *Astrophys. J.*, **430**, 834.
Moreno-Insertis, F. 1994, in *Solar Magnetic Fields*, ed. M. Schüssler and W. Schmidt (Cambridge: Cambridge Univ. Press), p. 117.
Parker, E. N. 1955, *Astrophys. J.*, **122**, 293.
Parker, E. N. 1979, *Cosmical Magnetic Fields* (Oxford: Clarendon), p. 141ff.
Thomas, J. H., Markiel, J. A., and Van Horn, H. M. 1995, *Astrophys. J.*, **453**, 403, Paper I
Winget, D. E., *et al.* 1994, *Astrophys. J.*, **430**, 839.

## Discussion

*M.T. Ruiz*: Could you comment on the possible relation between the existence of a fossil magnetic field of about $10^6$ G in the interior of the Sun and the strong ($10^6 - 10^9$ G) fields observed in white dwarfs?

*J.A. Markiel*: A process such as this could lead to a magnetic white dwarf, but I am not familiar with the evidence for a fossil field in the Sun, and the small fraction of white dwarfs observed to be strongly magnetized seems to indicate that such evolution would be rare.

*J. Kubat*: The dynamo process would result in more or less periodic energy storage and release (similar to the Sun). Would such "white dwarf flares" be observable? If so, may they serve as a tool for the detection of the magnetic field?

*J.A. Markiel*: I expect there would not be enough energy involved for flare activity to be observable. Indirect methods such as asteroseismology are likely to be the only ways to detect these fields.

# MAGNETIC FIELD SURVEY OF DC WHITE DWARFS

ANGELA PUTNEY

*California Institute of Technology*
*Mail Stop 105-24*
*Pasadena, CA 91125, USA*

**Abstract.** A survey of $\sim 50$ DC white dwarf stars was conducted in circular spectropolarimetry to search for magnetic fields $> 30$ kG. Four DC stars were discovered with magnetic fields above 30 kG: G 111-49 with $B_e \sim -220$ MG, G 183-35 with $B_e = +6.8 \pm 0.5$ MG, G 256-7 with $B_e = +4.9 \pm 0.5$ MG, and G 234-4 with $B_e = +39.6 \pm 11.6$ kG. A new magnetic DB white dwarf was also discovered, LB 8827 with $B_e = 1.0 \pm 0.5$ MG. A total of 17% of the white dwarfs in the survey have a magnetic field $> 30$ kG. This value is far larger than the 2% of DA stars, but more than half of the DC stars were originally misclassified. Only 5% of the re-classified DC stars have magnetic fields above 30 kG. Three magnetic stars were re-observed to investigate the possibility of rotation. Two are definitely rotating: LHS 1734 with 16 min $< P < 1$ yr and G 158-45 (=LHS 1044) with a probable period $P \sim 4.44$ hr but a definite period $P \leq 1$ d or $P \sim 4$ d. G 183-35 might be rotating with 50 min $< P < $ a few yr.

## 1. Introduction

When I became interested in magnetic white dwarfs (MWDs) several years ago there were about 25 isolated white dwarfs known to have magnetic fields greater than 1 MG (1 MG $= 10^6$ Gauss). Some appeared to rotate with periods of hours to weeks and others appeared not to rotate. This raised several questions, such as

- How many white dwarfs have magnetic fields $> 1$ MG?
- What are the field strengths of the rest of the white dwarfs?
- Why do some MWDs rotate and others do not?
- Where do the fields come from? Are they fossil fields?

*I. Isern et al. (eds.), White Dwarfs*, 413–419.

The most efficient way to answer these questions is to take circular spec-
tropolarimetric observations of as many white dwarfs as possible. I undertook a survey of DC white dwarfs to try and answer the first two questions.
I also investigated a few of the survey objects more closely to look at the
rotation question. I still cannot answer the final question with certainty,
but my data is consistent with the fossil field theory (i.e., the magnetic flux
is conserved from the main sequence).

## 2.  The survey

All of the survey objects were taken from the McCook & Sion 1987 catalog
of white dwarf stars. All stars were classified as DC, or at least one of its
classifications was DC. None were listed as a binary. All were magnitude
16.5 or brighter and had a declination of $-20°$ or greater (these two criteria allowed for reasonable observing times on the Palomar 200'' (5 m)
telescope). After applying these criteria to the list of 1279 white dwarfs,
I was left with 52 stars to observe. After completing the observations I
re-classified many of the objects. I ended up with 18 DA, 3 DB, 2 DQ, 1
DZ, 19 DC, 3 DC+other star, 1 hot subdwarf (sdOC), 2 BL Lac Objects
(Flemming et al., 1993), 1 was non-existent, and 2 were not observed.

## 3.  Magnetic Field Strength Determination

After the re-classification, I was left with 46 white dwarfs. I determined a
magnetic field strength for each of these stars. The method used for the field
strength determination varied for different field strength regimes. There are
four basic magnetic field strength regimes: $B \geq 50$ MG, $1$ MG$\leq B \leq 50$
MG, $B \leq 1$ MG, and B is indeterminable. Table 1 shows the breakdown of
measured magnetic field strength with spectra type.

TABLE 1.  Number of Stars per Spectral Type per Magnetic Field Range

| Spectral Type | Magnetic Field Strength Range | | | | |
| --- | --- | --- | --- | --- | --- |
| | < 10 kG | 10 kG – 1 MG | 1 MG – 100 MG | > 100 MG | Indeterminable |
| DA | 6 | 6 | 6 | 0 | 0 |
| DB | 2 | 1 | 0 | 0 | 0 |
| DC | 0 | 0 | 0 | 1 | 21 |
| DQ | 0 | 0 | 0 | 0 | 2 |
| DZ | 1 | 0 | 0 | 0 | 0 |

## 3.1.  B ≥ 50 MG

These stars have the most bizarre flux and circular polarization spectra of all four regimes. The absorption features in flux appear at unexpected wavelengths and typically the polarization spectrum has a non-zero, $\lambda$-dependent continuum and strong features. In this regime the magnetic and Coulombic forces are comparable, which causes the hydrogen (or helium or whichever element) atom to be very distorted in shape, yielding absorption features at wavelengths very different from the zero field case. Originally stars similar to these were classified DXP for unknown atmospheric constituents but observed polarization. A comparison of the flux and polarization spectra to the magnetic field as a function of wavelength curves of hydrogen (Wunner *et al.*, 1985) can give you a magnetic field range for hydrogen lines. No theoretical calculations yet exists for other elements or molecules in this magnetic field regime. One star in my survey fell in this category, G 111-49 with a field of −220 MG (see also Putney (1995), and Putney 1995, Ph.D. thesis, California Institute of Technology).

## 3.2.  1 MG ≤ B ≤ 50 MG

This is the easiest regime to recognize. The flux spectrum will exhibit the classical Zeeman split lines and the circular polarization spectrum will display the classical Zeeman signature of one positive feature for every negative feature. The continuum polarization is typically null or very close to it. The atoms in the stars in this regime are actually perturbed by the quadratic Zeeman effect. Six of the stars in the survey fell in this regime, G 158-45 (=LHS 1044), LHS 1734, G 43-54, G 256-7, PG 1312+098, and G 183-35. G 256-7 (B = +4.9 ± 0.5 MG) and G 183-35 (B = 6.8 ± 0.5 MG) were previously unknown to have strong magnetic fields (Putney, 1995); G 158-45, LHS 1734, and G 43-54 were discovered to have fields by Bergeron, Ruiz, & Leggett (1992; 1996); PG 1312+098 was first reported to have a magnetic field by Schmidt (1987).

## 3.3.  B ≤ 1 MG

This is the best understood regime, but in some ways it is the hardest one to work in. This is the linear Zeeman effect region, with the quadratic Zeeman effect being noticeable for the strongest fields. The transitions are understood for all elements, therefore, at the wavelength of each absorption line in flux, the circular polarization can be checked for the existence of the Zeeman signature. A magnetic field was positively detected in only one star in this regime, G 234-4 at +39.6 ± 11.6 kG.

## 3.4.  B INDETERMINABLE

Two types of objects fall in this category. The first has no absorption or
emission features in flux (a true DC) and no circular polarization, either
in the form or features or continuum. Thus there is nothing to measure.
The lack of continuum polarization places an upper limit on the magnetic
field of $\sim$ 20 MG. For the present survey, these objects are probably helium
white dwarfs which are too cool to show any helium absorption.

The second type are the stars with molecular carbon absorption features
in flux but null circular polarization. As has been demonstrated with G 99-
37 (Angel & Landstreet 1974; Schmidt, Bergeron, & Fegley 1995), the $C_2$
bands do not shift via the Zeeman effect until fairly high fields are reached
(value unknown, but $\gg$ 10 MG). G 99-37 has a continuum polarization
of $\geq$ 1 %, varying as a function of wavelength, and it has the tell-tale
Zeeman signature in circular polarization for the CH feature only. Angel &
Landstreet (1974) calculated a field strength of 3.6 MG from this feature,
but they expected that the dipolar field was closer to 10 MG. When the
transitions of $C_2$ in a strong magnetic field are known, a limit could be
placed on the field strength of these stars.

## 4.  Results

The 46 white dwarfs observed represent all (minus two) the DC white
dwarfs visible to the Northern Hemisphere with $m_v \leq 16.5$. This number is
not large nor statistically significant. However, Schmidt & Smith (1995;SS)
made a very similar survey of 169 DA stars. Combining the results of these
two surveys and the list of known MWDs (SS), more concrete statements
can be made than from my survey alone. The distribution of white dwarfs
per magnetic decade shows a bimodal distribution with one peak between
10 and 100 MG and the second below 10 kG (value unknown since fields
are not measured accurately below $\sim$30 kG). The distribution of magnetic
stars per temperature class ($\theta$) compared to all white dwarfs which have
been observed in circular spectropolarimetry per $\theta$ showed no preference
for any temperature. This implies that MWDs appear at all temperatures
in proportionally the same numbers (recently, many MWDs have been discovered at lower temperatures ($\theta = 8$ or 9, Bergeron $et$ $al.$ 1996), this is
probably due more to the focus of recent white dwarf surveys on cool white
dwarfs than on a real effect, but time will tell). A comparison of magnetic
field strength to temperature produced a very interesting result. Magnetic
stars with temperatures $>$ 16000 K all have fields below 50 MG, whereas
magnetic stars below that temperature can reach fields of 1000 MG and
tend to either be 100 MG and higher or below 50 MG (see Figure 1). The
peak in magnetic fields tantalizingly appears around the temperature of the

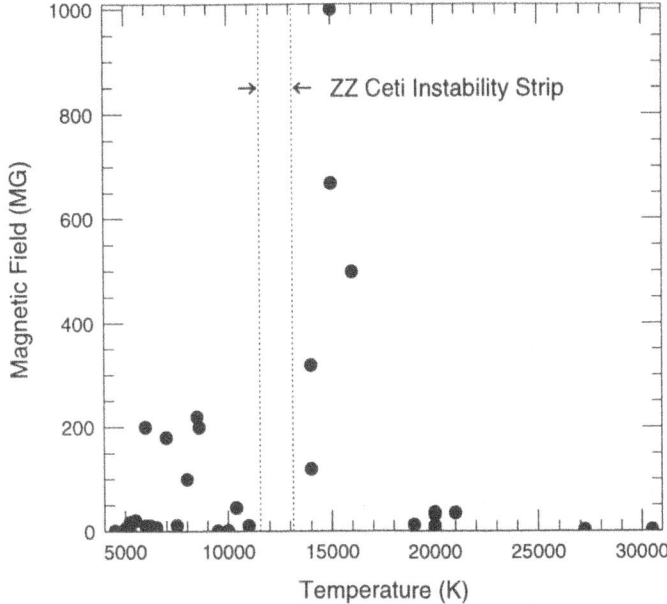

*Figure 1.* Magnetic field strength versus temperature for all stars in the MWD table of SS and the new stars discovered in the present survey. The temperature edges of the ZZ Ceti instability strip are from Kepler & Nelan (1993). GD 229 (16000 K, 500 MG) is given the more conservative field strength from Schmidt, Latter, & Foltz (1990), rather than the value in SS.

ZZ Ceti instability strip. Perhaps some process induces a magnetic field at this temperature. One interpretation of this figure is that there are two populations, one with magnetic fields $\leq 50$ MG and a second one where some process enhances the magnetic field at about 16000 K and then the enhanced field decays away. I leave this plot on the doorstep of the theorists.

The overall results are that 17% of all stars classified as DC pre-1987 have B > 30 kG, but falls to 5% after re-classification (= G 111-49). 4% of all white dwarfs have B > 30 kG when the present and SS surveys are combined.

## 5. Rotation

Three of the survey objects were observed repeatedly to look for rotation. Two, G 158-45 and LHS 1734, are rotating and the third, G 183-35, might be rotating. None were observed often enough to determine a rotation period, however, strong limits can be placed on G 158-45 ($\sim$ 11 hours or 3-4 days). This does bring up the whole question of rotation in white dwarfs and the effect of the magnetic field on rotation.

Conservation of momentum from the main sequence to the white dwarf phase would yield rotation faster than the break-up velocity, therefore angular momentum is lost during evolution to the white dwarf phase. Naively one would expect this loss to increase with increasing magnetic field (an enhancement of angular momentum transfer from core to envelope by magnetic fields). However, the fastest and the slowest white dwarfs also have the strongest magnetic fields. Rotation periods of MWDs tend to fall in either the $\sim 100$ yr range (or non-rotating, or "century club") or hours to weeks. These are measured by observing the change in polarization and absorption features over time. It is harder to measure rotation in non-magnetic white dwarfs because the features do not, typically, change over time. Pilachowski & Milkey (1987) measured rotation velocities from line cores for many white dwarfs; 15 of which were also in the SS survey, meaning there is a magnetic field determination for them. The rotation periods of these stars were all in the range of several hours. Comparisons of the distribution of period with magnetic field for MWDs and regular white dwarfs implies either (a) a magnetic field increases the angular momentum loss in white dwarfs except for 3 to 5 stars, or (b) it does not dramatically increase the angular momentum loss, except for 5 stars (the century club). If a MWD does not appear to rotate, it could be due to the rotation axis being aligned with our line-of-sight. When this problem was last studied (Schmidt & Norsworthy, 1991), about 1/3 of the MWDs monitored for rotation appeared not to rotate (5 stars). The idea that 1/3 of all MWDs have their rotation axis aligned with our line-of-sight is preposterous. With more observations, these 5 stars have moved to under 1/4 of MWDs with measured rotation periods. Many of the recently discovered MWDs have been observed to rotate, but have yet to have their rotation period measured. If this trend continues, it may just turn out that some or all of these five do indeed have their rotation axes aligned with our line-of-sight, which would solve our dilemma. However, it is much too early to be making any bold statements.

# References

Angel, J.R.P. & Landstreet, J.D. 1974, ApJ, 191, 457

Bergeron, P., Ruiz, M.-T., & Leggett, S.K. 1992, ApJ, 400, 315

Bergeron, P., Ruiz, M.-T., & Leggett, S.K. 1996, ApJSS, in press

Kepler, S.O. & Nelan, E.P. 1993, AJ, 105, 608

McCook, G.P. & Sion, E.M. 1987, ApJSS, 65, 603

Fleming, T.A., Green, R.F., Jannuzi, B.T., Liebert, J., Smith, P.S., & Fink, H. 1993, AJ, 106, 1729

Pilachowski, C.A. & Milkey, R.W. 1987, PASP, 99, 836

Putney, A. 1995, ApJ, 451, L67

Schmidt, G.D. 1987, Mem.Soc.Astron.Ital., 58, 77

Schmidt, G.D. & Norsworthy, J.E. 1991, ApJ, 366, 270

Schmidt, G.D. & Smith, P.S. 1995, ApJ, 448, 305 (SS)

Schmidt, G.D., Bergeron, P., & Fegley, B. 1995, ApJ, 443, 274

Schmidt, G.D., Latter, W.B., & Foltz, C.B. 1990, ApJ, 443,274

Wunner, G., Rösner, W., Herold, H., & Ruder, H. 1985, A&A, 149, 102

## Discussion

*P. Bradley*: Are the field morphologies of the non-rotating MWDs different from the rotating ones?

*A. Putney*: Thus far, all the rotating and non-rotating MWDs modeled are best fit with offset dipoles. So, no, the field morphologies are not different.

*P. Bradley*: More general question. Any idea of how close to pole-on you would have to be to pole-on before you cannot tell if the star is rotating?

*S. Jordan*: That all depends on the S/N ratio of the observation and on how fast the star is rotating. If the period is much shorter than the exposure time you may artificially produce "non-dipole" components which do not exist in reality.

*M. Wood*: For a magnetic field aligned with the rotation axis, you'd expect no sign of the rotation period in your observations, yes?

*A. Putney*: If the magnetic field is a centered dipole, that is indeed the case. However, if the field is an off-centered dipole the field is not symmetric and, except for a pole-on view, the rotation will be seen even in this case.

*M. Van Kerkwijk*: Is it possible that the "non-rotating" WDs are in fact rotating very fast? What kind of limits can one set?

*A. Putney*: Yes, it is possible that the "non-rotating" MWDs are rotating very fast. This could explain some of the discrepancies in the models, but not all. The limits are set by the observations. For a bright MWD, like Grw+70°8247, on the Keck 10-m, consecutive exposures of about 1 min have revealed no changes. For dimmer stars, the limit will be longer. On the Palomar 200-in, the limits are more like 10-15 min and no changes have been detected on the few we have observed.

*I. King*: It is dangerous to suggest that a large fraction of objects of any class are seen pole-on, because such a small fraction of the total solid angle of a sphere is near the pole.

*A. Putney*: I am not suggesting that these are all pole-on, but rather trying to point out that as more of the MWDs are found to be rotating, the idea that at least some of the non-rotating stars are pole-on, should be reconsidered. Other explanations should still be sought.

# NEW IUE-SPECTRA OF THE MAGNETIC WHITE DWARF GRW +70°8247 AND THEIR INTERPRETATION

DIETER ENGELHARDT AND IRMELA BUES

*Astronomisches Institut der Universität Erlangen-Nürnberg*
*Dr.Remeis Sternwarte,*
*Sternwartstr.7, 96049 Bamberg, Germany*

**Abstract.** The hydrogen model of quasi-Landau-resonances in a very strong magnetic field ($\approx 10^4$ Tesla) is applied to Fourier-transformed IUE spectra of Grw +70°8247.

## 1. Introduction

The quantitative analysis of spectra of hot white dwarfs with very strong polarizations indicating magnetic fields (B$\leq 10^5$Tesla) is still a quantum mechanical puzzle. For some of the white dwarfs with magnetic fields around $10^2$ Tesla the element composition in the atmospheres could be determined as hydrogen - rich, and even for Grw +70°8247 an identification of the strange 4135 feature as due to quasi-stationary transitions of Zeeman - shifted components of the Balmer lines is verified by modeling the flux in the visible by Wickramasinghe and Ferrario (1988). The hydrogen data of Rösner et al. (1984), used sucessfully by them do not predict any lines in the UV, with the exception of red-shifted Ly$_\alpha$. Nevertheless some UV-lines occur in the Fourier-transformed spectra of Grw +70°8247 as well as in spectra of BPM25114, PG1015 + 01 and PG1031 + 234. This can be understood within the framework of quasi-Landau-resonances. In this paper we analyze new IUE-spectra (swp56229 and swp56341, low resolution) of Grw+70°8247 taken by us (1995 Nov.26 and Dec.26).

## 2. Quasi-Landau-resonances versus adiabatic approximation

The 'classical' analysis of magnetic white dwarfs (MWDs) is based on detailed models of the hydrogen atom in magnetic fields (Schrödinger or Pauli-equation) and of the atmosphere of the MWD. Concerning the hydrogen atom two questions have never been answered satisfactorily:

*I. Isern et al. (eds.), White Dwarfs, 421–425.*

1.) How large is the effect of special relativity?

2.) Is it possible to shift Landau-resonances in fields of $10^5$ Tesla from $\approx 10$ eV to the red, e.g. into the visible spectrum? (This question was posed by Wickramasinghe at the 9th European workshop on white dwarfs in Kiel.)

The first question, neglected here, is relevant for fields of $\approx 4.7 \cdot 10^5$ Tesla.

The answer to the second one is 'Yes': the behaviour of high states in fields of **5 Tesla** can be understood within the framework of quasi-Landau-resonances (Main et al.1986). Due to scaling laws for the energy and the magnetic field this model is applicable to fields of $\approx 10^5$ Tesla and energies $< 1$ Rydberg: transitions occur in the UV.

The concept of the quasi-Landau-resonances is contradictory to the commonly used adiabatic approximation. There – within a classical picture!! – the motion of the electron in the field plane (homogeneous magnetic field) is treated as independent of the motion in direction of the magnetic vector. For very large fields, the speed of the electron (classically!) is much larger in the plane than perpendicular. In the direction of the magnetic vector there is no influence of the field upon the electron due to the right-hand-rule. For this reason it is usually assumed, that the hydrogen atom has the shape of a cigar.

Within the framework of quasi-Landau-resonances, however, the hydrogen atom is flat: this model makes sense in terms of quantum mechanics, because the f-values of transitions between Landau-alike states are much larger than those between cigar-alike states. In this picture, the electron is close to the proton in the direction of the magnetic field. The motion of the electron is restricted to the field plane.

## 3.  A new method of magnetic white dwarf analysis

The energy-eigen-values $E$ of quasi-Landau-resonances may be written as:

$$E = \hbar\omega_c \frac{1}{n - \delta},$$

where $\hbar$ is Planck's constant, $\omega_c$ the cyclotron frequency, n is an integer number and $0.3 < \delta < 0.4$. This formula was derived by Main et al. from the experiments mentioned above and can be applied to larger fields ($\approx 10^5$ Tesla).

The method of Fourier-transformation has been sucessfully applied to the analysis of states close to the continuum. For IUE-data it is an ideal tool. Reduced IUE data cannot be investigated directly with this method, but have to be transformed to frequencies by a Fourier-transformation.

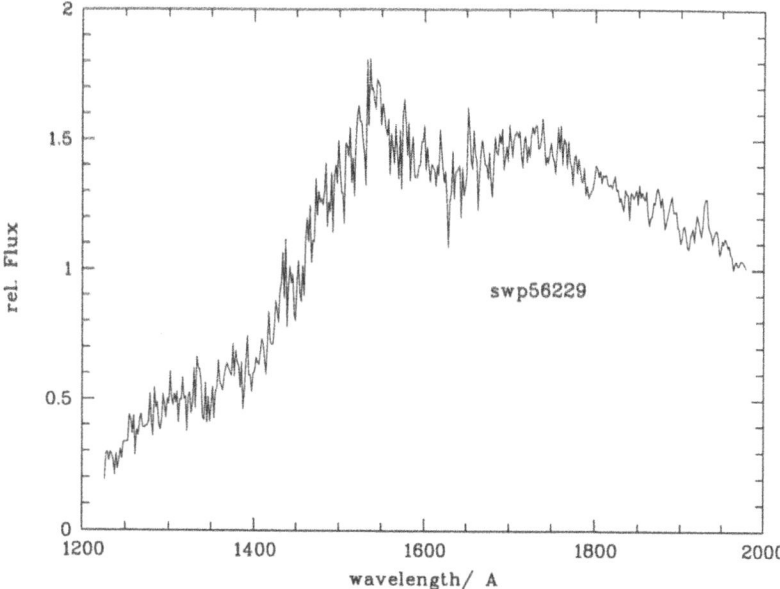

*Figure 1.* New IUE-spectra taken by I. Bues: Flux as a function of wavelength .

The analysis steps are:

1.) Flux points, apparently not due to the star must be removed.

2.) The IUE-spectrum – flux as a function of wavelength ($\lambda$) – is transformed to a function of frequency.

3.) The function of frequency is Fourier-transformed.

4.) If a mode in the Fourier transformed spetrum is identified as the ground mode –belonging to the cyclotron frequency– then one can derive directly the magnetic field strength of the star.

In the Fourier-transformed spectra of 'normal' MWDs many modes occur (e.g. BPM25114). Only two modes are found in Grw+70°8247 as well as for GD229 (Engelhardt 1996).

### 3.1. OLD SPECTRA

The spectra swp7490 and swp1660 taken by Greenstein et al. (1978) with IUE were analyzed by this method. The maximum mode can be identified: The magnetic field strength is $3.2 \cdot 10^4$ Tesla and $3.13 \cdot 10^4$ Tesla. The values are derived directly from the numerical results. The second mode may originate from an electric field (Engelhardt 1996). The result is in accordance with the sophisticated model (dipole assumption) of Wickramasinghe and Ferrario.

## 3.2.  NEW SPECTRA

In Fig. 1 the new spectrum swp56229 is plotted. The transformed spectra of swp56229 and swp56341 are shown in Fig.2 and 3. The magnetic field strength is $3.36 \cdot 10^4$ Tesla, in swp56229 and swp56341. The second mode is also seen.

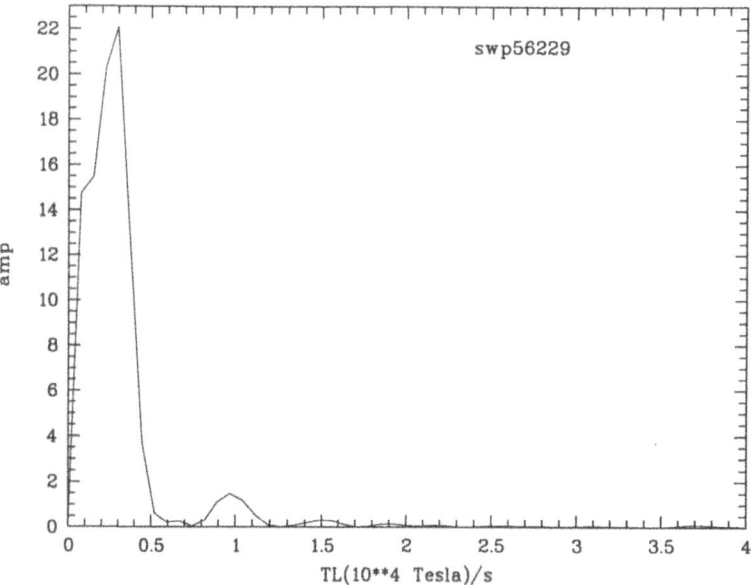

*Figure 2.*   Fourier spectrum of the UV-region observed by the IUE-satellite. The Fourier amplitude 'amp' versus time in units of the Larmor time $T_L$ for $10^4$ Tesla. The number of the analyzed spectrum is swp56229, the field strength ($B_{eff}$) derived from the periodicity is $3.36 \cdot 10^4$ Tesla.

## 4.  Conclusions

A new method for the detection of magnetic field strengths on magnetic white dwarfs has been shown. It was applied to 17 years old and new IUE-spectra of Grw+70°8247. The magnetic field strength of $3.2 \cdot 10^4$ Tesla is confirmed. In spite of the various look of the IUE-spectra the magnetic field strength remains nearly constant, thus demonstrating the long-term stability of the field and the definitely nondipolar structure.

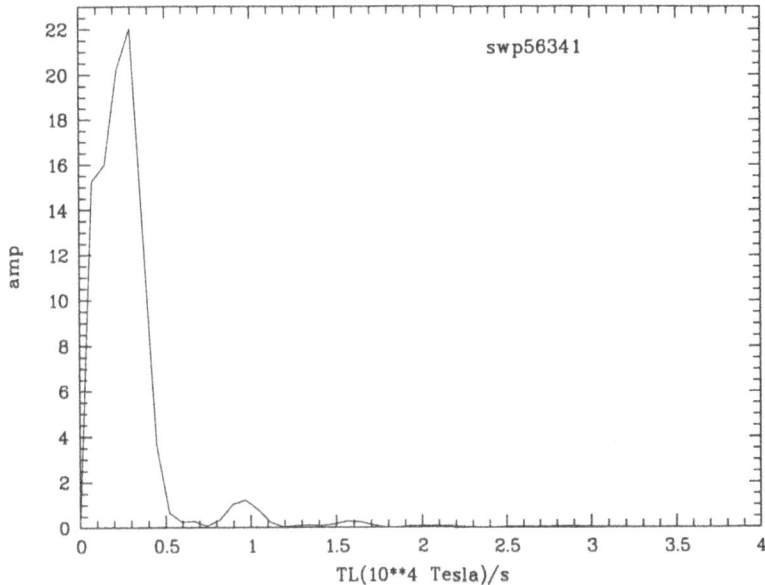

*Figure 3.* Fourier spectrum of the UV-region observed by the IUE-satellite. The Fourier amplitude 'amp' versus time in units of the Larmor time $T_L$ for $10^4$ Tesla. The number of spectrum is swp56341, the field strength $(B_{eff})$ derived from the periodicity is $3.36 \cdot 10^4$ Tesla.

## References

Engelhardt, D. 1996, PHD Thesis, University of Erlangen-Nürnberg
Greenstein, J.L., Boksenberg, A.: 1978, *Mon. Not., R., Astron. Soc.*, **185**, 823
Main,J., Wiebusch, G., Holle, A., Welge, K.H. 1986, *Phys.Rev.Lett.*, **57**, 2789
Rösner, W., Wunner, G., Herold, H., Ruder 1984, *J.Phys. B*, **17**, 29
Wickramasinghe, D., T., Ferrario, L. 1988, *Ap. J.*, **327**, 222

# Section VI:
# Pulsating White Dwarfs

# CONVECTION AND THE ZZ CETI INSTABILITY STRIP

D. KOESTER

*Institut für Astronomie und Astrophysik*
*Universität Kiel, D-23098 Kiel, Germany*

AND

G. VAUCLAIR

*Observatoire Midi-Pyrénées*
*14 Ave E. Belin, 31400 Toulouse, France*

## 1. Introduction

Theoretical predictions of the ZZ Ceti instability strip (Bradley and Winget 1994; Brassard et al. as cited in Bergeron et al. 1995 (B95)) have found a very strong dependence of the blue edge on the assumed efficiency of the convective energy transport. Convection is usually described within the framework of the mixing length approximation (MLT); in the following we will use the nomenclature introduced by Fontaine et al. (1981) and Koester et al. (1994), where ML1/$\alpha$=1.5 stands for a specific choice of the parameters a,b,c and a ratio of mixing length over presure scale height of 1.5.

This dependency of the results on the parametrization of MLT offers in principle the possibility, to "calibrate" MLT for DA white dwarfs by comparing the theoretical instability strip with empirical determinations of $T_{eff}$ and log g for ZZ Ceti stars. Unfortunately, however, this empirical determination has not been very successful. Published results for the effective temperature e.g. for two of the hottest ZZ Ceti, G117-B15A and G226-29, span a range of more than 2000 K. For GD165, B95 and Bragaglia et al. (1995) find a difference of more than 2600 K for $T_{eff}$, using the same models and fitting procedures. This indicates that a determination of parameters from optical spectra is extremely difficult, because of the well known effect that the Balmer lines reach maximum strength close to the ZZ Ceti range and change very little with a change of atmospheric parameters. We will come back to this problem below.

*I. Isern et al. (eds.), White Dwarfs, 429–435.*

An alternative path is the use of the UV spectra, obtained with the IUE satellite, or more recently with the Hubble Space Telescope. This method has had its own difficulties in the past, because the UV spectra are dominated by the quasi-molecular satellites in the wing of Ly $\alpha$ and a quantitative theory for the absorption coefficient has become available only recently (Allard and Koester 1992; Koester et al. 1994; Allard et al. 1994). Using these new calculations Koester et al. (1994) demonstrated that the HST spectrum of G117-B15A could only be fitted assuming an interme-diate efficiency of convective energy transport (e.g. ML1/$\alpha$=2). B95 in an important study of a large sample of ZZ Ceti stars arrived at a similar con-clusion; their prefered choice of parameters is ML2/$\alpha$=0.6. The resulting spectra for this choice are almost identical to those with the parameters chosen by Koester et al. (1994). B95 also point out that the UV spectra alone do not allow a unique solution: even if the parametrization of MLT is kept fixed, the UV spectra can be fit equally well along a line in the $T_{eff}$ – log g diagram running from low $T_{eff}$, low log g to higher $T_{eff}$ and higher log g. The basic physical reason for this is that the appearance of the Ly $\alpha$ line wing with its satellites is largely determined by the ionization balance of hydrogen and the relative numbers of neutral and charged perturbers. In order to arrive at a unique solution, B95 use simultaneously the optical *and* the UV spectra.

The determination of atmospheric parameters for ZZ Ceti stars is also necessary to make full use of the asteroseismologic information on the in-ner structure. Because the variable DA have much simpler pulsation spectra than the other classes of variable white dwarfs, with much smaller num-bers of observed frequencies, the identification of pulsation modes from the power spectra alone is extremely difficult. Accurate atmospheric parame-ters are needed to constrain the ranges of temperature and masses allowed by asteroseismology.

## 2.  New Observations and Analysis

In this paper we present new results for the six objects G226-29, GD165, G117-B15A, L19-2, R548, and G29-38. The first four of these are close to the blue edge of the instability strip, based on previous temperature determinations and on their short periods, small amplitudes, and simple power spectra.

For all these objects, with the exception of G117-B15A, new optical spectra with very high S/N (100 - 300) were obtained at the DSAZ (Calar Alto) or ESO (La Silla). G226-29, GD165, G117-B15A, and G29-38 have new UV spectra obtained with the HST/FOS, and for G117-B15A a new, very accurate gravitational redshift measurement has become available (Reid

1996). Optical and UV spectra were analysed with a very large grid of model atmospheres, extending over the range of parameters expected for the variable DA and using more than 10 different versions of MLT.

## 2.1. G226-29

G226-29 is the brightest and presently the best observed ZZ Ceti. Averaging the time-resolved HST spectra taken by Kepler et al. (1995) we have obtained a very high S/N spectrum. Fig. 1 illustrates an analysis similar to that of B95. Because the UV spectrum only defines a relation between $T_{eff}$ and log g, they first determined a solution for the optical spectrum using different MLT parametrizations. For each choice of parameters, they kept the surface gravity fixed from the optical solution, and determined the $T_{eff}$, which best fits the UV spectrum. Using this procedure for all objects with appropriate observations they find that on average the choice ML2/$\alpha$=0.6 provides the most consistent results for the "optical" and "UV" temperatures.

The circles in Fig. 1 indicate the results of our fits to the optical spectra, the crosses are fits to the UV spectra, obtained by keeping the log g value from the optical solution fixed, as in B95. Using ML2 gives results with log g < 8.20, with $\alpha$ = 0.5, 0.6, 0.7, 0.8, 0.9, 1.0 from top to bottom. Using the arguments of B95 we would conclude from this that ML1/$\alpha$ = 1.5 provides a consistent solution at $T_{eff}$ = 12100 K, log g = 8.30. However, closer inspection shows that the UV fit at these parameters is not very good; the solution is also incompatible with a constraint derived from the parallax and the V magnitude, given by the diagonal dotted lines.

Detailed inspection of all fits, and many experiments with changes in observational reductions and the fitting procedures, have led us to the conclusion that the real errors in the parameter determinations are much larger than the usually very small formal errors given by the $\chi^2$ fitting. The fit results for these high S/N spectra are completely determined by systematic errors of either models or observations. Fits over a wide range of $T_{eff}$ are practically indistinguishable for the eye, and even for the formal solution the results may differ largely for two spectra taken in the same night at almost identical conditions. Theoretical spectra, e.g. with ML1/$alpha$ = 1.5 and ML2$\alpha$ = 0.6 are almost identical; yet the formal fitting to high S/N spectra can lead to significantly different solutions, as in this case. We conclude from this that results based solely on the fitting of optical Balmer lines should be considered with extreme caution; this also means that the method used by B95 is problematic and results for individual objects may be very uncertain. However, we emphasize that this present study agrees with their main result concerning the best choice of MLT parameters: ML2/$\alpha$ =0.6,

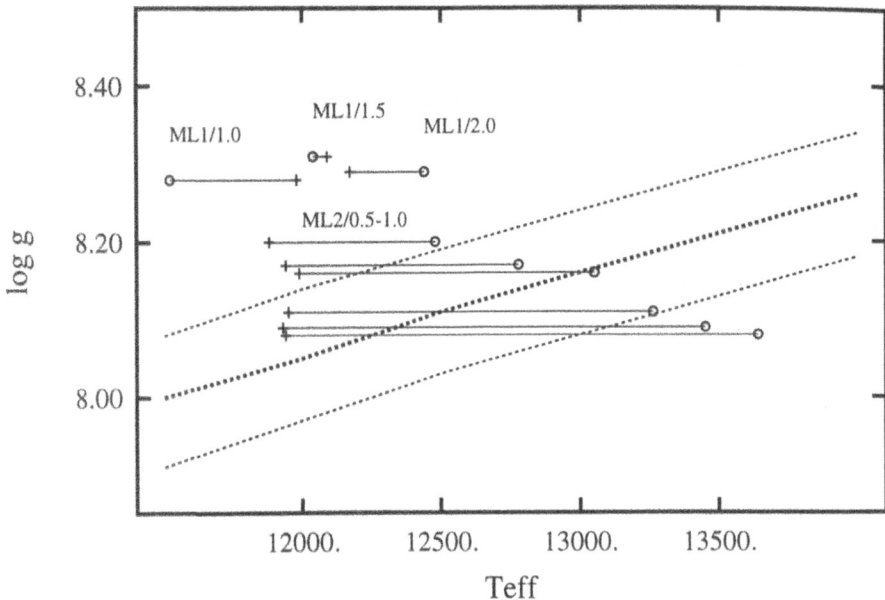

*Figure 1.* Fit to optical(circles) and UV spectra (crosses) of G226-29 using different versions of MLT. See text

or any choice with similar convective efficiency, gives the most consistent fit to all data.

The obvious conclusion for individual objects is that all available information needs to be used and that in case of conflicts the optical solution should be given a low weight. Applying this to G226-29 using the parallax 0.0819±0.0046 (van Altena et al. 1994) with the V magnitude, the absolute UV flux, and fits to the UV spectral shape leads to a consistent solution at $T_{eff} = 11830$ K, log g $= 8.05$.

## 2.2. THE OTHER VARIABLE DA

For GD165 we have obtained an UV spectrum with HST/FOS and found a consistent solution to optical, UV, and parallax at $T_{eff} = 11960$, log g $= 7.87$. The parallax, in this case, is not very accurate and does not give a very stringent constraint.

This is even more the case for G117-B15A, where the parallax is useless. However, Reid (1996) has obtained a highly accurate gravitational redshift of 24.2±0.8 km/s which demands log g $= 7.873±0.020$. The only solution, which is consistent with the UV spectrum, is then found at $T_{eff} = 11680$.

For G29-38 we obtain a consistent solution to the optical spectrum and the UV spectrum at $T_{eff} = 11600$, log g $= 8.05$. However, this solution is inconsistent with the parallax at the $4\sigma$ level, and we have not been able, even excluding the optical spectrum, to find a consistent solution for parallax, V, UV absolute flux, and UV spectral shape.

For R548 and L19-2 no UV spectra of good quality are available and the results are based on optical spectra only. They are thus subject to the uncertainties mentioned above and can only be considered as preliminary.

## 3. Results and Discussion

The final results for the six objects are summarized in Table 1. The last column shows the data consistent with this solution (optical or UV spectra, parallax, gravitational redshift. A : or :: means uncertain or very uncertain for the reasons discussed above. Fig. 2 shows these objects in a log g - $T_{eff}$

TABLE 1. Atmospheric parameters for 6 ZZ Ceti stars.

| object | $T_{eff}$ | log g | remarks |
|--------|-----------|-------|---------|
| GD165 | 11960 | 7.87 | opt, UV, $\pi$ |
| G226-29 | 11830 | 8.05 | UV, $\pi$ |
| G117-B15A | 11680 | 7.87 | UV, $V_{gr}$ |
| G29-38 | 11600 | 8.05 | : UV, opt |
| R548 | 12000 | 7.98 | :: opt |
| L19-2 | 12200 | 7.99 | :: opt |

diagram (squares), together with the results of B95 (circles) for the same objects. The solid lines are the theoretical blue edges (ML1, ML2, ML3 from left to right) of Brassard et al., taken from Fig. 14 in B95. While there are large differences between the results for individual objects, the general trend remains the same: the results agree with the theoretical prediction using ML2 (but $\alpha = 1.0$ !) for the envelope calculations. The surface gravities for our solutions are generally lower than in B95, and more in line with surface gravities for hotter DA. No stability analysis is available using ML2/$\alpha$=0.6 or a similar version, which is found to give the best fit to the spectra. As an approximation to its location we have used the criterion that the thermal time scale at the bottom of the convection zone is equal to the period, which results in a blue edge (dotted line) at much cooler $T_{eff}$ than the observed objects, when interpreted with the same MLT version. This confirms a conclusion reached already by B95: when MLT is used to describe convection, a more efficient version is necessary to describe the upper convection zone (emerging spectrum) than for the envelope stability analysis, which depends mostly on the depth of the convection zone.

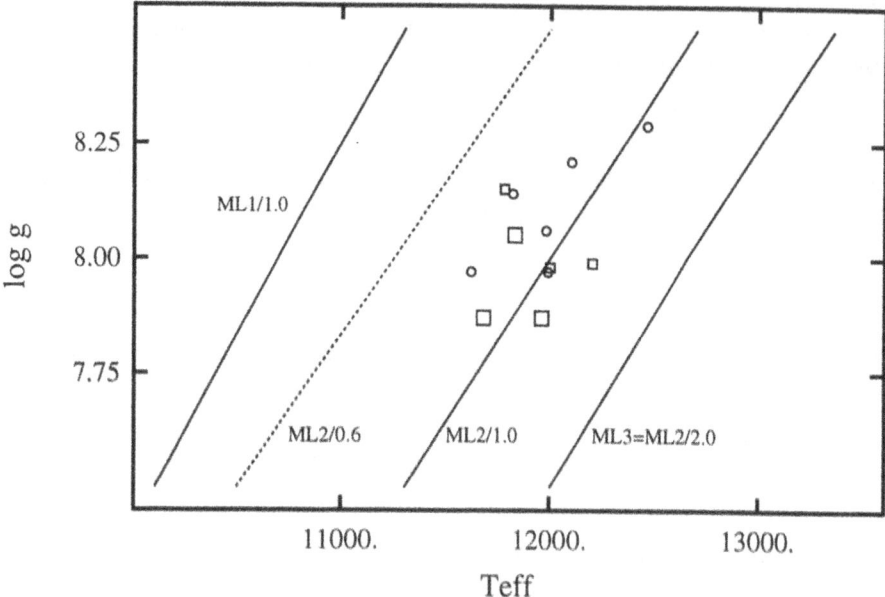

*Figure 2.*   Location of observed variables and theoretical blue edges. Squares are the results of this study, with larger squares for the more reliable results (first 3 in Table 1). Circles are results from B95 for the same objects. See text for explanations.

Finally we want to emphasize again that great caution is necessary when using atmospheric parameters based on optical spectral fitting in the ZZ Ceti region. These results can be subject to very large error and in general it is necessary to confirm the results with independent information.

## References

Allard N.F., Koester D. 1992, A&A 258, 464
Allard N.F., Koester D., Feautrier N., Spielfiedel A. 1994, A&AS 108, 417
Bergeron P., Wesemael F., Lamontagne R., Fontaine G., Saffer R.A., Allard N.F. 1995, ApJ 449, 258 (B95)
Bradley P.A., Winget D.E. 1994, ApJ 421, 236
Bragaglia A., Renzini A., Bergeron P. 1995, ApJ 443, 735
Daou D., Wesemael F., Bergeron P., Fontaine G., Holberg J.B. 1990, ApJ 364, 242
Fontaine G., Villeneuve B., Wilson J. 1981, ApJ 243, 550
Kepler O.S. et al. 1995, private communication
Koester D., Allard N.F., Vauclair G. 1994, A&A 291, L9
Reid I.N. 1996, AJ 111, 2000
van Altena W.F., Lee J.T., Hoffleit E.D. 1994, The General Catalogue of Trigonometric Parallaxes (New Haven: Yale University Observatory)

## Discussion

*D. O'Donoghue:* If you cannot help the asteroseismologists by determining $T_{eff}$ and log g, can they help you if, e.g., they determine the mass?

*D. Koester:* Every firm constraint we can use is a help. But for me the most basic and reliable data are spectra, colors, parallaxes, and gravitational redshifts. Asteroseismology may be more useful for the PG1159 and DBV objects, where you observe in excess of 100 excited modes.

*M.H. van Kerkwijk:* Have there been any attempts to use a mixing assumption that resembles reality better than ML theory?

*D. Koester:* There have been attempts by the hydrodynamics group in Kiel, for example. The problem is that you get a different answer depending on whether you want to describe the temperature structure, the energy flux, the velocity field (important for diffusion and mixing). The bottom line is that there is no flavor of MLT which correctly describes everything.

# WET OBSERVATIONS OF THE DAV VARIABLE HL TAU 76: FIRST RESULTS.

N.DOLEZ

*Observatoire Midi-Pyrénées*

AND

S.J.KLEINMAN

*University of Texas &*
*Big Bear Solar Observatory*
*California Institute of Technology*

**Abstract.** We present some preliminary results on the Whole Earth Telescope (WET) observations of HL Tau 76 obtained in February 1996 (Xcov13). We also re-analyzed the data of two previous monosite observations (obtained in 1989 and 1990). Six frequencies (in the range 1.6mHz to 2.6mHz) have been identified as independent modes, and many more frequencies present in the power spectra of the data which have been shown to be harmonic or linear combinations of the original six modes. We also present a comparison of these results with those on another group of DA white dwarfs with large amplitude oscillations and similar behaviour.

## 1. Introduction

DAV (or ZZ Ceti) white dwarfs have a fairly simple theoretical structure, due to gravitational settling of the chemical species ( "onion skins"). The number of known stars in this class (25) allows statistical interpretations so we can use both individual stars, and the group of stars to learn about the properties of DAs in general. Differences in spacing between successive radial overtones of the non-radial oscillations can be interpreted as a signature of the H/He and He/C transition zones. Hydrogen content in DA white dwarfs is a very important topic concerning the late stages of stellar evolution and the formation of planetary nebulae. All this makes the DAVs exciting targets for WET observations.

*I. Isern et al. (eds.), White Dwarfs, 437–443.*
© *1997 Kluwer Academic Publishers.*

HL Tau 76 is the first discovered ZZ Ceti (Landolt, 1968). It has been observed by several groups in the early seventies (Warner & Nather 1972, Pages 1972, Fitch 1973). The star belongs to a well-defined sub-group within the ZZ Ceti group, whose main features are:

- large number of observed frequencies.
- rather long periods (near 500 s)
- large amplitude
- cool temperatures

The stars in this group also tend to exhibit long-term variability in the pulsation amplitudes.

The choice of HL Tau 76 as a WET target was motivated by its previously known interesting behaviour : large number of observed frequencies, large amplitude (and large S/N expected), and possible amplitude changes in its modes.

## 2. Observations

The WET is a network of telescopes devoted to observing rapidly oscillating stars (Nather et al, 1990). The WET observed HL Tau 76 in the Xcov13 campaign (February, 1996) as a second priority target. The acquired light curve is presented in Figure 1 and the journal of observations in Table 1. Due to bad weather and the fact that the star was second priority, The coverage is nearly equivalent to what could be obtained in a monosite campaign.

TABLE 1. Xcov13 observations of HL Tau 76

| where | date | UT | duration (sec) |
|---|---|---|---|
| BAO 85-cm | 1996 Feb 16 | 12:22:40 | 4060 |
| NOT 2.5m * | 1996 Feb 17 | 21:19:40 | 6255 |
| BAO 85-cm * | 1996 Feb 20 | 11:29:40 | 16250 |
| BAO 85-cm * | 1996 Feb 21 | 11:06:00 | 17510 |
| BAO 85-cm * | 1996 Feb 22 | 11:08:50 | 16630 |
| OHP 193 | 1996 Feb 22 | 19:32:40 | 8480 |
| OHP 193 * | 1996 Feb 23 | 19:28:40 | 15330 |
| OHP 193 * | 1996 Feb 24 | 19:08:30 | 17220 |
| NOT 2.5m | 1996 Feb 25 | 22:21:00 | 9570 |

BAO :Beijing-XingLong Observatory
OHP :Haute Provence Observatory
NOT :La Palma Nordic Telescope
*Only the runs marked with an asterisk (*) were used in this study.*

Prior to the WET campaign, the star had been observed in several monosite campaigns. We present below analysis of the combined results of the WET observations, a 1989 monosite campaign (Haute Provence Observatory and Izana: see Auvergne et al 1992), and a 1990 monosite campaign (McDonald, see Kleinman 1995). Table 2 and Table 3 below present the journal of observations of those two campaigns.

TABLE 2.   November and December 1989 observations of HL Tau 76

| where | date | UT | duration (sec) |
|-------|------|-----|----------------|
| Izana-Teneriffe | 1989 Oct 29 | 00:29:00 | 19800 |
| Izana-Teneriffe | 1989 Nov 02 | 23:34:00 | 23280 |
| Izana-Teneriffe | 1989 Nov 04 | 00:07:00 | 17640 |
| Izana-Teneriffe | 1989 Nov 05 | 00:11:20 | 21600 |
| Izana-Teneriffe | 1989 Nov 07 | 01:49:20 | 15240 |
| Izana-Teneriffe | 1989 Nov 08 | 02:48:10 | 15000 |
| OHP-193 | 1989 Dec 26 | 18:03:00 | 29880 |
| OHP-193 | 1989 Dec 27 | 18:10:00 | 25320 |

*From this table, only the Izana data have been used in the present work.*

TABLE 3.   November 1990 observations of HL Tau 76

| where | date | UT | duration (sec) |
|-------|------|-----|----------------|
| McDonald 36-in | 1990 Oct 24 | 8:39:00 | 11710 |
| McDonald 36-in | 1990 Oct 25 | 5:22:00 | 25000 |
| McDonald 36-in | 1990 Oct 26 | 4:13:00 | 27220 |

The quality of the data obtained in these three data sets is extremely good. Unfortunately, the aliasing problems are still severe (even in the WET campaign, as the coverage was not sufficient) and prevent any positive identification of multiplets, and hence of the $\ell$ value of the modes.

## 3.  Analysis

We computed separate power spectra of the three data sets (Figure 2). The behaviour of the star is very similar in the three power spectra. Six frequencies appears to be independent modes and most of them are present in the three spectra. Many more frequencies are present which have been

shown to be harmonics or linear combinations of the six independent modes
(Table 4).

The periods of the six modes are in the range $1.4mHz$ to $2.6mHz$, (ie:
$1.44mHz, 1.52mHz, 1.68mHz, 1.85mHz, 2.02mHz, 2.61mHz$), putting the
star among the long period ZZ Cetis. The averaged period spacing based on
those six modes is 56 seconds. If they are $\ell = 1$ modes of successive radial
orders, this value is compatible with a mass of 0.6 solar mass (see Bradley
1994; Bradley & Kleinman 1996). However the spacing between the modes
shows discrepancies compared to the averaged value, which could be the
signature of trapping of oscillations in the composition transition regions
— "mode-trapping". This result is encouraging and we expect that a more
careful analysis of the data (with hopefully more observations and fitting
with theoretical models) could yield an accurate picture of the structure of
the star.

TABLE 4.  Xcov13 selected frequencies

| frequency $\mu Hz$ | amplitude $mmag$ | identification. | frequency $\mu Hz$ | amplitude $mmag$ | identification |
|---|---|---|---|---|---|
| 742 | 9.2 | | 2782 | 7.5 | |
| 939 | 18.3 | 5-2 | 3551 | 7.8 | 1+4 |
| 1072 | 23.2 | | 3677 | 7.8 | 2+4 |
| 1256 | 8.9 | | 3872 | 10.7 | 3+4 |
| 1280 | 9.1 | | 4029 | 4.5 | 4+4 |
| 1521 | 10.3 | 1 | 4483 | 5.0 | 3+5 |
| 1675 | 14.4 | 2 | 4638 | 3.0 | 4+5 |
| 1848 | 37.3 | 3 | 5364 | 4.1 | 4+2+2 |
| 2023 | 27.8 | 4 | 5547 | 3.3 | 4+2+3 |
| 2223 | 6.7 | | 5713 | 3.4 | 4+2+4 |
| 2614 | 20.1 | 5 | 5896 | 3.3 | 4+3+4 |

*Compare to fig.2 : most frequencies are present in the three data sets. Identification of
combinations can be uncertain due to aliasing problem.*

## 3.1.  DISCUSSION OF SOME UNIDENTIFIED MODES

Only one remaining high frequency mode could not be identified as a combination of the six "basic" modes. (ie: frequency at $6.197mHz$ in the 1989
data set). Although its amplitude is rather small, it seems to be real: the
fact that it fits fairly well the shape of the window function indicates that
it is present during most of the nights in the data set. At the other ex-
tremity of the spectra, the low frequencies can approximately be explained

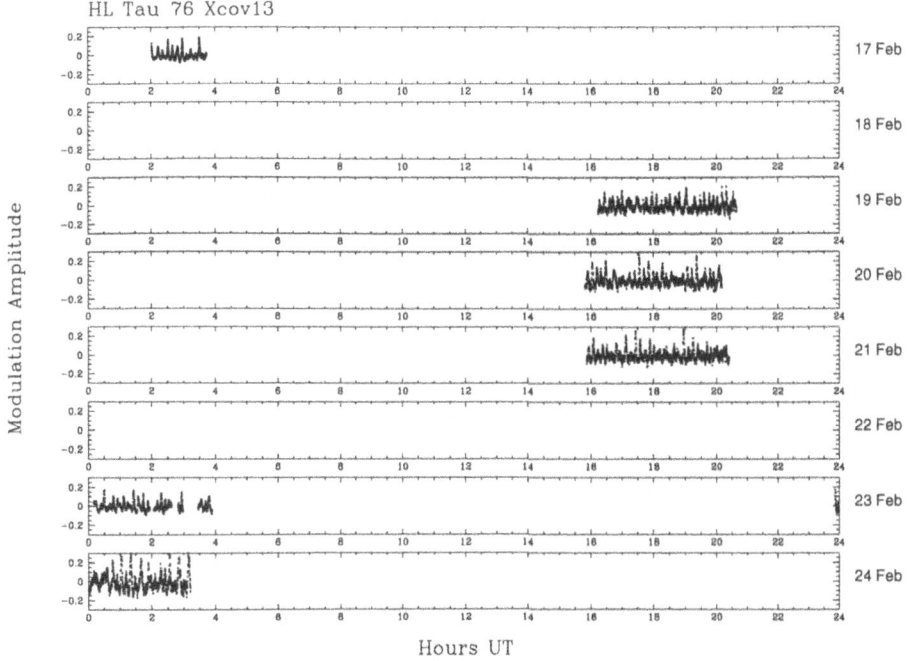

*Figure 1.*    Light curve from the Xcov13 WET campaign, February 1996. Included in this figure are only the data which have been effectively used for computing the power spectra.

as linear combinations of the six main modes, however the fit is only very approximate (for frequencies near $.92mHz$ and $1.05mHz$), the amplitude is rather large for difference frequencies, and the frequencies seem to change slightly from one data set to the other (for example, we get frequencies at $1.049mHz$ and $1.060mHz$ in December 1989, $1.090mHz$ in November 1989 and $1.072mHz$ in Xcov13). Some of these frequencies could very well be real modes of the star. The amplitudes of the oscillations in HL Tau 76 do show some variations when we compare the different data sets. However those variations are certainly much smaller than in comparable stars as G29–38 or G191–16, making this star an intriguing choice for further observations and analysis.

We performed some comparisons with G29–38 (Kleinman 1995) and found striking similarities: same period spacing and same range of periods for the largest amplitude modes. If we believe that the modes of similar frequencies are the same $k$ and $\ell$ in the two stars, we find a difference of 12 sec (plus or minus 4 sec) between the modes, the HL Tau 76 periods being

**HL Tau 76 Amplitude Spectra**

HL Tau 76 Amplitude Spectra

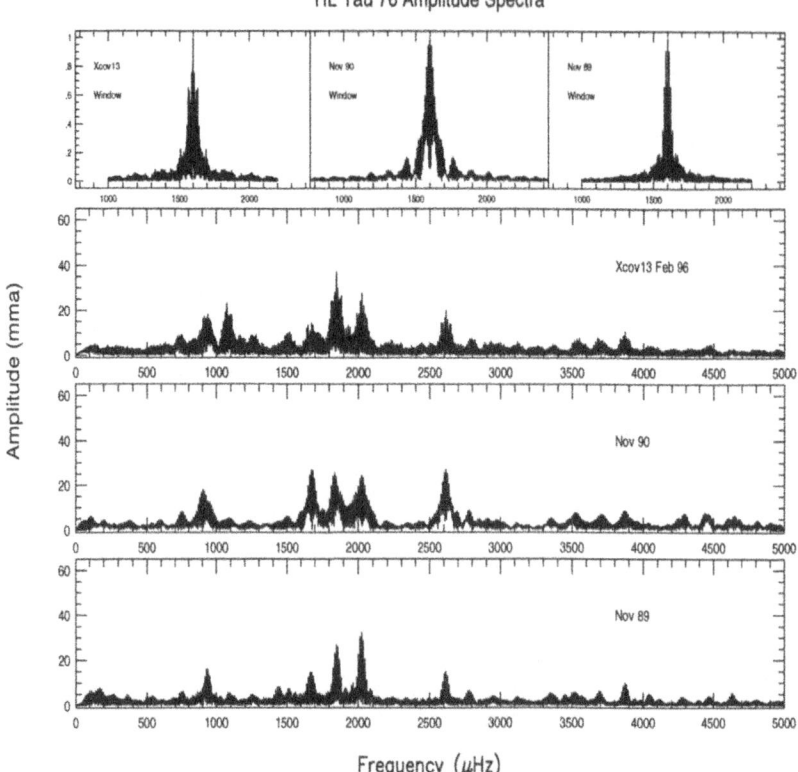

*Figure 2.* Power spectra of the three data sets. the first panel shows the three corresponding window functions. The following panels shows respectively (by group of three) the Xcov13, the October 1989 and the November 1990 power spectra.

the lowest (Fig 3 or Table 5). The similarity is also very clear with several other long periods DAs such as G191–16 , G255–2, and KUV8368+4026. Interestingly, several of those stars present a "gap" near 450sec. in the sequence of periods present in the spectra. As shown in Bradley & Kleinman (1996), in the case of G29–38, depending on the model, this gap can be filled by one or two modes, which in turn has large consequences on the possible internal structure of the star.

## 4. Conclusion

The WET campaign of observations on HL Tau 76has been extremely successful, and we obtained a very significant result. We can now, by combining all data obtained so far on HL Tau 76, present a much more accurate picture of the mode structure and long-term behaviour of the star.

The large amplitude, long period DAV white dwarfs seems to merge as a very homogeneous sub-group (including G29–38, HL Tau 76, G255–2, G19–

TABLE 5. Some non-combination modes
in HL Tau 76and G29–38

| period (sec) HL Tau 76 | period (sec) G29–38 | difference |
|---|---|---|
| 657 | 655 | -2 |
| 597 | 612 | 15 |
| 541 | 552 | 11 |
| 494 | 495 | 1 |
| 382 | 400 | 18 |

16 and possibly KUV8368–4026), and as such constitute a very promising class of objects to be studied.

Concerning HL Tau 76, we still have to reach a clear identification of degree and orders of the modes. For that we need a better coverage on a multisite campaign, to be able to resolve the multiplets structure of the modes, and thus, with the possible detection of more modes, be able to identify them. For several observational reasons, HL Tau 76 appears still to be one of the best candidates of its class for this kind of observations. Because of the homogeneity of the group, observing even one object belonging to it could give a great insight to the whole.

**List of WET participants:** Dolez N., Kleinman S., Vauclair G., Chevreton M., Winget D., Provencal J., Nitta A., O'Brien S., Breger M., Zola S., Krzesinski J, Moskalik P., Pajdosz G., Solheim J.E., Pfeiffer B., O'Donoghue D., Kanaan A., Clemens C., Nather E, Sullivan D., Marar T.M.K., Ashoka B.N., Leibowitz E., Hemar S.

## References

Auvergne M., Chevreton M., Belmonte J.A., Vauclair G., Dolez N., Goupil M.J. (1991), in "White Dwarfs", *Proceedings of the 7th European Workshop, G.Vauclair & E.Sion (Eds), Kluwer Academic Publisher,* p.167

Bradley P. (1994), in "White Dwarfs",*Proceedings of the 9th European Workshop, Koester D., Werner K. (Eds),* Springer Verlag, p. 285

Bradley P., Kleinman S. (1996),*These Proceedings*

Fitch W.S. (1973), *ApJ,* **181,** *L95*

Kleinman S.J. (1995), PhD Thesis

Landolt A.U. (1968), *ApJ,* **153,** *151*

Nather R.E., Winget D.E., Clemens J.C., Hansen C.J., Hine B.P., (1990),*ApJ,* **361,***309*

Pages C.G. (1972), *M.N.R.A.S.,* **159,** *25*

Warner B. Nather R.E. (1972), *M.N.R.A.S.,* **156,** *1*

# PRELIMINARY ASTEROSEISMOLOGY OF G 29–38

PAUL BRADLEY
*Los Alamos National Laboratory*
*Los Alamos, NM 87545*

AND

SCOT KLEINMAN
*Dept. of Astronomy and McDonald Observatory*
*University of Texas*
*Austin, TX 78712*

**Abstract.** G 29–38 is a cool pulsating DA white dwarf with a series of modes from 110 s to over 1000 s found by Kleinman et al.. We present the results of our attempts to find a seismological fit to G 29–38 by looking for a model that identifies most of the observed modes as $\ell = 1$ modes. G 29–38, like several other DAV stars, has a discrepancy between the spectroscopically favored mass of about $0.70 M_\odot$ and the parallactic mass of about $0.85 M_\odot$. Our preliminary model has a mass of $0.75 M_\odot$ and a hydrogen layer mass of $5 \times 10^{-7} M_\star$, and it requires that most modes be $\ell = 1$.

## 1. Introduction: G 29–38 and the ZZ Ceti Stars

A major controversy in the white dwarf field concerns the thickness of the hydrogen layer in DA white dwarfs. Fontaine & Wesemael (1997) and Shipman (1997) discuss the available evidence from two separate viewpoints and both agree they are consistent with a range of hydrogen layer masses from "thick" ones at $\sim 10^{-4} M_\star$ to "thin" ones that are closer to $10^{-10} M_\star$. However, they could not say what the relative fraction of thick versus thin DAs is or if there is actually a continuum of hydrogen layer masses between the two extremes.

One way to directly determine the hydrogen layer mass of a DA is through asteroseismology of the pulsators, called the ZZ Ceti white dwarfs.

*I. Isern et al. (eds.), White Dwarfs, 445–450.*
© *1997 Kluwer Academic Publishers.*

The ZZ Ceti white dwarfs lie between 12, 500 K and 11, 000 K and are driven
by a mixture of the $\kappa,\gamma$ mechanism and convection-pulsation interactions
(see Brassard & Fontaine 1997). A nice feature of the ZZ Ceti stars is their
normality; other than their pulsations, they do nothing to stand out from
the crowd of normal DA white dwarfs. Because of this, we believe that what
we learn about the ZZ Ceti stars should apply to the vast majority of DA
white dwarfs. DAs make up about 80% of all spectroscopically identified
white dwarfs, so ZZ Ceti star structure could shed light on the structure of
most white dwarfs. This potential is already being realized, as Fontaine &
Wesemael (1997), Pfeiffer et al. (1996), and Bradley (1995, 1997) describe.
Most of the white dwarfs they discuss are among the hottest ZZ Ceti stars,
and *all* of them have only a few observed linearly independent modes. What
we really need is a DA white dwarf that has many modes present, so we
can tightly constrain the hydrogen layer mass for at least one DA white
dwarf. This is where G 29–38 comes in.

## 2.  Meet G 29–38

Kleinman et al. (1997) describes the observed mode structure and observations, so here we will only mention the facts vital to this work. G 29–38 is
the first ZZ Ceti star for which we know there are a large number of modes,
with at least 17 modes that Kleinman et al. (1997) suggest are mostly
$\ell = 1$ modes. G 29–38 is one of the cooler ZZ Ceti stars, so it offers us a
chance to see how its hydrogen layer mass compares to the younger, hotter
ZZ Ceti stars that are typically $10^8$ yr younger. GD 154 is the only other
"cool" ZZ Ceti star with a tentative hydrogen layer mass; it is possibly the
thinnest, at $\sim 10^{-10} M_\star$ (Pfeiffer et al. 1996). Determining the hydrogen
layer mass for G 29–38 might help us decide if we need to seriously consider some unlikely evolution scenarios, such as the hydrogen layer mass
changing from thick to thin as a white dwarf cools through the ZZ Ceti
instability strip.

G 29–38 has some observational discrepancies and oddities that make
it especially interesting (or troublesome). First, G 29–38 is unique amongst
the known ZZ Ceti stars because it displays a strong infrared excess that
is currently believed to be re-radiation by circumstellar dust (Zuckerman
1993). G 29–38 also has a discrepancy between the spectroscopic mass
$(0.69 M_\odot)$ and the mass implied by trigonometric parallax $(0.85 M_\odot)$. This
problem is seen in other ZZ Ceti stars, with the most spectacular case
being R 548 $(0.59 M_\odot$ vs. $\sim 0.2 M_\odot)$. The most recent parallax value is
$0.''0734 \pm 0.''0040$ (van Altena, Lee, & Hoffleit 1994), which is not much
different from the previous value of $0.''0709 \pm 0.''0042$ reported by Harrington & Dahn (1980). We use Bergeron et al.'s (1995) temperature of

$11,800 \pm 250$ K and gravity of $8.14 \pm 0.05$ as our reference values.

To fit the observed periods of G 29–38, we use models generated with the white dwarf evolution code discussed by Bradley (1996) and Wood (1994) in our analysis. The pulsation periods of some models are in tables $1-12$ in Bradley (1996). We use a slightly modified version of the nonradial pulsation analysis code described by Kawaler, Hansen, & Winget (1985). Bradley (1996) discusses the application of this Runge-Kutta-Fehlberg code to DA white dwarf models. For our purposes, it is enough to know these codes compute theoretical periods with better than 0.1 s accuracy except for periods greater than $\sim 900$ s, where uncertainties in the outer layer structure of our models limits the accuracy.

## 3. Description of Seismological Fits

Here, we decribe our attempts to fit models to the suite of observed periods. At present, no one model fits all of the observed periods, but a number of them match at least some periods. At this point, we do not consider the existing mismatches serious, because we are more interested in seeing if an $\ell = 1$ mode interpretation is possible for G 29–38 and providing a guide for future observations. We describe how well we can match the observations for models of a given mass. We believe this will give the reader insight into how we perform seismology and make it clear why we prefer $0.75 M_\odot$ models. At $0.75 M_\odot$, we have some models that fit almost all the periods to within 10 s or 1%.

### 3.1. G 29–38 MAY BE $\leq 0.65 M_\odot$

Models at $0.60 M_\odot$ or less can only match the 110 s period as an $\ell = 2$, $k = 1$ mode, but only for hydrogen layer masses less than $10^{-5} M_\star$. When a model fits some periods, it tends to fit either the low period ($< 600$s) or longer period ($> 500$s) modes well, but not both. This indicates that the mean period spacing between modes is too large in the models. The best looking fit is at $M_H \sim 10^{-5} M_\star$, but it is not convincing. These models have seismological distances 5 to 6 pc farther than trigonometric parallax allows.

### 3.2. G 29–38 MAY BE $\approx 0.70 M_\odot$

These models have $\log g \approx 8.17$, which is within the errors of the observational value of 8.14. However, no model matches the 110 s or 237 s modes and most fail at 177 s. Some models do well between 400 to 800s and when this happens, the hydrogen layer mass tends to lie between $10^{-4} M_\star$ and

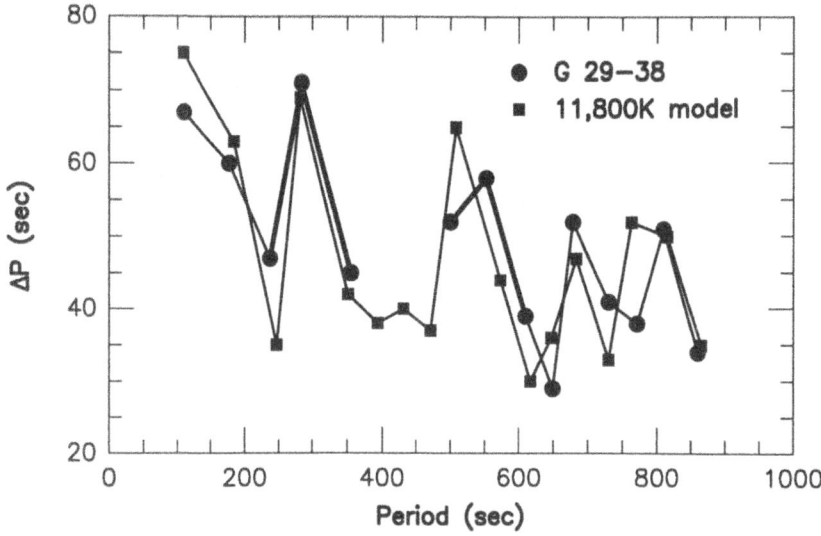

*Figure 1.* A comparison of the observed and theoretical period spacing ($\Delta P$) diagrams. The model (model B in Table 1 matches the general trends of the observed periods.

$10^{-6}M_\star$ . Here, the seismological and spectroscopic distance is about 3 pc greater than the parallax distance.

### 3.3.  G 29–38 IS $\approx 0.75 M_\odot$

The seismological $\log g \approx 8.26$ is a bit higher than model atmospheres predict. However, models with $M_H \approx 5 \times 10^{-7} M_\star$ fit quite well, predicting all oberved periods to within 10 s (save for the 552 s mode) out to 900 s (see Table 1 and Figure 1). The best model says the 110 s mode is an $\ell = 2$, $k = 1$ mode, while all the rest are $\ell = 1$ modes. The seismological distance is 15.2 pc, about $2\sigma$ higher than the parallax value of 13.6 pc.

Assuming model B (the best fit) represents the internal structure of G 29–38, we examine how closely we can constrain the surface layer masses and the location of the C/O interface. As Table 1 indicates, changing the helium or hydrogen layer mass by a factor of two (model C and D, repectively) causes a noticeably poorer fit. Even moving the location the C/O core transition from 20:80 C/O to pure C from $0.75 M_\star$ to $0.80 M_\star$ has a noticeable effect. This shows that once we are more certain of the mode identification and eliminate any other possible model choice, we are in an excellent position to constrain the structure of G 29–38.

TABLE 1.  Comparison of G 29–38 to Models

| Name | G 29–38 | ch751057 | y751057 | ch753557 | ch751056 |
|---|---|---|---|---|---|
| Alias | ZZ Psc | Model A | Model B | Model C | Model D |
| $T_{\mathrm{eff}}$ (K) | 11,820 | 11,830 | 12,100 | 11,910 | 11,620 |
| $\log g$ | 8.14 | 8.26 | 8.26 | 8.26 | 8.26 |
| $M_\star / M_\odot$ | 0.69 | 0.75 | 0.75 | 0.75 | 0.75 |
| $M_{\mathrm{H}}/M_\star$ | | $5 \times 10^{-7}$ | $5 \times 10^{-7}$ | $5 \times 10^{-7}$ | $10^{-6}$ |
| $M_{\mathrm{He}}/M_\star$ | | $10^{-2}$ | $10^{-2}$ | $5 \times 10^{-3}$ | $10^{-2}$ |
| dist. (pc) | 13.6 | 15.3 | 15.2 | 15.6 | 15.3 |
| $(\ell, k)$ | | Period (s) | | | |
| (2, 1) | 110 | 109 | 109 | 112 | 112 |
| (1, 1) | 177 | 184 | 184 | 191 | 181 |
| (1, 2) | — | 198 | 199 | 201 | 189 |
| (1, 3) | 237 | 238 | 247 | 233 | 238 |
| (1, 4) | 284 * | 280 | 282 | 287 | 283 |
| (1, 5) | 355 | 351 | 351 | 354 | 334 |
| (1, 6) | 400 * | 392 | 393 | 405 | 394 |
| (1, 7) | — | 436 | 431 | 430 | 437 |
| (1, 8) | — | 471 | 471 | 481 | 458 |
| (1, 9) | 500 | 507 | 508 | 518 | 502 |
| (1,10) | 552 | 565 | 573 | 567 | 564 |
| (1,11) | 610 * | 606 | 617 | 607 | 592 |
| (1,12) | 649 | 633 | 647 | 642 | 630 |
| (1,13) | 678 * | 684 | 683 | 691 | 680 |
| (1,14) | 730 | 728 | 730 | 728 | 714 |
| (1,15) | 771 | 754 | 763 | 762 | 748 |
| (1,16) | 809 | 798 | 815 | 806 | 795 |
| (1,17) | 860 | 851 | 865 | 857 | 833 |
| (1,18) | 894 | 891 | 900 | 902 | 883 |
| (1,19) | 915 | 945 | 947 | 960 | 932 |

## 3.4.  G 29–38 MAY BE $\approx 0.80 M_\odot$

When the mass increases to $0.80 M_\odot$, the model surface gravity is 8.32, and we still have a distance (14.8 pc) that is larger than the parallax value. We are finally able to have an $\ell = 1$, $k = 1$ period near 110s, when we allow $M_{\mathrm{H}} \approx 10^{-4} M_\star$. However, the fits to the other periods are not that good, and the best case occurs when $M_{\mathrm{H}} \approx 5 \times 10^{-5} M_\star$.

3.5.  G 29–38 MAY BE $\approx 0.85 M_\odot$

Only when we move to $0.85 M_\odot$ do we produce a model whose luminosity is consistent with trigonometric parallax, but at the expense of a whopping suface gravity near 8.40, a full 0.25 dex above the spectroscopic value. Here, only portions of the model period spectrum fits the observations; in this case, the mean period spacing of the models is too short to match.

## 4.  Verdict

Based on the available modeling effort and the requirement that most of the modes be $\ell = 1$, we believe the mass of G 29–38 lies near $0.75 M_\odot$, with a hydrogen layer mass near $5 \times 10^{-7} M_\star$, and a helium layer mass near $10^{-2} M_\star$. We urgently need additional observational constraints on the $\ell$ identity of the observed modes, and suggest that further searches for $m$-splitting be carried out. Detecting additional modes may also help, especially if we can find $\ell = 2$ modes. Finally, our hydrogen layer mass of $5 \times 10^{-7} M_\star$ falls in the middle of the extremes between thick and thin hydrogen layer masses. This provides one piece of evidence in support of a continuum of hydrogen layer masses, but we need more evidence before this assertion is significant.

## References

Bergeron, P., Wesemael, F., Lamontagne, R., Fontaine, G., Saffer, R.A., & Allard, N.F. 1995, ApJ, 449, 258

Bradley, P.A. 1995, Proc. 9th European Workshop on White Dwarf Stars, eds. D. Koester & K. Werner (Berlin: Springer), 284

Bradley, P.A. 1996, ApJ, 468, 350

Bradley, P.A. 1997, in preparation

Brassard, P., & Fontaine, G. 1997, these proceedings

Brickhill, A.J. 1991, MNRAS, 251, 673

Fontaine, G., & Wesemael, F. 1997, these proceedings

Harrington, R.S., & Dahn, C.C. 1980, AJ, 85, 454

Kawaler, S.D., Hansen, C.J., & Winget, D.E., 1985, ApJ, 295, 547

Kleinman, S.J., et al. 1997, these proceedings

Pfeiffer, B., et al. 1996, A&A, in press

Shipman, H.L. 1997, these proceedings

van Altena, W.F., Lee, J.T., & Hoffleit, E.D. 1994, The General Catalogue of Trigonometric Parallaxes, (New Haven: Yale University Observatory)

Wood, M.A. 1994, in Proc. of IAU Colloq. 147, The Equation of State in Astrophysics, ed. G. Chabrier & E. Schatzman, (Cambridge: Cambridge Univ.), 612

Zuckerman, B. 1993, in ASP Conf. Ser. in Planets Around Pulsars, eds. J.A. Phillips, J.E. Thorsett, & S.R. Kulkerni, ASP Conf. Ser., (San Francisco: ASP) 36, 303

# RECENT ADVANCES IN THE THEORETICAL DETERMINATION OF THE ZZ CETI INSTABILITY STRIP

P. BRASSARD AND G. FONTAINE

*Département de Physique, Université de Montréal*

## 1. Introduction

We present here the preliminary results of our efforts to improve the theoretical determination of the locus of the ZZ Ceti instability strip. The main goals of this project are: to find a blue edge compatible with the observations, to investigate the effects of various convection theories, and to find a red edge. To reach these goals we make many improvements over previous studies. These improvements concern state-of-the-art constitutive physics, realistic atmospheric structures, new pulsation codes, and additional terms in the equations of pulsation.

## 2. Input Physics

### 2.1. EQUATION OF STATE

The equation of state is made of three parts corresponding to different regions of the $(\rho, T)$ plane. For the first region, of primary relevance in the high atmosphere, the equation of state is obtained by the solution of the Saha equations with a small correction term for the Coulomb interaction. The second part is the region of partial ionization, located approximately at the base of the enveloppe. For the H and He species, we use the Saumon, Chabrier and Van Horn (1995; referred as SCV) equation of state. The SCV equation of state is complemented by an improved version of the Fontaine, Graboske and Van Horn (1977) equation of state from Fontaine (1993; referred as IFGV) to cover entirely that region. For the C species, we use the IFGV equation of state. The O equation of state is given by the solution of the Saha equations with a Coulomb term and a pressure ionization term from Eggleton, Faulkner and Flannery (1973). In the last

*I. Isern et al. (eds.), White Dwarfs, 451–457.*

region, corresponding to the deep interior with complete ionized material, we use the Lamb (1974) equation of state.

From these different sources of data we have built four large tables of data, one for each species (H, He, C and O). Special care has been taken to insure a smooth transition of the various quantities between the different regions and sources of data.

## 2.2. OPACITIES

For the radiative opacities, we use the new OPAL 1995 data from Iglesias and Rogers (1996). At low temperatures, for the H species we use the opacities with quasi-molecular absorption from Allard et al. (1994) and those kindly provided to us by P. Bergeron (1996) for He. The opacities used here are for pure composition ($Z = 0$).

We have developed a set of polynomial fits for the conductive opacities. Data are from Hubbard and Lampe (1969) for the low densities and from Itoh et al. (1983, 1984, 1993a, 1993b) and Mitake et al. (1984) for the high densities (solid and liquid phases). The purpose of these fits is to smooth the transition between the two regions and to act as an extrapolant for the regions in the $(\rho, T)$ plane not covered by tabulated data. We have allowed for a contribution from the heavy elements in the deep interior by setting $Z = 0.02$ in the determination of the conductive opacities.

## 2.3. CONVECTION

Convection is the main source of uncertainties in building stellar models. Despite many flaws, the mixing-length theory is the standard workhorse in stellar modelization. We have tested various parameterizations for the mixing-length. For comparisons purposes, we have used the three parameterizations of Tassoul, Fontaine and Winget (1990, hereafter TFW): ML1, ML2 and ML3. The ML1 parameterization corresponds to the standard version of the mixing-length theory according to Böhm-Vitense (1958), while the ML2 parameterization is the version of Böhm and Cassinelli (1971). The differences between the ML1 and ML2 version of the mixing-length theory arise only from the different geometrical factors associated to different assumed shapes for the eddies. The ML3 parameterization is actually a sub-variety of the ML2 parameterization assuming a higher convective efficiency: $\alpha$, the ratio of the mixing-length over the local pressure scale height is 2.0 instead of 1.0. More recently, Bergeron et al. (1995) found that a mixing-length parameterization assuming a ML2 geometry, but with $\alpha = 0.6$, brings the most consistent $T_{\text{eff}}$ and $\log g$ determinations in agreement with optical spectroscopy, ultraviolet spectroscopy, photometry, trigonometric parallax measurements, and gravitational redshift masses data. In the rest of this paper, we will refer to this parametrization as the ML6 version of the mixing-length theory.

We have also allowed our models to be build with the Canuto-Mazitelli (1991,1992) theory of convection. The main benefits of this new theory over the mixing-length theory are: a large spectrum in the size of the eddies, $\alpha$ is replaced by a non-local expression, and, according to the authors, this is essentially a parameter-free theory.

Because we deal with complete stellar models, we have modified the mixing-length and the Canuto-Mazitelli theories to take account for optically thin material.

## 2.4. ATMOSPHERIC STRUCTURES

At large optical depth ($\tau \gg 1$) the radiative gradient is given by the diffusion approximation:

$$\nabla_{\text{diff}} = \frac{3}{16\sigma} \frac{P\kappa}{gT^4} F. \tag{1}$$

For the optically thin regions, a first approximation is to use gray atmosphere. For a purely radiative grey atmosphere, the temperature stratification is given by the grey relation $T^4 = \frac{3}{4} T^4_{\text{eff}} [\tau + H(\tau)]$, where $H(\tau)$ is the Hopf function. The temperature gradient is thus given by

$$\nabla_{\text{grey}} = \frac{3}{16\sigma} \frac{P\kappa}{gT^4} F[1 + H'(\tau)]. \tag{2}$$

By analogy with these expressions, we define a realistic temperature gradient which is particularly of interest in presence of convection

$$\nabla_{\text{real}} = \frac{3}{16\sigma} \frac{P\kappa}{gT^4} W, \tag{3}$$

where $W = W(P, T, \tau)$ is a function extracted from detailed atmospheric computations. To build our models, we have extracted tabular data for the $W$ function (ML1, ML2, ML3 and ML6 convection) from the atmospheres models of Bergeron et al. (1995).

Our models are built with $\nabla_{\text{real}}$ as the radiative gradient along with a boundary condition also obtained from the models of Bergeron et al. (1995). For comparison, the models of TFW are built with $\nabla_{\text{diff}}$ as the radiative gradient and the boundary condition is obtained from the grey relation. The models of Wood (1990) are very similar, except that he uses $\nabla_{\text{gray}}$ for the radiative gradient from the top of the atmosphere down to the top of the convective zone where the gradient is switched to $\nabla_{\text{diff}}$. Figure 1 provides a comparison for the temperature stratification of one of our models with one from TFW.

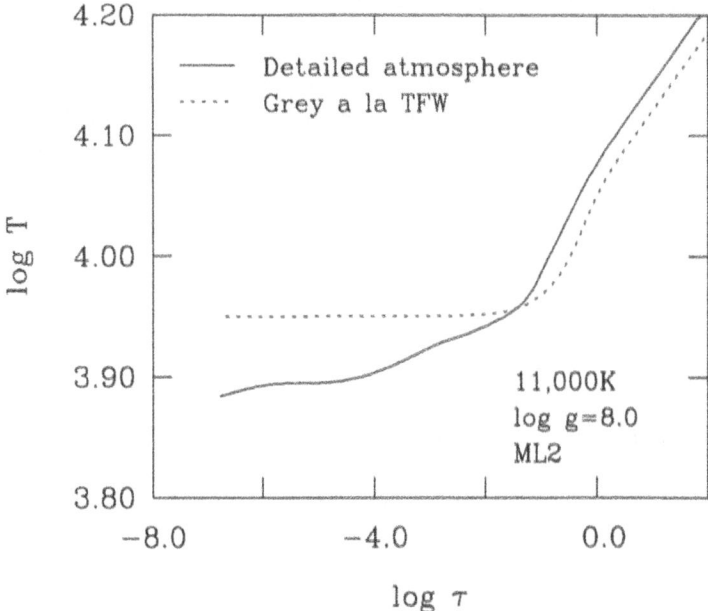

*Figure 1.*   Temperature stratification of a grey atmosphere compared to a detailed one, obtained from the solution of the transfert equation

## 3.  Method and Results

The differential equations for linear, nonadiabatic and nonradial oscillations are obtained after an application of a periodic perturbation on the usual equations of stellar structure (see, for example, Unno et al. 1989). In this derivation, many approximations are used. From these, three may be of importance for white dwarf seismology: the radiative gradient is given by $\nabla_{\text{diff}}$, thermal imbalance terms are neglected, and the perturbation of the convective flux is also neglected. This last approximation is usually referred to as the "frozen convection" approximation in the literature.

These three limitations have been taken care of in our pulsation equations: we use $\nabla_{\text{real}}$ for the radiative gradient, and perturbations of the thermal imbalance terms and of the convective flux have been included. Perturbations of the convective flux is obtained from the pertubations of the mixing-length (or Canuto-Mazitelli) theory equations. This procedure is equivalent to assuming that convection reacts *instantaneously* to the change in the physical conditions induced by the presence of oscillations. This is a good approximation because, for white dwarf stars, we find that the convective turn-over time scale *near the blue edge* is more than two orders of magnitude lower than a typical period of oscillation.

To numerically solve this new pulsation equations set, we use the finite element method of Brassard et al. (1992) together with our new generation of stellar models. Blue edges obtained for the ML1, ML2, ML3 and ML6 mixing-length theories are shown in Figure 2. The four dots in each panel

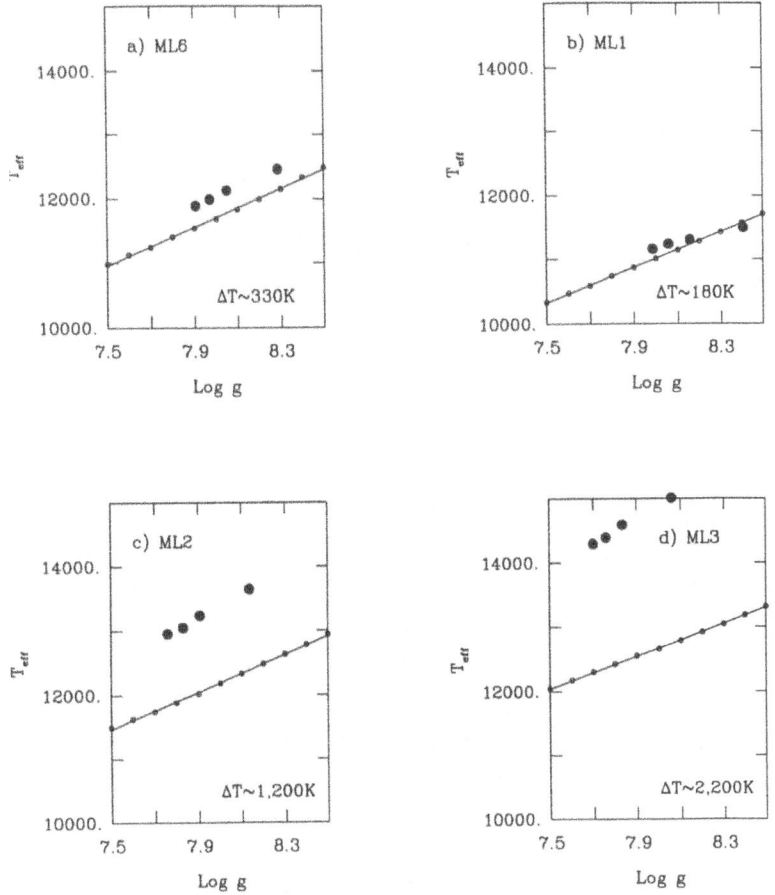

*Figure 2.* Theoretical blue edges obtained for different version of the mixing-length theory along with the observational blue edge derived from atmospheric parameters determination

of Figure 2 correspond to the observational blue edge. To get this blue edge, we have extracted the four stars that seem to be at or near the blue edge from the sample of the 22 ZZ Ceti stars analyzed by Bergeron et al. (1995). Also, we can note that these comparisons are self-consistent because our models reproduce the same temperature stratification, as well as the same convection zone localization, than those from Bergeron et al. (1995).

## 4. Conclusion

As the consequence of the improvements that we made in the description and analysis of our models, we find that it is possible to obtain theoretical blue edge localizations that are in good agreement with the observations. From Figure 2, we see that this is possible only for theories with low convective efficiencies (ML1, ML6, CM [not shown]). This result is contrary to previous theoretical determinations of the blue edge that require higher convective efficiency (ML2, ML3) to be compatible with the observations (see, for example, Wesemael et al. 1991). This result is also compatible with recent hydrodynamic computations (Ludwig et al. 1994) showing an effective low convective efficiency. Finally, at this time, we still do not find any red edge localization.

This work was supported in part by the NSERC Canada and by the fund FCAR (Québec).

## References

Allard, N.F., Koester, D., Feautrier, N., and Spielfiedel, A. 1994, *Astr.Ap.Suppl.*, **108**, 417

Bergeron, P., 1996, private communication

Bergeron, P., Wesemael, F., Lamontagne, R., Fontaine, G., Saffer, R.A., and Allard, N.F. 1995, *Ap.J.*, **449**, 258

Böhm, K.-H., and Cassinelli, J. 1971, *Astr. Ap.*, **12**, 21

Böhm-Vitense, E. 1958, *Zs. Ap.*, **46**, 108

Brassard, P., Pelletier, C., Fontaine, G., and Wesemael, F. 1992 *Ap.J.Suppl.*, **80**, 725

Canuto, V.M., and Mazzitelli, I. 1991 *Ap.J.*, **370**, 295

Canuto, V.M., and Mazzitelli, I. 1992 *Ap.J.*, **389**, 724

Eggleton, P., Faulkner, J., and Flannery, G.P., 1973 *Astr.Ap.*, **23**, 261

Fontaine, G., 1993, private communication

Fontaine, G., Graboske, H.C., Jr., and Van Horn, H.M. 1977, *Ap.J.Suppl.*, **35**, 293

Hubbard, W.B., and Lampe, M. 1969, *Ap.J.Suppl.*, **18**, 297

Iglesias, C.A., and Rogers, F.G. 1996, *Ap.J.*, **464**, 943

Itoh, N., Hayashi, H., and Kohyama, Y. 1993a, *Ap.J.*, **418**, 405

Itoh, N., and Kohyama, Y. 1993b, *Ap.J.*, **404**, 268

Itoh, N., Kohyama, Y., Matsumoto, N., and Seki, M. 1984, *Ap.J.*, **285**, 758

Itoh, N., Mitake, S., Iyetomi, H., and Ichimaru, S. 1983, *Ap.J.*, **273**, 774

Lamb, D.Q. 1974, Ph. D. Thesis, University of Rochester

Ludwig, H.-G., Jordan, S., and Steffen, M. 1994 *Astr.Ap.*, **294**, 105

Mitake, S., Ichimaru, S., and Itoh, N. *Ap.J.*, **277**, 375

Saumon, D., Chabrier, G. and Van Horn, H.M., 1995, *Ap.J.Suppl.*, **99**, 713

Tassoul, M., Fontaine, G. and Winget, D.E. 1990, *Ap.J.Suppl.*, **72**, 335

Unno, W., Osaki, Y., Ando, H., Saio, H. and Shibahashi, H. 1989, *Nonradial Oscillations of Stars* (Tokyo: University of Tokyo Press)

Wesemael, F., Bergeron, P., Fontaine, G., and Lamontagne, R. 1991, in *White Dwarfs*, NATO ASI Series, Vol. 336, eds. G. Vauclair and E. Sion (Dordrecht: Kluwer), p.159.

## Discussion

*S. Jordan*: I would be very careful saying that CM description is parameter free, because there are assumptions in it that are probably not valid in a white dwarf atmosphere. None of the hydrodynamic calculations of the Kiel group indicate a stratification of eddies of different sizes. They rather show downdrafts at some positions that sometimes merge into strong downward jets. Between them, the atmosphere is relatively calm!

*P. Brassard*: As I said, that is what they claim.

*J. Kubát*: 1) What does your "detailed" atmosphere mean? 2) Would you expect changes in your results if you use some more realistic model atmosphere (e.g. NLTE line-blanketed)?

*P. Brassard*: 1) I mean an atmosphere with the same temperature stratification as a complete model atmosphere. 2) No. For nonadiabatic computations, we need only the thermal and mechanical structure of the atmosphere and NLTE effects are negligible in this case.

*G. Chabrier*: Pierre, when you perturb the convection equations ("unfrozen convection"), which time-dependent convection theory do you use?

*P. Brassard*: None; we assume that convection adjusts instantaneously.

*M.H. van Kerkwijk*: What is the problem in finding the red edge? Any idea how to make progress?

*P. Brassard*: 1) We found that, near the red edge, the convective turn-over time scale becomes comparable to the pulsation periods. In this case, a complete time-dependent convection theory may be necessary. 2) We use a linear theory. If the red edge arise from non-linear effects, we cannot reproduce them.

*A. Talon*: Hop!

# COOL DAV ASTEROSEISMOLOGY

S.J. KLEINMAN
*University of Texas*
*Currently: Big Bear Solar Observatory*
*California Institute of Technology*

## 1. Introduction

Through their non-radial, g-mode pulsations, the variable white dwarfs provide us with a means to look beyond their surfaces and peer into their inner compositions and structures. Because of their pulsational complexity, however, we require multi-site campaigns (the Whole Earth Telescope, WET Nather et al. 1990, for example) to gather enough data to isolate and identify their modes of oscillation. The WET has made remarkable progress with the DOVs and DBVs (Winget et al. 1991, Winget et al. 1994), but the DAVs have not readily revealed their interiors, even to long multi-site campaigners.

Traditionally, the DAVs are divided into two separate groups: a cool group with large amplitude, long period, and complex (i.e., non-linear) oscillations and a hotter group with smaller amplitude, shorter period, and simpler oscillations. The large amplitude pulsators provide us with many periods of variability, desirable for precise asteroseismological measurements, but most are linear combinations of other modes and many are unstable in amplitude and are thus unsuited for typical asteroseismological analysis. The lower amplitude pulsators have only a few modes — too few to make a believable mode identification for an individual star. Clemens (1993, 1994) showed that by treating the ensemble of the hot DAVs as a single star, he could make consistent mode identifications, suggesting all the DAVs have a very similar structure and fairly thick Hydrogen layer masses (see elsewhere in these proceedings for discussion on the H-layer mass issue).

The cool DAVs, however, remained a problem. With their variable, non-linear behavior, we were not even sure if it was worth studying them for asteroseismological measurements. Are their variations normal-modes that

*I. Isern et al. (eds.), White Dwarfs, 459–465.*
© *1997 Kluwer Academic Publishers.*

can be used like those in the other white dwarf pulsators for glimpses of their interiors, or are they simply a hodge-podge of non-linear effects which may indeed be instructive, but not as easily or directly as the rest of the white dwarf pulsators?

Armed with an immense data set on a number of cool DAVs, I dare to answer the last question in the affirmative: the cool DAVs are normal-mode pulsators and may actually be our best hope at obtaining detailed structure information for DAs. The approach to the problem was two-pronged: 1) obtain a long timebase of data (over 10 years) on a single DAV (G29–38) and 2) obtain as much data as possible (at least a few observing seasons) on as many different DAVs as possible. In this paper, I will only have space to discuss the first part of the approach in some detail (an upcoming journal article will discuss it more), but Dolez and Kleinman (these proceedings) will comment on the second approach as well.

Zuckerman (private communication) has stated the observed infrared excess of G29–38 (Zuckerman & Becklin, 1987) has not been found in other DAs, particularly in other cool DAVs. This uniqueness raises the cautionary flag — is G29–38 simply a weird duck or can it be an example for other stars in its subclass? Observations of the other cool DAVs (see Dolez & Kleinman, these proceedings, and Kleinman, 1995) are therefore vital and currently indicate that except for the IR excess, G29–38 does indeed appear to be a "typical" member of its class.

## 2. Data

G29–38's power spectra change dramatically from year to year, with less dramatic changes occurring even during a given season. If the star oscillates in normal g-modes, but picks and chooses which particular modes to oscillate in at any given time, we should see modes that recur from season to season and distinct areas where no modes ever appear. Thus, by observing the star for many seasons, we can obtain a subset of modes which will come close to approximating (at least in theory) the set of all possible visible modes in which G29–38 oscillates. If the normal-mode theory of asteroseismology applies to this star, we should discover an underlying stable structure of modes emerging from the combined data.

I therefore gathered an extensive set of data on G29–38. Table 1 lists other sites that contributed data to the project. (The full list of participants will be included in an upcoming journal article.) Most of the data are from single-site runs at McDonald observatory using 3- or sometimes 2-star high-speed time-series photomultiplier-based photometers (Kleinman et al., 1996). Additionally, there are data from two WET runs and a two-site campaign in 1985 involving SAAO and McDonald observatory.

TABLE 1. Participating Observing Sites

| Observatory | Telescope(s) |
|---|---|
| CTIO | 1.5m |
| Itajuba, LNA | 1.6m |
| KPNO | 1.3m |
| La Palma (INT) | 2.5m |
| Maidanak | 1.0m |
| Mauna Kea | Air Force 24", CFHT 3.6m |
| McDonald | 30", 36", 82" |
| OHP | 1.93m |
| SAAO | 30",40",74" |
| Siding Spring | 24",40" |

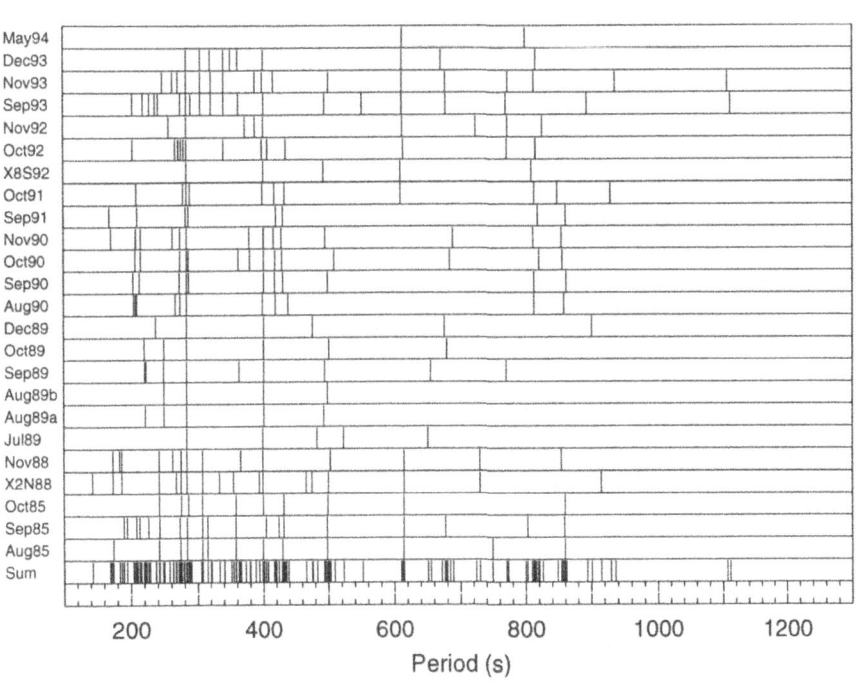

*Figure 1.* Schematic diagram of G29-38's periodicities for the entire data set.

Figure 1 is a schematic mode plot for all the G29–38 data. Ignoring the true amplitude of the modes, I plotted a single line for each observed pulsation period. A wealth of information is contained within this plot, but for now, note the bottom *Sum* panel and the conspicuous lack of groupings, which would be indicative of a subset of single-$\ell$ pulsations. The obvious conclusion here is that the modes we see are not explainable by single-$\ell$, normal-mode pulsations. While there appear to be some gaps in G29–38's schematic mode plot, there are few concise groupings; modes seem to be able to appear wherever they want to and are not confined to obey strict period spacings demanded of single-$\ell$ pulsation models.

Since it is already known that this star (and others of its class) have linear combination modes, I start by removing the linear combinations and see what the remaining modes look like. If we are indeed seeing normal g-mode pulsations, we expect to find series of nearly equally spaced periodicities, representing a succession of different-$k$, same-$\ell$ modes. If there is more than one $\ell$ present, we expect to see two such patterns. More likely, however, is a set of $\ell=1$ pulsations with perhaps, but not necessarily, a mode or two of $\ell=2$. The simplest approach is to assume only $\ell=1$ modes and see what, if anything, cannot be explained as either an $\ell=1$, or a linear combination mode.

Our ability to identify (and hence remove) the linear combinations depends greatly on the S/N ratio and the resolution of each transform. To help avoid uncertain and incorrect identifications, I now restrict my analysis to the best data sets available — one per year: Aug85 is the August, 1985 data set from a combined campaign by SAAO and McDonald. X2N88 is the data from the WET Xcov2 campaign on G29–38 in November, 1988. Sep89 is a 20-night data set from McDonald from September, 1989. Oct90, and Sep93 are also single-site data sets from McDonald during October, 1990 and September, 1993 respectively. X8S92 is the data from the second WET run (Xcov8) on G29–38 in September, 1992.

Figure 2 is the schematic period diagram minus the identified linear combinations from this selected data subset. This new schematic period diagram is substantially cleaner and qualitatively shows the mode groupings we would expect for normal-mode pulsations. The roughly equally-spaced groups seen in the *Sum* row suggest a mean period of roughly 50s. We see a near complete set of modes from 110s to 900s with two more at longer periods. With few exceptions, the groups are tight and have distinct gaps between them.

The data from September, 1993 have the most modes and nicely reproduce the expected $\approx 50s$ spacing. This season is unique in that there appear to be six consecutive $k$s in one period range. The other seasons also show roughly equally spaced groups, but are often missing one or more $k$s in between each observed mode. Most notable in this respect are the X2N88, Sep89, and X8S92 data sets. Almost half of the observed non-combination modes repeat at least once in the data set; four are present in four of the five data sets. We now have strong evidence for a series of successive-$k$, same-$\ell$ modes.

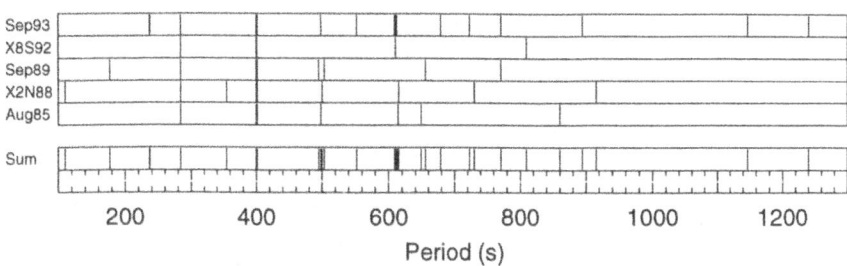

*Figure 2.*   Schematic diagram of G29-38's periodicities minus the linear combination modes.

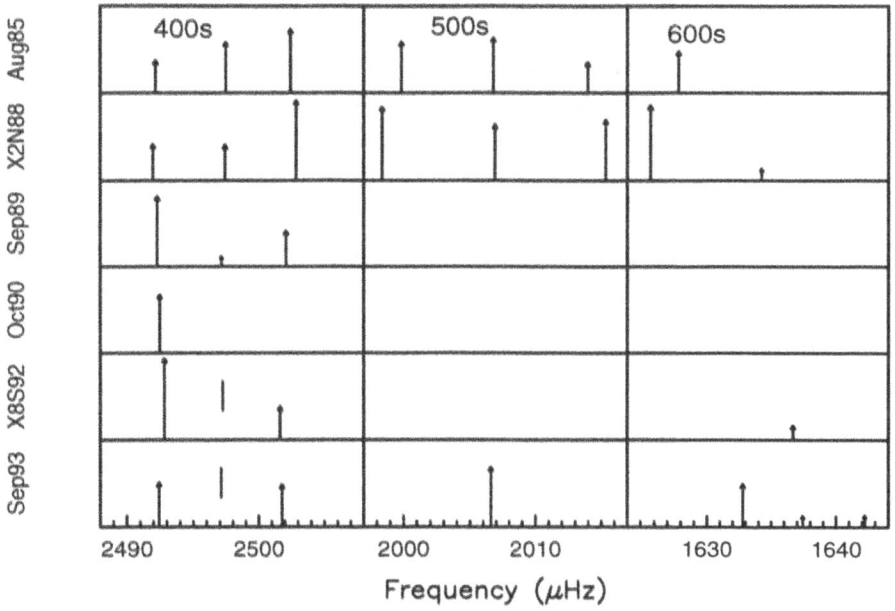

*Figure 3.*   The observed multiplets as a function of time. In some cases the central mode was not visible in the data but is represented by a shorter line segment at the average of the two flanking modes. Note the variable spacing see in all three sets of multiplets. The 600s modes, in particular are interesting: one year they match the 400s spacing, and another year, the 500s spacing. These measurements are not explainable by simple first-order rotational-splitting models.

## 3. Interpretation

Standard g-mode theory predicts that for large-$k$, the $k$-spacing between modes of a given $\ell$ is constant in period. For uniform slow rotation (and a negligible magnetic field) the $m$-spacing for modes of a given $\ell$ is constant in frequency. If we see modes of more than one $\ell$ in a star, we can use these spacings and their ratios to help identify the $\ell$ values. Unfortunately, we seem to have mostly a single-$\ell$ here and only have a few multiplets (which themselves exhibit strange behavior: see Figure 3 which shows apparently variable $m$-splittings with possible periods of a few years or less), so we cannot make solid mode identifications based on these criteria. We can search for equal period spacings in the data, but since we have modes with periods as low as 110s, we are not completely in the high-$k$ limit and therefore expect significant deviations from uniform period spacing. We have tried to search for equal period spacings using methods that should allow for departures from a strictly uniform spacing (which will also be present due to mode-trapping in the star's outer layers), but failed to find any convincing results. Nonetheless, we believe the pattern of modes in Figure 2 is NOT random and most likely represents a series of successive-$k$, $\ell=1$ modes. Clearly, with a period spacing on the order of 50s, the observed modes cannot be dominantly anything else other than $\ell=1$ while maintaining any reasonable match to G29–38's observed mass and distance. Bradley and Kleinman (these proceedings) discuss preliminary attempts to produce an asteroseismological model to this set of modes. The modeling must be considered preliminary because based on the evidence presented so far, we cannot yet solidly identify each mode with its $k$, $\ell$, and $m$ values. Their success, however, suggests the $\ell=1$ interpretation is plausible and provides additional impetus to look further for independent mode identifications. Once identified, the abundance of pulsation modes should provide very precise asteroseismological information about the insides of the DAVs.

I would like to thank Paul Bradley and Noel Dolez for helpful discussions during the preparation of this work.

## References

Clemens, J.C., 1993, Baltic Astronomy, 2, 407

Clemens, J.C., 1994, Ph.D. Thesis, University of Texas at Austin

Kleinman, S.J., 1995, Ph.D. Thesis, University of Texas at Austin

Kleinman, S.J., Nather, R.E., Phillips, T., 1996, PASP, 108, 356

Nather, R. E., Winget, D. E., Clemens, J. C., Hansen, C. J., Hine, B. P., 1990, ApJ, 361, 309

Winget, D.E., et al., 1991, ApJ, 378, 326

Winget, D.E., et al., 1994, ApJ, 430, 839

Zuckerman, B. and Becklin, E. E., 1987, Nature, 330, 138

## Discussion

*S. Kawaler*: How can you identify "real modes" as opposed to linear combination modes? I.e., if a+b=c, how do you say that a and b are real and c is the combination when it could be that c and a are real and b=c-a?

*S. Kleinman*: Using different data sets, we often see that the "real" modes appear without their combinations, whereas combinations cannot appear without their corresponding "real" mode components. You can also use arguments based on the number of combinations of each "real" modes and the relative amplitudes of each mode in question, but the clearest argument is the first one, of course.

*Y. Wu*: Do you see the same variations in the rotational splittings for other DAVs?

*S. Kleinman*: With current observations, no. There are some variations, but not as large, although this could simply due to the lack of data of this quality and timebase on the other stars.

*Y. Wu*: We think the amplitude variations are due to mode interactions. For the trapped modes, do you see larger variations in amplitudes than for other modes?

*S. Kleinman*: The largest two modes with the largest amplitude changes do appear to be trapped modes. It is not clear, however, if all trapped modes show such variations.

# OVERSTABILITY OF G-MODES IN ZZ CETI VARIABLES

*how to deal with the convection zone?*

YANQIN WU
*Theoretical Astrophysics, CALTECH*
*Pasadena, CA 91125, U.S.A.*

**Abstract.** Many types of variable stars have convection zones in the pulsation driving regions. To study the driving of pulsations, we need to consider how convection reacts to pulsation. A physically consistent treatment of time-dependent convection is not available. The conventional way around this difficulty is to use the 'frozen-in' approximation, valid in the limit that the pulsation time scale is much *shorter* than the time scale for convective re-adjustment. Following Brickhill, we introduce an approximation that is valid in the opposite limit: an instantaneously adjusting convection zone that keeps its isentropic profile during the pulsation cycle.

For variables that fall into this extreme, overstability of the pulsation modes can be studied analytically. The interaction between pulsation and convection may drive the mode. This new kind of driving mechanism merits the name 'convective driving' (Brickhill, 1983).

Applied to DA white dwarfs, we find overstability for low-order g-modes with periods similar to the observed ones. We derive the blue edge *and* the red edge for the ZZ Ceti instability strip, giving a width of $\sim 800\,\mathrm{K}$, comparable to what has been observed. The red edge of the instability strip results from the decrease in surface visibility of the driven modes with decreasing temperature. The location of the theoretical instability strip coincides roughly with the observed one.

## 1. Introduction

Variable DA white dwarfs (DA for hydrogen atmosphere) constitute one of the three variable white dwarf populations. They are called ZZ Cetis after the prototype, and are found in a narrow temperature range around $12,000\,\mathrm{K}$ (the ZZ Ceti instability strip). ZZ Cetis are characterized by multiperiodic flux variations, with periods ranging from 100 to 1200 s, amplitudes

*I. Isern et al. (eds.), White Dwarfs, 467–472.*
© *1997 Kluwer Academic Publishers.*

from 1 to 30 mmag. These periodicities represent stationary waves inside the star (eigen-modes). They cause surface flux variations by compressing and inflating different parts of the star, most of these actions happening in the upper atmosphere. The restoring force is buoyancy. These modes are non-radial gravity modes (g-mode). The entire star is organized into pulsations of only a few modes, and the energy stored in each of these modes is as large as $10^{30}$ erg. The excitation mechanism has not been well understood.

The fact that the ZZ Ceti instability strip is an apparent extension of the well known Cepheid strip (which includes pulsators like Cepheids, $\delta$ Scuti and oscillating Ap stars) suggests that the $\kappa$-mechanism (opacity mechanism) is involved in their excitation. Hydrogen is indeed partially ionizing in the surface layer of these white dwarfs, as is required for the $\kappa$-mechanism. But due to the high gravity, this partial ionization zone is very thin, and its thermal time is orders of magnitude shorter than the mode periods. This driving is in-sufficient when compared with the damping by the radiative interior.

The opacity maximum associated with the ionization causes convection in the upper atmosphere, and almost all the stellar flux is carried by convective motion in this region. The thermal time scale at the bottom of the convection zone is similar to the mode periods. This makes one wonder whether the convection zone provides the driving. Theoretically, dealing with perturbations to the convective flux is a problem yet to be solved. Baker & Kippenhahn (1963) proposed the 'frozen-in' approximation, in which the convective flux itself is assumed not to be affected by the perturbations. This assumption is valid only in the limit that the pulsation time scale is much shorter than the typical eddy turn-over time. And when this assumption is adopted, as shown by Presnell (1987), the bottom of the convection zone will contribute to the driving, the so-called 'convective blocking'.

For ZZ Ceti variables, the convective response time is shorter than a second throughout the convective layer, while the mode periods are longer than a hundred seconds. We either have to rely on numerical hydro-dynamical simulations (Gautschy et al.,1996) or to look for another simplifying assumption that is valid in this limit. Brickhill (1983, 1991a, 1991b) assumes *instantaneous response* of the convection to the state of pulsation. He presents the first physically consistent calculations of mode overstability, mode visibility, and the instability strip. Our investigation supports most of his conclusions.

## 2. The Instantaneous Assumption and the Consequences

An efficient convection zone will have an almost flat entropy profile, thanks to the labour of turbulent eddies (Schwarzschild 1966). When perturbed, for instance, when the energy flux injected from below is slowly varying over a time much longer than the typical eddy turn-over time, the convection zone maintains its *isentropic* profile by uniformly changing its entropy. To do so, it absorbs (releases) energy from (into) the incoming flux, and allows only part of the flux variations to show up at the stellar surface. Recall a Carnot-type engine: it absorbs heat when compressed, releases it when expanded, and after a whole cycle, net work is done. The convection zone drives g-modes in the same way, giving rise to what is called 'convective driving' (Brickhill, 1983).

Here, we sketch the main line of argument underlying our analytic study. Interested readers are referred to our forth-coming paper for a complete derivation.

A mode with frequency $\omega$ and energy $E$ has a growth rate $\gamma$ defined as,

$$\gamma \equiv \frac{\omega}{2\pi} \oint dt \frac{dE}{dt}. \tag{1}$$

For perturbations depending on time as $e^{-i\omega t}$, $\gamma$ can be written into the following useful forms,

$$\gamma = 2\omega R^2 \oint dt \int_0^R dz\, \rho\, \frac{k_B}{m_p} \delta T \frac{d\delta s}{dt} = \frac{\omega}{2\pi} L \oint dt \int_0^R dz \frac{\delta T}{T} \frac{d}{dz}\left(\frac{\delta F}{F}\right). \tag{2}$$

Here, $k_B$ is the Boltzman constant, $m_p$ the proton mass, $R$ the stellar radius, $L$ the stellar luminosity, $z$ the vertical depth counting down from the photosphere, and $\delta T$, $\delta s$, $\delta F$ are Lagrangian perturbations for temperature, entropy and flux respectively. In the convection zone, the first formula in equation (2) is more convenient, while in the radiative zone, the second is used. For our present discussion, we use adiabatic g-mode eigenfunctions, i.e., eigenfunctions calculated from static white dwarf models without considering heat leakage (non-adiabaticity) or convective viscosity.

To derive $\delta F$ in the radiative zone, we use the equation of radiative diffusion

$$F = \frac{ac}{3\kappa\rho} \frac{dT^4}{dz}, \tag{3}$$

to relate $\delta F$ to $\delta T$ and $\delta \rho$; $\kappa$, the radiative opacity, is a function of $\rho$ and $T$.

In the convection zone, $\delta s$ is derived from the isentropic assumption. The heat absorbed by the convection zone when the entropy level is raised uniformly by $\Delta s$ is

$$\Delta Q = \int_{cvz} dz\, \rho \frac{k_B}{m_p} T\, \delta s \approx \Delta s \int_{cvz} dz\, \rho \frac{k_B}{m_p} T = F\, \tau_b\, \Delta s, \qquad (4)$$

where

$$\tau_b \equiv \frac{1}{F} \int_{cvz} dz\, \rho \frac{k_B}{m_p} T. \qquad (5)$$

The conventional thermal relaxation time, $\tau_{th}$, is defined as

$$\tau_{th}(z) \equiv \frac{1}{F} \int_0^z dz\, \rho\, c_p T, \qquad (6)$$

where $c_p$, the specific heat per unit mass at constant pressure, is of order $k_B/m_p$. Obviously, $\tau_b$ is of order $\tau_{th}$ at the base of the convection zone ($z_b$).

We use the subscript $ph$ for photospheric quantities. The first law of thermodynamics yields,

$$\frac{d\Delta Q}{dt} = \delta F_b - \delta F_{ph}. \qquad (7)$$

$\delta F_b$ can be derived from equation (3).

We need to relate $\Delta s$ with $\delta F_{ph}$ to get a closed form and to solve for $\Delta s$. Given the time scales involved, we can regard these perturbative quantities as differences between two adjacent static models with slightly different luminosities. In these static models, the convection zone connects the photosphere with the radiative interior. The upper convective region is super-adiabatic due to the small sound speed there. Entropy at the photosphere ($s_{ph}$), can be easily related to the flux there ($F_{ph}$) using equations for stellar structure and the ideal gas law, while the super-adiabatic entropy jump ($s_b - s_{ph}$) can be related to $F_{ph}$ using mixing length theory. These allow us to derive a relation between $s_b$ and $F_{ph}$, as well as its differential form. Symbolically, we write

$$\Delta s = \delta s_b = (B + C)\frac{\delta F_{ph}}{F}, \qquad (8)$$

where $B$ and $C$ represent contributions from the photosphere and the super-adiabatic region, respectively. They can both be derived analytically when properties of the gas is known, and they are typically of order 10. We define a useful thermal time scale here,

$$\tau_c = (B + C)\tau_b. \qquad (9)$$

We call this 'thermal adjustment time' of the convection zone. Numerically, $\tau_c$ is of order 20 times longer than the thermal time at the bottom of the convection zone.

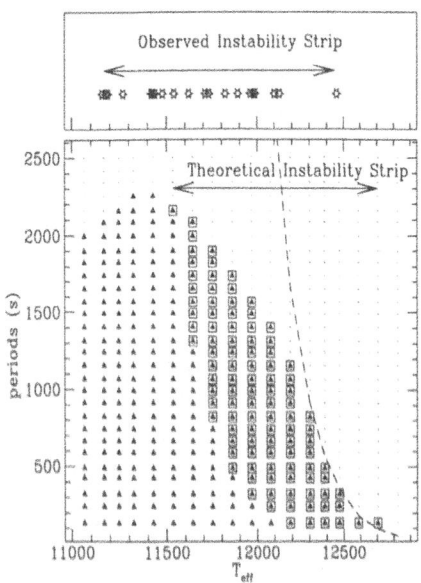

*Figure 1.* Observed and theoretical instability strips for DA white dwarfs. The inferred effective temperatures for all known pulsators are marked in the upper panel. The lower panel shows the main theoretical results. For a range of effective temperatures, the eigen-modes of the star (with spherical degree $l = 1$) are marked with small dots if they were calculated to be stable, and with triangles if they were overstable. The dashed line represents a crude estimate for the maximum overstable period as a function of $T_{eff}$. It ignores effects like non-adiabaticity and turbulent viscosity, both taken into account in the full calculations. For modes marked with a square, the reduction of surface visibility is not very severe, and they will be observable. The width of the theoretical instability strip is very similar to the observed one. The location is also consistent, if one takes into account possible offsets in the inferred temperatures due to uncertainties in, e.g., convective efficiency.

## 3. Models, Complications and Results

We build a series of models for the upper layers in white dwarfs, based on hydrogen opacity and equation of state. The entropy in the convection zone can be directly measured from the model and be compared with equation (8). Eigen-modes are calculated from full white dwarf models kindly provided to us by P. Bradley. Driving/damping rates are thereupon calculated with equation (2).

A summary of our main theoretical results is presented in Figure 1. In obtaining these results, we take into account two complicating factors: 1) the thermal time scale at the bottom of the convection zone is comparable or longer than the periods of overstable modes, this implies that the structure of the modes should be significantly altered by non-adiabatic effect, i.e., heat leakage; 2) an adiabatic eigen-function will give rise to a large viscous damping in the convection zone, but in reality, this damping is negligible – the large turbulent viscosity forces a realistic eigenfunction to have a flat velocity profile.

As the temperature of the white dwarf drops, the surface convection zone deepens, and $\tau_c$ increases rapidly. We find that only g-modes with periods shorter than $\tau_c$ can be overstable (not a sufficient condition, though, as is apparent from the figure). When a DA white dwarf first cools to about 12,500 K, $\tau_c$ becomes longer than the shortest period g-mode ($k = 1$). This defines the blue edge of the instability strip. As the star cools further, it will pulsate at longer and longer periods. This result is reminiscent of the observed trend reported by Clemens (1993).

Overstable modes have growth rates of order $\gamma \sim 1/(k\tau_1)$, where $k$ is the mode's radial order and $\tau_1$ is the thermal timescale evaluated at the top of the mode's propagation cavity. For low order g-modes ($l = 1, k = 1$), this could be of order $10^{-15}\,\mathrm{s}^{-1}$; while for a $k = 20$ mode, the driving rate is increased to $10^{-5}\,\mathrm{s}^{-1}$.

The surface visibility of the overstable modes is a critical factor in determining the red edge of the instability strip. Combining equations (4), ( 7) and (8), we arrive at,

$$\frac{\delta F_{\mathrm{ph}}}{F} = \frac{1}{1 - i\omega\tau_c}\frac{\delta F_b}{F}. \tag{10}$$

This equation describes the surface visibility of a mode affected by the thermal adjustment of the convection zone. Ignoring the phase information, we have

$$\left|\frac{\delta F_{\mathrm{ph}}}{\delta F_b}\right| = \frac{1}{\sqrt{1 + (\omega\tau_c)^2}}. \tag{11}$$

Amplitude of the flux variations at the surface (the observable quantity) is reduced from the interior amplitude by an amount depending on the mode period and the thickness of the convection zone. The convection zone acts like a low-pass filter: for similar amplitudes at the base of the convection zone, longer period modes show up more strongly at the surface than their shorter companions. This could *partially* explain the other trend found by Clemens (1993) from DA pulsation data, viz., longer period modes have larger amplitudes. The nonlinear resonances that limit the amplitude of modes will also contribute to this trend. Beyond the red edge of the instability strip, there are still many modes overstable, but they all have $\omega\tau_c \gg 1$, and are therefore invisible.

## References

Baker, N., Kippenhahn, R. 1965, ApJ, 142, 869
Brickhill, A. J. 1983, MNRAS, 204, 537
Brickhill, A. J. 1991, MNRAS, 251, 673
Brickhill, A. J. 1991, MNRAS, 252, 334
Clemens, J. C. 1993, Baltic Astronomy, 2, 407
Gautschy, A., Ludwig, H. G., Freytag, B. 1996, A&A, 311, 493
Pesnell, W. D. 1987, ApJ, 314, 598
Schwarzschild, M. 1958, *Structure and Evolution of the Stars*. Princeton University Press,

# A CONTINUING SPECTROSCOPIC STUDY OF THE INSTABILITY STRIP OF THE PULSATING DB (V777 HER) WHITE DWARFS: THE VIEW IN THE OPTICAL

F. WESEMAEL
*Département de Physique, Université de Montréal*

A. BEAUCHAMP
*CAE Électronique Ltée*

P. BERGERON
*Lockheed Martin Electronic Systems Canada*

G. FONTAINE
*Département de Physique, Université de Montréal*

R.A. SAFFER
*Space Telescope Science Institute*

AND

J. LIEBERT
*Steward Observatory, University of Arizona*

## 1. Introduction

The variable DB white dwarfs are the helium-rich analogs of the better-known, and extensively studied, ZZ Ceti stars. Both classes of objects are non-radial oscillators, whose pulsations are driven by the $\kappa$ mechanism which operates at the base of the hydrogen convection zone in the ZZ Ceti stars, and at the base of the helium zone in the helium-rich V777 Her stars. Each class defines its own instability strip along the white dwarf cooling sequence.

Non-adiabatic analyses of DB models (Winget *et al.* 1983; Bradley & Winget 1994a) show that the location of the blue edge of the DB instability strip is a sensitive function of the convective efficiency adopted in the envelope, as well as of the total stellar mass and, to a much lesser extent, of the helium envelope mass. Current constraints on the extent of the instability strip derive exclusively from *IUE* observations, last analyzed by Thejll et

473

*I. Isern et al. (eds.), White Dwarfs,* 473–476.

al. (1991), who place the coolest variables in their sample near 22,000 K, and the hottest one near 24,000 K. Given the efforts currently being made (Beauchamp et al. 1996) to update the effective temperature scale of DB stars on the basis of an extensive spectroscopic analysis of these objects, it appears of undeniable interest to look at the problem of the location and width of the instability strip of the V777 Her stars. We present here updated results of this continuing investigation (see Wesemael et al. 1996 for an earlier report), and summarize our joint analysis of the eight known V777 Her stars and of 21 other DB or DBA stars above 17,000 K.

## 2. Boundaries and Structure of the Strip

Our nominal fits are carried out with pure-helium models for the DB stars, and with a grid of helium-rich models with traces of hydrogen ($\log y \gtrsim 3.5$) for the DBA stars. All fits are restricted to the so-called ML2 convective efficiency — the most appropriate for the DB stars according to the results of Beauchamp et al. (1996). Typical total internal errors on $T_{\text{eff}}$ should be in the 500 K range for our fits with pure-helium atmospheres. However, the influence of small, invisible amounts of hydrogen in the atmosphere of DB stars remains a serious concern, as effective temperatures determined on the basis of helium models with a small admixture of hydrogen can often be cooler than those determined on the basis of pure-helium models by a few thousand Kelvin. To analyze this effect in some detail, we have redone all the fits of DB stars with models containing traces of hydrogen at the detection threshold. This threshold is a function of effective temperature, and was set at $\log y = 5.0$ below 20,000 K, $\log y = 4.0$ between 20,000K and 25,000 K, and $\log y = 3.5$ above 25,000 K. Note that, in the bright prototype GD 358 (V777 Her; $T_{\text{eff}} \sim 24,000$ K), Provencal & Shipman (1996) have recently imposed $\log y > 4.5$.

Results of these fits are summarized in Figure 1, where we show the structure of the instability strip for DB and DBA stars under the assumption of pure-helium DB models (top) and under the assumption of helium-rich DB models containing traces of hydrogen (bottom). The coolest of the variable stars we observed currently sits at 22,000 K, a temperature which marks the empirical red edge of the strip. This temperature is in good agreement with the red edge determined by Thejll et al. (1991) from the *IUE*. Both estimates are unaffected by the abundance dilemma discussed above, as the star marking the red edge (PG 1456+103) is a DBA star. Its hydrogen abundance is fixed in our analysis by the fit to the Balmer lines, while the sensitivity of the *IUE* temperature to the presence of trace amounts of hydrogen remains small.

The effective temperature we determine for GD 358 with pure-helium models, 25,323 K, is consistent with the 24,000± 1000 K derived by Koester *et al.* (1985) from optical and *IUE* data. Its spectroscopic gravity is normal, $\log g = 7.95$, and is consistent with the asteroseismological stellar mass inferred by Bradley and Winget (1994b).

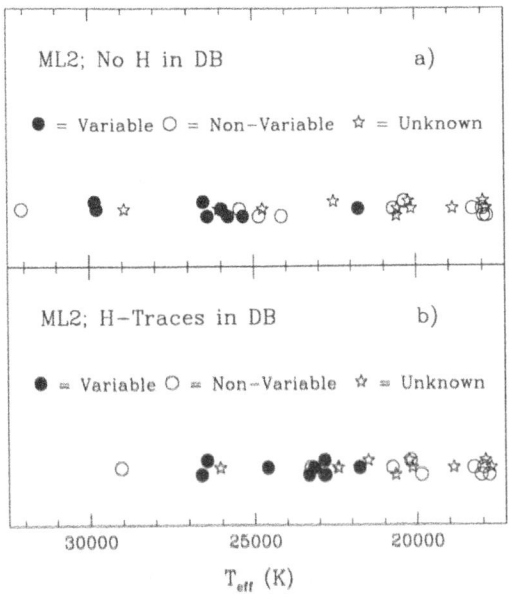

*Figure 1.* Schematic representation of the instability strip of DB and DBA stars. All effective temperatures are determined on the basis of optical synthetic spectra incorporating the ML2 version of the mixing-length theory. *a)*: Pure-helium models for the DB stars; *b)*: Models with traces of hydrogen for the DB stars.

Returning for now to the instability strip, the empirical blue edge determined on the basis of pure-helium models is near 29,800 K (both CBS 114 and PG 1654+160), more than 2000 K cooler than the hot, non-variable star PG 0112+104. However, should the variable DB stars contain large, but nevertheless undetectable, quantities of hydrogen, the blue edge temperature would be reduced to $T_{\text{eff}} = 26,600$ K, and the role of the hottest variable DB would then be assumed by PG 1351+489, with CBS 114 less than 200 K cooler. Because the red edge in anchored by the DBA star PG 1456+103, the net effect of using models with traces of hydrogen is thus to squeeze, rather than to shift, the instability strip to lower effective temperatures.

Equally noteworthy is the apparent structure observed in the strip, irrespective of our handling of the hydrogen abundance: in our analysis based on pure-helium models, the coolest variable appears separated from its closest variable neighbor by 3000 K — a span which includes 2 confirmed constant stars (PG 0921+091 and GD 205), as well as two additional stars

(PG 1540+681 and PG 2234+064) for which no published observations exist. If models with traces of hydrogen are preferred, the situation is similar, in that the constant stars GD 205 and PG 2254+159 both currently lie within the strip, together with PG 2234+064 and PG 2246+120, for which no published observations exist. Given the historical emphasis of photometric searches of new DB variables on broad-line stars culled from the PG survey, it is more than likely that the three undocumented PG stars discussed here have already been observed, and found constant.

While non-adiabatic investigations of the pulsational properties of the DB stars remain scarce, a preliminary comparison of our results with theoretical expectations is already possible. For the analysis with pure-helium models, our determinations are consistent with the predictions of envelope models based on ML2 convection only for the coolest variable, near 22,000 K. The location of the hotter objects — above 25,000 K — indicate, as was the case in the ZZ Ceti stars (Bergeron *et al.* 1995), that — in those stars — the convective efficiency in the deep, driving regions must be larger than in the shallow, photospheric layers, where ML2 was adopted. However, this requirement may not be as severe were we to adopt our analysis of DB stars based on models with traces of hydrogen, for which the instability strip is squeezed to lower effective temperatures. It seems, though, that the quality and homogeneity of the theoretical data available to interpret the observed boundaries of the instability strip now lags behind the observations. DB stars would clearly benefit from a detailed, homogeneous study in the line of that described by Brassard & Fontaine (1997) for the ZZ Ceti stars.

This work was supported in part by the NSERC Canada, by the Fund FCAR (Québec), and by the NSF grant AST 92-17961.

## References

Beauchamp, A., Wesemael, F., Bergeron, P., Liebert, J., & Saffer, R.A. 1996, in *Hydrogen-Deficient Stars*, eds. C.S. Jeffery & U. Heber, p. 295

Bergeron, P., Wesemael, F., Lamontagne, R., Fontaine, G., Saffer, R.A., & Allard, N. 1995, *Ap.J.*, **449**, 258

Bradley, P.A. & Winget, D.E. 1994a, *Ap.J.*, **421**, 236

Bradley, P.A. & Winget, D.E. 1994b, *Ap.J.*, **430**, 850

Brassard, P. & Fontaine, G. 1997, these Proceedings

Koester, D., Vauclair, G., Dolez, N., Oke, J.B., Greenstein, J.L., & Weidemann, V. 1985, *Ast.Ap.*, **149**, 423

Provencal, J.L. & Shipman, H.L. 1996, in *Hydrogen-Deficient Stars*, eds. C.S. Jeffery & U. Heber, p. 321

Thejll, P., Vennes, S., & Shipman, H.L. 1991, *Ap.J.*, **370**, 355

Wesemael, F., Beauchamp, A., Bergeron, P., Fontaine, G., Saffer, R.A., & Liebert, J. 1996, in *Hydrogen-Deficient Stars*, eds. C.S. Jeffery & U. Heber, p. 322

Winget, D.E., Van Horn, H.M., Tassoul, M., Hansen, C.J., & Fontaine, G. 1983, *Ap.J. (Letters)*, **268**, L33

# THE POTENTIAL OF ASTEROSEIMOLOGY FOR HOT, B SUBDWARFS: A NEW CLASS OF PULSATING STARS?

S. CHARPINET, G. FONTAINE, P. BRASSARD
*Département de Physique, Université de Montréal*
*C.P. 6128, Succ. Centre-Ville, Montréal,*
*Québec, Canada, H3C 3J7*

AND

B. DORMAN
*Laboratory for Astronomy and Solar Physics*
*NASA/GSFC, Greenbelt, Maryland*
*20771 USA*

## 1. Introduction

In the last two decades, considerable progress has been made in our understanding of the physical properties and evolutionary status of hot, hydrogen-rich subdwarfs B stars (see, e.g., Saffer et al. 1994). It is now currently believed (e.g., Heber et al. 1984) that sdB stars are $\sim 0.5\ M_\odot$ objects belonging to the so-called extended horizontal branch (EHB), which never evolve toward the asymptotic giant branch (AGB) after core helium exhaustion (their hydrogen envelope masses being too small) (see, e.g., Dorman 1995 for a review). They remain at high effective temperatures ($T_{eff} \geq 20000$ K) throughout their core-burning evolution and ultimately contribute to a small fraction of the total white dwarf population (Bergeron et al. 1994).

While asteroseismology is proving to be an extremely powerful tool in studying other types of stars, its potential has not yet been studied for subdwarf B stars partly because of the lack (until recently) of sufficiently realistic equilibrium structures. Furthermore, luminosity variations caused by pulsational instabilities have not been reported for these stars (but see the exciting paper presented by O'Donoghue in these proceedings). However, from a rapid inspection of envelope models, the potential for driving pulsation modes appears to exist in these stars, hidden behind an opacity bump (responsible for a HeII-HeIII convection zone) analogous to what is

*I. Isern et al. (eds.), White Dwarfs, 477–480.*
© *1997 Kluwer Academic Publishers.*

found in other pulsating stars whose instabilities are always driven by one form or another of an opacity mechanism.

Motivated in part by this observation (and see Van Horn 1991), we undertook a systematic exploration of the potential of asteroseismology for subdwarf B stars. We report here the first results of this investigation.

## 2.  Computations and results

The first batch of equilibrium models investigated in this study consists of full stellar models taken from five evolutionary sequences chosen to map a significant fraction of the $\log g - T_{eff}$ plane actually occupied by the known subdwarf B stars (see Saffer et al. 1994). Models were computed for solar metallicity ($X = 0.70388$, $Z = 0.01718$) as described in Dorman (1992a,b) and Dorman et al. (1993), using the recent OPAL opacities (Rogers & Iglesias 1992). About 25 models per sequences with fixed core mass of $0.4758 M_\odot$ and envelope masses respectively of 0.0002, 0.0012, 0.0022, 0.0032, and 0.0042 $M_\odot$ were used for detailed adiabatic and nonadiabatic calculations with the Galerkin finite-element codes of Brassard et al. 1992 (adiabatic) and Fontaine et al. 1994; Brassard et al. 1996 (nonadiabatic).

Adiabatic calculations for radial ($l = 0$) and nonradial ($l = 1$, 2, and 3) pulsations in the $80 - 1500$ s period window (the most easily detectable modes with present-day fast photometric techniques) show that sdB stars have rich period spectra, including both radial and nonradial p, f, and g modes in that range of periods (see Charpinet et al. 1996 for a typical exemple of such period spectra). Provided we can demonstrate that some of these modes can be excited, this should motivate observational searches for pulsations in these stars.

Nonadiabatic calculations were carried out specifically to explore the question of mode stability and identify the regions of driving (if any). We considered the variations with depth of the local thermal time scale and of the work integral. Figure 1 illustrates the results for one typical model with $\log g = 5.46$, $T_{eff} = 27500$ K, age $\sim 8.4 \times 10^7$ yrs (time ellapsed since ZAHB) which belongs to the sequence with $M_{core} = 0.4758$ $M_\odot$ and $M_{env} = 0.0012$ $M_\odot$. From panel (a), it is clear that thermonuclear burning regions are inefficient to drive pulsation modes because the thermal time scale $\tau_{th}$ is orders of magnitude larger than the periods of interest leading to strongly adiabatic oscillations in those regions (in addition to the fact that the H burning shell is extremely weak). Thus, the $\epsilon$-mechanism appears to be irrelevant for exciting pulsation modes in hot B subdwarfs. In contrast, $\tau_{th}$ is of the same order of magnitude as the interesting pulsation periods in the envelope convection zone. Hence, driving might be efficient there.

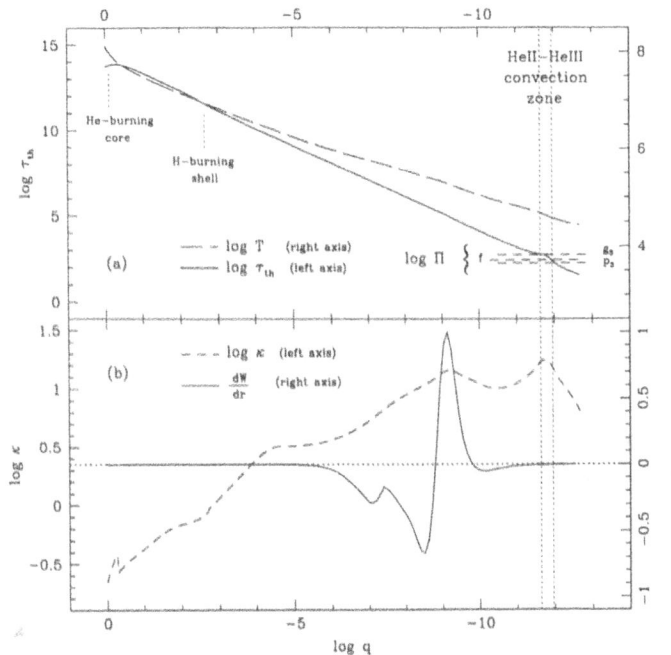

*Figure 1.* (a) Run of the thermal time scale $\tau_{th}$ (expressed in seconds) vs fractional mass depth $\log q \equiv \log(1 - M(r)/M)$ in a typical subdwarf B model. The locations of the helium burning core, the residual H burning shell, and the HeII-HeIII convection zone are indicated. (b) Run of the Rosseland opacity $\kappa$ and of the integrand of the work integral for the f-mode with $l = 2$ of our typical model.

Panel (b), however, reveals that significant driving (positive values of $\frac{dW}{dr}$) is not associated, as first expected, to the opacity bump caused by partial ionization of HeII-HeIII (retrospectively, a region too high in the envelope to carry enough weight in term of driving), but rather clearly corresponds to a secondary bump caused by heavy element ionization (near $\log q \sim -9.2$) which tends to disappear when metallicity is decreased. Hence, this figure leads to the important result that driving is closely related to metallicity through its effect on the opacity and, despite the fact that all our evolutionary models calculated with $Z \sim 0.02$ turned out to be stable, it gives us a precious clue for the search of instabilities in these stars.

Our last step was to compute 5 grids of static equilibrium envelope structures (we used a version of the code described by Brassard et al. 1996) taking into account variations of $Z$ (respectively $Z = 0.02$, 0.04, 0.06, 0.08, and 0.10). Each grid, made of 72 models, overlap the entire region in the $\log g - T_{eff}$ were hot B subdwarfs are found ($4.34 \leq \log T_{eff} \leq 4.62$ in steps of 0.04, and $4.8 \leq \log g \leq 6.4$ in steps of 0.2). We fixed the total mass of each model to 0.48 $M_{\odot}$ and the H-rich envelope mass to $\log q(H) = -4$.

*In agreement with our expectation concerning the effects of metallicity, we found unstable modes for models with $Z \geq 0.04$.* Radial as well as nonradial low radial order modes are predicted with e-folding times ($10^{-2}$ - $10^3$ yrs) significantly shorter than the typical evolutionary time scale of a hot B subdwarf ($\sim 10^8$ yrs). Hence, these instabilities would normally develop into observable amplitudes. We emphasize the fact that metal enrichment needs only occur in the driving region *itself* and not necessarily everywhere in the envelope (as we assumed in the relatively crude envelope models used in the present study). It is then worth noting that local enhancement and/or depletion of heavy elements resulting from a competition between radiative levitation, gravitational settling and weak stellar winds are strongly suspected in these stars (Michaud et al. 1989; Chayer et al. 1996). This greatly enhances the possibility that driving occurs as a result of a local enhancement of the metal content.

## 3. Conclusion

According to the relation we uncovered between driving and metallicity through the opacity mechanism, associated to the fact that local overabundances of metals in the driving region are expected in these stars, *we feel confident enough to risk the prediction that some subdwarf B stars should show luminosity variations resulting from pulsational instabilities.*

This work was supported in part by the NSREC of Canada and by the fund FCAR (Québec). B. D. acknowledges partial support from NASA RTOP 188-41-51-03

## References

Bergeron, P., Wesemael, F., Beauchamp, A., Wood, M.A., Lamontagne, R., Fontaine, G., & Liebert, J. 1994, ApJ, 432, 305
Brassard, P., Pelletier, C., Fontaine, G., & Wesemael, F. 1992, ApJS, 80, 725
Brassard, P., Fontaine, G., & Bergeron, P. 1996, in preparation
Charpinet, S., Fontaine, G., Brassard, P., & Dorman, B. 1996, ApJ Lett., 471, L103
Chayer P. et al. 1996, in preparation
Dorman, B. 1995, in Proc. 32nd Liège Astrophysical Colloq., Stellar Evolution: What should be done, ed. A. Noels, D. Fraipont-Caro, N. Grevesse, & P. Demarque (Liège: Institut d'Astrophysique), 291
Dorman B., Rood, R.T., O'Connell, R.W. 1993, ApJ, 415, 596
Dorman B. 1992a, ApJS, 80, 701
Dorman B. 1992b, ApJS, 81, 221
Fontaine, G., Brassard, P., Wesemael, F., & Tassoul, M. 1994, ApJ, 428, L61
Heber, U., Hunger, K., Jonas, G., & Kudritzki, R.P. 1984, A&A, 130, 119
Michaud, G., Bergeron, P., Heber, U., & Wesemael, F. 1989, ApJ, 338, 417
Rogers, F.J., & Iglesias, C.A. 1992, ApJ, 401, 361
Saffer , R.A., Bergeron, P., Koester, D, & Liebert, J. 1994, ApJ, 432, 351
Van Horn, H.M. 1991, in ASP Conf. Ser.20, Frontiers of Stellar Evolution, ed. D.L. Lambert (San francisco: ASP), 265

# MODELING TIME-RESOLVED SPECTRA OF ZZ CETI STARS

H. VÄTH AND D. KOESTER

*Institut für Astronomie und Astrophysik der Universität, 24098 Kiel, Germany (supas097@astrophysik.uni-kiel.d400.de)*

S. O. KEPLER

*Instituto de Fisica, Universidade Federal do Rio Grande do Sul, 91501-970 Porto Alegre, RS, Brazil*

AND

E. L. ROBINSON

*McDonald Observatory and Department of Astronomy, The University of Texas at Austin, Austin, TX 78712-1083, USA*

**Abstract.** We have developed a program with which time resolved spectra of oscillating white dwarfs can be calculated. This program automatically takes into account any non-linear variation of specific intensity with temperature. Here we apply this code to ZZ Ceti stars and present time resolved spectra and corresponding amplitude spectra. Finally, we analyze time resolved HST spectra of G 226−29.

## 1. Introduction

ZZ Ceti stars are variable DA white dwarfs (WDs) with effective temperatures $T_{eff} \sim 12000$ K. Their amplitudes range from $\sim 0.004$ mag to $\sim 0.3$ mag. The pulsations are non-radial $g$ modes with periods of the order of $\sim 100 - 1000$ s. A well-known property of $g$ modes is that radius variations are negligible because of the high surface gravity ($\log g$) of WDs (Robinson et al. 1982; hereafter RKN). Therefore, one can think of these oscillations as temperature waves.

Recently, Brassard et al. (1995; hereafter BFW) calculated spectra semi-analytically for temperature distributions across the surface expressed as $T_{eff} = T_{eff0}(1 + \epsilon\xi)$ where $\xi$ is a sum of spherical harmonics. The aim of this

481

*I. Isern et al. (eds.), White Dwarfs, 481–484.*

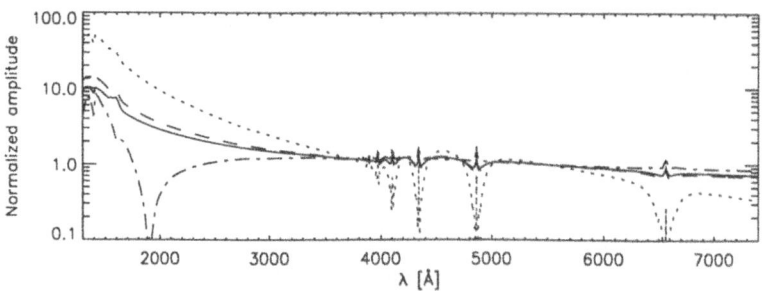

*Figure 1.* Normalized amplitude spectra for $l = 1$ (solid), $l = 2$ (dashed), $l = 3$ (dotted) and $l = 4$ (dash-dotted) modes ($T_{\text{eff}} = 12000$ K, $\log g = 8.25$, ML2/$\alpha$=0.6). Around 1900 Å the phase of the $l = 4$ mode changes by 180°, causing the amplitude to go to zero

analysis was to investigate non-linear effects that occur since the specific intensity $I$ is not a linear function of $T_{\text{eff}}$. We have developed a program that calculates the time-variable spectrum by integrating $I$ over the visible disk of the WD for a distribution of $T_{\text{eff}}$ across the WD surface. For now, we use for $T_{\text{eff}}$ an expression similar to that of BFW. However, we can adapt our program to non-linear variations of $T_{\text{eff}}$, which were suggested by Brickhill (1992) and could be caused by the response of the convective flux to linear pressure perturbations in deeper layers of these stars. We interpolate $I$ from a grid containing $I$ as a function of $\lambda$, $\mu$, $T_{\text{eff}}$, $\log g$ and different parametrizations of the mixing length theory (MLT). The grid was calculated with a stellar atmospheres code.

## 2. Time-resolved spectra and non-linear effects

There is more than one aim in our calculations. First, we want to be able to calculate amplitude spectra of oscillations. In Fig. 1 we show for $l = 1$ to 4 the normalized amplitude spectra (which are independent, for small amplitudes, of the often unknown inclination $i$; e.g., RKN, BFW). As one can see from Fig. 1, the slope varies strongly with $l$. By comparing the observed slope with the theoretical slope, one can therefore determine $l$ (the slope also depends on $T_{\text{eff}}$ and $\log g$, which may have to be constrained from other observations).

A second aim of our program is to calculate time-resolved spectra. This is shown in Fig. 2 for $l = 1$, $m = 0$ for a strong oscillation resulting in variations of $T_{\text{eff}}$ of $\pm 1000$ K. One can see that the spectra depend strongly on the inclination. Furthermore, one can see the effects of having a distribution of $T_{\text{eff}}$ at maximum and minimum instead of a uniform $T_{\text{eff}}$ across the WD surface.

A third aim is to properly take into account non-linear variations of $I$ with linear, i.e., sinusoidal, variations of $T_{\text{eff}}$. For the above case we show

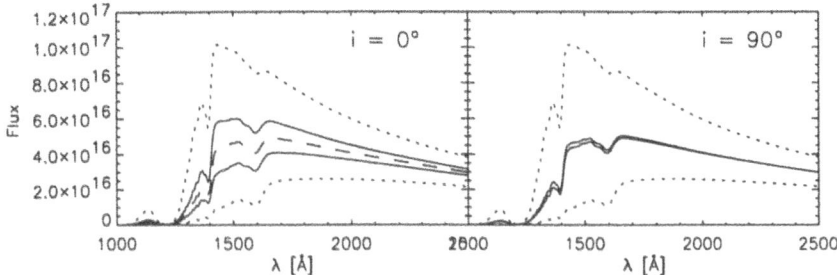

*Figure 2.* UV-spectra at maximum and minimum (solid) for $l = 1$, $m = 0$ oscillations with $\Delta T_{\text{eff}} = 1000$ K for $i = 0°$ (left) and $i = 90°$ (right). Also shown are the equilibrium spectrum ($T_{\text{eff}} = 12000$ K, $\log g = 8$, ML2/$\alpha$=0.6; dashed) and spectra at $T_{\text{eff}} = 12000 \pm 1000$ K (dotted)

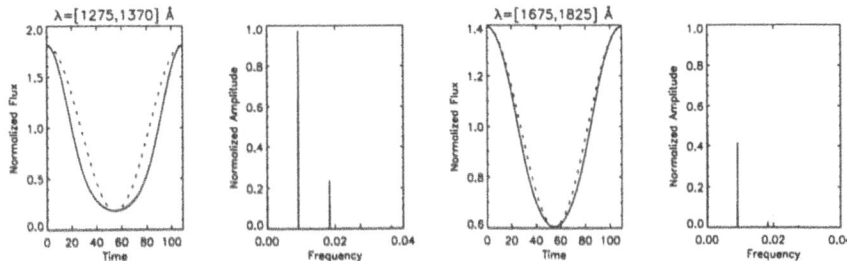

*Figure 3.* Light curves and corresponding Fourier spectra for oscillations as in Fig. 2 with $i = 0°$ close to the center of the Ly$_\alpha$ line (left panels) and just outside the Ly$_\alpha$ line (right panels). The underlying sinusoidal variation of $T_{\text{eff}}$ is overplotted (dotted)

in Fig. 3 the light curve for a filter located near the center of Ly$_\alpha$ and just outside this line. One can see that the light curve differs strongest from the underlying pure sinusoidal variation of $T_{\text{eff}}$ near the center of the spectral line. This causes strong amplitudes at higher harmonics in the Fourier spectrum.

## 3. Time-resolved spectra of G 226−29

This star is of great interest since it may well define the blue edge of the ZZ Ceti instability strip, which depends strongly on the a-priori unknown parametrization of the MLT (e.g., Bergeron et al. 1992). G 226−29 was observed with the HST using the FOS. Five observing runs were made, each about three hours long. The individual exposure times were about 9 s. Figure 4a shows the average spectra of the individual runs (these are corrected for a wavelength-dependent light loss modulated with the HST orbital period). Near $\sim 1500$ Å differences between the runs appear. However, the spectrum averaged over all runs (Fig. 4b) can be assumed to not be affected by this (it is also hardly affected by the light loss). We therefore fit this

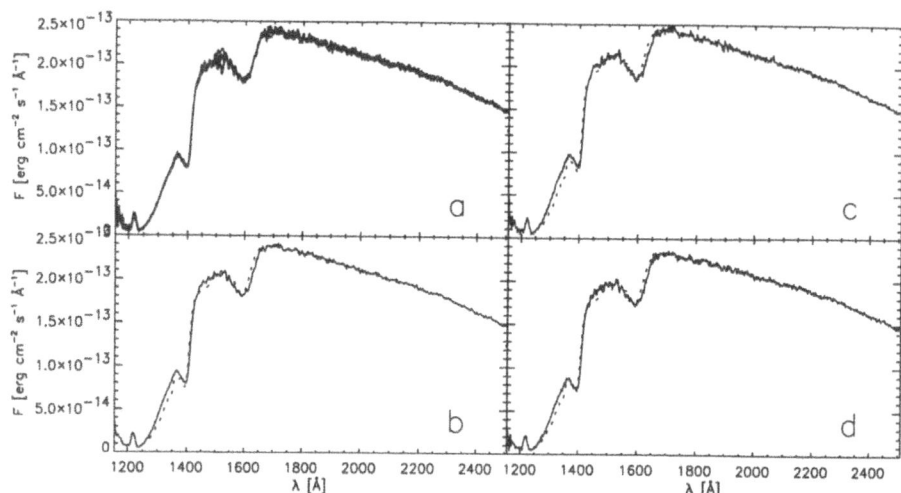

*Figure 4.* HST spectra of G 226−29. (a) average spectra of the 5 runs corrected for light losses. (b) spectrum averaged over all runs. (c) spectrum averaged at phases near maxima. (d) spectrum averaged at phases near minima. In (b) to (d) a fit is overplotted (dotted) using $T_{\rm eff} = 11820$ K, $\log g = 8.05$, ML2/$\alpha$=0.6, $l = 1$, $i = 72.5°$ and an amplitude such that $\left(\Delta T_{\rm eff}\right)_{\max} = 69$ K.

spectrum and obtain an average $T_{\rm eff} = 11820$ K (the error is dominated by systematic errors which are hard to estimate) at $\log g = 8.05$ (the latter adopted from Koester et al. 1996) using the ML2/$\alpha$=0.6 MLT, $\epsilon = 0.02$, $l = 1$ and $i = 72.5°$. The latter two parameters were adopted from results of WET observations by Kepler et al. (1995). These observations showed one triplet in the Fourier spectrum. It should be noted, however, that this interpretation of the triplet is not the only one possible. We exclude in the fit the region above $\approx 2300$ Å where the observed slope changes. This is evident also in other HST spectra and indicates some calibration problem. The high signal-to-noise HST spectrum reveals differences between observations and theory. These could be caused by continuing calibration problems, an imperfect line broadening theory for Ly$_\alpha$, an imperfect description of convection or a combination of those. Finally, we fit in Figs. 4c and d the average maximum and minimum spectra. A good fit is obtained indicating that $T_{\rm eff}$ varies on the WD surface by up to $69 \pm 3$ K.

## References

Bergeron P., Wesemael F., Fontaine G. 1992, ApJ 387, 288
Brassard P., Fontaine G., Wesemael F. 1995, ApJS 96, 545 (BFW)
Brickhill A.J. 1992, MNRAS 259, 529
Kepler S.O., Giovannini O., Wood M.A. et al. 1995, ApJ 447, 874
Koester D. et al. 1996, this conference
Robinson E.L., Kepler S.O., Nather R.E. 1982, ApJ 259, 219 (RKN)

# DISCOVERY OF TWO NEW ZZ CETI VARIABLES

N.DOLEZ, G.VAUCLAIR
*Observatoire Midi-Pyrénées*

J.N.FU
*Beijing Observatory*

AND

M.CHEVRETON
*Observatoire de Paris-Meudon*

## 1. Introduction

We report the discovery of two new pulsating DA variables . The candidates have been selected from the KISO survey (Nogushi et al. 1980). The selection has been based on the color indices of the objects.

The observations have been performed at the Haute Provence Observatory, in February 1996, on the 1.93m telescope. The instrument used was the Chevreton 4-channel photometer.

KUV11370+4222 shows an oscillations at a frequency of 3.89mHz (period = 257s), with a small amplitude of 4.7 mmag. two other peaks may be marginally significant at 2.16mHz and 3.41mHz.

## 2. Results

KUV08368+4026 has a dominant peak at a frequency of 1.62mHz (period 618s, amplitude 18mmag) and a secondary peak at 2.02 mHz (495s, 5.5mmag), These two low frequencies, as well as the large amplitude of the pulsation, are reminiscent of the group of large amplitude multiperiodic DAV (including several so called "cool DAV" as G29-38, Hltau76), and this discovery increases the possibility of a good understanding of those stars.

Due to the short duration of the observations, it was not possible to identify more modes in those stars. More observations are needed to confirm this detection, and find additional modes present in these non-radial pulsators.

*I. Isern et al. (eds.), White Dwarfs, 485–487.*

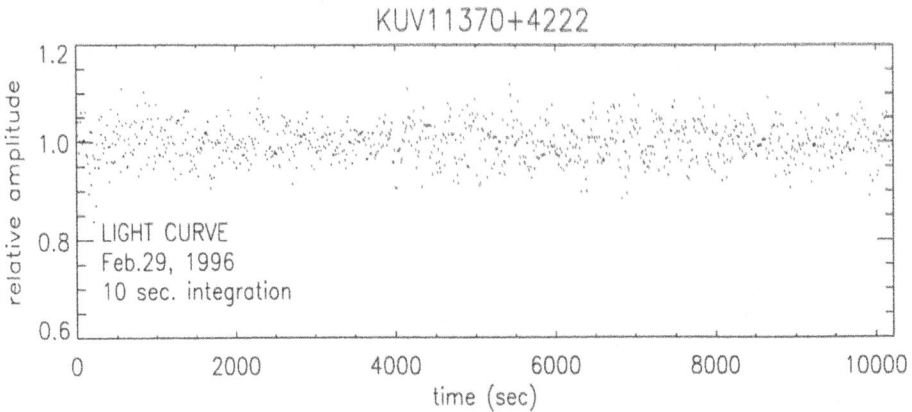

*Figure 1.* Light curve of KUV11370+4222, obtained at the Haute Provence Observatory on February 29, 1996

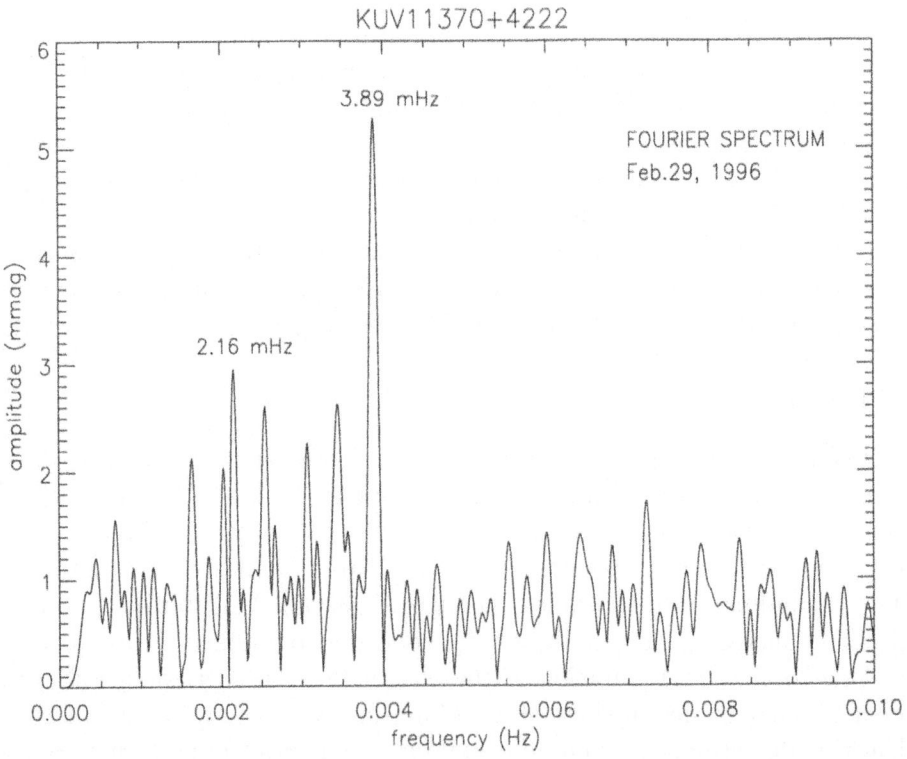

*Figure 2.* Amplitude spectrum of KUV11370+4222, Haute Provence Observatory, February 29, 1996

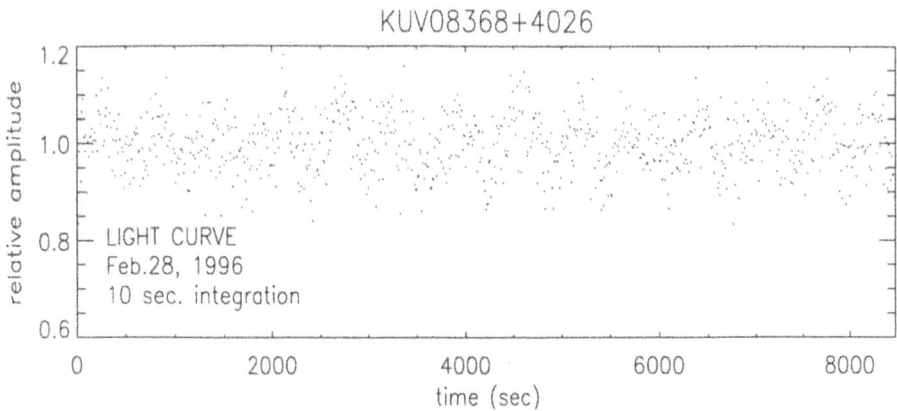

*Figure 3.* Light curve of KUV08368+4026, obtained at the Haute Provence Observatory on February 28, 1996

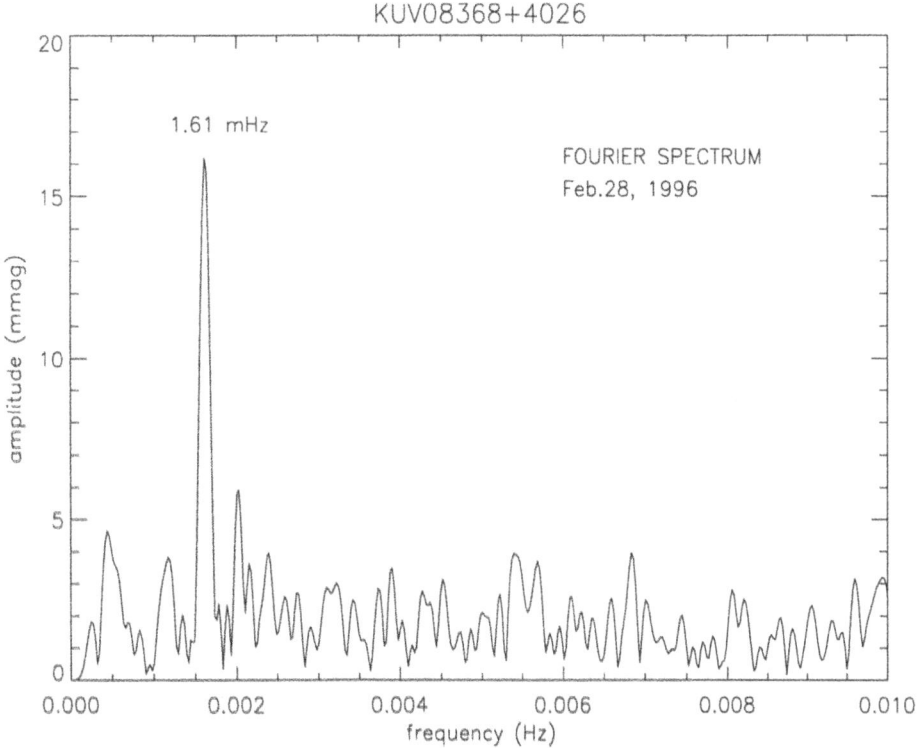

*Figure 4.* Amplitude spectrum of KUV08368+4026, Haute Provence Observatory, February 28, 1996

# NON-PULSATING DA WHITE DWARFS IN THE DAV REGION[1]

R. SILVOTTI
*Osservatorio Astronomico di Capodimonte*
*via Moiariello 16, I-80131 Napoli, Italy*

AND

C. BARTOLINI, G. COSENTINO, A. GUARNIERI AND A. PICCIONI
*Dipartimento di Astronomia*
*via Zamboni 33, I-40126 Bologna, Italy*

**Abstract.** In this paper we present the up to date results of our photoelectric monitoring program on DA White Dwarfs (WDs) near or inside the ZZ Ceti instability strip. From our data, none of the 12 targets listed below shows intrinsic luminosity variations within the observational limits. The results of this and previous works can be used as an observational check of the theoretical prediction that the blue edge of the strip depends mainly on the stellar mass (Bradley & Winget 1994).

## 1. Selection of the targets

The observational targets have been selected taking into account their effective temperature and surface gravity, their color indices (UBV, Strömgren, and MCSP (Greenstein 1982) systems), their H$\alpha$ and H$\beta$ equivalent widths (which are typically close to 40–45 and 50 Å respectively for DAV stars, Greenstein 1986) and their UV spectra (the contemporary presence of the $\lambda$ 1400 and $\lambda$ 1600 Å absorption features puts a strong constraint to the effective temperature and surface gravity, Allard & Koester 1992).

The observational edges of the DAV instability strip are (Bergeron et al. 1995): $11\,200 \leq T_{\text{eff}}/\text{K} \leq 12\,500$ and $7.91 \leq \log(g/\text{cm}\,\text{s}^{-2}) \leq 8.81$. For the color indices, the ranges of values for ZZ Ceti stars are:

---

[1] Based on observations obtained at the Loiano Observatory, Italy, and at the European Southern Observatory, Chile

*I. Isern et al. (eds.), White Dwarfs, 489–492.*
© *1997 Kluwer Academic Publishers.*

UBV      $0.14 \leq B-V \leq 0.25$      $-0.70 \leq U-B \leq -0.50$ [1]
STRÖ     $0.03 \leq b-y \leq 0.12$       $0.50 \leq u-b \leq 0.71$
MCSP     $-0.53 \leq V-I \leq -0.35$    $-0.40 \leq G-R \leq -0.29$

These limits are averages of different data published by McCook & Sion (1987), Wegner (1980), Fontaine et al. (1985), Greenstein (1984, 1986), Stobie et al. (1993, 1995). For the MCSP colors, we refer only to the AB79 calibration (Oke & Gunn 1983). Previous data, calibrated on the scale of AB69 (Oke & Schild 1970), have been corrected following Greenstein (1982, section II). Since the G–R color of DAV stars is strongly peaked around $-0.34$, G–R seems to be the most significant color index for studying the pulsational instability of DAV stars. Brickhill (1991, Fig. 4) has pointed out that, in fact, lines of constant G–R are nearly parallel to the slope of the theoretical instability strip in the $(T_{\mathrm{eff}}, \log g)$ plane.

## 2. Observations and results

The observations reported in this paper have been carried out between March 1991 and November 1995 using the two heads photometer (Piccioni et al. 1979 and Bartolini et al. 1993) of the 1.5 m Loiano reflector, and the one channel photometer (Lindgren and Gutiérrez 1991) of the 1.0 m La Silla telescope. We used two photomultipliers EMI 6256 SA at Loiano and a tube EMI 9789 QB at La Silla; both photomultipliers have the maximum sensitivity in the blue band. All the stars were observed with the Johnson B filter or without filter. During each observation we alternated the star monitoring with short measures of the sky background. Most stars have been observed more than one time (see Table 1). After a preliminary data reduction and Fourier analysis with the original time resolution (1–10 s), the data have been summed to an effective integration time of 9 or 10 s and re-analyzed for having more homogeneous results.

The log of the observations and the results obtained are presented in Table 1. For each star we report the effective temperature (column(2)) and the stellar mass (column(3)) from recent literature with relative reference (column(4)), the V magnitude (column (5)), the number of observations (column (6)), the duration in hours of the longest run (column (7)). In column (8) we report the upper limits to the pulsation amplitude in the period range between 40 and 1200 s. Finally, the mean noise-thresholds listed in column (9) are the averages of all the peaks present in each amplitude spectrum between 40 and 1200 s; these numbers give an estimate of the quality of the results, depending on both the quality and the length of the measurements.

[1]Exceptions are G 191-16 (B–V = 0.03, Greenstein 1974) and GD 99 (B–V = 0.28, U–B = –0.77, McCook & Sion 1987).

TABLE 1.  Log of observations and upper limits to the pulsation amplitude

| WD | $T_{eff}$ | $M/M_\odot$ | Ref | V | n | $d^h$ | $L^{mma}$ | $n^{mma}$ | Name |
|---|---|---|---|---|---|---|---|---|---|
| 0037−006 | 12400 | 0.70 | 4 | 14.7[1] | 1 | 3.2 | 3.2 | 1.2 | PB6089 |
| 0348+339 | 13000 | | 3 | 15.20 | 1 | 2.7 | 4.4 | 2.3 | GD52 |
| 0401+250 | 14100 | 0.48 | 1 | 13.80 | 2 | 2.1 | 7.4 | 1.8 | G8-8 |
| 1046+281 | | | | 15.40 | 1 | 2.3 | 10.4 | 5.2 | TON547 |
| 1116+026 | 13500 | 0.54 | 1 | 14.57 | 2 | 2.9 | 8.2 | 2.7 | GD133 |
| 1229−012 | | | | 14.24[1] | 2 | 4.0 | 4.6 | 1.2 | PG1229-013 |
| 1232+479 | 14700 | 0.53 | 2 | 14.52 | 2 | 6.9 | 8.6 | 1.8 | GD148 |
| 1241+235 | | | | 15.18[1] | 1 | 3.0 | 5.2 | 2.4 | LB16 |
| 1537+651 | 9800 | 0.68 | 1 | 14.64 | 4 | 2.3 | 14.4 | 6.9 | GD348 |
| 1600+369 | | | | 14.36 | 2 | 3.3 | 8.7 | 3.6 | EG115 |
| 2047+372 | 14800 | 0.71 | 1 | 12.93 | 3 | 2.8 | 3.5 | 1.9 | G210-36 |
| 2341+322 | 14000 | 0.47 | 1 | 12.92 | 2 | 6.5 | 2.7 | 0.9 | G130-5 |

Notes: [1] MCSP V magnitude.

Ref: (1) Kepler et al. 1995, (2) Bergeron, Saffer & Liebert 1992, (3) Dolez, Vauclair & Koester 1991, (4) Guseinov et al. 1983.

## 3. Discussion

The results of this study (summarized in Table 1) agree with the hypothesis that a few stars inside the ZZ Ceti instability strip are non-pulsating, as suggested by Kepler & Nelan (1993) and Kepler et al. (1995) (see also Dolez, Vauclair & Koester 1991). In particular PB 6089, GD 52, G 8-8, GD 133 and G 130-5, with effective temperatures very close to those of DAV stars, do not present pulsations with amplitudes higher than a few mmag. Taking into account only the spectro-photometric parameters of the observed stars (color indices, H$\alpha$ and H$\beta$ equivalent widths, and UV $\lambda$ 1400 and $\lambda$ 1600 Å absorption features), G 130-5 appears to be the best candidate for a non-pulsating DA WD inside the ZZ Ceti instability strip. On the other hand, if the effective temperature of the blue edge actually decreases with decreasing stellar masses (Bradley & Winget 1994), a mass of 0.47 $M_\odot$ (Kepler et al. 1995) would be sufficiently low to explain the non-variability of G 130-5. Another interesting object is PB 6089: it also has "good" (but not complete) spectro-photometric parameters; moreover, with effective temperature and stellar mass as given in Table 1, it should be inside the DAV strip. New accurate temperature and mass determinations for PB 6089 could help to better understand the behaviour of this star and also to check the connection between effective temperature of the blue edge and stellar mass.

With reference to the paper of Bartolini et al. (1993), we want to specify that for GD 348 and GD 148 new observations have not confirmed the possible variability of these stars. More exactly, we did not find any high-frequency variations in a short light-curve of GD 148. For GD 348 we would need at least one new longer observation (several hours) to clarify its photometric behaviour.

In addition to the stars reported in this paper, we want to emphasize that the following DA WDs have also been observed: GD 31, GD 47, GD 77, GD 294, GD 322, GD 340, G 21-15, G 157-82. They are not reported here because their observations, all shorter than 2 hours, are not enough significant. Nevertheless, even for them no periodic luminosity variations have been detected.

## Acknowledgements

We are grateful to Angela Bragaglia for having brought this poster to Blanes. This research was partially supported by the italian "Ministero per l'Università e la Ricerca Scientifica e Tecnologica".

## References

Allard N.F., Koester D., 1992, A&A 258, 464
Bartolini C., Cosentino G., Guarnieri A., Piccioni A., Silvotti R., 1993, in White Dwarfs: Advances in Observation and Theory, ed. M.A. Barstow (Kluwer: Dordrecht), p.543
Bergeron P., Saffer R.A., Liebert J., 1992, ApJ 394, 228
Bergeron P., Wesemael F., Lamontagne R., et al., 1995, ApJ 449, 258
Bradley P.A., Winget D.E., 1994, ApJ 421, 236
Brickhill A.J., 1991, MNRAS 252, 334
Dolez N., Vauclair G., Koester D., 1991, in White Dwarfs, eds. G. Vauclair and E. Sion (Kluwer: Dordrecht), p.361
Fontaine G., Bergeron P., Lacombe P., Lamontagne R., Talon A., 1985, AJ 90, 1094
Greenstein J.L., 1974, ApJ 189, L131
Greenstein J.L., 1982, ApJ 258, 661
Greenstein J.L., 1984, ApJ 276, 602
Greenstein J.L., 1986, ApJ 304, 334
Guseinov O.H., Novruzova H.I., Rustamov Y.S., 1983, Ap&SS 96, 1
Kepler S.O., Nelan E.P., 1993, AJ 105, 608
Kepler S.O., Giovannini O., Kanaan A., Wood M.A., Claver C.F., 1995, Proc. 3rd WET Workshop, eds. E.G. Meistas and J. Solheim, Baltic Astronomy vol.4, p.157
Lindgren H., Gutiérrez F.W., 1991, ESO operating manual n.13
McCook G.P., Sion E.M., 1987, ApJS 65, 603
Oke J.B., Gunn J.E. 1983, ApJ 266, 713
Oke J.B., Schild R.E., 1970, ApJ 161, 1015
Piccioni A., Bartolini C., Giovannelli F., Guarnieri A., 1979, Acta Astron. 29, 463
Stobie R.S., Chen A., O'Donoghue D., Kilkenny D., 1993, MNRAS 263, L13
Stobie R.S., O'Donoghue D., Ashley R., et al., 1995, MNRAS 272, L21
Wegner G., 1980, AJ 85, 538

The manufacturer's authorised representative in the EU is Springer
Nature Customer Service Centre GmbH, Europaplatz 3, 69115 Heidelberg,
Germany. If you have any concerns regarding our products, please
contact ProductSafety@springernature.com

Printed and bound by CPI Group (UK) Ltd, Croydon, CR0 4YY

23/04/2026

02095607-0010